AF718379

# Regulation of Cation Homeostasis in Plants

# Regulation of Cation Homeostasis in Plants

Edited by:

Carlos Tello Lacal

Delve Publishing

www.delvepublishing.com

**Regulation of Cation Homeostasis in Plants**

*Carlos Tello Lacal*

**Delve Publishing**

2010 Winston Park Drive,

2nd Floor

Oakville, ON L6H 5R7

Canada

www.delvepublishing.com

Tel: 001-289-291-7705

001-905-616-2116

Fax: 001-289-291-7601

Email: orders@arclereducation.com

**© 2018 Delve Publishing**

ISBN: 978-1-77361-043-6 (Hardcover)

This book contains information obtained from highly regarded resources. Reprinted material sources are indicated. Copyright for individual articles remains with the authors as indicated and published under Creative Commons License. A Wide variety of references are listed. Reasonable efforts have been made to publish reliable data and views articulated in the chapters are those of the individual contributors, and not necessarily those of the editors or publishers. Editors or publishers are not responsible for the accuracy of the information in the published chapters or consequences of their use. The publisher assumes no responsibility for any damage or grievance to the persons or property arising out of the use of any materials, instructions, methods or thoughts in the book. The editors and the publisher have at- tempted to trace the copyright holders of all material reproduced in this publication and apologize to copyright holders if permission has not been obtained. If any copyright holder has not been acknowledged, please write to us so we may rectify.

**Notice**: Registered trademark of products or corporate names are used only for explanation and identification without intent of infringement.

Delve Publishing publishes wide variety of books and eBooks. For more information about Delve Publishing and its products, visit our website at www.delvepublishing.com.

# DECLARATION

Some content or chapters in this book are open access copyright free published research work, which is published under Creative Commons License and are indicated with the citation. We are thankful to the publishers and authors of the content and chapters as without them this book wouldn't have been possible.

# ABOUT THE EDITOR

**Carlos Tello Lacal**

Carlos Tello (Valencia, Spain, 1981) studied Agricultural Engineering with specialization in Biotechnology at the Universidad Politécnica de Valencia, where he graduated with a BSc thesis on the characterization of an Arabidopsis mutant with increased tolerance to seed ageing and salt stress. He moved afterwards to Sevilla to complete his MSc and PhD at the Instituto de Recursos Naturales y Agrobiología, where he studied the mechanisms employed by plants in their adaptation to abiotic stress and more specifically the role of the Arabidopsis $Na^+/H^+$ antiporter SOS1 in the regulation of $Na^+$ and $K^+$ homeostasis. Since 2014, he is based in Zürich and develops several freelance science-related jobs. He has recently retrained himself in the field of cell culture in bioreactors by attending an advanced training course at the Zürcher Hochschule für Angewandte Wissenschaften

# TABLE OF CONTENTS

# LIST OF CONTRIBUTORS

**Carlos Tello**

**Manuel Nieves-Cordones**
Biochimie et Physiologie Moléculaire des Plantes, Institut de Biologie Intégrative des Plantes, UMR 5004 CNRS/UMR 0386 INRA/Montpellier SupAgro/Université Montpellier 2, Montpellier, France

**Vicente Martínez**
Departamento de Nutrición Vegetal, Centro de Edafología y Biología Aplicada del Segura – Consejo Superior de Investigaciones Científicas, Murcia, Spain

**Begoña Benito**
Centro de Biotecnología y Genómica de Plantas, Universidad Politécnica de Madrid, Madrid, Spain

**Francisco Rubio**
Departamento de Nutrición Vegetal, Centro de Edafología y Biología Aplicada del Segura – Consejo Superior de Investigaciones Científicas, Murcia, Spain

**Min Wang**
Agricultural Ministry Key Lab of Plant Nutrition and Fertilization in Low-Middle Reaches of the Yangtze River, Jiangsu Key Lab and Engineering Center for Solid Organic Waste Utilization, Nanjing Agricultural University, No. 1 Weigang, Nanjing 210095, China

**Qingsong Zheng**
Agricultural Ministry Key Lab of Plant Nutrition and Fertilization in Low-Middle Reaches of the Yangtze River, Jiangsu Key Lab and Engineering Center for Solid Organic Waste Utilization, Nanjing Agricultural University, No. 1 Weigang, Nanjing 210095, China

**Qirong Shen**
Agricultural Ministry Key Lab of Plant Nutrition and Fertilization in Low-Middle Reaches of the Yangtze River, Jiangsu Key Lab and Engineering Center for Solid Organic Waste Utilization, Nanjing Agricultural University, No. 1 Weigang, Nanjing 210095, China

**Shiwei Guo**
Agricultural Ministry Key Lab of Plant Nutrition and Fertilization in Low-Middle Reaches of the Yangtze River, Jiangsu Key Lab and Engineering Center for Solid Organic Waste Utilization, Nanjing Agricultural University, No. 1 Weigang, Nanjing 210095, China

**Toshio Yamaguchi**
Department of Microbiology, Faculty of Pharmacy, Niigata University of Pharmacy and Applied Life Sciences, Niigata, Japan

**Shin Hamamoto**
Department of Biomolecular Engineering, Graduate School of Engineering, Tohoku University, Sendai, Japan

**Nobuyuki Uozumi**
Department of Biomolecular Engineering, Graduate School of Engineering, Tohoku University, Sendai, Japan

**Fernando Alemán**
Centro de Edafología y Biología Aplicada del Segura-CSIC, Murcia, Spain

**Fernando Caballero**
Centro de Edafología y Biología Aplicada del Segura-CSIC, Murcia, Spain

**Reyes Ródenas**
Centro de Edafología y Biología Aplicada del Segura-CSIC, Murcia, Spain

**Rosa M. Rivero**
Centro de Edafología y Biología Aplicada del Segura-CSIC, Murcia, Spain

**Vicente Martínez**
Centro de Edafología y Biología Aplicada del Segura-CSIC, Murcia,

Spain

**Francisco Rubio**
Centro de Edafología y Biología Aplicada del Segura-CSIC, Murcia, Spain

**Tripti Sharma**
Molecular Biology, Institute for Biochemistry and Biology, University of Potsdam, Potsdam, Germany
IMPRS-PMPG, Max-Planck Institute of Molecular Plant Physiology, Potsdam, Germany

**Ingo Dreyer**
Centro de Biotecnologia y Genomica de Plantas, Universidad Politécnica de Madrid, Madrid, Spain

**Janin Riedelsberger**
Molecular Biology, Institute for Biochemistry and Biology, University of Potsdam, Potsdam, Germany
IMPRS-PMPG, Max-Planck Institute of Molecular Plant Physiology, Potsdam, Germany

**Cornelia Eisenach**
Laboratory of Plant Physiology and Biophysics, Institute of Molecular, Cell and Systems Biology, University of Glasgow, Bower Building, Glasgow G12 8QQ, UK

**Maria Papanatsiou**
Laboratory of Plant Physiology and Biophysics, Institute of Molecular, Cell and Systems Biology, University of Glasgow, Bower Building, Glasgow G12 8QQ, UK

**Ellin-Kristina Hillert**
Laboratory of Plant Physiology and Biophysics, Institute of Molecular, Cell and Systems Biology, University of Glasgow, Bower Building, Glasgow G12 8QQ, UK

**Michael R. Blatt**
Laboratory of Plant Physiology and Biophysics, Institute of Molecular, Cell and Systems Biology, University of Glasgow, Bower Building, Glasgow G12 8QQ, UK

**Annette Kugler**
Molecular Plant Physiology, Department Biology, University of Erlangen, Erlangen, Germany,

**Barbara Köhler**
Institute of Biochemistry and Biology, University of Potsdam, Potsdam-Golm, Germany and

**Klaus Palme**
Institute of Biology II/Botany, Faculty of Biology, University of Freiburg, Freiburg, Germany

**Patricia Wolff**
Institute of Biology II/Botany, Faculty of Biology, University of Freiburg, Freiburg, Germany

**Petra Dietrich**
Molecular Plant Physiology, Department Biology, University of Erlangen, Erlangen, Germany,

**Jun Ni**
College of Life and Environmental Sciences, Hangzhou Normal University

**Zhiming Yu**
College of Life and Environmental Sciences, Hangzhou Normal University

**Guankui Du**
Department of Biochemistry, Hainan Medical College

**Yanyan Zhang**
College of Life and Environmental Sciences, Hangzhou Normal University

**Jemma L. Taylor**
School of Life Sciences, Gibbet Hill Campus, University of Warwick

**Chenjia Shen**
College of Life and Environmental Sciences, Hangzhou Normal University

**Jing Xu**
College of Life and Environmental Sciences, Hangzhou Normal University

**Xunyan Liu**
College of Life and Environmental Sciences, Hangzhou Normal University

**Yifeng Wang**
State Key Laboratory of Rice Biology, China National Rice Research Institute, Chinese Academy of Agricultural Sciences

**Yunrong Wu**
State Key Laboratory of Plant Physiology and Biochemistry, College of Life Science, Zhejiang University

**Shane Waters**
Australian Centre for Plant Functional Genomics and Waite Research Institute, University of Adelaide, Glen Osmond, South Australia 5064, Australia

**Matthew Gilliham**
School of Agriculture, Food and Wine, Waite Research Institute and ARC Centre of Excellence in Plant Energy Biology, University of Adelaide, Glen Osmond, South Australia 5064, Australia

**Maria Hrmova**
Australian Centre for Plant Functional Genomics and Waite Research Institute, University of Adelaide, Glen Osmond, South Australia 5064, Australia

**Jon K. Pittman**
Faculty of Life Sciences, The University of Manches

**Kendal D. Hirschi**
USDA/ARS Children's Nutrition Research Center, Baylor College of Medicine

**Inês S Pires**
Instituto de Tecnologia Química e Biológica, Universidade Nova de Lis-

boa, Av. da República, 2780-157 Oeiras, Portugal and iBET
Department of Biology and Center for Genomics and Systems Biology, New York University

**Sónia Negrão**
Instituto de Tecnologia Química e Biológica, Universidade Nova de Lisboa, Av. da República, 2780-157 Oeiras, Portugal and iBET

**Melissa M Pentony**
Department of Biology and Center for Genomics and Systems Biology, New York University

**Isabel A Abreu**
Instituto de Tecnologia Química e Biológica, Universidade Nova de Lisboa, Av. da República, 2780-157 Oeiras, Portugal and iBET

**Margarida M Oliveira**
Instituto de Tecnologia Química e Biológica, Universidade Nova de Lisboa, Av. da República, 2780-157 Oeiras, Portugal and iBET

**Michael D Purugganan**
Department of Biology and Center for Genomics and Systems Biology, New York University

**Tyler McCubbin**
Department of Plant Sciences, University of California, Davis, CA 95616, USA

**Elias Bassil**
Department of Plant Sciences, University of California, Davis, CA 95616, USA

**Shiqi Zhang**
Department of Plant Sciences, University of California, Davis, CA 95616, USA

**Eduardo Blumwald**
Department of Plant Sciences, University of California, Davis, CA 95616, USA

**Brett A. Ford**
Department of Botany, La Trobe University, Melbourne, VIC, Australia

**Joanne R. Ernest**
Department of Botany, La Trobe University, Melbourne, VIC, Australia

**Anthony R. Gendall**
Department of Botany, La Trobe University, Melbourne, VIC, Australia

**Yang Zhou**
Institute of Tropical Bioscience and Biotechnology, Chinese Academy of Tropical Agricultural Sciences, Haikou, China
College of Plant Science & Technology, Huazhong Agricultural University, Wuhan, China
Hainan Key Laboratory for Sustainable Utilization of Tropical Bioresources/College of Agriculture, Hainan University, Haikou, China

**Xiaochang Yin**
Hainan Key Laboratory for Sustainable Utilization of Tropical Bioresources/College of Agriculture, Hainan University, Haikou, China

**Ruijun Duan**
Institute of Tropical Bioscience and Biotechnology, Chinese Academy of Tropical Agricultural Sciences, Haikou, China

**Gangping Hao**
Department of Biochemistry, Taishan Medical University, Tai'an, China

**Jianchun Guo**
Institute of Tropical Bioscience and Biotechnology, Chinese Academy of Tropical Agricultural Sciences, Haikou, China
College of Plant Science & Technology, Huazhong Agricultural University, Wuhan, China

**Xingyu Jiang**
Hainan Key Laboratory for Sustainable Utilization of Tropical Bioresources/College of Agriculture, Hainan University, Haikou, China

**Amanda L. Socha**
Department of Biological Sciences, Dartmouth College, Hanover, NH, USA

**Mary Lou Guerinot**
Department of Biological Sciences, Dartmouth College, Hanover, NH, USA

# PREFACE

Homeostasis can be defined in general terms as the active maintenance of the basal conditions of a system. The plant cell needs to control its cation homeostasis, that is to say tightly regulate the concentration of the main essential positively charged ions, such as K+, Ca2+ or Mg2+, as well as that of potentially toxic cations such as Na+, in order to maintain a balance allowing for the correct functioning of its metabolism, osmotic adjustment and nutrient acquisition. To this aim, plants employ a set of proteins including pumps, channels and transporters. This book contains a collection of peer-reviewed papers describing the main families of plant transporter proteins, with special emphasis on those regulating the uptake and distribution of K+ and its competitive inhibitor Na+.

In Chapter number 1, the editor of the present book describes the main effects caused by salt stress and the main adaptive responses that plants have developed against it, in addition to providing an insight into the main families of cation pumps, channels and transporters. Chapter 2 presents a detailed overview of the main systems controlling the uptake of Na+ and K+ through the roots in both Arabidopsis and rice. Chapter 3 looks into the importance of K+ as a plant macronutrient with special emphasis on its role in protection against stress conditions. In turn, Chapter 4 briefly outlines the main proteins catalyzing sodium transport in plants. The isolation of a point mutation increasing the affinity of HAK5, the most important member of the KT/KUP/HAK family, for K+ while reducing that for Na+ and Cs+ is presented in Chapter 5. In Chapter 6, the main Arabidopsis voltage-dependent and voltage-independent K+ channels are studied. In vivo distribution, traffic and response to ion concentration of voltage-dependent channels is illustrated in Chapter 7 through the example of K+ outward GORK channel. The two families of non-selective cation channels, cyclic-nucleotide gated channels (CNGC) and glutamate receptors (GLR), are dealt with in Chapters 8 and 9, describing the salt-dependent regulation of the CNGC19/CNGC20 subfamily and the functional characterization of rice GLR, respectively. Chapter 10 reviews the HKT family of Na+ and K+ transporters. The following eight chapters explore the main families of plant cation/H+ antiporters. In Chapter 11, the latest findings on the phylogenetic analysis and protein structure of the CaCA family of Ca2+/H+ antiporters in both Arabidopsis and rice are briefly presented. Chapter 12 describes the evolutionary divergence of NHX and SOS1 proteins within the CPA1 family. In Chapter 13, an Arabidopsis quadruple mutant lacking all the vacuolar NHX transporters is generated to investigate the role of these proteins in K+ homeostasis and microtubule dynamics. Chapter 14 describes the identification and characterization of orthologs in Brassica napus of the Arabidopsis endosomal NHX proteins AtNHX5 and AtNHX6. In Chapter 15, the genes for the Na+/H+ antiporter SOS1 and the H+-ATPase AHA1 of the halophyte Sesuvium portulacastrum are cloned and heterologously expressed in S. cerevisiae in order to study their coordinated function. Finally, Chapter 16 consists of a review of the proteins involved in the uptake and transport of Mn cations.

The present book intends to help college students, teachers, researchers and other readers interested in plant physiology better understand the mechanisms employed by plants to compensate transient imbalances in the levels of their main cations.

CHAPTER

1

# ION HOMEOSTASIS AND SALT TOLERANCE IN PLANTS

**Carlos Tello, PhD**

## INTRODUCTION

Soil salinity is an abiotic factor leading to serious consequences in agricultural production, since it limits both crop productivity and the potential agricultural use of previously uncultured soils (Munn, 2002; Flowers, 2004). Approximately $10^9$ ha of the total land area, covering 20% of total agricultural surface and 50% of irrigated land, are affected by salinity, being soils defined as saline when the electrical conductivity of the saturated extract exceeds 4 dS/m (Szábolcs, 1994; Flowers and Yeo, 1995; Yokoi *et al.*, 2002; Tester and Davenport, 2003). Soil salinization is naturally produced through the weathering of rocks or the deposition of salts carried by wind and water currents. In arid and semiarid regions, in which high agricultural productivity can only be achieved through irrigation, the occurrence of high insolation and low rainfall leads to the concentrations of salts in the soil. Unfortunately, irrigation tends to exacerbate this problem, especially when the drainage is insufficient and low quality water is used (Bressler and Hoffman, 1986; Ghassemi *et al.*, 1995; Serrano, 1996). Good agricultural practices, such as the irrigation with high quality water, installation of drainage systems or rectification of the soil sodium conductivity by adding calcium oxide or sulfate, can correct the problem (Tester and Davenport, 2003), being the employment of halotolerant cultivars another possible solution. Because both halophyte (salt-tolerant) and glycophyte (salt-sensitive) plants employ similar mechanisms to tackle salt stress and mostly differ in the expression profile of the genes involved, research in plant breeding against salt stress must focus on identifying both the elements conferring resistance to salt stress and their regulatory elements (Hasegawa *et al.*, 2000; Zhu, 2001).

## PHYSIOLOGICAL EFFECTS CAUSED BY SALT STRESS IN PLANTS

Unlike other multicellular organisms, plants are sessile and cannot escape when the growth conditions are inadequate. Therefore, they require the development of response mechanisms against all kinds of stress. The physiological effects produced by salt stress are caused by their two main components: osmotic imbalance and ion toxicity. A subset of secondary effects, such as oxidative stress, decline of cell growth, reduction of evapotranspiration, deficiencies in nutrient absorption and loss of photosynthesis efficiency, are derived from the above-mentioned primary effects (Niu et al., 1995; Zhu, 2001, 2002; Tester and Davenport, 2003; Munns and Tester, 2008).

### Osmotic Stress

The excess of salts originates a drastic reduction in osmotic potential of the soil and restricts water absorption through the roots. In few hours, the lower water content causes a transient growth reduction of leaves and roots. The water balance of the cells is altered, since they respond to the lack of water in the cytosol, where most metabolic reactions take place, by providing it from the vacuole. Thus, cells lose turgor and reduce their expansion. Additionally, plants minimize dehydration by closing the stomata to prevent water losses through evapotranspiration. Stomatal closure also diminishes $CO_2$ uptake, which reduces the fixation of this molecule through photosynthesis (Hasegawa *et al.*, 2000; Zhu, 2002; Munns *et al.*, 2002).

## Ion Toxicity and Electrochemical Imbalance

Homeostasis is the tendency of an organism to maintain its basal status despite changes in its environment. Under physiological conditions, the homeostatic concentrations of the main ions in the plant cytoplasm are 100-200 mM $K^+$, 1-10 mM $Na^+$ and $Cl^-$, and 100-200 nM $Ca^{2+}$ (Niu *et al.*, 1995). The excess of NaCl during salt stress episodes causes strong imbalances in ion homeostasis, which the cell needs to reestablish for its correct growth and metabolic functioning (Binzel *et al.*, 1998). Given the similarity of $Na^+$ and $K^+$ in terms of physicochemical properties, the former acts as a competitive inhibitor at the active site of enzymes requiring the latter as cofactor. The cell needs to maintain a high $K^+$:$Na^+$ ratio, since $Na^+$ concentrations above 100 mM in the cytoplasm inhibit biochemical and physiological processes (Bhandal and Malik, 1988; Marschner, 1995; Serrano, 1996). Essential cellular processes such as protein synthesis, RNA processing or sugar metabolism are strongly inhibited in the presence of high $Na^+$ concentrations, being the electrostatic equilibrium required to maintain the tertiary and quaternary structure of proteins, and therefore most of the enzyme activity, disrupted above 400 mM (Wyn-Jones and Pollard, 1983; Marschner, 1995; Blaha *et al.*, 2000; Forment *et al.*, 2002).

## Oxidative Stress

Salt stress causes oxidative damage due to the overaccumulation of reactive oxygen species (ROS), such as superoxide radicals ($\cdot O_2^-$), hydrogen peroxide ($H_2O_2$) or hydroxyl radicals ($\cdot OH^-$), which are naturally produced in aerobic processes such as electron transport at the chloroplasts and mitochondria or oxidation of metabolites. Excess of ROS causes oxidative damage to plasma membrane lipids, proteins and nucleic acids (Smirnoff *et al.*, 1993; Moran *et al.*, 1994; Borsani *et al.*, 2001; Hernández *et al.*, 2001). Plants degrade ROS through non-enzymatic scavengers such as ascorbate, glutathione, flavonoids, alkaloids and carotenoids, as well as through detoxifying enzymes such as superoxide dismutase, catalase or those belonging to the ascorbate-glutathione cycle (Chinnusamy *et al.*, 2004; Apel and Hirt, 2004).

### Defficiency in $K^+$ Uptake

$K^+$ is an essential element to the plant cell functioning in the osmotic adjustment, which additionally regulates stomatal movements and acts as a cofactor for several enzymes (Hasegawa *et al.*, 2000). $Na^+$ competes with $K^+$ for binding to the high and low affinity transport systems, contributing to the reduction of the $K^+$:$Na^+$ ratio in the cytoplasm (Rains and Epstein, 1967; Maathuis *et al.*, 1996). Cyclic nucleotide-gated channels (CNGC) have been proposed as a potential entry of $Na^+$ ions into the root (Amtmann *et al.*, 1999; Tyerman and Skerret, 1999; Zhu, 2003; Rus *et al.*, 2004). Additionally, $Na^+$ accumulation in the cell causes plasma membrane depolarization, preventing the passive entrance of $K^+$ ions through inward rectifier while favoring their exit through outward rectifier channels. Finally, the synthesis of compatible solutes used in cell osmoprotection consumes high amounts of ATP, which cannot be used for the high affinity $K^+$ uptake transport anymore (Shabala *et al.*, 2003; Shabala and Cuin, 2007).

### Other Secondary Effects of Salt Stress

They include photosynthesis deregulation, stomatal closure, cell growth arrest and activation of programmed cell death (PCD). Additionally, a relationship between salt stress and genotoxicity has been suggested, since the Arabidopsis mutant *uvs66* is sensitive to both genotoxic agents and salt stress (Albinsky *et al.*, 1999; Serrano *et al.*, 1999; Hasegawa *et al.*, 2000; Zhu, 2003).

## PLANT RESPONSES TO SALT STRESS

The enzymes of both glycophyte and halophyte plants are equally sensitive to high concentrations of $Na^+$ in the cytoplasm, having both types of species thus developed several adaptation processes that are triggered in the presence of salt stress (Hasegawa *et al.*, 2000; Zhu, 2003). Adaptive responses to high salinity can be divided into two groups: adaptations at the cellular and the whole plant level.

## Adaptive responses at the cellular level

### *Ion Homeostasis Regulation*

Sodium entry into the cell disrupts its ion and osmotic homeostasis. When the $Na^+$ concentration in the extracellular medium increases, a gradient at both sides of the plasma membrane is created. Together with the membrane potential, it favors the passive entry of $Na^+$ into the cell and can lead to apoptosis if this cation is overaccumulated (Niu *et al.*, 1995; Serrano *et al.*, 1999). In plants, ion homeostasis is mainly reestablished through $Na^+/H^+$ antiporters, which extrude $Na^+$ out of the cell or compartmentalize it into vacuoles (Xiong and Zhu, 2002). The energy required for cation transport comes from the proton gradient generated by the plasma membrane P-type $H^+$-ATPase or the vacuolar V-type $H^+$-ATPase and $H^+$-pyrophosphatase (Blumwald *et al.*, 2000). The first antiporter identified in plants was AtNHX1, which exhibits the same affinity for the $Na^+/H^+$ and $K^+/H^+$ exchange and high homology to ScNHX1 (Nass *et al.*, 1997; Venema *et al.*, 2002; Barragán *et al.*, 2012). Transgenic Arabidopsis and tomato plants overexpressing either this protein or the $H^+$-pyrophosphatase AVP1 showed increased tolerance to salt stress (Apse *et al.*, 1999; Gaxiola *et al.*, 2001; Zhang *et al.*, 2001). Extrusion of $Na^+$ out of the cell is carried out through the $Na^+/H^+$ antiporter SOS1, whose overexpression also results in an increased salt tolerance (Shi *et al.*, 2000, 2003; Qiu *et al.*, 2002; Yang *et al.*, 2009).

Ion accumulation in the vacuole lumen shifts its osmotic potential towards more negative values. In order to maintain the same potential at both sides of the tonoplast, plants synthesize compatible osmolites and accumulate them in the cytoplasm (Tester and Davenport, 2003). Compatible omolites can be accumulated in the cell without altering the intracellular biochemistry and include essential elements such as $K^+$, simple sugars such as glucose or fructose, complex sugars such as trehalose, raffinose and fructanes, amino acid derivatives such as proline, glycine betaine, β-alanine-betaine or proline-betaine, tertiary amines such as ectoin, and sulfonium compounds such as coline sulfate or dimethylsulfonium propionate (Bohnert and Jensen, 1996; Nuccio *et al.*, 1999). Finally, aquaporins are water channels located at the plasma membrane and tonoplast which allow the entry of ~$10^6$ water molecules per second in the presence of an osmotic pressure difference of 100 mM.

Their phosphorylation by $Ca^{2+}$-dependent protein kinases, favored at high hydric potentials, increases their permeability to promote water flux into the cell (Johansson *et al.*, 1996; Chrispeels *et al.*, 1999).

### Detoxification of Oxidative Stress

Salt and other types of abiotic stress generate ROS, which can be detoxified through the action of enzymes or the synthesis of protective osmolites (Hasegawa *et al.*, 2000). The production of transgenic plants overexpressing some of the detoxifying enzymes, such as glutathione peroxidase, superoxide dismutase, ascorbate peroxidase or glutathione reductase, resulted in improved salt tolerance (Allen *et al.*, 1997; Roxas *et al.*, 1997; Bohnert and Sheveleva, 1998). The synthesis of protective proteins such as dehydrines and osmotines, which maintain protein structure at high salinity, and that of late embryogenesis abundant (LEA) proteins, which counteract the osmotic component of salt stress and are involved in response to pathogens, is induced during stress episodes (Ingram and Bartels, 1996; Campbell and Close, 1997; Munns *et al.*, 2002; Brini *et al.*, 2011; Salleh *et al.*, 2012). Constitutive overexpression of barley LEA proteins increased salt tolerance in transgenic rice plants (Xu *et al.*, 1996).

Compatible osmolites are often accumulated at too low concentrations as to allow for their participation in osmotic readjustment and are therefore believed to be mostly employed in ROS scavenging and protection of cell structures against oxidative damage (Hong *et al.*, 2000; Zhu, 2001; Krassensky and Jonak, 2011). Some of these compatible solutes are amino acids. Proline is essential in the protection against cell damage caused by all types of stress, since it is accumulated as an osmolite, degrades ROS and acts as a chaperone to stabilize protein structures (Verbruggen and Hermans, 2008; Szabados and Savouré, 2010). Abiotic stress induces its synthesis through the reduction of glutamate to glutamate semialdehyde by P5C synthase (P5CS), followed by its reduction to proline by P5C reductase. In the absence of stress, its degradation to glutamate by proline dehydrogenase and P5C dehydrogenase is favored (reviewed by Szabados and Savouré, 2010). Overexpression of P5CS in tobacco and petunia produced plants with augmented tolerance to salt and drought due to increased proline accumulation, while the Arabidopsis *p5cs-1* mutant showed hypersensitivity to abiotic stress (Kishor *et al.*, 1995; Hong *et al.*,

2000; Yamada *et al.*, 2005; Szekely *et al.*, 2008). g-butyric acid (GABA) is a non-protein amino acid that is accumulated during stress situations. After its synthesis in the cytosol, it is transported into the mitochondria, where it is degraded to succinic acid and incorporated into the Krebs cycle (Shelp *et al.*, 1999; Fait *et al.*, 2008). Loss of function of GABA transaminase decreased tolerance to stress in Arabidopsis, while that of semialdehyde succinic dehydrogenase produced dwarf plants with hypersensitivity to UV-B light and thermal stress as a result of ROS overaccumulation (Bouche *et al.*, 2003; Renault *et al.*, 2010). Polyamine levels and tolerance to oxidative stress show a positive correlation. Although their specific function is poorly understood, they have been involved in membrane protection and ROS scavenging (Groppa and Benavides, 2008; Alcázar *et al.*, 2010; Hussain *et al.*, 2011). Low production of putrescine as a result of ADC1 and -2 loss of function reduced plant tolerance to all types of stress, while their overexpression produced hyper-resistant plants to drought, salt and cold stress (Capell *et al.*, 2004; Urano *et al.*, 2004; Cuevas *et al.*, 2008; Alcázar *et al.*, 2010). Some plants respond to abiotic stress by synthesizing glycine betaine, which protects membranes against oxidative damage and can participate in osmotic readjustment. Transgenic overexpression of this protein, especially when targeted to the chloroplast, improves abiotic stress tolerance (Rhodes and Hanson, 1993; Park *et al.*, 2007; Chen and Murata, 2011). Several carbohydrates can also function as compatible osmolites. Many angiosperms from cold regions employ fructanes, branched polymers formed through transfer of fructose to sucrose molecules, as storage sugars. Overexpression of fructosil transferases in tobacco and rice produced transgenic plants with increased tolerance to cold temperatures and drought (Pilon-Smits *et al.*, 1995; Vijn and Smeekens, 1999; Li *et al.*, 2007; Kawakami *et al.*, 2008). Starch hydrolysis through b-amylase increases in the presence of stress. Arabidopsis plants deficient in this degradation pathway showed increased sensitivity to cold and osmotic stress (Yano *et al.*, 2005; Valerio *et al.*, 2011). Trehalose is a non-reducing disaccharide which protects membranes and proteins and is accumulated in desiccation-resistant plants. Heterologous expression of trehalose synthesis genes from *E. coli* and *S. cerevisiae* improved tolerance to drought, salt and thermal stress in several crop species (Iordachescu and Imai, 2008; Paul *et al.*, 2008). In Arabidopsis, heterologous overexpression of several rice trehalose-6-phosphate isoforms produced plants tolerant to drought, salt and cold stress, while their loss of function reduced tolerance to heat stress (Avonce *et al*., 2004; Suzuki *et al*., 2008; Li *et al*., 2011). Transgenic Arabidopsis plants producing high levels of raffinose and galactinol also showed increased resistance to drought and salt stress (Taji *et al*., 2002; Nishizawa *et al*., 2008). Polyols are also involved in molecule stabilization and prevention of oxidative damage by hydroxyl radicals (Smirnoff and Cumbers, 1989; Shen *et al*., 1997). Overexpression of biosynthesis enzymes of myo-inositol and its cyclical derivatives increased salt tolerance in tobacco transgenic plants (Sheveleva *et al*., 1997; Majee *et al*., 2004; Patra *et al*., 2010).

## Adaptative Responses at the Whole Plant Level

### *$Na^+$ Transport Regulation*

The preservation of a high $K^+$:$Na^+$ ratio is essential for the survival of most species when subjected to salt stress (Gorham *et al*., 1990; Dubcovsky *et al*., 1998; Cuin *et al*., 2003) and plants must therefore regulate $Na^+$ transport in a very precise manner through processes such as loading from the root into the xylem, unloading in the leaves for vacuolar compartmentalization or recirculation from shoot to root for its extrusion (Munns *et al*., 2002). $Na^+$ enters the root passively through uniporters and channels, being the high affinity transporters HKT and non-selective cation channels (NSCC) the main candidates (Higinbotham, 1973; Rubio *et al*., 1995; Schachtman *et al*., 1997; Amtmann *et al*., 1999; Rus *et al*., 2001; Mäser *et al*., 2002; Garciadeblás *et al*., 2003; Rodríguez-Navarro and Rubio, 2006; Horie *et al*., 2007). The main mechanism by which glycophyte plants prevent $Na^+$ toxicity is the exclusion of this cation, being the $Na^+/H^+$ antiporter the main protein involved (Shi *et al*., 2000; Qiu *et al*., 2002). Once in the root, $Na^+$ diffuses radially through the apoplast towards the cortex until it reaches impermeable barriers such as the Casparian strip (Niu *et al*., 1995; Steudle, 2000). The presence of discontinuities on the endodermis of zones such as the root apex favor $Na^+$ entry into the apoplastic pathway. This mechanism is particularly important in certain species such as rice (Yeo *et al*., 1987; Yadav *et al*., 1996), while ion entry into the root pericycle occurs through the symplastic pathway in most species. In cotton, salt stress favors the quick maturation of the Casparian strip, while growth under high humidity and transpiration conditions produces roots with more permeable endodermis (Hose *et al.*, 2010; Enstone *et al.*, 2003). In Arabidopsis, $Na^+$ loading into the xylem is an active process carried out by the $Na^+/H^+$ antiporter SOS1 (Shi *et al.*, 2002; Qiu *et al.*, 2002). On the other hand, the uniporter HKT1;1 unloads $Na^+$ from the xylem and, probably in an indirect way, contributes to $Na^+$ compartmentalization in the vacuole, having its role in $Na^+$ recirculation through the phloem been speculated in the past (Rus *et al.*, 2001; Berthomieu *et al.*, 2003; Sunarpi *et al.*, 2005; Davenport *et al.*, 2007). Studies in tomato proposed a model in which the orthologues of these proteins coordinate their action to prevent $Na^+$ accumulation in the most vulnerable tissues. $Na^+$ that enters the root is loaded into the xylem by SlSOS1 and unloaded in the stem for its controlled accumulation by the tomato orthologue of HKT1;1. Likewise, the latter unloads this cation into mature leaves with a developed vasculature for its compartmentalization, while SlSOS1 excludes it from the cytosol of young leaves and favors its deposition in the apoplast (Olías *et al.*, 2009).

## *Salt Glands*

Some halophytes have developed these structures to adapt to saline media. Salt glands discharge the excess of salts by actively pumping them through the leave epidermis, where the waxy cuticle keeps them away from the mesophyll cells. Salt accumulation on the epidermis results in a drastic reduction of the osmotic potential, causing a water flux through these glands. For this reason, this strategy is only valid for plants belonging to habitats where water availability is not restricted, such as salty marshes (Tester and Davenport, 2003). Some grasses belonging to the *Chloroideae* subfamily, as well as the wild rice species *Porteresia coarctata*, have developed small bicellular glands which segregate salts to increase their halotolerance (Amarashinge and Watson, 1989; Flowers *et al.*, 1990; McWorther *et al.*, 1995; Marcum, 1999). Halophytes from very dry habitats have a special type of salt gland called salt trichomes. They are composed of two cells on the leaf surface, the outer of which accumulates water and salts until it bursts and releases $Na^+$ without producing excessive water losses (Tester and Davenport, 2003).

### *Plant Growth Inhibition*

Salt stress, like many other types of abiotic stress, inhibits plant growth, which is used to save energy resources that the plant can employ to survive adverse conditions (Zhu, 2001). Excess of $Na^+$ induces stomatal closure by repressing inward rectifying $K^+$ channels in guard cells. $Na^+$ accumulation in the shoot through evapotranspiration is thus prevented, being the impossibility to close the stomata an important cause of salt sensitivity in some species. Stomatal closure implies a lower $CO_2$ uptake and therefore a reduced photosynthetic efficiency, which ultimately diminishes plant growth (Robinson *et al.*, 1997; Véry *et al.*, 1998; Hasegawa *et al.*, 2000). Cyclin-dependent kinases (CDKs) are serine/threonine kinases involved in the cell cycle regulation, being the binding to positive regulators known as cyclins required (Evans *et al.*, 1983; Francis, 2007). Arabidopsis encompasses 7 classes of CDKs, termed CDKA to -F, and the CDK-like CDKL protein type (Menges *et al.*, 2005). Reduced CDK activity results in the cell cycle arrest at the G1/S or G2/M interphases and therefore, the inhibition of cell growth and elongation (Schuppler et al., 1998; Kitsios and Doonan, 2011). Water stress reduces cell expansion or division in pea leaves depending on the developmental stage, while it favors the formation of big cells in maize roots as a result of a delayed mitosis (Lecoeur *et al.*, 1995; Sacks *et al.*, 1997). In wheat mesophyll cells subjected to drought, the reduced cell division was observed to correlate with an inhibition in the activity of CDKA1 caused by posttranslational modifications and changes in the expression level of cyclins (Schuppler *et al.*, 1998; Granier *et al.*, 2000). In Arabidopsis, salt stress downregulates the expression of CDKA and –B types, causing the production of fewer and bigger cells in the shoot and root apex as a result of the mitosis transient inhibition (Burssens *et al.*, 2000; West *et al.*, 2004). ABA is accumulated upon salt and water stress and induces the expression of the CDK inhibitor ICK1, which arrests the cell cycle while enabling DNA endoreplication (Wang *et al.*, 1998; Weinl *et al.*, 2005). Overexpression of this inhibitor produces smaller plants with morphological alterations. Heterologous expression of ICK1 in *Brassica napus* plants produced alterations in both petal morphology and pollen fertility. In the Arabidopsis mutant *sto1*, selected for its increased tolerance to salt and osmotic stress, NaCl-mediated ICK1 accumulation is lower. By fusing ICK1 and deletion mutants of this protein to GFP, three functional domains responsible for its binding to CDKs (at the carboxy-terminal end), nuclear location (central region) and activity inhibition through increase of *in vivo* instability (amino-terminal end) were identified (Wang *et al.*, 2000; Zhou *et al.*, 2003; Ruggiero *et al.*, 2004). DELLA proteins form a nuclear protein subfamily belonging to the plant-specific GRAS family and are named after the conserved DELLA motif near the amino-terminal end (Peng *et al.*, 1997; Pysh *et al.*, 1999). The genome of Arabidopsis encodes 5 DELLA genes, *RGA*, *GAI*, *RGL1*, *RGL2* and *RGL3*, whose main function is the repression of the gibberellin response in the presence of stress (Tyler *et al.*, 2004). RGA1 was the first one to be identified in a screening in which the null mutant *rga1* partially suppressed the dwarf phenotype with delayed flowering of the gibberellin synthesis mutant *ga1-3*. The double mutant *rga1 gai* suppressed the phenotype completely, showing that RGA1 and GAI are the two primary repressors modulating this phenomenon of the gibberellin response, while further studies with the *rgl2* mutant implicated this protein in seed germination in the presence of light (Silverstone *et al.*, 1997; Dill and Sun, 2001; Lee *et al.*, 2002). Stabilized DELLA proteins repress gibberellin signaling, thus inhibiting plant growth under stress conditions. Gibberellin receptor Gibberellin-Insensitive Dwarf 1 (GID1) binds in the presence of this hormone to DELLA proteins and targets them for poliubiquitination-mediated degradation, process in which the positive regulator of gibberellin response SLEEPY1 also participates. The DELLA and VHYNP motifs at the amino-terminal end are necessary for the interaction with GID1, whereas the carboxy-terminal GRAS region is required for binding to SLEEPY1, favoring thus its integrity (Tyler *et al.*, 2004; Sun and Gubler, 2004; Hussain *et al.*, 2005; Ueguchi-Tanaka *et al.*, 2005; Griffiths *et al.*, 2006). A loss-of-function mutant for four of the five DELLA proteins (*gai rga1 rgl1 rgl2*) showed lower sensitivity to developmental delay caused by sublethal concentrations of $Na^+$ and higher content in bioactive gibberellins. Likewise, the response to ABA and ethylene was lower, proving the role of these proteins in integrating growth responses to different environmental signals (Achard *et al.*, 2006). A microarray for the detection of DELLA targets led to the identification of 14 genes, all of them induced by these proteins and repressed by gibberellins. Among them, several gibberellin biosynthesis genes, a gibberellin receptor, negative regulators of the gibberellin signaling pathway and XERICO, responsible for the accumulation of the antagonist to gibberellins ABA, were found (Zentella *et al.*, 2007). DELLA proteins also facilitate plant acclimation to cold temperature and improve their tolerance to oxidative stress. The cold-induced transcription factor CBF1, through the activation of the oxidase GA2, causes the degradation of gibberellins, which in turn favors the stabilization of DELLA proteins. Through the induction of ROS scavenging genes, these proteins protect cells from oxidative damage-mediated cell death (Achard *et al.*, 2008a, 2008b).

## ION TRANSPORT AND TOLERANCE TO STRESS

Ion transporters are essential to the plant, since they allow for both the uptake and intra and extracellular distribution of nutritents and the removal of harmful elements. The genome of Arabidopsis encodes approximately 800 membrane proteins, 65% of which can function as ion and organic compound transporters (Sussman and Harper, 1989). Pumps employ the energy obtained from the hydrolysis of ATP or pyrophosphate to transport ions against their electrochemical gradient. Unlike yeasts, plants lack $Na^+$ ATPases. Their pumps, P-type $H^+$-ATPase at the plasma membrane and V-type $H^+$-ATPase and $H^+$-pyrophosphatase at the tonoplast, transport $H^+$ against their gradient to generate a membrane potential allowing cation/$H^+$ antiporters to actively transport cations (Blumwald *et al.*, 2000; Serrano and Rodríguez-Navarro, 2001). Channels perform a fast and passive transport through the membrane at a rate of $10^7$ molecules per second. Transporters bind to one or two molecules of substrate at a time and transport them at a rate of $10^4$ molecules per second. They can be uniporters, which transport one molecule along its concentration gradient, or symporters and antiporters, which transport two molecules simultaneously, one along and one against the concentration gradient, in the same or opposite directions, respectively (Niu *et al.*, 1995; Chrispeels *et al.*, 1999).

The analysis of the Arabidopsis genome identified four subfamilies of cation/$H^+$ transporters with affinity for $Ca^{2+}$, $Na^+$, $K^+$ and $Mg^{2+}$: CaCA, NhaD, CPA1 and CPA2 (Mäser *et al.*, 2001). The CaCA subfamily is formed by 11 $Ca^{2+}$/$H^+$ transporters called CAX1-11, being some members capable of transporting $Mn^{2+}$ and $Cd^{2+}$ as well. AtMHX, phylogenetically related to the CAX subfamily, is involved in the compartmentalization of $Mg^{2+}$ and $Zn^{2+}$ into the vacuole (Shaul *et al.*, 1999; Cheng *et al.*, 2002;

Gaash et al., 2013). Both members of the NhaD subfamily, NHD1 and NHD2, show similarity with bacterial $Na^+/H^+$ antiporters, having only NHD1 been characterized as chloroplastic $Na^+/H^+$ antiporter involved in the $Na^+$-dependent uptake of pyruvate carried out by the BASS2 transporter (Pardo *et al*., 2006; Furumoto *et al*., 2011). The CPA1 subfamily consists of eight members: NHX1-6, SOS1 and NHX8. They are vacuolar cation/$H^+$ antiporters, a plasma membrane $Na^+/H^+$ antiporter and a plasma membrane $Li^+/H^+$ antiporter, respectively (Blumwald, 2000; Qiu *et al*., 2002; An *et al*., 2007; Bassil *et al*., 2011; Barragán *et al*., 2012). Finally, the CPA2 subfamily includes 28 cation/$H^+$ exchangers CHX and five KEA antiporters (Pardo *et al*., 2006). CHX13, CHX14, CHX17, CHX20 and functionally redundant CHX21 and CHX23, have been characterized *in planta*. They all have been suggested to be $K^+/H^+$ exchangers with roles in $K^+$ uptake, $K^+$ redistribution throughout the plant, $K^+$ uptake and homeostasis, regulation of guard cell movements and $K^+$ homeostasis in the female gametophyte, respectively (Cellier *et al*., 2004; Padmanaban *et al*., 2007; Zhao *et al*., 2008, 2015; Lu *et al*., 2011; Evans *et al*., 2012). Likewise, KEA1-3 have been characterized and demonstrated to constitute chloroplast $K^+/H^+$ antiporters with overlapping functions in chloroplast development, photosynthesis efficiency and rosette growth (Aranda-Sicilia *et al*., 2012; Dana *et al*., 2016)

The involvement of HKT transporters in $K^+$ and $Na^+$ high affinity symport, as well as $Na^+$ low affinity uptake in the root, has been proven. Additionally, their role in $Na^+$ uptake and recirculation through the phloem has been suggested in the past (Rubio *et al*., 1995; Rus *et al*., 2001; Laurie *et al*., 2002; Apse and Blumwald, 2007). The different HKT transporters display changing ionic selectivity when expressed in heterologous systems. The phylogenetic analysis carried out through comparison of amino acid sequences in transporters from monocots and dicots led to the establishment of two subfamilies (Platten *et al*., 2006). Members of subfamily 1 (HKT1;n) are present in both monocots and dicots and show specificity for $Na^+$, possibly due to the presence of a serine residue at the first pore loop. Transporters of subfamily 2 (HKT2;n) belong exclusively to the monocots, transport $Na^+$ and $K^+$, and the serine residue is replaced by glycine (except in OsHKT2;1, which acts as a $Na^+$ uniporter). The function of the only member of this family in Arabidopsis, AtHKT1;1, has been suggested to be long distance transport of $Na^+$.

Under salt stress, HKT1;1 would unload $Na^+$ from the xylem and favor the preferential accumulation of this cation in the root, thus protecting the leaf tissue from the damage caused by its overaccumulation in the shoot. Additionally, it indirectly favors the accumulation of $K^+$ in the shoot through depolarization-mediated activation of $K^+$ outward channels (Rus *et al*., 2004; Sunarpi *et al*., 2005; Davenport *et al*., 2007; Horie *et al*., 2009; Yao *et al*., 2010).

The excess of $Na^+$ interferes with ion uptake, being $K^+$ the most important given its functions within the cell and the similarity between both cations (Rains and Epstein, 1967; Binzel *et al*., 1988; Marschner, 1995; Maathuis *et al*., 1996; Blaha *et al*., 2000; Forment *et al*., 2002). The main family of $K^+$ transporters in Arabidopsis is KT/HAK/KUP, which was initially identified in *E. coli* and termed KT (K$^+$ Transporters), HAK (High Affinity K$^+$ Transporters) or KUP (K$^+$ Uptake Permeases) depending on the research groups originally describing them. The family includes 13 members involved in high and low affinity $K^+$ transport with localization at the plasma membrane and tonoplast (Senn *et al*., 2001; Véry and Sentenac, 2003). Nine of these proteins, KUP1, KUP4, KUP5, KUP6, KUP7, KUP9, KUP10, KUP11 and HAK5, have been proven to transport $K^+$ given their ability to complement $K^+$ uptake mutants of *E. coli*. Among them, KUP1, KUP4, KUP6, KUP7 and HAK5 have also been characterized *in planta*, playing HAK5 the most essential role in $K^+$ uptake under limiting concentrations of this nutrient (Ahn *et al*., 2004; Gierth *et al*., 2005; Kobayashi *et al*., 2010; Osakabe *et al*., 2013; Han *et al*., 2016).

Three families of $K^+$ channels exist in Arabidopsis: "Shaker"-type rectifying channels, two pore outward rectifying channels and non-selective cation channels (Shabala and Cuin, 2007). "Shaker"-type channels are located at the plasma membrane and allow the voltage-dependent $K^+$ transport (Véry and Sentenac, 2002). This family includes hyperpolarization-gated inward rectifying $K^+$ channels such as AKT1, KAT1, KAT2 and SPIK, depolarization-gated outward rectifying channels such as SKOR and GORK or weakly voltage-dependent channels such as AKT2 (Shabala and Cuin, 2007). "Shaker"-type channels control processes such as low affinity $K^+$ uptake through the root, loading of this cation into the xylem, its loading into and unloading from the phloem or its content in guard cells to regulate stomatal movements (Gaymard *et al.*, 1998; Hirsch *et al.*, 1998; Marten *et al.*, 1999; Lacombe *et al.*, 2000; Szyroki *et al.*, 2001). Two pore TPK channels are mostly localized at the tonoplast, where they can act as osmosensors, with the sole exception of TPK4, located at the plasma membrane and other compartments of the secretory pathway. Their $K^+$ outward rectifying ability is $Ca^{2+}$-dependent, having few members been functionally characterized (Schönknecht *et al.*, 2002; Becker *et al.*, 2004; Dunkel *et al.*, 2008; Maathuis, 2011). AtTPK1 is ubiquitously expressed in Arabidopsis and has been involved in vacuolar $K^+$ release during stomatal closure (Czempinski *et al.*, 2002; Gobert *et al.*, 2007). AtTPK4 exhibits low transcript abundance and is predominantly expressed in pollen, where it has been suggested to contribute to the $K^+$ conductance of the pollen tube (Becker *et al.*, 2004). AtTPK5, most abundantly expressed in senescent leaves and petals, is downregulated in the presence of salt stress, suggesting its role in maintaining $K^+$ levels at the cytoplasm or exchanging this cation for $Na^+$ at the vacuole (Maathuis *et al.*, 2006; Voelker *et al.*, 2010). Heterologous expression of AtTPK1, AtTPK2 and ATPK5 restored the growth of a $K^+$ uptake deficient *E. coli* strain in $K^+$-deprived medium (Isayenkov and Maathuis, 2013). Non selective cation channels (NSCC) are located at both the plasma membrane and tonoplast, having 40 candidate members been identified in the genome of Arabidopsis (Mäser *et al.*, 2001; Shabala *et al.*, 2007). They show high affinity for cations versus anions but they do not discriminate between $Na^+$ and $K^+$, being their role as main cause for $Na^+$ entry into the plant and NaCl-induced $K^+$ loss generally accepted (Shabala *et al.*, 2006; Demidchik and Maathuis, 2007; Zelman et al., 2013). This family includes cyclic nucleotide-gated channels (CNGC) and glutamate receptor-like proteins (GLR) (Leng *et al.*, 2002; Demidchik *et al.*, 2004). The CNGC members characterized in Arabidopsis are CNGC2, CNGC3, CNGC4 CNGC7, CNGC8, CNGC10, CNGC11, CNGC12, CNGC16, CNGC18, CNGC19 and CNGC20, suggested to participate in jasmonic acid-induced $Ca^{2+}$ uptake during the immune response (CNGC2 and CNGC4), indiscriminated uptake of $K^+$ and other monovalent cations (CNGC3 and CNGC10), induction of pollen tube tip growth (CNGC7, CNGC8 and CNGC18), transport of $Ca^{2+}$ and $Mg^{2+}$ (CNGC10), several $Ca^{2+}$-dependent physiological responses such as programmed cell death, (CNGC11 and CNGC12), stress tolerance in pollen (CNGC16) and reallocation of $Na^+$ within the plant during salt stress episodes (CNGC19 and CNGC20) (Leng *et al.*,

1999; Clough *et al.*, 2000; Balague *et al.*, 2003; Li *et al.*, 2005; Gobert *et al.*, 2006; Ali *et al.*, 2007; Frietsch *et al.*, 2007; Ma *et al.*, 2007; Urguhart *et al.*, 2007, 2011; Kugler *et al.*, 2009; Guo *et al.*, 2010; Caballero *et al.*, 2012; Tunc-Ozdemir *et al.*, 2013a, 2013b; Jin *et al.*, 2015; Lu *et al.*, 2016). The 20 Arabidopsis GLRs can be grouped into three clades, having only members of clade 3 been functionally characterized so far (Davenport, 2002). These proteins have been shown to constitute $Ca^{2+}$ channels and suggested to participate in the regulation of stomatal aperture (GLR3.1 and GLR3.5), plant defense signaling and response to pathogens (GLR3.3), transmission of environmental stress (GLR3.4) and activation (GLR3.3 and GLR3.6) or deactivation (GLR3.5) of wound-induced signaling (Meyerhoff *et al.*, 2005; Qi *et al.*, 2006; Manzoor *et al.*, 2013; Kong *et al.*, 2016; Salvador-Recatalá, 2016; Yoshida *et al.*, 2016).

## REFERENCES


1. Achard, P., Cheng, H., De Grauwe, L., Decat, J., Schoutteten, H., Moritz, T., Van der Straeten, D., Peng, J., Harberd, N.P. (2006) Integration to plant responses to environmentally activated pytohormonal signals. Science. 311, 91
2. Achard, P., Gong, F., Cheminant, S., Alioua, M., Hedden, P. Genschik, P. (2008a) The coldinducible CBF1 factor-dependent signaling pathway modulates the accumulation of the growth-repressing DELLA proteins via its effect on gibberellin metabolism. Plant Cell. 20, 2117–2129
3. Achard, P., Renou, J.P., Berthome, R., Harberd, N.P. and Genschik, P. (2008b) Plant DELLAs restrain growth and promote survival of adversity by reducing the levels of reactive oxygen species. Curr. Biol. 18, 656–660
4. Ahn, S.J., Shin, R., Schachtman, D.P. (2004) Expression of *KT/KUP* genes in *Arabidopsis* and the role of root hairs in $K^+$ uptake. Plant Physiol. 134, 1135-1145
5. Albinsky, D., Masson, J., Bogucki, A., Afsar, K., Vass, I., Nagy, F., Paszkowski, J. (1999) Plant responses to genotoxic stress are linked to an ABA/salinity signalling pathway. Plant J. 17, 73-82
6. Alcázar, R., Planas, J., Saxena, T., Zarza, X., Bortolotti, C., Cuevas,

J., Bitrian, M., Tiburcio, A.F., Altabella, T. (2010) Putrescine accumulation confers drought tolerance in transgenic Arabidopsis plants overexpressing the homologous arginine decarboxylase 2 gene. Plant Physiology and Biochemistry 48, 547–552

7. Ali, R., Ma, W., Lemtiri-Chlieh, F., Tsaltas, D., Leng, Q., von Bodman, S., Berkowitz, G.A. (2007) Death don't have no mercy and neither does calcium: Arabidopsis CYCLIC NUCLEOTIDE GATED CHANNEL2 and innate immunity. Plant Cell. 19, 1081-1095

8. Allen, R.D., Webb, R.P., Schake, S.A. (1997) Use of transgenic plants to study antioxidant defenses. Free Radic Biol Med. 23, 473-479.

9. Amarashinge, V., Watson, L. (1989) Variation in salt secretory activity of microhairs in grasses. Australian J. Plant Physiol. 16, 219-229

10. Amtmann, A., Jelitto, T.C., Sanders, D. (1999) $K^+$-selective inward-rectifying channels and apoplastic pH in barley roots. Plant Physiol. 120, 331-338

11. An, R., Chen, Q.J., Chai, M.F., Lu, P.L., Su, Z., Qin, Z.X., Chen, J., Wang, X.C. (2007) AtNHX8, a member of the monovalent cation: proton antiporter-1 family in *Arabidopsis thaliana*, encodes a putative $Li^+/H^+$ antiporter. Plant Journal. 49, 718-728

12. Apel, K., Hirt, H. (2004) Reactive oxygen species: metabolism, oxidative stress, and signal transduction. Annu. Rev. Plant Biol. 55, 373-399

13. Apse, M.P., Aharon, G.S., Snedden, W.A., Blumwald, E. (1999) Salt tolerance conferred by overexpression of a vacuolar $Na^+/H^+$ antiport in *Arabidopsis*. Science. 285, 1256-1258

14. Apse, M.P., Blumwald, E. (2007) $Na^+$ transport in plants. FEBS Letters. 25, 2247-2254

15. Aranda-Sicilia, M.N., Cagnac, O., Chanroj, S., Sze, H., Rodríguez-Rosales, M.P., Venema, K. (2012) Arabidopsis KEA2, a homolog of bacterial KefC, encodes a $K^+/H^+$ antiporter with a chloroplast transit peptide. Biochim. Biophys. Acta. 1818, 2362-2371

16. Avonce, N., Leyman, B., Mascorro-Gallardo, J.O., Van Dijck, P., Thevelein, J.M., Iturriaga, G. (2004) The Arabidopsis trehalose-6-P

synthase AtTPS1 gene is a regulator of glucose, abscisic acid, and stress signaling. Plant Physiology 136, 3649–3659

17. Balague, C., Lin, B.Q., Alcon, C., Flottes, G., Malmstrom, S., Kohler, C., Neuhaus, G., Pelletier, G., Gaymard, F., Roby, D. (2003) HLM1, an essential signalling component in the hypersensitive response, is a member of the cyclic nucleotide-gated channel ion channel family. The Plant Cell. 15, 365–379
18. Barragán, V., Leidi, E.O., Andrés, Z., Rubio, L., De Luca, A., Fernández, J.A., Cubero, B., Pardo, J.M. (2012) Ion exchangers NHX1 and NHX2 mediate active potassium uptake into vacuoles to regulate cell turgor and stomatal function in *Arabidopsis*. Plant Cell
19. Bassil, E., Ohto, M.A., Esumi, T., Tajima, H., Zhu, Z., Cagnac, O., Belmonte, M., Peleg, Z., Yamaguchi, T., Blumwald, E. (2011) The Arabidopsis intracellular $Na^+/H^+$ antiporters NHX5 and NHX6 are endosome associated and necessary for plant growth and development. Plant Cell. 23, 224-239
20. Becker, D., Geiger, D., Dunkel, M., Roller, A., Bertl, A., Latz, A., Carpaneto, A., Dietrich, P., Roelfsema, M.R.G., Voelker, Schmidt, D., Müller-Röber, B., Czempinski, K., Hedrich, R. (2004) AtTPK4, an Arabidopsis tandem-pore $K^+$ channel, poised to control the pollen membrane voltage in a pH- and $Ca^{2+}$-dependent manner. Proc Natl Acad Sci USA 101, 15621–15626
21. Berthomieu, P., Conéjéro, G., Nublat, A., Brackenbury, W.J., Lambert, C., Savio, C., Uozumi, N., Oiki, S., Yamada, K., Cellier, F., Gosti, F., Simonneau, T., Essah, P.A., Tester, M., Very, A.A., Sentenac, H., Casse, F. (2003) Functional analysis of AtHKT1 in *Arabidopsis* shows that Na+ recirculation by the phloem is crucial for salt tolerance. Embo J. 22, 2004-2014
22. Bhandal, I.S., Malik, C.P. (1988) Potassium estimation, uptake, and its role in the physiology and metabolism of flowering plants. Intern Rev Cytol. 110, 205-254
23. Binzel, M.L., Hess, F.D., Bressan, R.A., Hasegawa, P.M. (1988) Intracellular compartmentation of ions in salt adapted tobacco cells. Plant Physiol. 86, 607-614
24. Blaha, G., Stelzl, U., Spahn, C.M., Agrawal, R.K., Frank, J., Nierhaus, K.H. (2000) Preparation of functional ribosomal

complexes and effect of buffer conditions on tRNA positions observed by cryoelectron microscopy. Methods Enzymol. 317, 292-309

25. Blumwald, E. (2000). Sodium transport and salt tolerance in plants. Curr Opin Cell Biol. 12, 431-434
26. Bohnert, H.J., Jensen, R.G. (1996) Strategies for engineering water stress tolerance in plants. Trends Biotech. 14, 89-97
27. Bohnert, H.J., Sheveleva, E. (1998) Plant stress adaptations, making metabolism move. Curr Opin Plant Biol. 1, 267-274
28. Borsani, O., Valpuesta, V., Botella, M.A. (2001) Evidence for a role of salicylic acid in the oxidative damage generated by NaCl and osmotic stress in *Arabidopsis* seedlings. Plant Physiol. 126, 1024-1030
29. Bouche, N., Fait, A., Bouchez, D., Moller, S.G., Fromm, H. (2003) Mitochondrial succinicsemialdehyde dehydrogenase of the gammaaminobutyrate shunt is required to restrict levels of reactive oxygen intermediates in plants. Proceedings of the National Academy of Sciences, USA 100, 6843–6848
30. Bressler, E., Hoffman, G.C. (1986) Irrigation management for soil salinity control: theories and tests. Soil.Sci.Am.J. 50, 1552-1560
31. Brini, F., Hanin, M., Lumbreras, V., Amara, I., Khoudi, H., Hassairi, A., Pagés, M., Masmoudi, K. (2007) Overexpression of wheat dehydrin DHN-5 enhances tolerance to salt and osmotic stress in Arabidopsis thaliana. Plant Cell Rep. 26, 2017-2026
32. Burssens, S., Himanen, K., van de Cotte, B., Beeckman, T., Van Montagu, M., Inze, D. (2000) Expression of cell cycle regulatory genes and morphological alterations in response to salt tress in *Arabidopsis thaliana*. Planta 211, 632-40
33. Caballero, F., Botella, M.A., Rubio, L., Fernández, J.A., Martínez, V., Rubio, F. (2012) A Ca(2+)-sensitive system mediates low-affinity K(+) uptake in the absence of AKT1 in Arabidopsis plants. Plant Cell Physiol. 53, 2047-2059
34. Campbell, S.A., Close, T.J. (1997) Dehydrins: genes, proteins and association with phenotypic traits. New Phytologist. 137, 61-74
35. Capell, T., Bassie, L., Christou, P. (2004) Modulation of the

polyamine biosynthetic pathway in transgenic rice confers tolerance to drought stress. Proceedings of the National Academy of Sciences, USA 101, 9909–9914

36. Cellier, F., Conéjéro, G., Ricaud, L., Luu, D.T., Lepetit, M., Gosti, F., Casse, F. (2004) Characterization of AtCHX17, a member of the cation/$H^+$ exchangers, CHX family, from *Arabidopsis thaliana* suggests a role in $K^+$ homeostasis. The Plant Journal. 39, 834–846
37. Chen, T.H., Murata, N. (2011) Glycinebetaine protects plants against abiotic stress: mechanisms and biotechnological applications. Plant,Cell and Environment 34, 1–20
38. Cheng, N.H., Pittman, J.K., Shigaki, T., Hirschi, K.D. (2002) Characterization of CAX4, an *Arabidopsis* $H^+$/cation antiporter. Plant Physiol. 128, 1245-1254
39. Chinnusamy, V., Schumaker, K., Zhu, J.K. (2004) Molecular genetic perspectives on cross-talk and specificity in abiotic stress signalling in plants.J.Exp. Bot. 55, 225-236
40. Chrispeels, M.J., Crawford, N.M., Schroeder, J.I. (1999) Proteins for transport of water and mineral nutrients across the membranes of plant cells. Plant Cell. 11, 661-676.
41. Clough, S.J., Fengler, K.A., Yu, I., Lippok, B., Smith, R.K., Bent, A. (2000) The Arabidopsis dnd1 'defense, no death' gene encodes a mutated cyclic nucleotide-gated ion channel. Proceedings of the National Academy of Sciences, USA. 97, 9323–9328
42. Cuevas, J.C., López-Cobollo, R., Alcázar, R., Zarza, X., Koncz, C., Altabella, T., Salinas, J., Tiburcio, A.F., Ferrando, A. (2008) Putrescine is involved in Arabidopsis freezing tolerance and cold acclimation by regulating abscisic acid levels in response to low temperature. Plant Physiology 148, 1094–1105
43. Cuin, T.A., Miller, A.J., Laurie, S.A., Leigh, R.A. (2003) Potassium activities in cell compartments of salt-grown barley leaves. J. Exp. Bot. 54, 657-661
44. Czempinski, K., Zimmermann, S., Ehrhardt, T., Müller-Röber, B. (1997) New structure and function in plant $K^+$ channels: KCO1, an outward rectifier with a steep Ca2+ dependency. EMBO J. 16, 2565–2575
45. Dana, S., Herdean, A., Lundin, B., Spetea, C. (2016) Each of the

chloroplast potassium efflux antiporters affects photosynthesis and growth of fully developed Arabidopsis rosettes under short-day photoperiod. Physiol. Plant. 158, 483-491

46. Davenport, R. (2002) Glutamate receptors in plants. Annals of Botany. 90, 549-57

47. Davenport, R.J., Muñoz-Mayor, A., Jha, D., Essah, P.A., Rus, A., Tester, M. (2007) The $Na^+$ transporter AtHKT1;1 controls retrieval of $Na^+$ from the xylem in *Arabidopsis*. Plant, Cell and Environment. 30, 497-507

48. Demidchik, V., Essah, P.A., Tester, M. (2004) Glutamate activates cation currents in the plasma membrane of Arabidopsis root cells. Planta. 219, 167-175

49. Demidchik, V., Maathuis, F.J. (2007) Physiological roles of nonselective cation channels in plants: from salt stress to signalling and development. New Phytol. 175, 387-404

50. Dill, A., Sun, T.-P. (2001) Synergistic de-repression of gibberellin signaling by removing RGA and GAI function in *Arabidopsis thaliana*. Genetics 159, 777-785

51. Dubcovsky, J., Santa María, G., Epstein, E., Luo, M.C., Dvorák, J. (1996) Mapping of the $K^+$-$Na^+$ discrimination locus *Kna1* in wheat. Theor. Appl. Genet. 92, 448-454

52. Dunkel, M., Latz, A., Schumacher, K., Müller, T., Becker, D., Hedrich, R. (2008) Targeting of vacuolar membrane localized members of the TPK channel family. Mol Plant 1, 938–949

53. Enstone, D.E., Peterson, C.A., Ma, F. (2003) Root endodermis and exodermis: structure, function and responses to the environment. J. Plant Growth Regul. 21, 335-351

54. Evans, T., Rosenthal, E.T., Youngblom, J., Distel, D., Hunt, T. (1983) Cyclin: a protein specified by maternal mRNA in sea urchin eggs that is destroyed at each cleavage division. Cell 33:389-96

55. Evans, A.R., Hall, D., Pritchard, J., Newbury, H.J. (2012) The roles of the cation transporters CHX21 and CHX23 in the development of *Arabidopsis thaliana*. J Exp Bot 63, 59-67

56. Fait, A., Fromm, H., Walter, D., Galili, G., Fernie, A.R. (2008) Highway or by way: the metabolic role of the GABA shunt in plants.

Trends in Plant Science 13, 14–19

57. Flowers, T.J. (2004) Improving crop salt tolerance. J Exp Bot. 55, 307-319
58. Flowers, T.J., Flowers, S.A., Hajibagheri, M.A., Yeo, A.R. (1990) Salt tolerance in the halophytic wild rice *Porteresia coarctata cv Tateoka*. New Phytologist. 114, 675-684
59. Flowers, T.J., Yeo, A.R. (1995) Breeding for salinity resistance in crop plants. Where next? Australian J. Plant Physiol. 22, 875-884.
60. Forment, J., Naranjo, M.A., Roldan, M., Serrano, R., Vicente, O. (2002) Expression of *Arabidopsis* SR-like splicing proteins confers salt tolerance to yeast and transgenic plants. Plant J. 30, 511-519
61. Francis D. (2007) The plant cell cycle—15 years on. NewPhytol 174, 261-78
62. Frietsch, S., Wang, Y.F., Sladek, C., Poulsen, L.R., Romanowsky, S.M., Schroeder, J.I., Harper, J.F. (2007) A cyclic nucleotide-gated channel is essential for polarized tip growth of pollen. Proc Natl. Acad. Sci. USA. 104, 14531-14536
63. Furumoto T, Yamaguchi T, Ohshima-Ichie Y, Nakamura M, Tsuchida-Iwata Y, Shimamura M, Ohnishi J, Hata S, Gowik U, Westhoff P,Bräutigam A, Weber AP, Izui K. (2011) A plastidial sodium-dependent pyruvate transporter. Nature. 476, 472-475
64. Gaash, R., Elazar, M., Mizrahi, K., Avramov-Mor, M., Berezin, I., Shaul, O. (2013) Phylogeny and a structural model of plant MHX transporters. BMC Plant Biol. 13, 75
65. Garciadeblás, B., Senn, M.E., Bañuelos, M.A., Rodríguez-Navarro, A (2003) Sodium transporters and HKT transporters: the rice model. The Plant Journal. 34, 788-801
66. Gaxiola, R.A., Li, J., Undurraga, S., Dang, L.M., Allen, G.J., Alper, S.L., Fink, G.R. (2001) Drought- and salt tolerant plants result from the overexpression of the AVP1 $H^+$-pump. Proc. Natl. Acad. Sci. USA. 98, 11444-11449
67. Gaymard, F., Pilot, G., Lacombe, B., Bouchez, D., Bruneau, D., Boucherez, J., Michaux-Ferriere, N., Thibaud, J.B., Sentenac, H. (1998). Identification and disruption of a plant shaker-like outward channel involved in K+ release into the xylem sap. Cell. 94, 647-

655

68. **Ghassemi, F.**, Jakeman, A.J., Nix, H.A. (1995) Salinization of land and water resources. Human causes, extent, management and case studies. Sydney: University of New South Wales
69. Gierth, M., Mäser, P., Schroeder, J.I. (2005) The potassium transporter AtHAK5 functions in $K^+$ deprivation-induced high-affinity K+ uptake and AKT1 $K^+$ channel contribution to $K^+$ uptake kinetics in *Arabidopsis* roots. Plant Physiol. 137, 1105-1114
70. Gobert, A., Isayenkov, S., Voelker, C., Czempinski, K., Maathuis, F.J.M. (2007) The two-pore channel TPK1 gene encodes the vacuolar $K^+$ conductance and plays a role in K+ homeostasis. Proc. Natl. Acad. Sci. U.S.A. 104, 10726–10731
71. Gorham, J., Wyn-Jones, R.G., Bristol, A. (1990) Partial characterization of the trait of enhanced $K^+$-$Na^+$ discrimination in the D genome of the wheat plant. Planta. 180, 590-597
72. Granier, C., Inze, D., Tardieu, F. (2000) Spatial distribution of cell division rate can be deduced from that of p34(cdc2) kinase activity in maize leaves grown at contrasting temperatures and soil water conditions. Plant Physiol 124, 1393-402
73. Griffiths, J., Murase, K., Rieu, I., Zentella, R., Zhang, Z.L., Powers, S.J., Gong, F., Phillips, A.L., Hedden, P., Sun, T.P., and Thomas, S.G. (2006) Genetic Characterization and Functional Analysis of the GID1 Gibberellin Receptors in Arabidopsis. Plant Cell 18, 3399-3414
74. Groppa, M.D., Benavides, M.P. (2008) Polyamines and abiotic stress: recent advances. Amino Acids 34, 35–45.
75. Guo, K.M., Babourina, O., Christopher, D.A., Borsic, T., Rengel, Z. (2010) The cyclic nucleotide-gated channel AtCNGC10 transports $Ca^{2+}$ and $Mg^{2+}$ in Arabidopsis. Physiol Plant. 139, 303-312
76. Han, M., Wu, W., Wu, W.H., Wang, Y. (2016) Potassium Transporter KUP7 Is involved in $K^+$ acquisition and translocation in Arabidopsis root under $K^+$-limited conditions. Mol Plant. 7, 437-446
77. Hasegawa, P.M., Bressan, R.A., Zhu, J.K., Bohnert, H.J. (2000) Plant Cellular And Molecular Responses To High Salinity. Annu Rev Plant Physiol Plant Mol Biol. 51, 463-499

78. Hernández, J.A., Ferrer, M.A., Jiménez, A., Barceló, A.R., Sevilla, F. (2001) Antioxidant systems and (·O2$^-$)/ $H_2O_2$ production in the apoplast of pea leaves. Its relation with salt-induced necrotic lesions in minor veins. Plant Physiol. 127, 817-831

79. Higinbotham, N. (1973) Conceptual developments in membrane transport, 1924-1974. Plant Physiol. 54, 454-462

80. Hirsch, R.E., Lewis, B.D., Spalding, E.P., Sussman, M.R. (1998) A role for the AKT1 potassium channel in plant nutrition. Science. 280, 918-921

81. Hong, Z., Lakkineni, K., Zhang, Z., Verma, D.P. (2000) Removal of feedback inhibition of delta(1)-pyrroline-5-carboxylate synthetase results in increased proline accumulation and protection of plants from osmotic stress. Plant Physiol. 122, 1129-1136

82. Horie, T., Costa, A., Kim, T.H., Han, M.J., Horie, R., Leung, H.Y., Miyao, A., Hirochika, H., An, G., Schroeder, J.I. (2007) Rice OsHKT2;1 transporter mediates large $Na^+$ influx component into $K^+$-starved roots for growth. EMBO J. 26, 3003-3014

83. Horie, T., Hauser, F., Schroeder, J.I. (2009) HKT transporter-mediated salinity resistance mechanisms in Arabidopsis and monocot plants. Trends Plant Sci 14, 660-668

84. Hose, E., Clarkson, D.T., Steudle, E., Schreiber, L., Hartung, W. (2001) The exodermis: a variale apoplastic barrier. J. Exp. Bot. 52, 2245-2264

85. Hussain, S.S., Ali, M., Ahmad, M., Siddique, K.H. (2011) Polyamines: natural and engineered

86. abiotic and biotic stress tolerance in plants. Biotechnology Advances. 29, 300–311

87. Hussain, A., Cao, D., Cheng, H., Wen, Z., and Peng, J. (2005) Identification of the conserved serine/threonine residues important for gibberellin-sensitivity of Arabidopsis RGL2 protein. Plant J. 44: 88-99

88. Ingram, J., Bartels, D. (1996) The molecular basis of dehydration tolerance in plants: metabolic limitations. Plant Physiol. 122, 747-756

89. Iordachescu, M., Imai, R. (2008) Trehalose biosynthesis in response

to abiotic stresses. Journal of Integrative Plant Biology 50, 1223–1229

90. Isayenkov, S., Maathuis, F.J (2013) Arabidopsis thaliana vacuolar TPK channels form functional $K^+$ uptake pathways in *Escherichia coli*. Plant Signal Behav. 8(7), e24665
91. Jin, Y., Jing, W., Zhang, Q., Zhang, W. (2015) Cyclic nucleotide gated channel 10 negatively regulates salt tolerance by mediating Na+ transport in Arabidopsis. J Plant Res. 128, 211-220
92. **Johansson, I.**, Larsson, C., Ek, B., Kjellbom, P. (1996) Water transport activity of the plasma membrane aquaporin PM28A is regulated by phosphorylation. Plant Cell. 10, 451-459.
93. Kawakami, A., Sato, Y., Yoshida, M. (2008) Genetic engineering of rice capable of synthesizing fructans and enhancing chilling tolerance. Journal of Experimental Botany 59, 793–802
94. Kishor, P., Hong, Z., Miao, G.H., Hu, C., Verma, D. (1995) Overexpression of D-pyrroline-5-carboxylate synthetase increases proline production and confers osmotolerance in transgenic plants.. Plant Physiol. 108, 1387-1394
95. Kitsios, G., Doonan, J.H. (2011) Cyclin dependent protein kinases and stress responses in plants. Plant Signaling & Behavior 6, 204-209
96. Kobayashi, D., Uozumi, N., Hisamatsu, S., Yamagami, M. (2010) AtKUP/HAK/KT9, a $K^+$ transporter from *Arabidopsis thaliana*, mediates $Cs^+$ uptake in *Escherichia coli*. Biosci Biotechnol Biochem. 74, 203-205
97. Kong, D., Hu, H.C., Okuma, E., Lee, Y., Lee, H.S., Munemasa, S., Cho, D., Ju, C., Pedoeim, L., Rodríguez, B., Wang, J., Im, W., Murata, Y., Pei, Z.M., Kwak, J.M. (2016) L-Met Activates Arabidopsis GLR $Ca^{2+}$ Channels Upstream of ROS Production and Regulates Stomatal Movement. Cell Rep. 17, 2553-2561
98. Krasensky, J., Jonak, C. (2011) Drought, salt and temperature stress-induced metabolic rearrengements and regulatory networks. J. Exp. Bot. 63, 1593-1608
99. Kugler, A., Köhler, B., Palme, K., Wolff, P., Dietrich, P. (2009) Salt-dependent regulation of a CNG channel subfamily in Arabidopsis. BMC Plant Biol. 9, 140-150

100. Lacombe, B., Pilot, G., Michard, E., Gaymard, F., Sentenac, H., Thibaud, J.B. (2000) A shaker-like $K^+$ channel with weak rectification is expressed in both source and sink phloem tissues of *Arabidopsis*. Plant Cell. 12, 837-851

101. Laurie, S., Feeney, K.A., Maathuis, F.J., Heard, P.J., Brown, S.J., Leigh, R.A. (2002) A role for HKT1 in sodium uptake by wheat roots. Plant Journal. 32, 139-149

102. Lecoeur, Jr. M., Wery, J., Turc, O., Tardieu, F.O. (1995) Expansion of pea leaves subjected to short water deficit: cell number and cell size are sensitive to stress at different periods of leaf development. J Exp Bot 46, 1093-101

103. Lee, S., Cheng, H., King, K.E., Wang, W., He, Y., Hussain, A., Lo, J., Harberd, N.P., Peng, J. (2002) Gibberellin regulates Arabidopsis seed germination via RGL2, a GAI/RGAlike gene whose expression is up-regulated following imbibition. Genes Dev. 16, 646-658

104. Leng, Q., Mercier, R.W., Hua, B.G., Fromm, H., Berkowitz, G.A. (2002) Electrophysiological analysis of cloned cyclic nucleotide-gated ion channels. Plant Physiol. 128, 400-410

105. Leng, Q., Mercier, R.W., Yao, W.Z., Berkowitz, G.A. (1999) Cloning and first functional characterization of a plant cyclic nucleotide-gated cation channel. Plant Physiology 121, 753–761

106. Li, X.L., Borsics, T., Harrington, H.M., Christopher, DA. (2005) Arabidopsis AtCNGC10 rescues potassium channel mutants of *E. coli*, yeast and Arabidopsis and is regulated by calcium/calmodulin and cyclic GMP in *E. coli*. Functional Plant Biology. 32, 643–653

107. Li, H.J., Yang, A.F., Zhang, X.C., Gao, F., Zhang, J.R. (2007) Improving freezing tolerance of transgenic tobacco expressing sucrose: sucrose 1-fructosyltransferase gene from *Lactuca sativa*. Cell Tisssue and Organ Culture 89, 37–48

108. Li, H.W., Zang, B.S., Deng, X.W., Wang, X.P. (2011) Overexpression of the trehalose-6-phosphate synthase gene OsTPS1 enhances abiotic stress tolerance in rice. Planta 234, 1007–1018

109. Lu, Y., Chanroj, S., Zulkifli, L., Johnson, M.A., Uozumi, N., Cheung, A., Sze, H. (2011) Pollen tubes lacking a pair of $K^+$ transporters fail to target ovules in Arabidopsis. Plant Cell. 23, 81-93

110. Lu, M., Zhang, Y., Tang, S., Pan, J., Yu, Y., Han, J., Li, Y., Du, X.,

Nan, Z., Sun, Q. (2016) AtCNGC2 is involved in jasmonic acid-induced calcium mobilization. J Exp Bot. 67, 809-819

111. Ma, W., Yoshioka, K., Berkowitz, G.A. (2007) Cyclic nucleotide gated channels and ca-mediated signal transduction during plant innate immune response to pathogens. Plant Signal Behav. 2, 548-550
112. Maathuis, F. (2006) The role of monovalent cation transporters in plant responses to salinity. Journal of Experimental Botany. 57(5), 1137-1147
113. Maathuis, F. (2011) Vacuolar two-pore $K^+$ channels act as vacuolar osmosensors. New Phytol. 191, 84-91
114. Maathuis, F., Verlin, D., Smith, F., Sanders, D., Fernandez, J., Walker, N. (1996) The physiological relevance of $Na^+$-coupled $K^+$-transport. Plant Physiol. 112, 1609-1616.
115. Majee, M., Maitra, S., Dastidar, K.G., Pattnaik, S., Chatterjee, A., Hait, N.C., Das, K.P., Majumder, A.L. (2004) A novel salt-tolerant L-myo-inositol-1-phosphate synthase from *Porteresia coarctata* (Roxb.) Tateoka, a halophytic wild rice: molecular cloning, bacterial overexpression, characterization, and functional introgression into tobacco-conferring salt tolerance phenotype. Journal of Biological Chemistry 279, 28539–28552
116. Manzoor, H., Kelloniemi, J., Chiltz, A., Wendehenne, D., Pugin, A., Poinssot, B., Garcia-Brugger, A. (2013) Involvement of the glutamate receptor AtGLR3.3 in plant defense signaling and resistance to *Hyaloperonospora arabidopsidis*. Plant J. 76, 466-480
117. Marcum, K.B. (1999) Salinity tolerance mechanisms of grasses in the subfamily chloridoideae. Crop Sci. 39, 1153-1160
118. Marschner, H. (1995) 2nd Ed. Mineral nutrition of higher plants. Academic Press, London
119. Marten, I., Hoth, S., Deeken, R., Ache, P., Ketchum, K.A., Hoshi, T., Hedrich, R. (1999) AKT3, a phloem-localized K+ channel, is blocked by protons. Proc Natl Acad Sci USA. 96, 7581-7586
120. Mäser, P., Thomine, S., Schroeder, J.I., Ward, J.M., Hirschi, K., Sze, H., Talke, I.N., Amtmann, A., Maathuis, F.J., Sanders, D., Harper, J.F., Tchieu, J., Gribskov, M., Persans, M.W., Salt, D.E., Kim, S.A., Guerinot, M.L. (2001) Phylogenetic relationships within cation

transporter families of *Arabidopsis*. Plant Physiol. 126, 1646-1667.

121. Mäser, P., Eckelman, B., Vaidyanathan, R., Horie, T., Fairbairn, D.J., Kubo, M., Yamagami, M., Yamaguchi, K., Nishimura, M., Uozumi, N., Robertson, W., Sussman, M.R., Schroeder, J.I. (2002) Altered shoot/root $Na^+$ distribution and bifurcating salt sensitivity in *Arabidopsis* by genetic disruption of the $Na^+$ transporter AtHKT1. FEBS Lett. 531, 157-161
122. McWorther, C.G., Paul, R.N., Outzs, J.C. (1995) Bicellular trichomes of johnson grass (*Sorghum halepense*) leaves. Morphology, histochemistry and function. Weed Science. 43, 201-208
123. Menges M, de Jager SM, Gruissem W, Murray JA. (2005) Global analysis of the core cell cycle regulators of Arabidopsis identifies novel genes, reveals multiple and highly specific profiles of expression and provides a coherent model for plant cell cycle control. Plant J. 41, 546-66
124. Meyerhoff, O., Müller, K., Roelfsema, M.R., Latz, A., Lacombe, B., Hedrich, R., Dietrich, P., Becker, D. (2005) AtGLR3.4, a glutamate receptor channel-like gene is sensitive to touch and cold. Planta. 222, 418-427
125. Moran, J.F., Becana, M., Iturbe-Ormatexe, I., Fechilla, S., Klucas, R.V., Aparicio-Tejo, P. (1994) Drought induces oxidative stress in pea plants. Planta. 194, 346-352
126. Munns, R. (2002) Comparative physiology of salt and water stress. Plant Cell Environ. 25, 239-250
127. Munns, R., Tester, M. (2008) Mechanisms of salinity tolerance. Annu. Rev. Plant Biol. 59, 651-681
128. Nass, R., Cunningham, K.W., Rao, R. (1997) Intracellular sequestration of sodium by a novel $Na^+/H^+$ exchanger in yeast is enhanced by mutations in the plasma membrane $H^+$-ATPase. J. Biol. Chem. 272, 26145-26152.
129. Nishizawa, A., Yabuta, Y., Shigeoka, S. (2008) Galactinol and raffinose constitute a novel function to protect plants from oxidative damage. Plant Physiology 147, 1251–1263
130. Niu, X., Bressan, R.A., Hasegawa, P.M., Pardo, J.M. (1995) Ion Homeostasis in NaCl Stress Environments. Plant Physiol. 109, 735-742

131. Nuccio, M.L., Rhodes, D., McNeil, S.D., Hanson, A.D. (1999) Metabolic engineering of plants for osmotic stress resistance. Curr. Opin. Plant Biol. 2, 128-134

132. Olías, R., Eljakaoui, Z., Pardo, J.M., Belver, A. (2009) The $Na^+$/ $H^+$ exchanger SOS1 controls extrusion and distribution of Na+ in tomato plants under salinity stress. Plant Signaling & Behavior. 4:10, 973-976

133. Osakabe, Y., Arinaga, N., Umezawa, T., Katsura, S., Nagamachi, K., Tanaka, H., Ohiraki, H., Yamada, K., Seo, S.U., Abo, M., Yoshimura, E., Shinozaki, K., Yamaguchi-Shinozaki, K. (2013) Osmotic stress responses and plant growth controlled by potassium transporters in Arabidopsis. Plant Cell. 25, 609-624

134. Padmanaban S., Chanroj S., Kwak J. M., Li X., Ward J. M., Sze H. (2007) Participation of endomembrane cation/$H^+$ exchanger CHX20 in osmoregulation of guard cells. Plant Physiol. 144, 82–93

135. Pardo, J.M., Cubero, B., Leidi, E.O., Quintero, F.J. (2006) Alkali cation exchangers: roles on cellular homeostasis and stress tolerance. J Exp Bot. 57, 1181-1199

136. Park, E.J., Jeknic, Z., Chen, T.H., Murata, N. (2007) The codA transgene for glycinebetaine synthesis increases the size of flowers and fruits in tomato. Plant Biotechnology Journal 5, 422–430

137. Patra, B., Ray, S., Richter, A., Majumder, A.L. (2010) Enhanced salt tolerance of transgenic tobacco plants by co-expression of PcINO1and McIMT1 is accompanied by increased level of myo-inositol and methylated inositol. Protoplasma 245, 143–152

138. Paul, M.J., Primavesi, L.F., Jhurreea, D., Zhang, Y.H. (2008) Trehalose metabolism and signaling. Annual Review of Plant Biology 59, 417–441

139. Peng, J., Carol, P., Richards, D.E., King, K.E., Cowling, R.J., Murphy, G.P., and Harberd, N.P. (1997) The Arabidopsis GAI gene defines a signalling pathway that negatively regulates gibberellin responses. Genes Dev. 11: 3194-3205

140. Pilon-Smits, E., Ebskamp, M., Paul, M.J., Jeuken, M., Weisbeek, P.J., Smeekens, S. (1995) Improved performance of transgenic fructan-accumulating tobacco under drought stress. Plant Physiol. 107, 125-130

141. Platten, J.D., Cotsaftis, O., Berthomieu, P., Bohnert, H., Davenport, R.J., Fairbairn, D.J., Horie, T., Leigh, R.A., Lin, H.X., Luan, S., Mäser, P., Pantoja, O., Rodríguez-Navarro, A., Schachtman, D.P., Schroeder, J.I., Sentenac, H., Uozumi, N., Véry, A.A., Zhu, J.K., Dennis, E.S., Tester, M. (2006) Nomenclature for HKT transporters, key determinants of plant salinity tolerance. Trends Plant Sci. 11(8), 372-374

142. Pysh, L.D., Wysocka-Diller, J.W., Camilleri, C., Bouchez, D., Benfey, P.N. (1999) The GRAS gene family in Arabidopsis: sequence characterization and basic expression analysis of the SCARECROW-LIKE genes. Plant J. 18: 111-119

143. Qi, Z., Stephens, N.R., Spalding, E.P. (2006) Calcium entry mediated by GLR3.3, an Arabidopsis glutamate receptor with a broad agonist profile. Plant Physiol. 142, 963-971

144. Qiu, Q.S., Guo, Y., Dietrich, M.A., Schumaker, K.S., Zhu, J.K. (2002) Regulation of SOS1, a plasma membrane $Na^+/H^+$ exchanger in *Arabidopsis thaliana*, by SOS2 and SOS3. Proc Nat Acad Sci USA. 99, 8436-8441.

145. Rains, D., Epstein, E. (1967) Sodium absorption bybarley roots. Its mediation by mechanisms 2 of alkali cation transport. Plant Physiol. 42, 224-228

146. Renault, H., Roussel, V., El Amrani, A., Arzel, M., Renault, D., Bouchereau, A., Deleu, C. (2010) The Arabidopsis pop2-1 mutant reveals the involvement of GABA transaminase in salt stress tolerance.BMC Plant Biology 10, 20

147. Rhodes, D., Hanson, A.D. (1993) Quaternary ammonium and tertiary sulfonium compounds in higher plants. Ann. Rew Plant Physiol Plant Mol Biol. 44, 357-384

148. Rodríguez-Navarro, A., Rubio, F. (2006) High-affinity potassium and sodium transport systems in plants. Journal of Experimental Botany 57, 1149–1160

149. Robinson, M.F., Véry, A.A., Sanders, D., Mansfield, T.A. (1997) How can stomata contribute to salt tolerance? Ann. Bot. 80, 387-393

150. Roxas, V.P., Smith, R.K., Jr., Allen, E.R., Allen, R.D. (1997) Overexpression of glutathione S-transferase/glutathione peroxidase

enhances the growth of transgenic tobacco seedlings during stress. Nat Biotechnol.15, 988-991

151. Rubio, F., Gassmann, W., Schroeder, J.I. (1995) Sodium-driven potassium uptake by the plant potassium transporter HKT1 and mutations conferring salt tolerance. Science. 270, 1660-1663

152. Ruggiero, B., Koiwa, H., Manabe, Y., Quist, T.M., Inan, G., Saccardo, F., Joly, R.J., Hasegawa, P.M., Bressan, R.A., Maggio, A. (2004) Uncoupling the effects of abscisic acid on plant growth and water relations. Analysis of *sto1/nced3*, an abscisic acid-deficient but salt stress-tolerant mutant in Arabidopsis. Plant Physiol. 136, 3134-3147

153. Rus, A., Lee, B.H., Munoz-Mayor, A., Sharkhuu, A., Miura, K., Zhu, J.K., Bressan, R.A., Hasegawa, P.M. (2004) AtHKT1 facilitates $Na^+$ homeostasis and $K^+$ nutrition in planta. Plant Physiol. 136, 2500-2511

154. Rus, A., Yokoi, S., Sharkhuu, A., Reddy, M., Lee, B.H., Matsumoto, T.K., Koiwa, H., Zhu, J.K., Bressan, R.A., Hasegawa, P.M. (2001) AtHKT1 is a salt tolerance determinant that controls $Na^+$ entry into plant roots. Proc Natl Acad Sci USA. 98, 14150-14155

155. Sacks, M.M., Silk, W.K., Burman, P. (1997) Effect of Water Stress on Cortical Cell Division Rates within the Apical Meristem of Primary Roots of Maize. Plant Physiol 114, 519-27

156. Salleh, F.M., Evans, K., Goodall, B., Machin, H., Mowla, S.B., Mur, L.A., Runions, J., Theodoulou, F.L., Foyer, C.H., Rogers, H.J. (2012) A novel function for a redox-related LEA protein (SAG21/AtLEA5) in root development and biotic stress responses. Plant Cell Environ. 35, 418-429

157. Salvador-Recatalà, V. (2016) New roles for the GLUTAMATE RECEPTOR-LIKE 3.3, 3.5, and 3.6 genes as on/off switches of wound-induced systemic electrical signals. Plant Signal Behav. 11, e1161879

158. Schachtman, D.P., Kumar, R., Schröeder, J.I., Marsh, E.L. (1997) Molecular and functional characterisation of a novel low-affinity cation transporter (LCT1) in higher plants. Proc. Natl. Acad. Sci. USA. 94, 11079-11084

159. Schönknecht, G., Spoormaker, P., Steinmeyer, R., Brüggeman,

L., Ache, P., Dutta, R., Reintanz, B., Godde, M., Hedrich, R., and Palme, K. (2002) KCO1 is a component of the slow-vacuolar SV ion channel. FEBS Lett. 511, 28-32

160. Schuppler, U., He, P.H., John, P.C., Munns, R. (1998) Effect of water stress on cell division and cell-division-cycle 2-like cell-cycle kinase activity in wheat leaves. Plant Physiol. 117, 667-678
161. Senn, M.E., Rubio, F., Banuelos, M.A., Rodríguez-Navarro, A. (2001) Comparative functional features of plant potassium HvHAK1 and HvHAK2 transporters. J Biol Chem. 276, 44563-44569
162. Serrano, R. (1996) Salt tolerance in plants and microorganisms: toxicity targets and defense responses. Int Rev Cytol. 165, 1-52
163. Serrano, R., Mulet, J.M., Ríos, G., Márquez, J.A., Larrinoa, I.F., Leube, M.P., Mendizábal, I., Pascual-Ahuir, A., Proft, M., Ros, R., Montesinos, C. (1999) A glimpse of the mechanisms of ion homeostasis during salt stress. J. Exp. Botany 50, 1023-1036
164. Serrano, R., Rodríguez-Navarro, A. (2001) Ion homeostasis during salt stress in plants. Curr. Opin. Cell Biol. 13, 399-404
165. Shabala, S. (2003) Regulation of potassium transport in leaves:from molecular to tissue level. Ann Bot, London. 92, 627-634.
166. Shabala, S., Demidchik, V., Shabala, L., Cuin, T.A., Smith, S.J., Miller, A.J., Davies, J.M., Newman, I.A. (2006) Extracellular $Ca^{2+}$ ameliorates NaCl-induced $K^+$ loss from Arabidopsis root and leaf cells by controlling plasma membrane $K^+$ -permeable channels. Plant Physiol. 141, 1653-1665.
167. Shabala, S., Cuin, T.A. (2007) Potassium transport and plant salt tolerance. Physiologia Plantarum 133, 651-669
168. Shaul, O., Hilgemann, D.W., Almeida-Engler, J., Van Montagu, M., Inz, D., Galili, G. (1999) Cloning and characterization of a novel $Mg^{2+}/H^+$ exchanger. EMBO J. 18, 3973-3980
169. Shelp, B.J., Bown, A.W., McLean, M.D. (1999) Metabolism and functions of gammaaminobutyric acid. Trends in Plant Science 4, 446–452
170. Shen, B., Jensen, R.G., Bohnert, H.J. (1997) Increased resistance to oxidative stress in transgenic plants by targeting mannitol biosynthesis to chloroplasts. Plant Physiol. 113, 1177-1183

171. Sheveleva, E., Chmara, W., Bonhert, H.J., Jensen, R.G. (1997) Increased salt and drought tolerance by D-ononitol production in transgenic *Nicotiana tabacum* L. Plant Physiol. 115, 1211-1219

172. Shi, H., Ishitani, M., Kim, C., Zhu, J.K. (2000) The *Arabidopsis* thaliana salt tolerance gene *SOS1* encodes a putative $Na^+/H^+$ antiporter. Proc Natl Acad Sci USA. 97, 6896-6901.

173. Shi, H., Lee, B.H., Wu, S.J., Zhu, J.K. (2003) Overexpression of a plasma membrane $Na^+/H^+$ antiporter gene improves salt tolerance in *Arabidopsis* thaliana. Nat Biotechnol. 21, 81-85

174. Silverstone, A.L., Mak, P.Y.A., Casamitjana-Martínez, E., Sun, T.-P. (1997) The new RGA locus encodes a negative regulator of gibberellin response in *Arabidopsis thaliana*. Genetics 146, 1087-1099

175. Smirnoff, N. (1993) The role of active oxygen in the response of plants to water-deficit and desiccation. New Phytol. 125, 27-58

176. Smirnoff, N., Cumbes, Q.J. (1989) Hydroxyl radical scavenging activity of compatible solutes. Phytochemistry 28, 1057–1060

177. Steudle, E. (2000) Water uptake by roots: effects of water deficit. J. Exp. Bot. 51, 1531-1542

178. Sun, T.-P., Gubler, F. (2004) Molecular mechanism of gibberellin signaling in plants. Annu. Rev. Plant Biol. 55,197-223

179. Sunarpi, Horie, T., Motoda, J., Kubo, M., Yang, H., Yoda, K., Horie, R., Chan, W.Y., Leung, H.Y., Hattori, K., Konomi, M., Osumi, M., Yamagami, M., Schroeder, J.I., Uozumi, N. (2005) Enhanced salt tolerance mediated by AtHKT1 transporter-induced $Na^+$ unloading from xylem vessels to xylem parenchyma cells. Plant J. 44, 928-938

180. Sussman, M.R., Harper, J.F. (1989) Molecular biology of the plasma membrane of higher plants. Plant Cell. 1, 953-960

181. Suzuki, N., Bajad, S., Shuman, J., Shulaev, V., Mittler, R. (2008) The transcriptional coactivatorM BF1c is a key regulator of thermotolerance in Arabidopsis thaliana. Journal of Biological Chemistry 283, 9269–9275

182. Szabados, L., Savoure, A. (2010) Proline: a multifunctional amino acid. Trends in Plant Science 15, 89–97

183. Szábolcs, I. (1994). Soil salinization. Handbook of plant crop stress.

Pressarkli, M. Ed. Marcel Dekker, New York, pp: 3-11

184. Szyroki, A., Ivashikina, N., Dietrich, P., Roelfsema, M.R., Ache, P., Reintanz, B., Deeken, R., Godde, M., Felle, H., Steinmeyer, R., Palme, K., Hedrich, R. (2001) KAT1 is not essential for stomatal opening. Proc Natl Acad Sci USA. 98, 2917-2921
185. Taji, T., Ohsumi, C., Iuchi, S., Seki, M., Kasuga, M., Kobayashi, M., Yamaguchi-Shinozaki, K., Shinozaki, K. (2002) Important roles of drought- and cold-inducible genes for galactinol synthase in stress tolerance in Arabidopsis thaliana. The Plant Journal 29, 417–426
186. Tester, M., Davenport, R. (2003) $Na^+$ tolerance and $Na^+$ transport in higher plants. Ann Bot ,London. 91, 503-527
187. Tunc-Ozdemir, M., Rato, C., Brown, E., Rogers, S., Mooneyham, A., Frietsch, S., Myers, C.T., Poulsen, L.R., Malhó, R., Harper, J.F. (2013a) Cyclic nucleotide gated channels 7 and 8 are essential for male reproductive fertility. PLoS One. 8, e55277
188. Tunc-Ozdemir, M., Tang, C., Ishka, M.R., Brown, E., Groves, N.R., Myers, C.T., Rato, C., Poulsen, L.R., McDowell, S., Miller, G., Mittler, R., Harper, J.F. (2013b) A cyclic nucleotide-gated channel (CNGC16) in pollen is critical for stress tolerance in pollen reproductive development. Plant Physiol. 161, 1010-1020
189. Tyerman, S.D., Skerret, M. (1999) Root ion channels and salinity. Scientia Horticulturae. 78, 175-235
190. Tyler, L., Thomas, S.G., Hu, J., Dill, A., Alonso, J.M., Ecker, J.R., and Sun, T.-p. (2004) DELLA proteins and gibberellin-regulated seed germination and floral development in Arabidopsis. Plant Physiol. 135, 1008-1019
191. Ueguchi-Tanaka, M., Ashikari, M., Nakajima, M., Itoh, H., Katoh, E., Kobayashi, M., Chow, T.Y., Hsing, Y.I., Kitano, H., Yamaguchi, I., Matsuoka, M. (2005) GIBBERELLIN INSENSITIVE DWARF1 encodes a soluble receptor for gibberellin. Nature 437: 693-698
192. Urano, K., Yoshiba, Y., Nanjo, T., Ito, T., Yamaguchi-Shinozaki, K., Shinozaki, K. (2004) Arabidopsis stress-inducible gene for arginine decarboxylase AtADC2 is required for accumulation of putrescine in salt tolerance. Biochemical and Biophysical Research Communications 313, 369–375
193. Urquhart, W., Chin, K., Ung, H., Moeder, W., Yoshioka, K. (2011)

The cyclic nucleotide-gated channels AtCNGC11 and 12 are involved in multiple $Ca^{2+}$-dependent physiological responses and act in a synergistic manner. J Exp. Bot. 62, 3671-3682

194. Urquhart, W., Gunawardena, A.H., Moeder, W., Ali, R., Berkowitz, G.A., Yoshioka, K. (2007) The chimeric cyclic nucleotide-gated ion channel ATCNGC11/12 constitutively induces programmed cell death in a $Ca^{2+}$-dependent manner. Plant Mol. Biol. 65, 747-761
195. Valerio, C., Costa, A., Marri, L., Issakidis-Bourguet, E., Pupillo, P., Trost, P., Sparla, F. (2011) Thioredoxin-regulated beta-amylase (BAM1) triggers diurnal starch degradation in guard cells, and in mesophyll cells under osmotic stress. Journal of Experimental Botany 62, 545–555
196. Venema, K., Quintero, F.J., Pardo, J.M., Donaire, J.P. (2002) The *Arabidopsis* $Na^+/H^+$ exchanger AtNHX1 catalyzes low affinity $Na^+$ and $K^+$ transport in reconstituted liposomes. J Biol Chem. 277, 2413-2418
197. Verbruggen, N., Hermans, C. (2008) Proline accumulation in plants: a review. Amino Acids 35, 753–759
198. Véry, A., Robinson, M., Mansfield, T., Sanders, D. (1998) Guard cell cation channels are involved in $Na^+$-induced stomatal closure in a halophyte. Plant J. 14, 509-521
199. Véry, A.A., Sentenac, H. (2003). Molecular mechanisms and regulation of $K^+$ transport in higher plants. Annu Rev Plant Biol. 54, 575-603
200. Vijn, I., Smeekens, S. (1999) Fructan: more than a reserve carbohydrate? Plant Physiology 120, 351–359
201. Voelker, C., Gómez-Porras, J.L., Becker, D., Hamamoto, S., Uozumi, N., Gambale, F., Müller-Röber, B., Czempinski, K., Dreyer, I. (2010) Roles of tandem-pore $K^+$ channels in plants - a puzzle still to be solved. Plant Biol. 12(1), 56-63
202. Voelker, C., Schmidt, D., Müller-Röber, B., Czempinski, K. (2006) Members of the *Arabidopsis* AtTPK/KCO family form homomeric vacuolar channels *in planta*. Plant J. 48, 296–306
203. Wang, H., Qi, Q., Schorr, P., Cutler, A.J., Crosby, W.L., Fowke, L.C. (1998) ICK1, a cyclindependent protein kinase inhibitor from *Arabidopsis* thaliana interacts with both Cdc2a and CycD3, and its

expression is induced by abscisic acid. Plant J. 15, 501-510

204. Wang, H., Zhou, Y., Gilmer, S., Whitwill, S., Fowke, L.C. (2000) Expression of the plant cyclin-dependent kinase inhibitor ICK1 affects cell division, plant growth and morphology. Plant J. 24, 613-623

205. Weinl, C., Marquardt, S., Kuijt, S.J., Nowack, M.K., Jakoby, M.J., Hülskamp, M., Schnittger, A. (2005) Novel functions of plant cyclin-dependent kinase inhibitors, ICK1/KRP1, can act non-cell-autonomously and inhibit entry into mitosis. Plant Cell. 17, 1704-1722

206. West, G., Inze, D., Beemster, G.T. (2004) Cell cycle modulation in the response of the primary root of Arabidopsis to salt stress. Plant Physiol 135, 1050-8

207. Wyn-Jones, R.G., Pollard, A. (1983) Proteins, enzymes and inorganic ions. En: Encyclopedia of Plant Physiology. New Series., A. Läuchli and R.L. Bieleskio, eds. Springer-Verlag., Berlin, pp. 528-562

208. Xiong, L., Zhu, J.K. (2002) Molecular and genetic aspects of plant responses to osmotic stress. Plant Cell Environ. 25, 131-139

209. Xu, D., Duan, X., Wang, B., Hong, B., Ho, T., Wu, R. (1996). Expression of a late embryogenesis abundant protein gene, *HVA1*, from barley confers tolerance to water deficit and salt stress in transgenic rice. Plant Physiol. 110, 249-257

210. Yadav, R., Flowers, T., Yeo, A. (1996) The involvement of the transpirational bypass flow in sodium uptake by high- and low-sodium-transporting lines of rice developed through intravarietal selection. Plant Cell Environ. 19, 329-336

211. Yamada, M., Morishita, H., Urano, K., Shiozaki, N., Yamaguchi-Shinozaki, K., Shinozaki, K., Yoshiba, Y. (2005) Effects of free proline accumulation in petunias under drought stress. Journal of Experimental Botany 56, 1975–1981.

212. Yang, Q., Cheng., Z.Z., Zhou, X.F., Yin, H.B., Li, X., Xin, X.F., Hong, X.H., Zhu, J.K., Gong, Z. (2009) Overexpression of SOS (Salt Overly Sensitive) genes increases salt tolerance in transgenic *Arabidopsis*. Molecular Plant. 2(1), 22-31.

213. Yano, R., Nakamura, M., Yoneyama, T., Nishida, I. (2005) Starchrelated alpha-glucan/water dikinase is involved in the cold-induced development of freezing tolerance in Arabidopsis. Plant Physiology 138, 837–846

214. Yao, X., Horie, T., Xue, S., Leung, H.Y., Katsuhara, M., Brodsky, D.E., Wu, Y., Schroeder, J.I. (2010) Differential sodium and potassium transport selectivities of the rice OsHKT2;1 and OsHKT2;2 transporters in plant cells. Plant Physiol. 152(1), 341-355

215. Yeo, A.R., Yeo, M.E., Flowers, T.J. (1987) The contribution of an apoplastic pathway to sodium uptake by rice roots in saline conditions. J. Exp. Botany. 38, 1141-1153

216. Yokoi, S., Bressan, R.A., Hasegawa, P.M. (2002) Salt Stress Tolerance of Plants. JIRCAS Working Report, 25-33

217. Yoshida, R., Mori, I.C., Kamizono, N., Shichiri, Y., Shimatani, T., Miyata, F., Honda, K., Iwai, S. (2016) Glutamate functions in stomatal closure in Arabidopsis and fava bean. J Plant Res. 129, 39-49

218. Zelman, A.K., Dawe, A., Berkowitz, G.A. (2013) Identification of cyclic nucleotide gated channels using regular expressions. Methods Mol. Biol. 1016, 207-224

219. Zentella, R., Zhang, Z.L., Park, M., Thomas, S.G., Endo, A., Murase, K., Fleet, C.M., Jikumaru, Y., Nambara, E., Kamiya, Y., Sun, T.P. (2007) Global analysis of DELLA direct targets in early gibberellin signaling in Arabidopsis. Plant Cell. 19, 3037-3057

220. Zhang, H.X., Hodson, J.N., Williams, J.P., Blumwald, E. (2001) Engineering salt-tolerant Brassica plants: characterization of yield and seed oil quality in transgenic plants with increased vacuolar sodium accumulation. Proc Natl Acad Sci USA. 98, 12832-12836

221. Zhao, J., Cheng, N.H., Motes, C.M., Blancaflor, E.B., Moore, M., Gonzales, N., Padmanaban, S., Sze, H., Ward, J.M., Hirschi, K.D. (2008) AtCHX13 is a plasma membrane $K^+$ transporter. Plant Physiol. 48, 796-807

222. Zhao, J., Li, P., Motes, C.M., Park, S., Hirschi, K.D. (2015) CHX14 is a plasma membrane $K^+$-efflux transporter that regulates $K^+$

redistribution in *Arabidopsis thaliana*. Plant Cell Environ. 38, 2223, 2238

223. Zhou, Y., Li, G., Brandizzi, F., Fowke, L.C., Wang, H. (2003) The plant cyclin-dependent kinase inhibitor ICK1 has distinct functional domains for in vivo kinase inhibition, protein instability and nuclear localization. Plant J. 35, 476-489
224. Zhu, J.K. (2001) Plant salt tolerance. Trends Plant Sci. 6, 66-71
225. Zhu, J.K. (2002) Salt and drought stress signal transduction in plants. Annu Rev Plant Biol. 53, 247-273
226. Zhu, J.K. (2003) Regulation of ion homeostasis under salt stress. Curr Opin Plant Biol. 6, 441-445

CHAPTER

2

# COMPARISON BETWEEN ARABIDOPSIS AND RICE FOR MAIN PATHWAYS OF $K^+$ AND $NA^+$ UPTAKE BY ROOTS

Manuel Nieves-Cordones[1], Vicente Martínez[2], Begoña Benito[3] and Francisco Rubio[2]*

[1]Biochimie et Physiologie Moléculaire des Plantes, Institut de Biologie Intégrative des Plantes, UMR 5004 CNRS/UMR 0386 INRA/Montpellier SupAgro/Université Montpellier 2, Montpellier, France

[2]Departamento de Nutrición Vegetal, Centro de Edafología y Biología Aplicada del Segura – Consejo Superior de Investigaciones Científicas, Murcia, Spain

[3]Centro de Biotecnología y Genómica de Plantas, Universidad Politécnica de Madrid, Madrid, Spain

**Keywords**: potassium, sodium, uptake, roots, *Arabidopsis*, rice

$K^+$ is an essential macronutrient for plants. It is acquired by specific uptake systems located in roots. Although the concentrations of $K^+$ in the soil solution are widely variable, $K^+$ nutrition is secured by uptake systems that exhibit different affinities for $K^+$. Two main systems have been described for root $K^+$ uptake in several species: the high-affinity HAK5-like transporter and the inward-rectifier AKT1-like channel. Other unidentified systems may be also involved in root $K^+$uptake, although they only seem to operate when $K^+$ is not limiting. The use of knock-out lines has allowed demonstrating their role in root $K^+$ uptake in *Arabidopsis* and rice. Plant adaptation to the different $K^+$ supplies relies on the finely tuned regulation of these systems. Low $K^+$-induced transcriptional up-regulation of the genes encoding HAK5-like transporters occurs through a signal cascade that includes changes in the membrane potential of root cells and increases in ethylene and reactive oxygen species concentrations. Activation of AKT1 channels occurs through phosphorylation by the CIPK23/CBL1 complex. Recently, activation of the *Arabidopsis* HAK5 by the same complex has been reported, pointing to CIPK23/CBL as a central regulator of the plant's adaptation to low $K^+$. $Na^+$ is not an essential plant nutrient but it may be beneficial for some plants. At low concentrations, $Na^+$ improves growth, especially under $K^+$ deficiency. Thus, high-affinity $Na^+$ uptake systems have been described that belong to the HKT and HAK families of transporters. At high concentrations, typical of saline environments, $Na^+$ accumulates in plant tissues at high concentrations, producing alterations that include toxicity, water deficit and $K^+$ deficiency. Data concerning pathways for $Na^+$ uptake into roots under saline conditions are still scarce, although several possibilities have been proposed. The apoplast is a significant pathway for $Na^+$ uptake in rice grown under salinity conditions, but in other plant species different mechanisms involving non-selective cation channels or transporters are under discussion.

## INTRODUCTION

Given the constant increase in world population, high-yield crop production has become a necessity for agriculture. As that the nutrient sources of the land are limited, the input of nutrients by the addition of fertilizers ensures a continuous supply for plants, circumventing reductions in plant yield. The use of fertilizers has raised crop yield considerably, for example, from 50 to 80% of wheat and corn grain yield is attributable to nutrient fertilization (Stewart et al., 2005). However, this practice comes with high economic and environmental costs.

Potassium ($K^+$) is an essential macronutrient that is required by plants to complete their life cycle (Taiz and Zeiger, 1991). $K^+$ can make up to 10% of the total plant dry weight (Leigh and Wyn Jones, 1984) and fulfills important functions for metabolism, growth, and stress adaptation. Specifically, it acts as an enzyme activator, protein synthesis stabilizer, neutralization of protein negative charges, and it participates in cytoplasmic pH homeostasis as well (Marschner, 2012). An optimal $K^+$ concentration in the cytosol of around 100 mM is required for the performing of the functions mentioned above (Wyn Jones and Pollard, 1983), and plant cells maintain the cytosolic $K^+$ concentration around this value (Walker et al., 1996).

$K^+$ constitutes about 2.9% of the earth's crust but the concentration of $K^+$ in the soil solution is highly variable, in the $10^{-5}$ to $10^{-3}$ M range (Barraclough, 1989; Marschner, 2012). Since roots are able to take up $K^+$ at a higher rate than this cation can diffuse from the bulk soil solution, a $K^+$ depletion zone near the root surface can be formed with $K^+$ concentrations of just a few micromolar (Baldwin et al., 1973; Claassen and Jungk, 1982). More importantly, increasing areas of the world are currently described as being $K^+$deficient for agricultural practices (Mengel et al., 2001; Moody and Bell, 2006; Römheld and Kirkby, 2010; Kirkby and Schubert, 2013).

$K^+$ deficiency has a negative impact on plant growth since cellular expansion and photosynthesis are severely affected under these conditions (Bednarz and Oosterhuis, 1999; Hafsi et al., 2014). This deficiency also correlates with a decrease in protein synthesis and subsequent decline in growth (Walker et al., 1996, 1998). $K^+$ deficiency has been shown to inhibit lateral root development in *Arabidopsis* (Armengaud et al., 2004; Shin and Schachtman, 2004; Kellermeier et al., 2014) and in barley (Drew, 1975) and in the up-regulation of genes involved in $K^+$ uptake (Ashley et al., 2006; Nieves-Cordones et al., 2014). In addition, $K^+$-deficient plants are more sensitive to abiotic and biotic stresses such as drought, cold, salinity, or fungal attacks (Marschner, 2012; Zörb et al., 2014).

Sodium ($Na^+$) is not an essential element for plants but, for some species it can be a beneficial element that stimulates growth (Wakeel et al., 2010, 2011; Kronzucker et al., 2013). In these cases, $Na^+$ can be regarded as a functional nutrient (Subbarao et al., 2003), that can partly replace $K^+$ in some functions such as osmotic adjustment of the large central vacuole, cell turgor regulation leading to cell enlargement, or long-distance transport of anions (Subbarao et al., 2003; Horie et al., 2007; Gattward et al., 2012; Battie-Laclau et al., 2013).

On the other hand, $Na^+$ has been extensively associated to its negative impact on crop yield. Excess of $Na^+$salts in the soil results in both reduced soil water availability (due to the decrease in water potential) and ionic toxicity. When accumulated at high concentrations in the cytoplasm, $Na^+$ results in deleterious effects on cell biology, e.g., on photosynthetic activity or on membrane integrity (due to displacement of membrane-bound $Ca^{2+}$ ions) (Cramer et al., 1985). Thus, $Na^+$ is usually compartmentalized outside the cytoplasm (Morgan et al., 2014), in vesicles such as the vacuole, where it is used as an osmoticum. Estimates of the area of salt-affected soils vary widely, ranging from 6 to 10% of the earth's land area (Eynard et al., 2005; Munns and Tester, 2008). Importantly, 20% of irrigated lands are affected by secondary salinization, limiting agriculture worldwide.

In the present review, we summarize recent advances in the field of $K^+$ and $Na^+$ uptake in the plant root, with special attention to the transport systems and their regulation mechanisms. We believe that the studies performed on the model plant *Arabidopsis* and the results of recent research in crops such as rice suggest that the results obtained with model species cannot be fully extended to other plant species.

## $K^+$ AND $NA^+$ UPTAKE BY ROOTS: KINETIC FEATURES AND SENSITIVITY TO OTHER CATIONS

$K^+$ and $Na^+$ can enter the root apoplast and diffuse toward inner cell layers (Sattelmacher et al., 1998). However, this pathway is interrupted by the endodermis, where the Casparian strip, which is impermeable to water and ions, is located (Schreiber et al., 1999; Tester and Leigh, 2001; Marschner, 2012; Geldner, 2013; Barberon and Geldner, 2014). To cross this impermeable barrier, nutrient ions enter the cytosol of a root peripheral cell either from the epidermis, cortex or endodermis and move from cell to cell (symplastic pathway) through plasmodesmata (Burch-Smith and Zambryski, 2012). Diffusion within the symplasm beyond the endodermic barrier allows nutrient ions to reach the stele, where they will initiate their travel toward the aerial parts within the xylem vessels (Lauchli, 1972).

It is worth noting that the Casparian strip may be absent in some places (Maathuis, 2014) allowing ions to reach the root stele and xylem vessels through the apoplastic pathway via bypass flow (Kronzucker and Britto, 2011). Since this flow is relatively low, most of the ions that reach the root xylem vessels are probably taken up across the plasma membrane of a root peripheral cell (Tester and Leigh, 2001). Thus, their entry into the root symplasm would have been mediated by membrane transport systems, channels, transporters or cotransporters. It should be noted that the bypass flow was observed in rice at $Na^+$ external concentrations as low as 25 mM (Yeo et al., 1987) and it may contribute to salt stressing effects under high salinity by having an effect in shoot $Na^+$ content (Yeo et al., 1987; Faiyue et al., 2012; Maathuis, 2014). $Na^+$bypass flow has been described in other species, besides rice, such as mangroves (Krishnamurthy et al., 2014), maize, and broad bean (Peterson et al., 1981), but not in *Arabidopsis* (Essah et al., 2003).

More than 60 years ago, through the application of the concept of enzyme kinetics for the study of root $K^+$absorption (Epstein and Hagen, 1952), Epstein et al. (1963) suggested that at least two transport systems were involved in root $K^+$ uptake: a high-affinity system that operates at low external concentrations and a low-affinity system at higher concentrations. A similar scheme was also described for $Na^+$ uptake (Rains and Epstein, 1967a). This biphasic behavior has since been observed in many plant species, with some exceptions. Maize, for example, shows a linear, non-saturating response to $K^+$ in the low-affinity range (Kochian and Lucas, 1982). More recently, it has been shown that this linear response is dominated by the apoplastic movement of $K^+$, and the "true" transmembrane flux saturates at modest rates (Coskun et al., 2016). In the high-affinity range of concentrations, $K^+$ uptake is an active process that takes place against the $K^+$ electrochemical potential, most likely by a $K^+/H^+$ symport, whilst the low-affinity uptake can take place by passive transport through inwardly rectifying $K^+$ channels (Maathuis and Sanders, 1996a; Maathuis et al., 1997; Rodríguez-Navarro, 2000). It should also be noted that the limits in $K^+$concentrations for the operation of a symporter or a channel depend on the plasma membrane potential and the cytoplasmic pH and $K^+$ concentrations. Thus, assuming, for example, that a cytoplasmic $K^+$concentration of 100 mM and a membrane potential of -240 mV exists, $K^+$ uptake could take place through a channel from an external $K^+$ concentration as low as 10 μM, which falls within the high-affinity system described by Epstein et al. (1963), Hirsch et al. (1998) and Spalding et al. (1999). High-affinity $K^+$uptake becomes apparent when $K^+$ tissue concentrations decrease due to $K^+$ starvation (Glass, 1976; Kochian and Lucas, 1982; Siddiqi and Glass, 1986; Martínez-

Cordero et al., 2005). Providing $NH_4^+$ to the nutrient solution used to grow the plants, has a large influence on the $NH_4^+$ sensitivity of high-affinity $K^+$uptake. In some species such as barley (Santa-María et al., 2000), pepper (Martínez-Cordero et al., 2005), or *Arabidopsis* (Rubio et al., 2008), the presence of $NH_4^+$ in the growth solution induced an $NH_4^+$-insensitive high-affinity $K^+$ uptake component. In others, such as tomato (Nieves-Cordones et al., 2007), high-affinity $K^+$ uptake was dominated by an $NH_4^+$-sensitive component, irrespectively of the presence or the absence of $NH_4^+$ in the growth solution. By contrast, both $Na^+$ (Martínez-Cordero et al., 2005; Kronzucker et al., 2006, 2008; Nieves-Cordones et al., 2007, 2010; Alemán et al., 2009; Cheng et al., 2015) and $Cs^+$ (Rubio et al., 2000; White and Broadley, 2000; Qi et al., 2008) usually inhibit high-affinity $K^+$ uptake.

Regarding low-affinity $K^+$ transport, the role of $TEA^+$, $Cs^+$, and $Ba^{2+}$, in blocking animal and plant $K^+$channels is well known, as these inhibit low-affinity $K^+$ uptake, supporting the idea of the channel-mediated nature of this transport (Ketchum and Poole, 1991; Blatt, 1992; Hille, 1992; Very and Sentenac, 2002; Hoopen et al., 2010; Coskun et al., 2013). Unlike high-affinity $K^+$ uptake, low-affinity $K^+$ uptake is not down-regulated at high external $K^+$ (Maathuis and Sanders, 1996b; Szczerba et al., 2006) and it is $NH_4^+$-insensitive (Spalding et al., 1999; Santa-María et al., 2000; Kronzucker et al., 2003; Szczerba et al., 2006). $Na^+$ suppresses $K^+$ uptake both in the low- and the high-affinity ranges, with low-affinity $K^+$ uptake being more sensitive to this inhibition (Epstein et al., 1963; Rains and Epstein, 1967a; Kronzucker et al., 2006, 2008). It is worth noting that $K^+$ uptake seems to be insensitive to $Ca^{2+}$ in barley and *Arabidopsis*(Cramer et al., 1989; Caballero et al., 2012; Coskun et al., 2013).

As for $Na^+$ uptake, it has been shown that root high-affinity $Na^+$ uptake is significant when plants are starved of $K^+$ (Garciadeblas et al., 2003; Haro et al., 2010). Under these conditions, $Na^+$ is able to partially replace $K^+$ and support plant growth (Maathuis and Sanders, 1993; Garciadeblas et al., 2003; Horie et al., 2007; Wakeel et al., 2011; Wakeel, 2013). High-affinity $Na^+$ uptake has been shown to be sensitive to $K^+$and $Ca^{2+}$ in barley (Rains and Epstein, 1967b) and to $K^+$ and $Ba^{2+}$ in rice (Garciadeblas et al., 2003). By contrast, low-affinity $Na^+$ uptake seems to be insensitive to $Ca^{2+}$ and $K^+$ in rice (Malagoli et al., 2008) while it is sensitive to $Ca^{2+}$ in *Arabidopsis*, barley, and wheat (Cramer et al., 1987, 1989; Essah et al., 2003; D'Onofrio et al., 2005). $Na^+$ uptake can be reduced by the application of $K^+$, and such reduction can alleviate salt stress effects to some extent (Wakeel, 2013). For example, a reduction in tissue $Na^+$ content due to increased $K^+$ application was observed in strawberry or in *Jatropha curcas* (Khayyat et al., 2009; Rodrigues et al., 2012).

## IDENTIFICATION OF K⁺ TRANSPORT SYSTEMS IN PLANTS

The molecular approaches developed in the last 25 years have led to the characterization of many $K^+$ and $Na^+$ transport systems in plants. The first $K^+$ uptake system identified in plants, AKT1, was isolated by complementing a $K^+$-uptake deficient yeast strain with an *Arabidopsis* cDNA library (Sentenac et al., 1992). Sequence analysis and heterologous expression in Sf9-insect cells showed that the AKT1 cDNA encoded an inward-rectifier $K^+$ channel belonging to the Shaker family (Gaymard et al., 1996). The gene encoding this channel showed specific constitutive expression in epidermal root cells (Lagarde et al., 1996) and AKT1 was proposed as the system mediating low-affinity $K^+$ uptake in the roots.

Later, a PCR-based approach led to the identification of a cDNA from barley, HvHAK1, that mediated high-affinity $K^+$ uptake in yeast (Santa-María et al., 1997). Specific expression of its mRNA in $K^+$-starved roots and its kinetic properties in yeast prompted researchers to propose it as the system mediating the high-affinity $K^+$ uptake observed in barley roots (Epstein et al., 1963). Subsequent studies led to the identification of an *Arabidopsis* homolog of HvHAK1, that was named AtHAK5 (Rubio et al., 2000).

In addition to AKT1 and AtHAK5-like transporters, other $K^+$ uptake systems could be involved in root $K^+$uptake and in trans-membrane $K^+$ movements in other plant organs. Moreover, the sequence of whole genomes of plants evidenced the existence of large gene families encoding putative $K^+$ transport systems (Grabov, 2007; Véry et al., 2014; Nieves-Cordones et al., 2016).

## Initial Characterizations in *Arabidopsis*

The studies on heterologous systems and on gene expression patterns for AKT1 and AtHAK5 produced data suggesting that these two systems played important roles in $K^+$ acquisition by the root. However, a demonstration for the proposed roles was only possible when *Arabidopsis* knock-out mutants for these two genes became available (Spalding et al., 1999; Gierth et al., 2005; Rubio et al., 2008, 2010; Pyo et al., 2010). The studies showed that while *Arabidopsis* wild-type (WT) and *akt1* plants could deplete external $K^+$ ($Rb^+$) to values around 1 μM, the *athak5* plants did not diminish its concentration below 30 μM. In agreement with this, reduced growth was observed in *athak5* plants grown in the presence of 1 μM $K^+$ (Qi et al., 2008) or 10 μM $K^+$ (Pyo et al., 2010; Ragel et al., 2015). Moreover, $K^+$ ($Rb^+$) uptake in *athak5* plants was completely inhibited by the presence of $Ba^{2+}$ in the external solution. By contrast, a complete inhibition of $K^+$ ($Rb^+$) depletion was observed in the presence of $NH_4^+$ in the *akt1* line. An *athak5 akt1* double mutant did not show $K^+$ ($Rb^+$) uptake at external concentrations below 50 μM, and it could only promote $K^+$ ($Rb^+$) uptake at concentrations higher than 100 μM. All these results demonstrated that in *Arabidopsis* plants, AtHAK5 was the only system mediating $K^+$ uptake at concentrations below 10 μM and that this system was inhibited by $NH_4^+$. At concentrations between 10 and 200 μM, both AtHAK5 and AKT1 contributed to $K^+$ uptake, defining AKT1 as a $Ba^{2+}$-sensitive component of $K^+$ uptake. Above 500 μM AtHAK5 contribution was negligible as the *AtHAK5* gene was repressed at this external $K^+$ concentration, and AKT1 role became more relevant. Unidentified systems could compensate for the lack of AKT1 since the *akt1* and the *athak5 akt1* lines grew at similar rates than the WT line when the external $K^+$ concentration was sufficiently high (~10 mM).

The studies described for *Arabidopsis* allowed for depicting a model for the contribution of AtHAK5 and AKT1 to root $K^+$ uptake (Alemán et al., 2011). They also allowed for extending the studies to species where knock-out mutants were not available, through the use of $NH_4^+$ and $Ba^{2+}$ as specific inhibitors of HAK5 and AKT1, respectively. Thus, it was shown that in tomato and pepper plants grown in the absence of $NH_4^+$, an $NH_4^+$-sensitive component, probably mediated by SlHAK5 (formerly LeHAK5) and CaHAK1, respectively, dominated $K^+$ uptake from concentrations corresponding to the high-affinity component (inhibition of $K^+$ uptake by $NH_4^+$ was close to 80% in tomato, for example; Martínez-Cordero et al., 2005; Nieves-Cordones et al., 2007). These results contrast with those obtained for *Arabidopsis*, where both AtHAK5 and AKT1 contribute to $K^+$ uptake within the range of the $K^+$ concentration assigned to the high-affinity component, as demonstrated by the use of single (Rubio et al., 2008), and double KO mutants (Rubio et al., 2010) in AtHAK5 and AKT1. Therefore, it can be concluded that the *Arabidopsis* model cannot be completely extended to other plant species, crops included, and highlights the need for the characterization of knock-out lines in each particular species to address the relevance of the different $K^+$ transport systems. In addition, the availability of whole genomes for many plant species revealed some differences with respect to *Arabidopsis*. In *Arabidopsis*, AtHAK5 is the only member that belongs to the cluster Ia of HAK transporters, which are involved in high-affinity $K^+$ uptake in roots (Nieves-Cordones et al., 2016). In tomato, a highly homologous gene to *SlHAK5* is located just 2580 bp downstream from it in the tomato genome (Fernandez-Pozo et al., 2015). In rice, the *OsHAK21* gene, also belonging to cluster Ia is induced by salinity, something that has not been described in other species for members of this cluster. This transporter has been linked to rice tolerance to salinity through the maintenance of $Na^+/K^+$ homeostasis, although the physiological mechanism remains unclear (Shen et al., 2015).

## Rice Transport Systems Contributing to Root $K^+$ Uptake

The recent characterization of T-DNA insertion rice lines knocked-out for $K^+$ uptake systems such as OsHAK1, OsHAK5, and OsAKT1, has importantly contributed to increase our understanding of the relative contribution of such systems to $K^+$ uptake in a species different from *Arabidopsis*, which is of great importance in agriculture.

A T-DNA insertion mutant with OsAKT1 knocked-out (Golldack et al., 2003) showed reduced growth and decreased root and shoot $K^+$ concentrations when grown in the presence of 1 and 0.1 mM $K^+$ (Li et al., 2014). $K^+$ flux experiments and electrophysiological approaches showed an impairment of $K^+$ uptake in the *osakt1* line. Strong expression of *OsAKT1* was detected in epidermal root cells, but it was also found in cortex, endodermis, and vascular bundles, suggesting a direct or indirect role in $K^+$ translocation. In addition, slight expression was detected in shoots.

Knock-out mutants of the gene encoding the high-affinity $K^+$ transporter OsHAK1 (Bañuelos et al., 2002) were also characterized. The studies with *oshak1* lines showed that OsHAK1 contributed about 50–55% of high-affinity $K^+$ uptake in the range of 0.05–0.1 mM external $K^+$ and about 30% of $K^+$ uptake at 1 mM external $K^+$ (Chen et al., 2015). Root and shoot growth, cell size and internal $K^+$ concentrations were reduced in the *oshak1* mutant at both 0.1 and 1 mM $K^+$ and this deficient phenotype could not be rescued at high external $K^+$ (5 mM $K^+$). Transcripts of *OsHAK1* are preferentially accumulated in roots of $K^+$-starved plants, as it is observed with *AtHAK5* in *Arabidopsis*. In addition, *OsHAK1* is strongly expressed at the xylem parenchyma and phloem of root vascular tissues, shoot meristems and vascular bundles of leaf sheaths. Moreover, *oshak1* plants show reduced $K^+$ translocation from root to shoot. In addition, the *osakt1* and *oshak1* lines were inhibited throughout development, showing delayed grain filling and reduced grain yield, suggesting that they may play an important role in rice productivity.

OsHAK5 was isolated and characterized as a high-affinity transport system in heterologous systems (Horie et al., 2011). Expression studies showed that OsHAK5 localized to the plasma membrane and that under normal $K^+$ supply its transcripts were detected in root, root–shoot junction and leaf sheath. $K^+$ starvation enhanced its expression in root epidermis, parenchyma of stele tissue, primordials of lateral roots, mesophyll, and parenchyma cells of the vascular bundle. The expression pattern of *OsHAK5* supported its role in $K^+$ uptake, but also in $K^+$ distribution between roots and shoots, especially at low external $K^+$. The lower accumulation of $K^+$ in roots of overexpressing lines and the higher $K^+$ accumulation in the knock-out lines, grown in low $K^+$, supported this idea (Yang et al., 2014). These authors suggested that OsHAK5 may mediate $K^+$ accumulation in xylem parenchyma cells to enable $K^+$ channels to release $K^+$ efficiently into the xylem sap. A role for OsHAK5 in $K^+$ signaling is also proposed, as $K^+$ in the phloem may act as a signal to convey the shoot demand of $K^+$ and $K^+$ xylem loading (Engels and Marschner, 1992), and *OsHAK5* is abundantly expressed in phloem tissue. A role for the AKT2 channel in $K^+$ signaling by modulating $K^+$ in the phloem has been also proposed (Deeken et al., 2002) and AKT2 has been recently suggested as a pathway for $Na^+$ entry into the roots (Salvador-Recatala, 2016). Since AKT2-mediated $K^+$ transport is $Ca^{2+}$-sensitive (Latz et al., 2007), interesting interactions between salt stress, $Ca^{2+}$, and AKT2 may emerge.

The differential abundance of *OsHAK1*, *OsHAK5*, and *OsAKT1* transcripts suggests that they play non-redundant functions. *OsHAK1* and *OsAKT1* are highly expressed in all cell types of roots and at low levels in shoots. However, while the *OsHAK1* gene was induced 8- to 12-fold by $K^+$ starvation (Chen et al., 2015), *OsAKT1* expression was not affected by the external $K^+$ concentrations (Li et al., 2014). By contrast, *OsHAK5* was less expressed in roots and strongly in shoots (Yang et al., 2014). It has been proposed that both OsHAK1 and OsAKT1 contribute to $K^+$ acquisition at low and high external concentrations. At low $K^+$, OsHAK1 dominates high-affinity $K^+$ uptake over OsAKT1 and OsHAK5. By contrast, OsHAK5 dominates $K^+$ transport from root to shoots (Yang et al., 2014).

The studies described above suggest that OsHAK1, OsHAK5, and OsAKT1 are involved in $K^+$ uptake at low and high concentrations as well as in $K^+$ translocation from root to shoot. Contribution to $K^+$ uptake over a wide range of $K^+$ concentrations by a unique system has been demonstrated for the *Arabidopsis* AKT1 (Alemán et al., 2011). For transporters of the HAK family, the overexpression of AtKUP1 in *Arabidopsis*suspension cells produced enhanced $K^+$ uptake at micromolar and millimolar $K^+$ concentrations (Kim et al., 1998). The rice studies reviewed here suggest that OsHAK1 and OsHAK5 may contribute to $K^+$ uptake at both low and high concentrations, which would explain why the growth of *oshak1* and *oshak5* lines is not rescued at high external $K^+$. However, the idea of dual affinity for HAK transporters should be taken with caution. It is possible that, in addition to $K^+$ uptake, the knock-out lines are affected in other processes. In fact, Yang et al. (2014) highlight the role of some HAK/KUP/KT transporters in auxin distribution, irrespective of $K^+$ supply. At this stage, it cannot be ruled out that OsHAK1 and OsHAK5 play a part in this process.

## Role of HAK Transporters in $K^+/Na^+$ Homeostasis under NaCl Stress

$K^+/Na^+$ homeostasis has been shown to be crucial for tolerance of plants to salinity (Maathuis and Amtmann, 1999). Maintaining $K^+$ uptake rates at high external $Na^+$ is crucial for $K^+/Na^+$ homeostasis and salt tolerance (Munns and Tester, 2008; Cuin et al., 2012; Cheng et al., 2015). However, in the presence of high $Na^+$ concentrations, the low-$K^+$ induction of genes encoding high-affinity $K^+$ transporters is not observed (Nieves-Cordones et al., 2008, 2010). In addition, under salinity, $K^+$ transport through high-affinity HAK transporters is competitively inhibited (Santa-María et al., 1997; Rubio et al., 2000). In fact, OsHAK1 has been defined as the $Na^+$-sensitive high-affinity $K^+$ pathway in rice (Chen et al., 2015). Nonetheless, studies in low-$K^+$-grown *Arabidopsis* and rice plants showed that AtHAK5 and OsHAK1 function was still pivotal in maintaining $K^+$ uptake and plant growth in the presence of high $Na^+$ (Nieves-Cordones et al., 2010; Chen et al., 2015). In addition to OsHAK1, OsHAK5 may also play a role in salt tolerance, as high salt transiently enhances *OsHAK5* expression (Yang et al., 2014) and the transporter mediates $Na^+$-insensitive $K^+$ uptake (Horie et al., 2011). In the presence of salinity, OsHAK5-overexpressing lines accumulated more $K^+$ in shoots and showed enhanced growth as compared to WT. In contrast, *oshak5* lines accumulated more $Na^+$ in shoots and grew less than WT plants (Yang et al., 2014). The authors propose that the higher accumulation of $Na^+$ in shoots in *oshak5* may be due to a hyperpolarized membrane potential of mesophyll cells in knock-out mutants that would favor $Na^+$accumulation in shoots. Therefore, the $K^+$ transport systems that contribute to maintaining a depolarized membrane potential of mesophyll cells to evade excessive $Na^+$ accumulation under salinity may play a role in salt tolerance. The function of another rice HAK transporter, OsHAK21, seems to be important for salt tolerance (Shen et al., 2015). The gene encoding this transporter is enhanced by salinity and the protein is localized to the plasma membrane of xylem parenchyma and endodermal cells (putative passage cells). It has been shown that the knock-out *oshak21* line is more salt sensitive because of a higher and a lower accumulation of $Na^+$ and $K^+$ respectively, which points to OsHAK21 as a key transporter needed for the maintenance of $Na^+/K^+$ homeostasis in rice under salt stress.

## Other Systems That May Be Involved in Root K⁺ Uptake

Recently, a T-DNA insertion mutant in the *Arabidopsis* KUP7, which belongs to cluster V of the HAK family (Nieves-Cordones et al., 2016) has been characterized (Han et al., 2016). The results showed that the *kup7* line was sensitive to low $K^+$ (<100 μM), showing lower internal $K^+$ concentrations. It could be rescued by higher $K^+$ concentrations or by complementing the mutant with the WT gene. The *KUP7* gene was ubiquitously expressed in many organs and the KUP7 protein was localized to the plasma membrane. KUP7 could complement a yeast strain deficient in $K^+$ uptake. $K^+$ transport studies showed that KUP7 was involved in root $K^+$ uptake and $K^+$ translocation to the shoot. It seemed to operate at higher concentrations than AtHAK5, and may be an alternative system involved in $K^+$ uptake in the *athak5 akt1 Arabidopsis* line (Caballero et al., 2012). The observed effect on $K^+$ translocation may be an indirect effect of a reduced uptake. However, Han et al. (2016) speculated that KUP7 may mediate $K^+$ release into the xylem sap. It should be noted that besides $K^+$ uptake, some HAK transporters have been shown to mediate $K^+$ efflux (Bañuelos et al., 2002; Garciadeblas et al., 2002; Osakabe et al., 2013). Thus, $K^+$ release into the xylem sap could take place through this type of transporters, if the electrochemical potential for $K^+$ allows for this movement. It is well known that $K^+$ loading of the xylem is mainly mediated by SKOR channels and the possible specific contribution of HAK transporters to $K^+$ loading is yet to be determined.

All of the above described results can be summarized into two different models for the $K^+$ uptake systems in *Arabidopsis* and rice roots, shown in Figure 1 and Table 1.

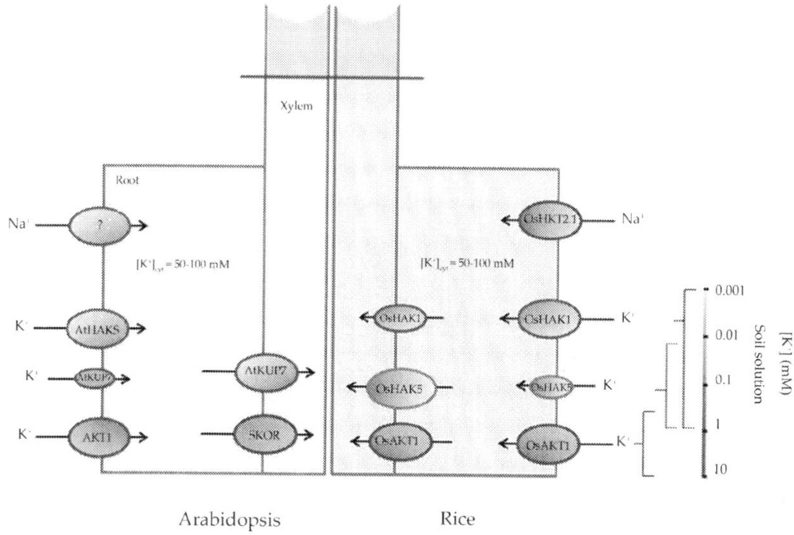


**Figure 1:** Schematic comparison of the systems involved in $K^+$ **and** $Na^+$ **movements in** ***Arabidopsis*** **and rice roots.** The availability of knock-out mutants in *Arabidopsis* and rice plants for specific transport systems has allowed for elucidating their roles in $K^+$ transport. The figure shows the predicted function of each system for which knock-out mutants have been studied. AtHAK5 and AKT1 are the main systems for $K^+$ uptake in *Arabidopsis* plants. In addition, a member of the KT/HAK/KUP family, AtKUP7, seems to also be involved in $K^+$ uptake. $K^+$ release into the xylem is mainly mediated by SKOR and also partially by AtKUP7. In rice, AtHAK5 and AKT1 functions are fulfilled by the rice homologs OsHAK1 and OsAKT1. An additional system, OsHAK5, partially contributes to high-affinity $K^+$uptake, but at higher concentrations than OsHAK1. These three rice systems may directly or indirectly facilitate $K^+$ release into the xylem, with the contribution by OsHAK5 to $K^+$ release into the xylem being more relevant. It is not clear if such contribution is a direct (by mediating $K^+$ efflux into xylem vessels) or indirect (by favoring $K^+$ accumulation in endodermal cells) outcome of the aforementioned transporters. Regarding $Na^+$ uptake, the genetic identity of $Na^+$ uptake systems in *Arabidopsis* remains to be elucidated. GLRs, CNGCs or other non-selective cation channels could be involved. In rice, OsHKT2;1 has been shown to mediate $Na^+$ uptake during $K^+$ deficiency. In the presence of $K^+$ or high external $Na^+$ concentrations other unknown systems should take part in $Na^+$ uptake.

**TABLE 1:** Summary for $K^+$ **and** $Na^+$ **uptake features observed in** *Arabidopsis* **and rice roots.**

| Species | Cation | Type of uptake[1] | $K_m$ (µM) | Sensitivity | Transport systems involved | Reference |
|---|---|---|---|---|---|---|
| *Arabidopsis* | $K^+$ | High affinity | 24* | (–) $NH_4^+$, $Ba^{+2}$, $Cs^+$ (=) $Ca^{+2}$ | AtHAK5, AKT1 | Spalding et al., 1999; Gierth et al., 2005; Rubio et al., 2008; Caballero et al., 2010 |
| | | Low affinity | 3,991* | (–) $Ba^{+2}$, $Cs^+$, TEA, $La^{+3}$, $Na^+$ (=) $NH_4^+$, $Ca^{+2}$ | AKT1 | Spalding et al., 1999; Gierth et al., 2005; Caballero et al., 2012 |
| | $Na^+$ | High-affinity | n.d. | n.d. | – | – |
| | | Low-affinity | Linear response | (–) $Ca^{+2}$, $Ba^{+2}$, cyclic nucleotides (+) $La^{+3}$, GABA (=) TEA, $Cs^+$ | | Maathuis and Sanders, 2001; Essah et al., 2003 |
| Rice | $K^+$ | High-affinity | 11–18 | (–) $NH_4^+$ | OsHAK1, OsHAK5, OsAKT1 | Bañuelos et al., 2002; Li et al., 2014; Yang et al., 2014; Chen et al., 2015 |
| | | Low-affinity | n.d. | n.d. | OsHAK1, OsAKT1 | Li et al., 2014; Chen et al., 2015 |
| | $Na^+$ | High-affinity | 60 or 477–655¶ | (–) $K^+$, $Ba^{+2}$ | OsHKT2;1 | Garciadeblás et al., 2003; Horie et al., 2007; Han et al., 2010 |
| | | Low-affinity | n.d. | (=) $K^+$, $Ca^{2+}$ | OsHKT2;1 | Horie et al., 2007; Malagoli et al., 2008 |

[1]High-affinity uptake takes into account cation uptake observed at external concentrations < 0.5 mM while the low affinity one does so at > 0.5 mM. *Obtained with $Rb^+$ as an analog for $K^+$. (+), (–), and (=) stand for activation, inhibition, or no effect on cation uptake, respectively. n.d., not determined. ¶ Only one component was identified in Horie et al. (2007) for external $Na^+$ concentrations up to 5 mM.

## REGULATION OF K+ UPTAKE

In general terms, the genes encoding HAK transporters that mediate high-affinity $K^+$ uptake in roots are strongly induced by $K^+$ deprivation, whereas the genes encoding AKT1 channels are not. Several elements in the signal transduction cascade that results in the activation of HAK transcription have been identified. One of the first events when a root faces $K^+$ deprivation is the hyperpolarization of the cell's plasma membrane (Amtmann and Blatt, 2009). A positive correlation has been found between the membrane potential and the expression levels of *SlHAK5* and *AtHAK5*, independent of the root's $K^+$ concentration (Nieves-Cordones et al., 2008; Rubio et al., 2014). Thus, it has been proposed that the hyperpolarization of the membrane potential may be the first element in the low-$K^+$ signal cascade. The mechanisms linking membrane potential and gene expression are unknown but, changes in cytoplasmic $Ca^{2+}$ derived from the activity of hyperpolarization-activated $Ca^{2+}$ channels, could provide a connecting mechanism (Véry and Davies, 2000). Increases in ethylene (Jung et al., 2009) and reactive oxygen species (ROS; Shin and Schachtman, 2004; Hernandez et al., 2012) are also involved, probably acting following the hyperpolarization of the membrane potential. Other hormones such as jasmonic acid (Armengaud et al., 2004, 2010; Takehisa et al., 2013; Schachtman, 2015) and cytokinins (Nam et al., 2012) may also play a role in $K^+$ signaling.

At the end of the cascade, transcription factors such as DDF2, JLO, bHLH121, TFII_A for *Arabidopsis AtHAK5* (Kim et al., 2012; Hong et al., 2013), bind the gene's promoter to activate its expression.

Interestingly, some environmental conditions such as the lack of N, P, or S, that hyperpolarize root cell membrane potential (Rubio et al., 2014) and elevate ROS levels (Shin et al., 2005), also activate the transcription of *AtHAK5*-type genes. However, under such conditions, no HAK-mediated uptake is observed, suggesting a post-transcriptional regulation for these HAK transporters that is elicited specifically by $K^+$ starvation (Rubio et al., 2014). Recently, it has been shown that the *Arabidopsis* AtHAK5 transporter is activated by complexes that contain the CIPK23 kinase and CBL1, CBL8, CBL9, or CBL10 $Ca^{2+}$ sensors. AtHAK5 phosphorylation by CIPK23 leads to increases in the maximal rate of transport ($V_{max}$) and the affinity for $K^+$ transport (Ragel et al., 2015). It can be assumed that a specific low-$K^+$-induced $Ca^{2+}$ signal is registered by the CBL $Ca^{2+}$ sensor, that in turn promotes phosphorylation of the transporter by CIPK23.

As a voltage-dependent inward-rectifier $K^+$ channel, AKT1 activity is regulated by the membrane potential (Gaymard et al., 1996; Xu et al., 2006). In addition, its activity is modulated by interaction with other proteins. AKT1 forms tetrameric channels by interacting with the AtKC1 subunit (Daram et al., 1997; Pilot et al., 2003; Duby et al., 2008; Geiger et al., 2009). Upon interaction with AtKC1 in root cells, the activation potential for AKT1-containing channels becomes more negative, in comparison with AKT1 homotetramers (Reintanz et al., 2002; Wang et al., 2010). Moreover, AtKC1 interacts, in turn, with the SNARE proteins SYP121 (Honsbein et al., 2009) and VAMP721 (Zhang et al., 2015). SYP121 was shown to activate $K^+$ uptake through AKT1/AtKC1 channels while VAMP721 negatively regulated this process. Phosphorylation also plays a role in AKT1 regulation. AKT1 activity is enhanced at low $K^+$ by the same regulators as AtHAK5, i.e., the CIPK23/CBL1/9 complex, supporting the putative role of $Ca^{2+}$ as a secondary messenger involved in low-$K^+$ signaling (Li et al., 2006; Xu et al., 2006). Inactivation of the channel is achieved by the AIP1 phosphatase (Luan, 2009). It is worth to note that CBL proteins can also modify AKT1 activity in absence of CIPKs, as it is the case of CBL10 which is a negative regulator of AKT1 (Ren et al., 2013). Recently it has been shown that CIPK23 and

AtKC1 act synergistically and balance $K^+$uptake/leakage to modulate AKT1-mediated responses of *Arabidopsis* to low $K^+$ (Wang et al., 2016). Finally, nitric oxide has recently been shown to lower AKT1 activity by modulating vitamin B6 biosynthesis, constituting a new mechanism for the regulation of $K^+$ uptake (Xia et al., 2014).

The current model for the regulation of these two main systems involved in $K^+$ uptake, i.e., AtHAK5 and AKT1, is depicted in Figure 2.

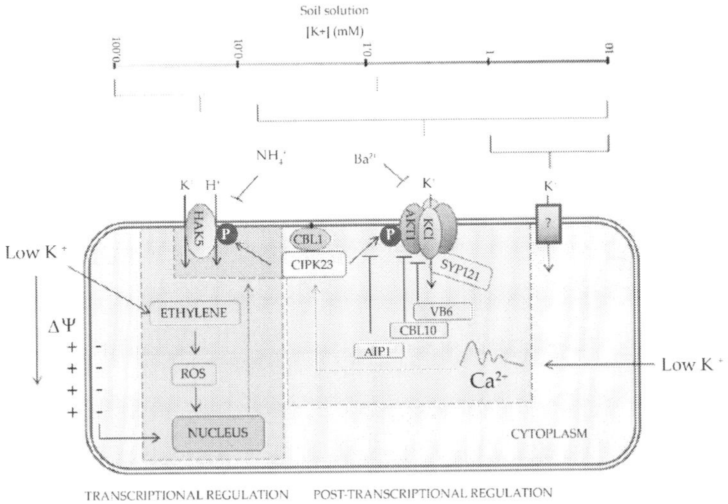


**Figure 2:** Main pathways for root $K^+$ **uptake and their regulatory mechanisms.** HAK5-type transporters are high-affinity $K^+$ transporters involved in $K^+$ uptake at very low concentrations. When the external $K^+$ concentration increases, the inward-rectifier $K^+$ channel AKT1 together with HAK5, contributes to $K^+$ uptake. At $K^+$ concentrations above 200 μM, HAK5 is not present and AKT1 is the main system for low-affinity $K^+$ uptake. At very high concentrations, other unknown systems can secure $K^+$ supply if AKT1 is not functional. The HAK5 and AKT1 uptake systems are subjected to finely tuned regulation. At low $K^+$ concentrations, a hyperpolarization of the plasma membrane induces *HAK5* transcription. The signal cascade of HAK5 regulation is dependent on ethylene and ROS production. In addition, low external $K^+$ likely produces a specific cytoplasmic $Ca^{2+}$ signal that is registered by the $Ca^{2+}$ sensor CBL1, which induces CIPK23 recruitment to the plasma membrane resulting in the phosphorylation and subsequent activation of HAK5 and AKT1. The channel activity is downregulated through dephosphorylation by the AIP1 phosphatase, and interaction with CBL10 and vitamin B6. Other subunits such as KC1 and the SNARE protein SYP121 also cooperate in AKT1 regulation. It

can be concluded that whereas HAK5 is subjected to transcriptional and post-transcriptional regulation, $K^+$ uptake through AKT1 is mainly regulated post-transcriptionally. The CIPK23/CBL1 complex emerges as a key regulator of $K^+$ nutrition.

It is worth mentioning that CIPK23/CBL is also involved in the regulation of $NO_3^-$ uptake (Ho et al., 2009; Tsay et al., 2011; Léran et al., 2015), and that $K^+$ starvation significantly reduced the $NO_3^-$ concentrations in tomato plants (Rubio et al., 2014) and induced several genes for $NO_3^-$ uptake (Armengaud et al., 2004). This indicates that a cross-regulation between $K^+$ and $NO_3^-$ nutrition exists and that the CIPK23/CBL complex may constitute one of the key elements for such regulation.

## UPTAKE OF $NA^+$ AND CELL MECHANISMS INVOLVED

As for the identities of genes involved in $Na^+$ uptake from external solutions, the scenario is far less clear than that for $K^+$. With respect to high-affinity $Na^+$ uptake, despite this activity being widely observed in roots from several species (Garciadeblas et al., 2003; Haro et al., 2010), only a few transport systems belonging to the HKT and HAK transporter families have been shown to take up $Na^+$ within this range of concentrations. High-affinity $K^+$ transporters (HKT) are related to fungal and bacterial $K^+$ transporters from the Trk/Ktr families (Corratgé-Faillie et al., 2010). In plants, however, HKT transporters display varying $Na^+$/$K^+$ permeabilities. Phylogenetic and functional analyses have led to the identification of two HKT subfamilies (Platten et al., 2006): subfamily I, present in both monocotyledonous and dicotyledonous species, and subfamily II, identified only in monocotyledonous species so far. Subfamily II HKT transporters are expected to be all $K^+$-permeable and can operate as $Na^+$/$K^+$ symporters (Rubio et al., 1995; Jabnoune et al., 2009; Yao et al., 2010; Oomen et al., 2012) or $K^+$-selective uniporters (Horie et al., 2011; Sassi et al., 2012) when heterologously expressed in yeast and/or *Xenopus* oocytes. Subfamily I HKT transporters are $Na^+$-selective in *Arabidopsis* and rice and are mostly involved in $Na^+$ recirculation through vascular tissues (Maathuis, 2014; Véry et al., 2014), thus falling beyond the scope of the present review. In the rice cultivar Nipponbare, OsHKT2;1 provides a major pathway for root high-affinity $Na^+$uptake that supports plant growth under limiting $K^+$ supply

(Garciadeblas et al., 2003; Horie et al., 2007; Figure 1). Plants lacking a functional *OsHKT2;1* gene have shown reduced growth and lower $Na^+$ content when starved of $K^+$ in the presence of 0.5 mM $Na^+$, and under such conditions $Na^+$ can partially compensate $K^+$ demand (Horie et al., 2007). Besides $Na^+$, OsHKT2;1 can also transport $K^+$ when expressed in *Xenopus* oocytes (Jabnoune et al., 2009; Oomen et al., 2012), but $K^+$ transport is not detected when it was expressed in yeast or in tobacco BY2 cells (Horie et al., 2001; Yao et al., 2010). In OsHKT2;1 expressing oocytes, the shifts in reversal potentials induced by $K^+$ depended on the pre-treatment of oocytes. When the oocytes were pre-treated in low-$Na^+$ (0.5 mM $Na^+$) they showed smaller shifts that when pretreated with high-$Na^+$ (96 mM $Na^+$; Yao et al., 2010). Comparisons of $Rb^+$ influx between WT and *oshkt2;1* roots did not reveal significant differences between these genotypes (Horie et al., 2007). Thus, the possible involvement of OsHKT2;1 in root $K^+$ uptake remains to be verified. Another subfamily II HKT transporter, OsHKT2;2, is absent in the Nipponbare (japonica) cultivar but present in the indica cultivar Pokkali. The transporter obtained from the latter cultivar is permeable to both $Na^+$ and $K^+$ at low external concentrations when expressed in tobacco BY2 cells, yeast and *Xenopus* oocytes (Horie et al., 2001; Yao et al., 2010; Oomen et al., 2012). It is important to note that a natural chimera OsHKT2;2/1 present in the Nona Bokra (indica) cultivar maintains high-affinity $K^+$ uptake even at high $Na^+$concentrations, something that is also observed for Pokkali OsHKT2;2 but not for Nipponbare OsHKT2;1 (Oomen et al., 2012). Despite the lack of data concerning *oshkt2;2* knock-out mutants, it is tempting to speculate that OsHKT2;2 contributes to both $K^+$ and $Na^+$ high-affinity uptake in rice roots. As for HAK transporters, two members, none of them from higher plants, have been shown to mediate high-affinity $Na^+$ uptake: PpHAK13 from the moss *Physcomitrella patens* and YlHAK1 from the yeast *Yarrowia lipolytica* (Benito et al., 2012). PpHAK13, which belongs to cluster IV (Nieves-Cordones et al., 2016), transports $Na^+$, but not $K^+$, and high-affinity $Na^+$ uptake is abolished in *pphak13* mutants plants which evidences that PpHAK13 forms the major pathway for $Na^+$ entry at low external concentrations in *P. patens* plants. On the other hand, YlHAK1 is able to transport $Na^+$ and $K^+$ when expressed in yeast, but the latter cation is only transported when $Na^+$ is not added to the experimental solution. High-affinity $Na^+$transporters from the HAK family in higher plants are still to be identified.

Concerning low-affinity $Na^+$ uptake, it is generally accepted that $Na^+$ can enter the plant through ion channels (Maathuis, 2014). $Na^+$-permeable channels include glutamate-like receptors (GLRs; Davenport, 2002) and cyclic nucleotide gated channels (CNGCs; Assmann, 1995; Bolwell, 1995; Trewavas, 1997; Newton and Smith, 2004) and possibly other, non-identified, non-selective cation channels (NSCCs; Maathuis and Sanders, 1993; Tyerman et al., 1997; Demidchik and Tester, 2002; Essah et al., 2003). Voltage-independent NSCCs (VI-NSCC) may constitute the main class of NSCCs involved in $Na^+$ entry since they are highly sensitive to $Ca^{2+}$ as observed for $Na^+$ uptake in roots (Demidchik and Maathuis, 2007). Moreover, VI-NSCC blockers such as quinine or lanthanides also inhibited root $Na^+$ influx (Essah et al., 2003; Wang et al., 2006). Despite these encouraging observations linking VI-NSCCs and root $Na^+$uptake, the molecular identity of these channels remains obscure at present. Several HAK and HKT transporters, which are expressed in roots, are also permeable at millimolar $Na^+$ concentrations (Santa-María et al., 1997; Horie et al., 2001; Takahashi et al., 2007; Mian et al., 2011; Oomen et al., 2012). It is worth to note that OsHKT2;1 and HvHKT2;1 contribute to $Na^+$ uptake in the millimolar range but they are downregulated in the presence of salt stress or $K^+$. Thus, when taking into account other experimental conditions and plant species, it remains unclear which other type/family of transport systems constitute the major pathway for low-affinity $Na^+$ uptake. It is likely that there is a large redundancy between the aforementioned channels and transporters. Insights into the identification of the contributing transport proteins would be of extraordinary biotechnological value since low-affinity $Na^+$ uptake allows for the massive entry of $Na^+$ within the plant that gives rise to toxicity. Interestingly, the secondary messengers cyclic AMP and GMP affect $Na^+$ influx. Studies on *Arabidopsis* seedlings (Maathuis and Sanders, 2001; Essah et al., 2003) and on pepper plants (Rubio et al., 2003) have shown that unidirectional $Na^+$ influx is reduced by cGMP addition. CNGCs have a cyclic-nucleotide binding domain and their activity is modulated by cyclic-nucleotides (Gao et al., 2014, 2016). It can be expected that the cyclic-nucleotide regulation of $Na^+$ fluxes occurs through direct cGMP (or cAMP) binding to this domain as it is the case of animal CNGCs (Craven and Zagotta, 2006).

Recently, it has been shown that salt stress triggers the formation of endocytic vesicles via a clathrin-independent mechanism (Baral et al., 2015). Such vesicles lead to the formation of vacuole-like structures that may help plants to better cope with salt stress. Endocytosis can modify the transporter complement of the plasma membrane (Sutter et al., 2007), thus affecting the $Na^+$ uptake pathways. But more interestingly, the endocytic process involves bulk-flow entry into root cells that may transport $Na^+$ ions from the apoplast to the vacuole-like structures. If this were true, it would constitute a parallel pathway for $Na^+$ uptake, independent from that mediated by transmembrane proteins. Moreover, this direct $Na^+$transport into vacuoles would prevent $Na^+$ accumulation in the cytosol which leads to cell toxicity. Interestingly, vesicle trafficking has been recently suggested to play a role in plant adaptation to salt stress (Garcia de la Garma et al., 2015).

## CONCLUDING REMARKS

$K^+$ is an essential macronutrient for plants while $Na^+$ may be beneficial or detrimental at low or high concentrations, respectively. Plant roots possess specific $K^+$ transport systems that can function under a wide range of concentrations to secure $K^+$ ions. The studies with knock-out mutants of the model plant *Arabidopsis* have led to the identification of two major pathways for $K^+$ uptake: the high-affinity $K^+$transporter AtHAK5 and the inward-rectifying $K^+$ channel AKT1. These systems operate at low (micromolar) and high (millimolar) external $K^+$ concentrations, although an overlap in their operation is observed in the 10 200 μM $K^+$ range. Different mechanisms that include transcriptional and post-transcriptional regulation modulate the activity of these two systems in response to $K^+$ supply. Importantly, the CIPK23/CBL1 complex activates AtHAK5 as well as AKT1, pointing to its role as a central regulator of $K^+$ nutrition. Homologs of AtHAK5 and AKT1 have been found in many plant species, and in some of them, paralog genes exist which suggest function redundancy. This precludes assigning a function by only using sequence homology or heterologous expression studies. The recent characterization of rice knock-out plants has shed light on this matter. OsHAK1, OsHAK5, and OsAKT1 seem to contribute to root $K^+$ uptake as well as $K^+$ release into xylem, and they probably play additional unknown functions in the shoot. However, they play non-redundant roles:

(i) OsHAK1 is mainly involved in root $K^+$ uptake at low concentrations, (ii) OsAKT1 mostly at high $K^+$ concentrations, and (iii) OsHAK5 is more relevant for $K^+$translocation to the shoot. These systems may also play a role in salinity tolerance by maintaining the $K^+/Na^+$ homeostasis.

Regarding $Na^+$ transport systems, the information is scarcer. High-affinity $Na^+$ uptake and high-affinity $Na^+$ transporters have been described in some species, but they are lacking in many others. The pathways for low-affinity $Na^+$ uptake are not clearly identified yet, and several families of transporters contain members that could be good candidates. Recently, it has been described that salt stress induced a new endocytic pathway that is clathrin-independent, non-discriminatory in its choice of cargo, and that operates across all layers of the root. This new pathway may contribute to the bulk $Na^+$ uptake and distribution across the root cells under saline stress conditions.

All of the above highlight the importance of characterizing the function of each transporter in each particular species. The studies with knock-out lines in *Arabidopsis* and rice evidence that the conclusions drawn in model species cannot be always fully extended to other, non-model species.

## AUTHOR CONTRIBUTIONS

All authors listed, have made substantial, direct and intellectual contribution to the work, and approved it for publication.

## REFERENCES

1. Alemán, F., Nieves-Cordones, M., Martínez, V., and Rubio, F. (2009). Differential regulation of the HAK5 genes encoding the high-affinity K+ transporters of Thellungiella halophila and Arabidopsis thaliana. Environ. Exp. Bot. 65, 263–269. doi: 10.1016/j.envexpbot.2008.09.011
2. Alemán, F., Nieves-Cordones, M., Martínez, V., and Rubio, F. (2011). Root K+ acquisition in plants: the Arabidopsis thaliana model. Plant Cell Physiol. 52, 1603–1612. doi: 10.1093/pcp/pcr096
3. Amtmann, A., and Blatt, M. R. (2009). Regulation of macronutrient transport. New Phytol. 181, 35–52. doi: 10.1111/j.1469-8137.2008.02666.x

4. Armengaud, P., Breitling, R., and Amtmann, A. (2004). The potassium-dependent transcriptome of Arabidopsis reveals a prominent role of jasmonic acid in nutrient signaling. Plant Physiol. 136, 2556–2576. doi: 10.1104/pp.104.046482
5. Armengaud, P., Breitling, R., and Amtmann, A. (2010). Coronatine-insensitive 1 (COI1) mediates transcriptional responses of Arabidopsis thaliana to external potassium supply. Mol. Plant 3, 390–405. doi: 10.1093/mp/ssq012
6. Ashley, M. K., Grant, M., and Grabov, A. (2006). Plant responses to potassium deficiencies: a role for potassium transport proteins. J. Exp. Bot. 57, 425–436. doi: 10.1093/jxb/erj034
7. Assmann, S. M. (1995). Cyclic AMP as a second messenger in higher plants (status and future prospects). Plant Physiol. 108, 885–889.
8. Baldwin, J. P., Nye, P. H., and Tinker, P. B. (1973). Uptake of solutes by multiple root systems from soil. 3. Model for calculating solute uptake by a randomly dispersed root system developing in a finite volume of soil. Plant Soil 38, 621–635. doi: 10.1007/BF00010701
9. Bañuelos, M. A., Garciadeblas, B., Cubero, B., and Rodríguez-Navarro, A. (2002). Inventory and functional characterization of the HAK potassium transporters of rice. Plant Physiol. 130, 784–795. doi: 10.1104/pp.007781
10. Baral, A., Irani, N. G., Fujimoto, M., Nakano, A., Mayor, S., and Mathew, M. K. (2015). Salt-induced remodeling of spatially restricted clathrin-independent endocytic pathways in Arabidopsis root. Plant Cell 27, 1297–1315. doi: 10.1105/tpc.15.00154
11. Barberon, M., and Geldner, N. (2014). Radial transport of nutrients: the plant root as a polarized epithelium. Plant Physiol. 166, 528–537. doi: 10.1104/pp.114.246124
12. Barraclough, P. B. (1989). Root-growth, macro-nutrient uptake dynamics and soil fertility requirements of a high-yielding winter oilseed rape crop. Plant Soil 119, 59–70. doi: 10.1007/BF02370269
13. Battie-Laclau, P., Laclau, J.-P., Piccolo, M. D. C., Arenque, B. C., Beri, C., Mietton, L., et al. (2013). Influence of potassium and sodium nutrition on leaf area components in Eucalyptus grandis trees. Plant Soil 371, 19–35. doi: 10.1007/s11104-013-1663-7

14. Bednarz, C. W., and Oosterhuis, D. M. (1999). Physiological changes associated with potassium deficiency in cotton. J. Plant Nutr. 22, 303–313. doi: 10.1016/j.jphotobiol.2012.02.002
15. Benito, B., Garciadeblas, B., and Rodriguez-Navarro, A. (2012). HAK transporters from Physcomitrella patens and Yarrowia lipolytica mediate sodium uptake. Plant Cell Physiol. 53, 1117–1123. doi: 10.1093/pcp/pcs056
16. Blatt, M. R. (1992). K+ channels of stomatal guard cells. Characteristics of the inward rectifier and its control by pH. J. Gen. Physiol. 99, 615–644. doi: 10.1085/jgp.99.4.615
17. Bolwell, G. P. (1995). Cyclic AMP, the reluctant messenger in plants. Trends Biochem. Sci. 20, 492–495. doi: 10.1016/S0968-0004(00)89114-8
18. Burch-Smith, T. M., and Zambryski, P. C. (2012). Plasmodesmata paradigm shift: regulation from without versus within. Annu. Rev. Plant Biol. 63, 239–260. doi: 10.1146/annurev-arplant-042811-105453
19. Caballero, F., Botella, M. A., Rubio, L., Fernández, J. A., Martínez, V., and Rubio, F. (2012). A Ca2+-sensitive system mediates low-affinity K+ uptake in the absence of AKT1 in Arabidopsis plants. Plant Cell Physiol. 53, 2047–2059. doi: 10.1093/pcp/pcs140
20. Chen, G., Hu, Q., Luo, L., Yang, T., Zhang, S., Hu, Y., et al. (2015). Rice potassium transporter OsHAK1 is essential for maintaining potassium-mediated growth and functions in salt tolerance over low and high potassium concentration ranges. Plant Cell Environ. 38, 2747–2765. doi: 10.1111/pce.12585
21. Cheng, D., Wu, G., and Zheng, Y. (2015). Positive correlation between potassium uptake and salt tolerance in wheat. Photosynthetica 53, 447–454. doi: 10.1007/s11099-015-0124-3
22. Claassen, N., and Jungk, A. (1982). Potassium dynamics at the soil-root interface in relation to the uptake of potassium by maize plants. Z. Pflanzenernahr. Bodenkd. 145, 513–525. doi: 10.1002/jpln.19821450603
23. Corratgé-Faillie, C., Jabnoune, M., Zimmermann, S., Véry, A. A., Fizames, C., and Sentenac, H. (2010). Potassium and sodium transport in non-animal cells: the Trk/Ktr/HKT transporter family.

Cell. Mol. Life Sci. 67, 2511–2532. doi: 10.1007/s00018-010-0317-7

24. Coskun, D., Britto, D. T., Kochian, L. V., and Kronzucker, H. J. (2016). How high do ion fluxes go? A re-evaluation of the two-mechanism model of K(+) transport in plant roots. Plant Sci. 243, 96–104. doi: 10.1016/j.plantsci.2015.12.003
25. Coskun, D., Britto, D. T., Li, M., Oh, S., and Kronzucker, H. J. (2013). Capacity and plasticity of potassium channels and high-affinity transporters in roots of barley and Arabidopsis. Plant Physiol. 162, 496–511. doi: 10.1104/pp.113.215913
26. Cramer, G., Epstein, E., and Lauchli, A. (1989). Na-Ca interactions in barley seedlings - relationship to ion-transport and growth. Plant Cell Environ. 12, 551–558. doi: 10.1111/j.1365-3040.1989.tb02128.x
27. Cramer, G. R., Lauchli, A., and Polito, V. S. (1985). Displacement of Ca2+ by Na+ from the plasmalemma of root-cells - A primary response to salt stress. Plant Physiol. 79, 207–211. doi: 10.1104/pp.79.1.207
28. Cramer, G. R., Lynch, J., Lauchli, A., and Epstein, E. (1987). Influx of Na+, K+ and Ca2+ into roots of alt-stressed cotton seedlings. Plant Physiol. 79, 207–211. doi: 10.1104/pp.79.1.207
29. Craven, K. B., and Zagotta, W. N. (2006). CNG and HCN channels: two peas, one pod. Annu. Rev. Physiol. 68, 375–401. doi: 10.1146/annurev.physiol.68.040104.134728
30. Cuin, T. A., Zhou, M., Parsons, D., and Shabala, S. (2012). Genetic behaviour of physiological traits conferring cytosolic K+/Na+ homeostasis in wheat. Plant Biol. (Stuttg) 14, 438–446. doi: 10.1111/j.1438-8677.2011.00526.x
31. Daram, P., Urbach, S., Gaymard, F., Sentenac, H., and Cherel, I. (1997). Tetramerization of the AKT1 plant potassium channel involves its C- terminal cytoplasmic domain. EMBO J. 16, 3455–3463. doi: 10.1093/emboj/16.12.3455
32. Davenport, R. (2002). Glutamate receptors in plants. Ann. Bot. 90, 549–557. doi: 10.1093/aob/mcf228
33. Deeken, R., Geiger, D., Fromm, J., Koroleva, O., Ache, P., Langenfeld-Heyser, R., et al. (2002). Loss of the AKT2/3 potassium

channel affects sugar loading into the phloem of Arabidopsis. Planta 216, 334–344. doi: 10.1007/s00425-002-0895-1

34. Demidchik, V., and Maathuis, F. (2007). Physiological roles of nonselective cation channels in plants: from salt stress to signalling and development. New Phytol. 175, 387–404. doi: 10.1111/j.1469-8137.2007.02128.x
35. Demidchik, V., and Tester, M. (2002). Sodium fluxes through nonselective cation channels in the plasma membrane of protoplasts from Arabidopsis roots. Plant Physiol. 128, 379–387. doi: 10.1104/pp.010524
36. D'Onofrio, C., Kader, A., and Lindberg, S. (2005). Uptake of sodium in quince, sugar beet, and wheat protoplasts determined by the fluorescent sodium-binding dye benzofuran isophthalate. J. Plant Physiol. 162, 421–428. doi: 10.1016/j.jplph.2004.07.017
37. Drew, M. C. (1975). Comparison of effects of a localized supply of phosphate, nitrate, ammonium and potassium on growth of seminal root system, and shoot, in barley. New Phytol. 75, 479–490. doi: 10.1111/j.1469-8137.1975.tb01409.x
38. Duby, G., Hosy, E., Fizames, C., Alcon, C., Costa, A., Sentenac, H., et al. (2008). AtKC1, a conditionally targeted Shaker-type subunit, regulates the activity of plant K+ channels. Plant J. 53, 115–123. doi: 10.1111/j.1365-313X.2007.03324.x
39. Engels, C., and Marschner, H. (1992). Adaptation of potassium translocation into the shoot of maize (Zea mays) to shoot demand: evidence for xylem loading as a regulating step. Physiol. Plant. 86, 263–268. doi: 10.1034/j.1399-3054.1992.860211.x
40. Epstein, E., and Hagen, C. E. (1952). A kinetic study of the absorption of alkali cations by barley roots. Plant Pshyiol. 27, 457–474. doi: 10.1104/pp.27.3.457
41. Epstein, E., Rains, D. W., and Elzam, O. E. (1963). Resolution of dual mechanisms of potassium absorption by barley roots. Proc. Natl. Acad. Sci. U.S.A. 49, 684–692. doi: 10.1073/pnas.49.5.684
42. Essah, P. A., Davenport, R., and Tester, M. (2003). Sodium influx and accumulation in Arabidopsis. Plant Physiol. 133, 307–318. doi: 10.1104/pp.103.022178
43. Eynard, A., Lal, R., and Wiebe, K. (2005). Crop response in salt-

affected soils. J. Sustain. Agric. 27, 5–50. doi: 10.1007/BF00223914

44. Faiyue, B., Al-Azzawi, M. J., and Flowers, T. J. (2012). A new screening technique for salinity resistance in rice (Oryza sativa L.) seedlings using bypass flow. Plant Cell Environ. 35, 1099–1108. doi: 10.1111/j.1365-3040.2011.02475.x
45. Fernandez-Pozo, N., Menda, N., Edwards, J. D., Saha, S., Tecle, I. Y., Strickler, S. R., et al. (2015). The sol genomics network (SGN)– from genotype to phenotype to breeding. Nucleic Acids Res. 43, D1036–D1041. doi: 10.1093/nar/gku1195
46. Gao, Q. F., Fei, C. F., Dong, J. Y., Gu, L. L., and Wang, Y. F. (2014). Arabidopsis CNGC18 is a Ca2+-permeable channel. Mol. Plant 7, 739–743. doi: 10.1093/mp/sst174
47. Gao, Q. F., Gu, L. L., Wang, H. Q., Fei, C. F., Fang, X., Hussain, J., et al. (2016). Cyclic nucleotide-gated channel 18 is an essential Ca2+ channel in pollen tube tips for pollen tube guidance to ovules in Arabidopsis. Proc. Natl. Acad. Sci. U.S.A. 113, 3096–3101. doi: 10.1073/pnas.1524629113
48. Garcia de la Garma, J., Fernandez-Garcia, N., Bardisi, E., Pallol, B., Asensio-Rubio, J. S., Bru, R., et al. (2015). New insights into plant salt acclimation: the roles of vesicle trafficking and reactive oxygen species signalling in mitochondria and the endomembrane system. New Phytol. 205, 216–239. doi: 10.1111/nph.12997
49. Garciadeblas, B., Benito, B., and Rodriguez-Navarro, A. (2002). Molecular cloning and functional expression in bacteria of the potassium transporters CnHAK1 and CnHAK2 of the seagrass Cymodocea nodosa. Plant Mol. Biol. 50, 623–633. doi: 10.1023/A:1019951023362
50. Garciadeblas, B., Senn, M. E., Bañuelos, M. A., and Rodriguez-Navarro, A. (2003). Sodium transport and HKT transporters: the rice model. Plant J. 34, 788–801. doi: 10.1046/j.1365-313X.2003.01764.x
51. Gattward, J. N., Almeida, A.-A. F., Souza, J. O. Jr., Gomes, F. P., and Kronzucker, H. J. (2012). Sodium-potassium synergism in Theobroma cacao: stimulation of photosynthesis, water-use efficiency and mineral nutrition. Physiol. Plant. 146, 350–362. doi: 10.1111/j.1399-3054.2012.01621.x

52. Gaymard, F., Cerutti, M., Horeau, C., Lemaillet, G., Urbach, S., Ravallec, M., et al. (1996). The baculovirus/insect cell system as an alternative to xenopus oocytes. J. Biol. Chem. 271, 22863–22870. doi: 10.1074/jbc.271.37.22863
53. Geiger, D., Becker, D., Vosloh, D., Gambale, F., Palme, K., Rehers, M., et al. (2009). Heteromeric AtKC1/AKT1 channels in Arabidopsis roots facilitate growth under K+ limiting conditions. J. Biol. Chem. 284, 21288–21295. doi: 10.1074/jbc.M109.017574
54. Geldner, N. (2013). The endodermis. Annu. Rev. Plant Biol. 64, 531–558. doi: 10.1146/annurev-arplant-050312-120050
55. Gierth, M., Maser, P., and Schroeder, J. I. (2005). The potassium transporter AtHAK5 functions in K+ deprivation-induced high-affinity K+ uptake and AKT1 K+ channel contribution to K+ uptake kinetics in Arabidopsis roots. Plant Physiol. 137, 1105–1114. doi: 10.1104/pp.104.057216
56. Glass, A. (1976). Regulation of potassium absorption in barley roots. An allosteric model. Plant Physiol. 58, 33–37. doi: 10.1104/pp.58.1.33
57. Golldack, D., Quigley, F., Michalowski, C. B., Kamasani, U. R., and Bohnert, H. J. (2003). Salinity stress-tolerant and -sensitive rice (Oryza sativa L.) regulate AKT1-type potassium channel transcripts differently. Plant Mol. Biol. 51, 71–81. doi: 10.1023/A:1020763218045
58. Grabov, A. (2007). Plant KT/KUP/HAK potassium transporters: single family - multiple functions. Anna. Bot. 99, 1035–1041. doi: 10.1093/aob/mcm066
59. Hafsi, C., Debez, A., and Abdelly, C. (2014). Potassium deficiency in plants: effects and signaling cascades. Acta Physiol. Plant. 36, 1055–1070. doi: 10.1007/s00709-015-0845-y
60. Han, M., Wu, W., Wu, W.-H., and Wang, Y. (2016). Potassium transporter KUP7 is involved in K+ acquisition and translocation in Arabidopsis root under K+-limited conditions. Mol. Plant 9, 437–446. doi: 10.1016/j.molp.2016.01.012
61. Haro, R., Banuelos, M. A., and Rodriguez-Navarro, A. (2010). High-affinity sodium uptake in land plants. Plant Cell Physiol. 51, 68–79. doi: 10.1093/pcp/pcp168

62. Hernandez, M., Fernandez-Garcia, N., Garcia-Garma, J., Rubio-Asensio, J. S., Rubio, F., and Olmos, E. (2012). Potassium starvation induces oxidative stress in Solanum lycopersicum L. roots. J. Plant Physiol. 169, 1366–1374. doi: 10.1016/j.jplph.2012.05.015

63. Hille, B. (1992). Ionic Channels of Excitable Membranes. Sunderland, MA: Sinauer Associates, Inc.

64. Hirsch, R. E., Lewis, B. D., Spalding, E. P., and Sussman, M. R. (1998). A role for the AKT1 potassium channel in plant nutrition. Science 280, 918–921. doi: 10.1126/science.280.5365.918

65. Ho, C. H., Lin, S. H., Hu, H. C., and Tsay, Y. F. (2009). CHL1 functions as a nitrate sensor in plants. Cell 138, 1184–1194. doi: 10.1016/j.cell.2009.07.004

66. Hong, J. P., Takeshi, Y., Kondou, Y., Schachtman, D. P., Matsui, M., and Shin, R. (2013). Identification and characterization of transcription factors regulating Arabidopsis HAK5. Plant Cell Physiol. 54, 1478–1490. doi: 10.1093/pcp/pct094

67. Honsbein, A., Sokolovski, S., Grefen, C., Campanoni, P., Pratelli, R., Paneque, M., et al. (2009). A tripartite snare-K+ channel complex mediates in channel-dependent K+ nutrition in Arabidopsis. Plant Cell 21, 2859–2877. doi: 10.1105/tpc.109.066118

68. Hoopen, F. T., Cuin, T. A., Pedas, P., Hegelund, J. N., Shabala, S., Schjoerring, J. K., et al. (2010). Competition between uptake of ammonium and potassium in barley and Arabidopsis roots: molecular mechanisms and physiological consequences. J. Exp. Bot. 61, 2303–2315. doi: 10.1093/jxb/erq057

69. Horie, T., Costa, A., Kim, T. H., Han, M. J., Horie, R., Leung, H. Y., et al. (2007). Rice OsHKT2;1 transporter mediates large Na+ influx component into K+-starved roots for growth. EMBO J. 26, 3003–3014. doi: 10.1038/sj.emboj.7601732

70. Horie, T., Sugawara, M., Okada, T., Taira, K., Kaothien-Nakayama, P., Katsuhara, M., et al. (2011). Rice sodium-insensitive potassium transporter, OsHAK5, confers increased salt tolerance in tobacco BY2 cells. J. Biosci. Bioeng. 111, 346–356. doi: 10.1016/j.jbiosc.2010.10.014

71. Horie, T., Yoshida, K., Nakayama, H., Yamada, K., Oiki, S., and Shinmyo, A. (2001). Two types of HKT transporters with different

properties of Na+ and K+ transport in Oryza sativa. Plant J. 27, 129–138. doi: 10.1046/j.1365-313x.2001.01077.x

72. Jabnoune, M., Espeout, S., Mieulet, D., Fizames, C., Verdeil, J. L., Conejero, G., et al. (2009). Diversity in expression patterns and functional properties in the rice HKT transporter family. Plant Physiol. 150, 1955–1971. doi: 10.1104/pp.109.138008
73. Jung, J. Y., Shin, R., and Schachtman, D. P. (2009). Ethylene mediates response and tolerance to potassium deprivation in Arabidopsis. Plant Cell 21, 607–621. doi: 10.1105/tpc.108.063099
74. Kellermeier, F., Armengaud, P., Seditas, T. J., Danku, J., Salt, D. E., and Amtmann, A. (2014). Analysis of the root system architecture of Arabidopsis provides a quantitative readout of crosstalk between nutritional signals. Plant Cell 26, 1480–1496. doi: 10.1105/tpc.113.122101
75. Ketchum, K. A., and Poole, R. J. (1991). Cytosolic calcium regulates a potassium current in corn (Zea mays) protoplasts. J. Membr. Biol. 119, 277–288. doi: 10.1007/BF01868732
76. Khayyat, M., Tafazoli, E., Rajaee, S., Vazifeshenas, M., Mahmoo-Dabadi, M. R., and Sajjadinia, A. (2009). Effects of NaCl and supple- mentary potassium on gas exchange, ionic content, and growth of salt-stressed strawberry plants. J. Plant Nutr. 32, 907–918. doi: 10.1080/01904160902870689
77. Kim, E. J., Kwak, J. M., Uozumi, N., and Schroeder, J. I. (1998). AtKUP1: an Arabidopsis gene encoding high-affinity potassium transport activity. Plant Cell 10, 51–62. doi: 10.2307/3870628
78. Kim, M. J., Ruzicka, D., Shin, R., and Schachtman, D. P. (2012). The Arabidopsis AP2/ERF transcription factor RAP2.11 modulates plant response to low-potassium conditions. Mol. Plant 5, 1042–1057. doi: 10.1093/mp/sss003
79. Kirkby, E., and Schubert, S. (2013). Management of potassium in plant and soil systems in China. J. Plant Nutr. Soil Sci. 176, 303–304. doi: 10.1002/jpln.201390018
80. Kochian, L. V., and Lucas, W. J. (1982). Potassium transport in corn roots: I. Resolution of kinetics into a saturable and linear component. Plant Physiol. 70, 1723–1731. doi: 10.1104/pp.70.6.1723
81. Krishnamurthy, P., Jyothi-Prakash, P. A., Qin, L., He, J., Lin, Q.,

Loh, C.-S., et al. (2014). Role of root hydrophobic barriers in salt exclusion of a mangrove plant Avicennia officinalis. Plant Cell Environ. 37, 1656–1671. doi: 10.1111/pce.12272

82. Kronzucker, H. J., and Britto, D. T. (2011). Sodium transport in plants: a critical review. New Phytol. 189, 54–81. doi: 10.1111/j.1469-8137.2010.03540.x
83. Kronzucker, H. J., Coskun, D., Schulze, L. M., Wong, J. R., and Britto, D. T. (2013). Sodium as nutrient and toxicant. Plant Soil 369, 1–23. doi: 10.1007/s11104-013-1801-2
84. Kronzucker, H. J., Szczerba, M. W., and Britto, D. T. (2003). Cytosolic potassium homeostasis revisited: K-42-tracer analysis in Hordeum vulgare L. reveals set-point variations in [K+]. Planta 217, 540–546. doi: 10.1007/s00425-003-1032-5
85. Kronzucker, H. J., Szczerba, M. W., Moazami-Goudarzi, M., and Britto, D. T. (2006). The cytosolic Na+:K+ ratio does not explain salinity-induced growth impairment in barley: a dual-tracer study using 42K+ and 24Na+. Plant Cell Environ. 29, 2228–2237. doi: 10.1111/j.1365-3040.2006.01597.x
86. Kronzucker, H. J., Szczerba, M. W., Schulze, L. M., and Britto, D. T. (2008). Non-reciprocal interactions between K+ and Na+ ions in barley (Hordeum vulgare L.). J. Exp. Bot. 59, 2793–2801. doi: 10.1093/jxb/ern139
87. Lagarde, D., Basset, M., Lepetit, M., Conejero, G., Gaymard, F., Astruc, S., et al. (1996). Tissue-specific expression of Arabidopsis AKT1 gene is consistent with a role in K+ nutrition. Plant J. 9, 195–203. doi: 10.1046/j.1365-313X.1996.09020195.x
88. Latz, A., Ivashikina, N., Fischer, S., Ache, P., Sano, T., Becker, D., et al. (2007). In planta AKT2 subunits constitute a pH- and Ca2+-sensitive inward rectifying K+ channel. Planta 225, 1179–1191. doi: 10.1007/s00425-006-0428-4
89. Lauchli, A. (1972). Translocation of inorganic solutes. Ann. Rev. Plant Physiol. 23, 197–218. doi: 10.1146/annurev.pp.23.060172.001213
90. Leigh, R. A., and Wyn Jones, R. G. (1984). A hypothesis relating critical potassium concentrations for growth to the distribution and functions of this ion in the plant-cell. New Phytol. 97, 1–13. doi: 10.1111/j.1469-8137.1984.tb04103.x

91. Léran, S., Edel, K. H., Pervent, M., Hashimoto, K., Corratgé-Faillie, C., Offenborn, J. N., et al. (2015). Nitrate sensing and uptake in Arabidopsis are enhanced by ABI2, a phosphatase inactivated by the stress hormone abscisic acid. Sci. Signal. 8:ra43. doi: 10.1126/scisignal.aaa4829

92. Li, J., Long, Y., Qi, G.-N., Li, J., Xu, Z.-J., Wu, W.-H., et al. (2014). The os-akt1 channel is critical for K+ uptake in rice roots and is modulated by the rice CBL1-CIPK23 complex. Plant Cell 26, 3387–3402. doi: 10.1105/tpc.114.123455

93. Li, L., Kim, B. G., Cheong, Y. H., Pandey, G. K., and Luan, S. (2006). A Ca2+ signaling pathway regulates a K+ channel for low-K+ response in Arabidopsis. Proc. Natl. Acad. Sci. U.S.A. 103, 12625–12630. doi: 10.1073/pnas.0605129103

94. Luan, S. (2009). The CBL-CIPK network in plant calcium signaling. Trends Plant Sci. 14, 37–42. doi: 10.1016/j.tplants.2008.10.005

95. Maathuis, F. J., Ichida, A. M., Sanders, D., and Schroeder, J. I. (1997). Roles of higher plant K+ channels. Plant Physiol. 114, 1141–1149. doi: 10.1104/pp.114.4.1141

96. Maathuis, F. J., and Sanders, D. (2001). Sodium uptake in Arabidopsis roots is regulated by cyclic nucleotides. Plant Physiol. 127, 1617–1625. doi: 10.1104/pp.010502

97. Maathuis, F. J. M. (2014). Sodium in plants: perception, signalling, and regulation of sodium fluxes. J. Exp. Bot. 65, 849–858. doi: 10.1093/jxb/ert326

98. Maathuis, F. J. M., and Amtmann, A. (1999). K+ Nutrition and Na+ toxicity: the basis of cellular K+/Na+ ratios. Ann. Bot. 84, 123–133. doi: 10.1006/anbo.1999.0912

99. Maathuis, F. J. M., and Sanders, D. (1993). Energization of potassium uptake in Arabidopsis thaliana. Planta 191, 302–307. doi: 10.1007/BF00195686

100. Maathuis, F. J. M., and Sanders, D. (1996a). Characterization of csi52, a Cs+ resistant mutant of Arabidopsis thaliana altered in K+ transport. Plant J. 10, 579–589. doi: 10.1046/j.1365-313X.1996.10040579.x

101. Maathuis, F. J. M., and Sanders, D. (1996b). Mechanisms of

potassium absorption by higher plant roots. Physiol. Plant. 96, 158–168. doi: 10.1111/j.1399-3054.1996.tb00197.x

102. Malagoli, P., Britto, D. T., Schulze, L. M., and Kronzucker, H. J. (2008). Futile Na+ cycling at the root plasma membrane in rice (Oryza sativa L.): kinetics, energetics, and relationship to salinity tolerance. J. Exp. Bot. 59, 4109–4117. doi: 10.1093/jxb/ern249

103. Marschner, P. (2012). Marschner's Mineral Nutrition of Higher Plants 3rd Edn. San Diego, CA: Academic Press.

104. Martínez-Cordero, M. A., Martinez, V., and Rubio, F. (2005). High-affinity K+ uptake in pepper plants. J. Exp. Bot. 56, 1553–1562. doi: 10.1093/jxb/eri150

105. Mengel, K., Kirkby, E. A., Kosegarten, H., and Appel, T. (2001). Principles on Plant Nutrition. Dordrecht: Kluwer.

106. Mian, A., Oomen, R. J., Isayenkov, S., Sentenac, H., Maathuis, F. J. M., and Very, A.-A. (2011). Over-expression of an Na+- and K+-permeable HKT transporter in barley improves salt tolerance. Plant J. 68, 468–479. doi: 10.1111/j.1365-313X.2011.04701.x

107. Moody, P. W., and Bell, M. J. (2006). Availability of soil potassium and diagnostic soil tests. Aust. J. Soil Res. 44, 265–275. doi: 10.1071/SR05154

108. Morgan, S. H., Maity, P. J., Geilfus, C. M., Lindberg, S., and Muehling, K. H. (2014). Leaf ion homeostasis and plasma membrane H+-ATPase activity in Vicia faba change after extra calcium and potassium supply under salinity. Plant Physiol. Biochem. 82, 244–253. doi: 10.1016/j.plaphy.2014.06.010

109. Munns, R., and Tester, M. (2008). Mechanisms of salinity tolerance. Ann. Rev. Plant Biol. 59, 651–681. doi: 10.1146/annurev.arplant.59.032607.092911

110. Nam, Y.-J., Tran, L.-S. P., Kojima, M., Sakakibara, H., Nishiyama, R., and Shin, R. (2012). Regulatory roles of cytokinins and cytokinin signaling in response to potassium deficiency in Arabidopsis. PLoS ONE 7:e47797. doi: 10.1371/journal.pone.0047797

111. Newton, R. P., and Smith, C. J. (2004). Cyclic nucleotides. Phytochemistry 65, 2423–2437. doi: 10.1016/j.phytochem.2004.07.026

112. Nieves-Cordones, M., Aleman, F., Martinez, V., and Rubio, F. (2010). The Arabidopsis thaliana HAK5 K+ transporter is required for plant growth and K+ acquisition from low K+ solutions under saline conditions. Mol. Plant 3, 326–333. doi: 10.1093/mp/ssp102

113. Nieves-Cordones, M., Alemán, F., Martínez, V., and Rubio, F. (2014). K+ uptake in plant roots. The systems involved, their regulation and parallels in other organisms. J. Plant Physiol. 171, 688–695. doi: 10.1016/j.jplph.2013.09.021

114. Nieves-Cordones, M., Martinez-Cordero, M. A., Martinez, V., and Rubio, F. (2007). An NH4+-sensitive component dominates high-affinity K+ uptake in tomato plants. Plant Sci. 172, 273–280. doi: 10.1016/j.plantsci.2006.09.003

115. Nieves-Cordones, M., Miller, A., Alemán, F., Martínez, V., and Rubio, F. (2008). A putative role for the plasma membrane potential in the control of the expression of the gene encoding the tomato high-affinity potassium transporter HAK5. Plant Mol. Biol. 68, 521–532. doi: 10.1007/s11103-008-9388-3

116. Nieves-Cordones, M., Rodenas, R., Chavanieu, A., Rivero, R. M., Martinez, V., Gaillard, I., et al. (2016). Uneven HAK/KUP/KT protein diversity among angiosperms: species distribution and perspectives. Front. Plant Sci. 7:127. doi: 10.3389/fpls.2016.00127

117. Oomen, R. J., Benito, B., Sentenac, H., Rodriguez-Navarro, A., Talon, M., Very, A.-A., et al. (2012). HKT2;2/1, a K+-permeable transporter identified in a salt-tolerant rice cultivar through surveys of natural genetic polymorphism. Plant J. 71, 750–762. doi: 10.1111/j.1365-313X.2012.05031.x

118. Osakabe, Y., Arinaga, N., Umezawa, T., Katsura, S., Nagamachi, K., Tanaka, H., et al. (2013). Osmotic stress responses and plant growth controlled by potassium transporters in Arabidopsis. Plant Cell 25, 609–624. doi: 10.1105/tpc.112.105700

119. Peterson, C. A., Emanuel, M. E., and Humphreys, G. B. (1981). Pathway of movement of apoplastic fluorescent dye tracers through the endodermis at the site of secondary root-formation in corn (Zea mays) and broad bean (Vicia faba). Can. J. Bot. 59:618. doi: 10.1139/b81-087

120. Pilot, G., Gaymard, F., Mouline, K., Cherel, I., and Sentenac, H.

(2003). Regulated expression of Arabidopsis Shaker K+ channel genes involved in K+ uptake and distribution in the plant. Plant Mol. Biol. 51, 773–787. doi: 10.1023/A:1022597102282

121. Platten, J. D., Cotsaftis, O., Berthomieu, P., Bohnert, H., Davenport, R. J., Fairbairn, D. J., et al. (2006). Nomenclature for HKT transporters, key determinants of plant salinity tolerance. Trends Plant Sci. 11, 372–374. doi: 10.1016/j.tplants.2006.06.001
122. Pyo, Y. J., Gierth, M., Schroeder, J. I., and Cho, M. H. (2010). High-affinity K+ transport in Arabidopsis: AtHAK5 and AKT1 are vital for seedling establishment and postgermination growth under low-potassium conditions. Plant Physiol. 153, 863–875. doi: 10.1104/pp.110.154369
123. Qi, Z., Hampton, C. R., Shin, R., Barkla, B. J., White, P. J., and Schachtman, D. P. (2008). The high affinity K+ transporter AtHAK5 plays a physiological role in planta at very low K+ concentrations and provides a caesium uptake pathway in Arabidopsis. J. Exp. Bot. 59, 595–607. doi: 10.1093/jxb/erm330
124. Ragel, P., Rodenas, R., Garcia-Martin, E., Andres, Z., Villalta, I., Nieves-Cordones, M., et al. (2015). The CBL-interacting protein kinase CIPK23 regulates HAK5-mediated high-affinity K+ uptake in Arabidopsis roots. Plant Physiol. 169, 2863–2873. doi: 10.1104/pp.15.01401
125. Rains, D. W., and Epstein, E. (1967a). Sodium absorption by barley roots: its mediation by mechanism 2 of alkali cation transport. Plant Physiol. 42, 319–323. doi: 10.1104/pp.42.3.314
126. Rains, D. W., and Epstein, E. (1967b). Sodium absorption by barley roots: role of the dual mechanisms of alkali cation transport. Plant Physiol. 42, 314–318. doi: 10.1104/pp.42.3.314
127. Reintanz, B., Szyroki, A., Ivashikina, N., Ache, P., Godde, M., Becker, D., et al. (2002). AtKC1, a silent Arabidopsis potassium channel alpha -subunit modulates root hair K+ influx. Proc. Natl. Acad. Sci. U.S.A. 99, 4079–4084. doi: 10.1073/pnas.052677799
128. Ren, X.-L., Qi, G.-N., Feng, H.-Q., Zhao, S., Zhao, S.-S., Wang, Y., et al. (2013). Calcineurin B-like protein CBL10 directly interacts with AKT1 and modulates K+ homeostasis in Arabidopsis. Plant J. 74, 258–266. doi: 10.1111/tpj.12123

129. Rodrigues, C. R. F., Silveira, J. A. G., Silva, E. N., Dutra, A. T. B., and Viegas, R. A. (2012). Potassium transport and partitioning alleviates toxic effects of sodium on young physic nut plants. Rev. Bras. Cien. Solo 36, 223–232. doi: 10.1590/S0100-06832012000100023

130. Rodríguez-Navarro, A. (2000). Potassium transport in fungi and plants. Biochim. Biophys. Acta 1469, 1–30. doi: 10.1016/S0304-4157(99)00013-1

131. Römheld, V., and Kirkby, E. (2010). Research on potassium in agriculture: needs and prospects. Plant Soil 335, 155–180. doi: 10.1007/s11104-010-0520-1

132. Rubio, F., Alemán, F., Nieves-Cordones, M., and Martínez, V. (2010). Studies on Arabidopsis athak5, atakt1 double mutants disclose the range of concentrations at which AtHAK5, AtAKT1 and unknown systems mediate K+ uptake. Physiol. Plant. 139, 220–228. doi: 10.1111/j.1399-3054.2010.01354.x

133. Rubio, F., Flores, P., Navarro, J. M., and Martinez, V. (2003). Effects of Ca2+, K+ and cGMP on Na+ uptake in pepper plants. Plant Sci. 165, 1043–1049. doi: 10.1016/S0168-9452(03)00297-8

134. Rubio, F., Fon, M., Rodenas, R., Nieves-Cordones, M., Aleman, F., Rivero, R. M., et al. (2014). A low K+ signal is required for functional high-affinity K+ uptake through HAK5 transporters. Physiol. Plant. 152, 558–570. doi: 10.1111/ppl.12205

135. Rubio, F., Gassmann, W., and Schroeder, J. I. (1995). Sodium-driven potassium uptake by the plant potassium transporter HKT1 and mutations conferring salt tolerance. Science 270, 1660–1663. doi: 10.1126/science.270.5242.1660

136. Rubio, F., Nieves-Cordones, M., Alemán, F., and Martinez, V. (2008). Relative contribution of AtHAK5 and AtAKT1 to K+ uptake in the high-affinity range of concentrations. Physiol. Plant. 134, 598–608. doi: 10.1111/j.1399-3054.2008.01168.x

137. Rubio, F., Santa-Maria, G. E., and Rodríguez-Navarro, A. (2000). Cloning of Arabidopsis and barley cDNAs encoding HAK potassium transporters in root and shoot cells. Physiol. Plant. 109, 34–43. doi: 10.1034/j.1399-3054.2000.100106.x

138. Salvador-Recatala, V. (2016). The AKT2 potassium channel mediates

NaCl induced depolarization in the root of Arabidopsis thaliana. Plant Signal. Behav. 11:e1165381. doi: 10.1080/15592324.2016.1165381

139. Santa-María, G. E., Danna, C. H., and Czibener, C. (2000). High-affinity potassium transport in barley roots. Ammonium-sensitive and -insensitive pathways. Plant Physiol. 123, 297 306. doi: 10.1104/pp.123.1.297
140. Santa-María, G. E., Rubio, F., Dubcovsky, J., and Rodriguez-Navarro, A. (1997). The hak1 gene of barley is a member of a large gene family and encodes a high-affinity potassium transporter. Plant Cell 9, 2281–2289. doi: 10.2307/3870585
141. Sassi, A., Mieulet, D., Khan, I., Moreau, B., Gaillard, I., Sentenac, H., et al. (2012). The rice monovalent cation transporter OsHKT2;4: revisited ionic selectivity. Plant Physiol. 160, 498–510. doi: 10.1104/pp.112.194936
142. Sattelmacher, B., Muhling, K. H., and Pennewiss, K. (1998). The apoplast - its significance for the nutrition of higher plants. Z. Pflanzen. Boden. 161, 485–498. doi: 10.1002/jpln.1998.3581610502
143. Schachtman, D. P. (2015). The role of ethylene in plant response to K+ deficiency. Front. Plant Sci. 6:1153. doi: 10.3389/fpls.2015.01153
144. Schreiber, L., Hartmann, K., Skrabs, M., and Zeier, J. (1999). Apoplastic barriers in roots: chemical composition of endodermal and hypodermal cell walls. J. Exp. Bot. 50, 1267–1280. doi: 10.1093/jexbot/50.337.1267
145. Sentenac, H., Bonneaud, N., Minet, M., Lacroute, F., Salmon, J. M., Gaymard, F., et al. (1992). Cloning and expression in yeast of a plant potassium ion transport system. Science 256, 663–665. doi: 10.1126/science.1585180
146. Shen, Y., Shen, L., Shen, Z., Jing, W., Ge, H., Zhao, J., et al. (2015). The potassium transporter OsHAK21 functions in the maintenance of ion homeostasis and tolerance to salt stress in rice. Plant Cell Environ. 38, 2766–2779. doi: 10.1111/pce.12586
147. Shin, R., Berg, R. H., and Schachtman, D. P. (2005). Reactive oxygen species and root hairs in Arabidopsis root response to nitrogen, phosphorus and potassium deficiency. Plant Cell Physiol. 46, 1350–1357. doi: 10.1093/pcp/pci145

148. Shin, R., and Schachtman, D. P. (2004). Hydrogen peroxide mediates plant root cell response to nutrient deprivation. Proc. Natl. Acad. Sci. U.S.A. 101, 8827–8832. doi: 10.1073/pnas.0401707101

149. Siddiqi, M. Y., and Glass, A. D. (1986). A model for the regulation of k influx, and tissue potassium concentrations by negative feedback effects upon plasmalemma influx. Plant Physiol. 81, 1–7. doi: 10.1104/pp.81.1.1

150. Spalding, E. P., Hirsch, R. E., Lewis, D. R., Qi, Z., Sussman, M. R., and Lewis, B. D. (1999). Potassium uptake supporting plant growth in the absence of AKT1 channel activity: inhibition by ammonium and stimulation by sodium. J. Gen. Physiol. 113, 909–918. doi: 10.1085/jgp.113.6.909

151. Stewart, W. M., Dibb, D. W., Johnston, A. E., and Smyth, T. J. (2005). The contribution of commercial fertilizer nutrients to food production. Agron. J. 97, 1–6. doi: 10.2134/agronj2005.0001

152. Subbarao, G. V., Ito, O., Berry, W. L., and Wheeler, R. M. (2003). Sodium - A functional plant nutrient. Crit. Rev. Plant Sci. 22, 391–416. doi: 10.1080/07352680390243495

153. Sutter, J.-U., Sieben, C., Hartel, A., Eisenach, C., Thiel, G., and Blatt, M. R. (2007). Abscisic acid triggers the endocytosis of the Arabidopsis KAT1 K+ channel and its recycling to the plasma membrane. Curr. Biol. 17, 1396–1402. doi: 10.1016/j.cub.2007.07.020

154. Szczerba, M. W., Britto, D. T., and Kronzucker, H. J. (2006). Rapid, futile K+ cycling and pool-size dynamics define low-affinity potassium transport in barley. Plant Physiol. 141, 1494–1507. doi: 10.1104/pp.106.082701

155. Taiz, L., and Zeiger, E. (1991). Plant Physiology, Chapters 8, 9, 10, 2nd Edn. Sunderland, MA: Sinauer.

156. Takahashi, R., Liu, S., and Takano, T. (2007). Cloning and functional comparison of a high-affinity K+ transporter gene PhaHKT1 of salt-tolerant and salt-sensitive reed plants. J. Exp. Bot. 58, 4387–4395. doi: 10.1093/jxb/erm306

157. Takehisa, H., Sato, Y., Antonio, B., and Nagamura, Y. (2013). Global transcriptome profile of rice root in response to essential

macronutrient deficiency. Plant Signal. Behav. 8:e24409. doi: 10.4161/psb.24409

158. Tester, M., and Leigh, R. A. (2001). Partitioning of nutrient transport processes in roots. J. Exp. Bot. 52, 445–457. doi: 10.1093/jexbot/52.356.653

159. Trewavas, A. J. (1997). Plant cyclic AMP comes in from the cold. Nature 390, 657–658. doi: 10.1038/37720

160. Tsay, Y.-F., Ho, C.-H., Chen, H.-Y., and Lin, S.-H. (2011). Integration of nitrogen and potassium signaling. Ann. Rev. Plant Biol 62, 207–226. doi: 10.1146/annurev-arplant-042110-103837

161. Tyerman, S. D., Skerrett, M., Garrill, A., Findlay, G. P., and Leigh, R. A. (1997). Pathways for the permeation of Na+ and Cl- into protoplasts derived from the cortex of wheat roots. J. Exp. Bot. 48, 459–480. doi: 10.1093/jxb/48.Special_Issue.459

162. Véry, A.-A., and Davies, J. M. (2000). Hyperpolarization-activated calcium channels at the tip of Arabidopsis root hairs. Proc. Natl. Acad. Sci. U.S.A. 97, 9801–9806. doi: 10.1073/pnas.160250397

163. Véry, A.-A., Nieves-Cordones, M., Daly, M., Khan, I., Fizames, C., and Sentenac, H. (2014). Molecular biology of K+ transport across the plant cell membrane: what do we learn from comparison between plant species? J. Plant Physiol. 171, 748–769. doi: 10.1016/j.jplph.2014.01.011

164. Very, A. A., and Sentenac, H. (2002). Cation channels in the Arabidopsis plasma membrane. Trends Plant Sci. 7, 168–175. doi: 10.1016/S1360-1385(02)02262-8

165. Wakeel, A. (2013). Potassium-sodium interactions in soil and plant under saline-sodic conditions. J. Plant Nutr. Soil Sci. 176, 344–354. doi: 10.1002/jpln.201200417

166. Wakeel, A., Farooq, M., Qadir, M., and Schubert, S. (2011). Potassium substitution by sodium in plants. Crit. Rev. Plant Sci. 30, 401–413. doi: 10.1080/07352689.2011.587728

167. Wakeel, A., Steffens, D., and Schubert, S. (2010). Potassium substitution by sodium in sugar beet (Beta vulgaris) nutrition on K-fixing soils. J. Plant Nutr. Soil Sci. 173, 127–134. doi: 10.1002/jpln.200900270

168. Walker, D. J., Black, C. R., and Miller, A. J. (1998). The role of cytosolic potassium and ph in the growth of barley roots. Plant Physiol. 118, 957–964. doi: 10.1104/pp.118.3.957

169. Walker, D. J., Leigh, R. A., and Miller, A. J. (1996). Potassium homeostasis in vacuolate plant cells. Proc. Natl. Acad. Sci. U.S.A. 93, 10510–10514. doi: 10.1073/pnas.93.19.10510

170. Wang, B., Davenport, R. J., Volkov, V., and Amtmann, A. (2006). Low unidirectional sodium influx into root cells restricts net sodium accumulation in Thellungiella halophila, a salt-tolerant relative of Arabidopsis thaliana. J. Exp. Bot. 57, 1161–1170. doi: 10.1093/jxb/erj116

171. Wang, X. P., Chen, L. M., Liu, W. X., Shen, L., Wang, F. L., Zhou, Y., et al. (2016). AtKC1 and CIPK23 synergistically modulate AKT1-mediated low potassium stress responses in Arabidopsis. Plant Physiol. 170, 2264–2277. doi: 10.1104/pp.15.01493

172. Wang, Y., He, L., Li, H. D., Xu, J. A., and Wu, W. H. (2010). Potassium channel alpha-subunit AtKC1 negatively regulates AKT1-mediated K+ uptake in Arabidopsis roots under low-K+ stress. Cell Res. 20, 826–837. doi: 10.1038/cr.2010.74

173. White, P. J., and Broadley, M. R. (2000). Mechanisms of caesium uptake by plants. New Phytol. 147, 241–256. doi: 10.1046/j.1469-8137.2000.00704.x

174. Wyn Jones, R. G., and Pollard, A. (1983). "Proteins, enzymes and inorganic ions," in Inorganic Plant Nutrition-Encyclopedia of Plant Physiology New Series 15, eds A. Lauchli and R. L. Bieleski (Berlin: Springer-Verlag), 528–562.

175. Xia, J., Kong, D., Xue, S., Tian, W., Li, N., Bao, F., et al. (2014). Nitric oxide negatively regulates AKT1-mediated potassium uptake through modulating vitamin B6 homeostasis in Arabidopsis. Proc. Natl. Acad. Sci. U.S.A. 111, 16196–16201. doi: 10.1073/pnas.1417473111

176. Xu, J., Li, H. D., Chen, L. Q., Wang, Y., Liu, L. L., He, L., et al. (2006). A protein kinase, interacting with two calcineurin B-like proteins, regulates K+ transporter AKT1 in Arabidopsis. Cell 125, 1347–1360. doi: 10.1016/j.cell.2006.06.011

177. Yang, T., Zhang, S., Hu, Y., Wu, F., Hu, Q., Chen, G., et al. (2014). The role of a potassium transporter oshak5 in potassium acquisition and transport from roots to shoots in rice at low potassium supply levels. Plant Physiol. 166, 945–959. doi: 10.1104/pp.114.246520

178. Yao, X., Horie, T., Xue, S., Leung, H. Y., Katsuhara, M., Brodsky, D. E., et al. (2010). Differential sodium and potassium transport selectivities of the rice OsHKT2;1 and OsHKT2;2 transporters in plant cells. Plant Physiol. 152, 341–355. doi: 10.1104/pp.109.145722

179. Yeo, A. R., Yeo, M. E., and Flowers, T. J. (1987). The contribution of an apoplastic pathway to sodium uptake by rice roots in saline conditions. J. Exp. Bot. 38, 1141–1153. doi: 10.1093/jxb/38.7.1141

180. Zhang, B., Karnik, R., Wang, Y., Wallmeroth, N., Blatt, M. R., and Grefen, C. (2015). The Arabidopsis R-SNARE VAMP721 Interacts with KAT1 and KC1 K+ channels to moderate k+ current at the plasma membrane. Plant Cell 27, 1697–1717. doi: 10.1105/tpc.15.00305

181. Zörb, C., Senbayram, M., and Peiter, E. (2014). Potassium in agriculture – Status and perspectives. J. Plant Physiol. 171, 656–669. doi: 10.1016/j.jplph.2013.08.008

CHAPTER

# 3

# THE CRITICAL ROLE OF POTASSIUM IN PLANT STRESS RESPONSE

Min Wang, Qingsong Zheng, Qirong Shen and Shiwei Guo *

Agricultural Ministry Key Lab of Plant Nutrition and Fertilization in Low-Middle Reaches of the Yangtze River, Jiangsu Key Lab and Engineering Center for Solid Organic Waste Utilization, Nanjing Agricultural University, No. 1 Weigang, Nanjing 210095, China

**Keywords**: biotic stress; abiotic stress; potassium; physiological and molecular mechanisms; plant resistance

## ABSTRACT

Agricultural production continues to be constrained by a number of biotic and abiotic factors that can reduce crop yield quantity and quality. Potassium (K) is an essential nutrient that affects most of the biochemical and physiological processes that influence plant growth and metabolism. It also contributes to the survival of plants exposed to various biotic and abiotic stresses. The following review focuses on the emerging role of K in defending against a number of biotic and abiotic stresses, including diseases, pests, drought, salinity, cold and frost and waterlogging. The availability of K and its effects on plant growth, anatomy, morphology and plant metabolism are discussed. The physiological and molecular mechanisms of K function in plant stress resistance are reviewed. This article also evaluates the potential for improving plant stress resistance by modifying K fertilizer inputs and highlights the future needs for research about the role of K in agriculture.

## INTRODUCTION

The world population is expanding rapidly and will pass from its current number of 7.0 billion to 9.4 billion by the year 2050 [1]. To provide enough food for an expanding world population, a massive increase in crop production is required to meet the food demands of future generations, while preserving the ecological and energy-related resources of our planet. However, agricultural production continues to be constrained by a variety of biotic (e.g., pathogens, insects and weeds) and abiotic (e.g., drought, salinity, cold, frost and waterlogging) factors that can significantly reduce the quantity and quality of crop production. Evidence indicates that biotic stress can cause a 28.2% yield loss of wheat, 37.4% loss of rice, 31.2% loss of maize, 40.3% loss of potatoes, 26.3% loss of soybeans and 28.8% loss of cotton [2]. Meanwhile, yield losses from abiotic stress were estimated at 65.8% for maize, 82.1% for wheat, 69.3% for soybeans and 54.1% for potatoes [3].

During their evolution, plants have developed a wide range of mechanisms to resist a variety of stressed conditions. Increasing evidence suggests that mineral nutrients play a critical role in plant stress resistance [4–8]. Out of all the mineral nutrients, potassium (K) plays a particularly critical role in plant growth and metabolism, and it contributes greatly to the survival of plants that are under various biotic and abiotic stresses. The importance of K fertilizer for the formation of crop production and its quality is known. As a consequence, potash consumption has increased dramatically in most regions of the world [9]. A strong positive relationship between K fertilizer input and grain yield has been shown [10].

K is an essential nutrient and is also the most abundant cation in plants. The concentration of $K^+$ in the cytoplasm has consistently been found to be between 100 and 200 mM [11], and apoplastic $K^+$ concentration may vary between 10 and 200 or even reach up to 500 mM [12]. K plays essential roles in enzyme activation, protein synthesis, photosynthesis, osmoregulation, stomatal movement, energy transfer, phloem transport, cation-anion balance and stress resistance [4].

This review is focused on the effects of K on plant resistance to various biotic (pathogen and insects) and abiotic (drought, salinity, cold and frost and waterlogging) stresses. K availability for plant growth, anatomy and morphology, as well as plant metabolism are discussed. This review also discusses the roles of K in stress-resistant mechanisms and evaluates the potential for improving plant resistance by modifying K fertilizer input and selecting appropriate plant species or varieties.

## THE ROLE OF POTASSIUM IN BIOTIC STRESS RESISTANCE

Increased evidence has shown that crop production is significantly restricted by biotic stresses. Oerke and Dehne [13] estimated that weeds produce the highest potential loss (32%), followed by animal pests (18%), fungi and bacteria (15%) and viruses (3%) from 1996 to 1998. These numbers reflect the total attainable production for eight major crops (wheat, rice, maize, barley, potatoes, soybeans, sugar beets and cotton). In many cases, K-deficient plants tend to be more susceptible to infection than those with an adequate supply of K. For example, the rate of rice borer infestation was greatest when there was no supply of K, but decreased rapidly as the K concentration increased [14] (Table 1). Similar results were found with a *Discula destructiva* Redlin infection in *Cornus florida* L. [15]. Williams and Smith [16] also reported that increased K fertilizer significantly reduced the disease incidence of stem rot and aggregate sheath spot (AgSS), and negative correlations were found between the percentage of K in leaf blades and disease severity. K fertilizer is widely reported to decrease insect infestation and disease incidence in many host plants. Perrenoud [17] reviewed 2449 references and found that the use of K significantly decreased the incidence of fungal diseases by 70%, bacteria by 69%, insects and mites by 63%, viruses by 41% and nematodes by 33%. Meanwhile, K increased the yield of plants infested with fungal diseases by 42%, bacteria by 57%, insects and mites by 36%, viruses by 78% and nematodes by 19%.

**Table 1:** Impacts of soil potassium build-up on rice stem borers and grain yield within a rice field [14].

| Serial number | Potassium treatments (kg ha$^{-1}$) | Stem borer infestation (%) | | Yield/plot (g/3 m$^2$) | Yield (kg ha$^{-1}$) |
|---|---|---|---|---|---|
| | | Dead heart | White heads | | |
| 1 | 40 kg | 3.05 b | 5.37 b | 1913.00 b | 6376.66 |
| 2 | 50 kg | 2.64 bc | 3.58 c | 2287.00 a | 7623.33 |
| 3 | 60 kg | 2.40 c | 3.37 c | 2317.00 a | 7723.33 |
| 4 | Control | 4.33 a | 7.12 a | 1690.00 c | 5633.33 |
| LSD value | | 0.619 | 0.561 | 219.4 | |

The means in each column are followed by at least one letter in common and are not significantly different at the 5% level of probability ($p \geq 0.05$). LSD, least significant difference.

K fertilizer application decreased the incidence of diseases in most cases, but sometimes had no effect or even the opposite effect. Prabhu *et al.*[18] noted that the effect of K on disease incidence can be classified as "increased", "decreased" and having "no effect" or "variable effect" (Table 2). The variable effects of K on disease incidence could be affected by the amount and source of K, plant and pathogen species and trial type. Nam *et al.*[19] found that strawberries that were grown with excess K were very susceptible to infection by the anthracnose pathogen, *Colletotrichum gloeosporioides*, but its resistance was greatly enhanced when no K was supplied. This result was observed because the low plant K status induces the synthesis of molecules, including reactive oxygen species (ROS) and phytohormones, such as auxin, ethylene and jasmonic acid (JA), as a result of its enhanced plant stress tolerance [5,20].

**Table 2:** Number of published papers reporting the effects of potassium on disease [18].

| | Decrease in disease | Increase in disease | No effect | Total |
|---|---|---|---|---|
| Fungi | 89 | 33 | 8 | 130 |
| Bacteria | 18 | 5 | 0 | 23 |
| Viruses | 9 | 5 | 3 | 17 |
| Nematode | 3 | 6 | 1 | 10 |

The mechanistic influences of K on plant disease resistance have been reported by several researchers. Higher $K^+$concentrations decreased the internal competition of pathogens for nutrient resources [15]. This nutritional status enables plants to allocate more resources to developing stronger cell walls for preventing pathogen infection and insect attack and to obtain more nutrients to be used for plant defense and damage repair [21]. DeDatta and Mikkelson [22] reported that the culm and stalk strength of rice were increased in the presence of adequate K concentrations as a result of increasing plant resistance. During airborne pathogen infections (especially from bacteria and viruses), the stomata were able to function properly when there was sufficient K, thus preventing pathogen invasion by rapid stomata closing [23,24].

K is also essential to the performance of multiple plant enzyme functions, and it regulates the metabolite pattern of higher plants, ultimately changing metabolite concentrations [4,21]. In a K-sufficient plant, the synthesis of high-molecular-weight compounds (such as proteins, starches and cellulose) was markedly increased, thereby depressing the concentrations of low-molecular-weight compounds, such as soluble sugars organic acids, amino acids and amides, in the plant tissues. These low-molecular-weight compounds are important for the development of infections and insect infestations, so lower concentrations, thereby, leave plants less vulnerable to disease and pest attacks in K-sufficient plants [4]. Adequate K increases phenol concentrations, which play a critical role in plant resistance [25]. Furthermore, Sarwar [14] concluded that less pest damage in higher K plants can be attributed to a lack of pest preference under sufficient nutrient concentrations, as well as the synthesis of defensive compounds leading to higher pest mortality. Figure 1 summarizes the main roles of K in plant resistance to biotic stress.

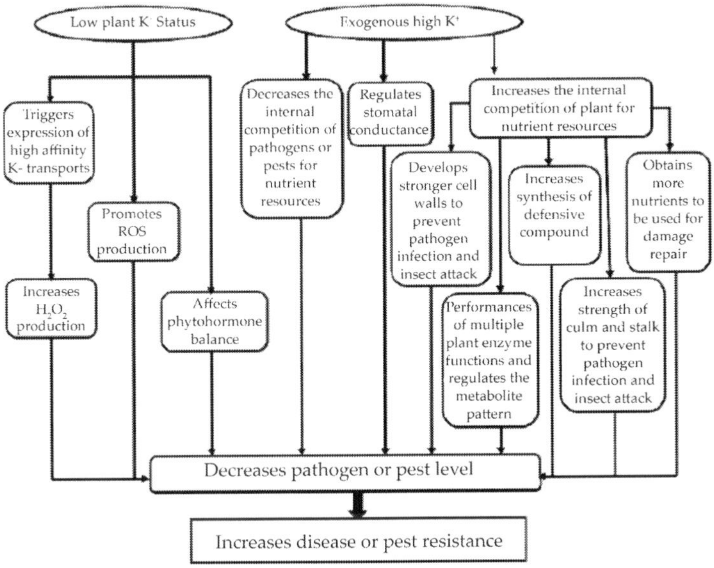


**Figure 1:** Role of K under biotic stress.

# THE ROLE OF POTASSIUM IN ABIOTIC STRESS RESISTANCE

## Potassium and Drought Stress

The major limitation for plant growth and crop production in arid and semi-arid regions is soil water availability. Plants that are continuously exposed to drought stress can form ROS, which leads to leaf damage [7,13,26] and, ultimately, decreases crop yield. During drought stress, root growth and the rates of $K^+$ diffusion in the soil towards the roots were both restricted, thus limiting K acquisition. The resulting lower K concentrations can further depress the plant resistance to drought stress, as well as K absorption. Maintaining adequate plant K is, therefore, critical for plant drought resistance. A close relationship between K nutritional status and plant drought resistance has been demonstrated. The roles of K in physiological and molecular mechanisms of plant drought resistance have been explored.

## Cell Elongation and Cell Membrane Stability

It is worthwhile to try to improve crop tolerance to stress in low-moisture soils by inducing deeper rooting, larger absorption surfaces and greater water retention in plant tissues. Deeper rooting could be achieved by deep placement of K fertilizer that is associated with other mineral nutrients, such as P and N, which both have root signaling functions [27]. Adequate amounts of K can enhance the total dry mass accumulation of crop plants under drought stress in comparison to lower K concentrations [28]. This finding might be attributable to stomatal regulation by $K^+$ and corresponding higher rates of photosynthesis [4]. Furthermore, K is also essential for the translocation of photoassimilates in root growth [6]. Root growth promotion by increased appropriate K supply under K-deficient soil was found to increase the root surface that was exposed to soil as a result of increased root water uptake [6]. Lindhauer [29] reported that fine K nutrition not only increased plant total dry mass and leaf area, but also improved the water retention in plant tissues under drought stress.

Increased evidence shows that the maintenance of membrane integrity and stability under drought stress is also essential for plant drought tolerance [30]. Cell membrane stability was significantly declined under drought stress [31]. In a study by Premachandra *et al.*[32], maize plants with higher K applications showed greater adaptation to water stress. This improvement was mainly attributed to the role of K in improving cell membrane stability and osmotic adjustment ability. An adequate K supply is essential to enhancing drought resistance by increasing root elongation and maintaining cell membrane stability.

## Aquaporins and Water Uptake

Aquaporins are channel proteins that are present in the plasma and intracellular membranes of plant cells. They play a crucial role in plant water relations by regulating the osmotic potential and hydraulic conductivity of membranes and make changes in plant water permeability [33,34]. Under drought stress conditions, aquaporin gene expression can be regulated [35,36] to help plants maintain their water balance [36–38].

During water stress, roots regulated their water and ion uptake capacities by modifying *PIPs* (plasma membrane intrinsic proteins) and $K^+$ channel at the transcription level to cope with the water deficiency

[37,39–42]. Liu *et al.*[41] observed that transcription expression levels of the *PIPs* and $K^+$ channel-encoding genes was induced by $K^+$ starvation and could be downregulated by a polyethylene glycol (PEG)-mediated water deficit in rice, which may result in a reduction in the membrane water permeability and then promote cellular water conservation during drought stress. However, it should be mentioned that the expression level of some other water channels, such as OsPIP1;1, OsPIP1;2, AtPIP1;4 and AtPIP2;5, could be induced by a relative long-term water-deficit treatment, which should result in greater osmotic water permeability and facilitated water flux [40,41]. Recently, Kanai *et al.*[43] also observed close coupling between aquaporin activities and K-channel transporters. The initial response of K deficiency was perceived in the form of a change in K-channel activity, which altered root hydraulic conductance, and transduction of the follow up signal resulted in a shift of aquaporin activity. They found that aquaporin activities might have been suppressed by K deficiency and result in a reduction of root hydraulic conductance and water supply to the growing stem for diameter expansion and the leaf for transpiration.

In *Arabidopsis* roots, the transcripts encoding water channels, PIP1;2 (PIP1b), PIP2;2 (PIP2b) and TIP1;2 (TIP), and the $K^+$transporter, HAK5, were reduced under treatment of the $K^+$-channel blocker (CsCl) [44]. Furthermore, water transport in onion roots was sensitive to inhibitors of the water channel and $K^+$-channel, and the reduction in hydraulic conductivity (*Lp*) by treatment with a $K^+$-channel inhibitor suggests that $K^+$ fluxes are linked to water channel activity in the plasma membrane [45]. Water channels and $K^+$ channel/transporters are functionally co-regulated as a part of plant osmoregulation to maintain appropriate cytosolic osmolarity and acclimate the plant to drought or other stresses [41]. Guo *et al.*[46] also showed a positive correlation between K absorption and water uptake in *Phaseolus vulgaris* plants. The loading of $K^+$ into the xylem most likely mediated the xylem hydraulic conductance that aided plants in maintaining cell turgor, stomatal aperture and gas exchange rates as part of their drought adaptations [47,48].

## Osmotic Adjustment

The maintenance of a favorable water status is critical for plant survival under drought stress. Osmotic adjustment is a major trait that is associated with maintaining high cellular turgor potential and water retention in response to drought stress. Many studies have shown that osmotic adjustment of leaves is positively correlated with drought tolerance in various plant species [49]. As one of the most prominent inorganic osmotica in plants, $K^+$ plays a key role in formation of the osmotic adjustment ability, even under drought conditions [4]. Cell turgor recovery in osmotically-generated stress was regulated by increasing $K^+$, $Cl^-$ and $Na^+$ uptake by root cells, which was partly mediated by voltage-gated $K^+$ transporters at the cellular plasma membrane [50]. Furthermore, sufficient K induces solute accumulation, thus lowering osmotic potential and helping to maintain plant cell turgor under osmotic stress. In summary, an adequate K status may facilitate osmotic adjustment, which maintains higher turgor pressure, relative water content and lower osmotic potential, thus improving the ability of plants to tolerate drought stress [8,51].

## *Stomatal Regulation*

One of the major functions of the stomata is to control plant water loss via transpiration. During drought stress, quick stomatal closure and internal moisture preservation are essential for plant adaptation to drought conditions. K plays a crucial role in turgor regulation within the guard cells during stomatal movement [4]. As stomatal closure is preceded by a rapid release of $K^+$ from the guard cells into the leaf apoplast, it is reasonable to think that stomata would be difficult to remain open under K-deficient conditions. Some studies also stated that K deficiency may induce stomatal closure and inhibit photosynthetic rates in several crop plants [52,53]. Conversely, many studies suggest that K had no effect on stomatal conductance and photosynthetic rates under well-watered conditions, but K starvation could favor stomatal opening and promote transpiration, compared with K sufficiency in several plants under drought stress [54–56]. Furthermore, photosynthetic rate was decreased under drought stress in K-deficient plants [51,54,57]. This discrepancy may be related to the plant species, experimental system and environmental factors within the experimental field or interspecific differences.

The effects of drought stress on stomata closure in olive trees and sunflower plants were found to be dependent on the $K^+$nutrient status [55,56]. When plants were supplied with different $K^+$ concentrations and then subjected to drought stress, their stomatal conductance was more markedly reduced in normal K plants than in low K plants (Table 3). Benlloch-Gonzalez *et al.*[56] explained that the low plant K status could inhibit water-stress-induced stomatal closure via ethylene synthesis, and stomatal conductance could be significantly reduced in $K^+$-starved plants after the adding of an ethylene synthesis inhibitor (cobalt). $K^+$ starvation increases the transcription of genes involved in ethylene production and signaling and stimulates ethylene production [56,58]. Then, the increased ethylene could inhibit the action of abscisic acid (ABA) on stomata and delay stomata closure [59,60]. During drought stress, the stomata cannot function properly in $K^+$-deficient plants, resulting in greater water loss. Drought stress did not decrease water use efficiency (WUE), whereas it did increase WUE by rapid stomata closing during water deficit [51]. Adequate levels of K nutrition enhanced plant drought resistance, water relations, WUE and plant growth under drought conditions [51].

**Table 3:** Effect of $K^+$ concentration in irrigation water (normal *versus* low K) and water availability in the growth medium (irrigation *versus* drought) on $K^+$ accumulation and stomatal conductance in leaves [56].

| Treatment | $K^+$ content in leaves (μmol/gFW) | Stomatal conductance (mmol/m$^2$/s) |
|---|---|---|
| Normal K: Irrigated | 133.6 ± 7.3 | 456 ± 5.6 |
| Normal K: Drought | 119.4 ± 3.8 | 281 ± 27.9 |
| Low K: Irrigated | 36.3 ± 1.4 | 462 ± 4.0 |
| Low K: Drought | 25.7 ± 0.8 | 351 ± 15.2 |

## Detoxification of Reactive Oxygen Species

Stomatal closing in response to drought stress leads to a reduction in photosynthetic efficiency as a consequence of chloroplast dehydration [7]. Photosynthesis inhibition can further disturb the balance between ROS production and antioxidant defense [61–63], resulting in ROS accumulation. The ROS have a dual action in biotic and abiotic stresses that depends on their cellular concentration [64]. Low levels of ROS could be involved in the stress-signaling pathway by triggering stress defense/acclimation responses [64,65]. However, ROS became extremely injurious to cellular membranes and other cellular components when its concentrations reached the point of phytotoxicity, resulting in oxidative stress and, eventually, cell death [64,66].

Drought stress-induced ROS production can additionally be enhanced in K-deficient plants [7]. Under drought stress, photosynthetic $CO_2$ fixation in K-deficient plants is substantially limited by impairment in stomata regulation, conversion of light energy into chemical energy and phloem export of photosynthates from source leaves into sink organs [51]. As the impairment in photosynthetic $CO_2$ fixation occurs, molecular $O_2$ is activated, leading to extensive generation of ROS [67] and, thereby, oxidative degradation of chlorophyll and membranes. The maintenance of adequate K nutrition is critical for mitigating or preventing damage by drought stress and controlling the water balance [68]. Egilla [51] suggested that increasing extrachloroplastic $K^+$ concentrations in plant cells with an excess $K^+$ supply could prevent photosynthesis inhibition under drought stress. An adaptive K requirement for drought-stressed plants could be related to the role of K in enhancing photosynthetic $CO_2$ fixation and transport of photosynthates into sink organs and inhibiting the transfer of photosynthetic electrons to $O_2$, thus reducing ROS production [7].

Beside the photosynthetic electron transport, nicotinamide adenine dinucleotide phosphate (NADPH)-dependent oxidase activation represents another major source for production of ROS in plant cells by a number of biotic and abiotic stress factors [65]. NADPH-oxidizing enzymes catalyze one-electron reduction of $O_2$ to $O_2^{\bullet-}$ by using NADPH as an electron donor [7]. Cakmak [7] reported that activity of NADPH oxidase was increased in cytosolic fractions of bean roots with increasing severity of K deficiency, resulting in an increase in NADPH-dependent $O_2^{\bullet-}$ generation. The reason for the increase of NADPH oxidase by K deficiency is probably that K deficiency induced ABA accumulation [69]. Furthermore, ABA has also been shown to be effective in increasing $H_2O_2$ and $O_2^{\bullet-}$ accumulations in roots or leaves [70,71], but this point needs to be clarified in future studies. An improvement in the plant K supply can inhibit ROS production under drought stress by reducing NADPH oxidase activity and maintaining photosynthetic electron transport [7]. In addition to K, various micronutrients, including Zn, B, Cu and Mn, have also been shown to be involved in detoxifying oxygen radicals [72]. The K supply is thus associated with other mineral nutrients and is essential for the detoxification of active oxygen under drought stress.

In brief, a sufficient K status increased cell membrane stability, root growth, leaf area and total dry mass for plants living under drought conditions and also improved water uptake and water conservation. Maintaining an adequate K nutritional status is critical for plant osmotic adjustment and for mitigating ROS damage as induced by drought stress. In Figure 2, we summarize the role of K in plants that are living under drought stress.

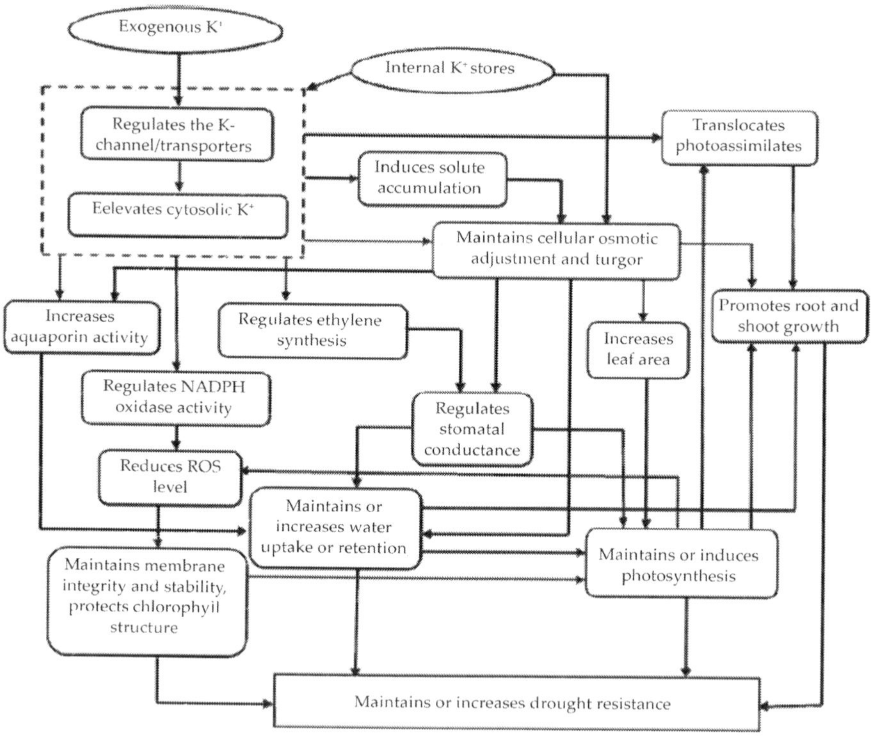


**Figure 2:** Role of K under drought stress.

## Potassium and Salt Stress

Salinity is a major abiotic stress that affects approximately 7% of the world's total land area. More than 800 million hectares of land around the world are affected by salinity [73], which results in billions of dollars in crop production losses. The accumulation of high salt concentrations in the soil makes it harder for plant roots to take up water and, thereby, disturbs a plant's water balance, while high concentrations of salts in plant tissue may be toxic. Salinity inhibits seed germination and plant growth, affects the leaf anatomy and physiology of plants and, thereby, influences their photosynthesis, water relations, protein synthesis, energy production and lipid metabolism [74]. Plant growth responds to salinity in the following two phases [75]: a rapid osmotic phase that decreases water availability to plants and inhibits the growth of young leaves, followed by a slow ionic phase that results in salt toxicity and accelerates the senescence of mature leaves. Plants have developed diverse strategies to resist salt stress, such as restricting $Na^+$ uptake, activating $Na^+$ exclusion or cellular compartmentalization of excessive $Na^+$ into the vacuole [76,77].

Salt-stressed root growth is restricted by osmotic effects and toxic effects of ions, which results in lower nutrient uptake and inhibits the translocation of mineral nutrients, especially $K^+$. As a result of the similarities in physicochemical properties between $Na^+$ and $K^+$, $Na^+$ could compete with $K^+$ for major binding sites in key metabolic processes, including both low-affinity (e.g., non-selective cation channels (NSCC)) and high-affinity (e.g., KUP and high-affinity $K^+$ transporter (HKT)) transporters and could also disturb plant metabolism [4,78]. $K^+$ deficiency can usually be observed under salinity stress. First, high levels of $Na^+$ inhibit $K^+$ activity in the soil solution, resulting in a reduction of $K^+$ availability. Second, $Na^+$not only interferes with $K^+$ translocation from root to shoot (especially in low $K^+$ status) [79], but also competes with $K^+$ for uptake sites at the plasma membrane, resulting in lower $K^+$ uptake. Third, salinity stress leads to plasma membrane dis-integrity and favors $K^+$ leaking, resulting in a rapid decline in cytosolic $K^+$[80]. Also, salinity induces significant membrane depolarization and favors $K^+$ leaking through depolarization-activated outward-rectifying (KOR) $K^+$ channels [78]. Therefore, keeping cellular $K^+$ content above a certain threshold and maintaining a high cytosolic $K^+/Na^+$ ratio (either by retaining $K^+$ or preventing $Na^+$ from accumulating in the leaves) is critical for plant growth and salt tolerance. An increasing K supply corresponded with higher $K^+$ accumulation in plant tissue, which reduced the $Na^+$ concentration and resulted in a higher $K^+/Na^+$ ratio. Members of the HKT transporter (high-affinity $K^+$ transporter) family that mediate $Na^+$-specific transport or $Na^+$-$K^+$ co-transport play a key role in plant $Na^+$

tolerance mechanisms [81,82]. HKT represents a primary mechanism in the regulation of $Na^+$ and $K^+$ homeostasis, as well as $Na^+$ exclusion [83,84].

Plant growth and salt tolerance were sharply reduced when exposed to a combination of salt stress and K-deficiency stress. $K^+$ deficiency significantly increased the negative effects that were induced by salt in the photosynthesis of barley and was accompanied by an increase in salt sensitivity [85]. Similar results were found by Qu *et al.*[86,87], in which $K^+$deficiency significantly inhibited nitrogen and photosynthetic carbon assimilation and also impaired the light reaction pathways of PS I and PS II in maize under salt stress. In a study by Chen *et al.*[88], $K^+$ flux from barley root in response to NaCl treatment was highly positively correlated with net $CO_2$ assimilation, plant growth, survival rate, relative grain yield and tolerance to salt stress (Table 4).

**Table 4:** Linear correlation (*r* values) between various physiological characteristics from 62 barley genotypes in glasshouse and laboratory experiments [88].

| Parameter | Grain yield | $K^+$ flux | Shoot biomass | Survival rate | Plant height | $[CO_2]$ass | TSW | Tillering |
|---|---|---|---|---|---|---|---|---|
| $K^+$ flux | 0.67 ** | - | - | - | - | - | - | - |
| Shoot biomass | 0.96 ** | 0.69 ** | - | - | - | - | - | - |
| Survival rate | 0.65 ** | 0.70 ** | 0.74 ** | - | - | - | - | - |
| Plant height | 0.70 ** | 0.69 ** | 0.61 ** | 0.51 ** | - | - | - | - |
| $[CO_2]$ass | 0.68 ** | 0.69 ** | 0.65 ** | 0.48 ** | 0.50 ** | - | - | - |
| TSW | 0.72 ** | 0.70 ** | 0.70 ** | 0.63 ** | 0.74 ** | 0.48 ** | - | - |
| Tillering | 0.48 ** | 0.26 * | 0.51 ** | 0.16 | 0.23 | 0.25 * | 0.33 * | - |
| Germination | 0.29 * | 0.21 | 0.31 * | 0.33 ** | 0.16 | 0.02 | 0.38 ** | 0.20 |

** $p < 0.01$; * $p < 0.05$; TSW: thousand-seed weight; $[CO_2]$ass: $CO_2$ assimilation.

Increased evidence has shown that K can involve osmotic adjustment of salt-stressed plants. During salt stress conditions, increased $Na^+$ concentrations were accumulated in the vacuole and a substantial osmotic potential gradient was established between the vacuole and the cytosol by depressing the cytosol's water activity. This change requires a coordinated increase in compatible solutes in the cytosol to balance out the osmotic pressure. Munns and Tester [75] reviewed that plants have a $Na^+$ exclusion mechanism that maintains a low level of $Na^+$ in the leaves during salt stress; thus, the major osmoticum in leaves was $K^+$. $K^+$ plays an important role in maintaining cell turgor and osmotic adjustment. The vacuole and the cytosol are the two major pools of K in plant cells. Cytosolic $K^+$ concentrations are maintained at a constant level and are essential for plant metabolism, while vacuolar $K^+$ concentrations may vary dramatically. Under $K^+$-deficient conditions, a constant cytosolic $K^+$ concentration was attributed to the consumption of vacuolar potassium [89].

Low $K^+$ status might induce the formation of ROS and related cell damage under saline conditions, which was attributed to the effects of $K^+$ deficiency and/or $Na^+$ toxicity on stomatal closing and the inhibition of photosynthetic activity and ultimately inhibits plant growth and reduces crop production [90]. Previous articles have shown that salinity-induced ROS formation can lead to programmed cell death (PCD), and a high cytosolic $K^+/Na^+$ ratio is essential for triggering salinity-induced PCD [91]. A decrease in the cytosolic $K^+$ pool would activate caspase-like proteases and lead to PCD under saline conditions. The ability of plants to satisfy their metabolic requirements for $K^+$ in the presence of salinity by using higher $K^+$fluxes and lower $Na^+$ fluxes that result in a higher $K^+/Na^+$ selectivity ratio is essential for salt tolerance. The addition of $K^+$to a saline culture solution has been found to increase $K^+$ concentrations in plant tissue that corresponds with a decrease in $Na^+$ content, with a further increase in plant growth and salt tolerance. Increased evidence shows that it is not the absolute quantity of $Na^+$*per se* that influences salt resistance, but rather the cytosolic $K^+/Na^+$ ratio that determines plant salt tolerance [11,78]. Figure 3 summarizes the role of K in plants living under salt stress.

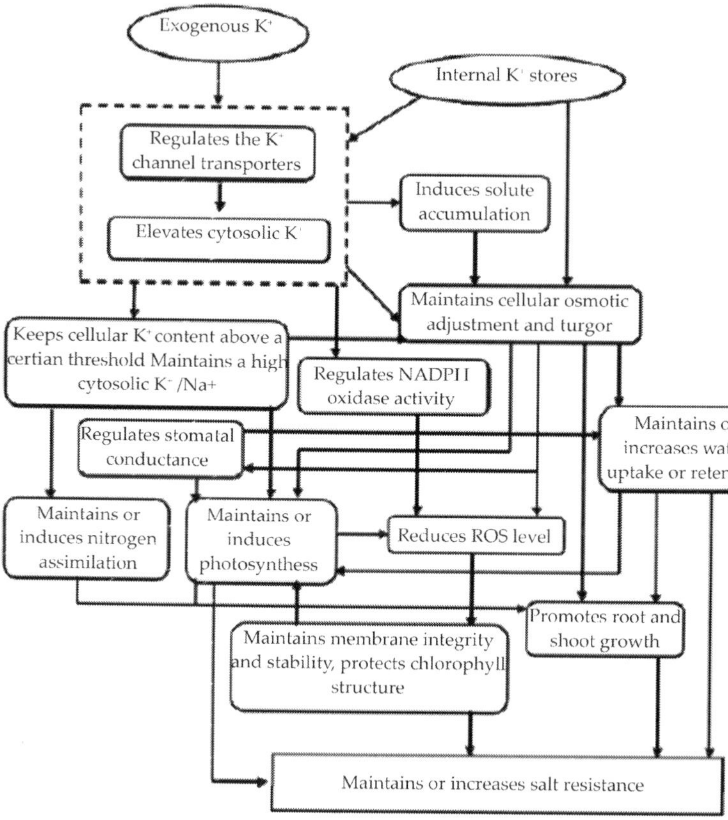


**Figure 3:** Role of K under salt stress.

## Potassium and Low-Temperature Stress

Cold stress inhibits plant growth and development, which results in limited crop productivity. It affects plants by directly inhibiting metabolic reactions and indirectly influencing cold-induced osmotic, oxidative and other stresses. The effect of increasing $K^+$ applications on yield and cold tolerance studied by Devi *et al.*[92] in *Panax ginseng* showed that a high $K^+$ concentration activated the plant's antioxidant system and increased levels of ginsenoside-related secondary metabolite transcripts, which are associated with cold tolerance. Cold stress may destroy photosynthetic processes and reduce the effectiveness of antioxidant enzymes, resulting in ROS accumulation [66,93,94]. K improved plant survival under cold stress by increasing antioxidant levels and reducing ROS production [7,92].

Greater frost damage in K-deficient plants is related to water deficiency from the chilling-induced inhibition of water uptake and freezing-induced cellular dehydration [95]. A significant negative correlation was found between frost damage and leaf K concentration, and an adequate K supply can effectively increase frost resistance [6,8]. Bogdevitch [96] found that oats that were supplied with sufficient K could survive late frost without obvious damage, whereas much of the crop that was grown on K-deficient soil did not survive. This finding could be attributed to a regulation of osmotic and water potential and a reduction of electrolyte leakage caused by cold stress [8,97]. High concentrations of $K^+$ protected against freezing by lowering the freezing point of the plant's cell solution. Furthermore, an adapted cytosol $K^+$ concentration is also essential for enzyme activities that are involved in regulating frost resistance [8].

Because the plasma membrane is the primary site for perceiving changes in temperature, membrane fluidity can be decreased by cold stress as a result of changes in fatty acid unsaturation and the lipid-protein composition of the cell membrane [98]. The ratio of unsaturated/saturated fatty acids in the cell membrane was essential for plant cold tolerance, and the higher the ratio in the cell membrane, the more tolerant the tissue is to cold stress [99]. A decrease in membrane fluidity could further affect the transport of ions, water and metabolites. The effects of nitrogen and potassium on spikelet sterility induced by low temperature at the reproductive stage of rice were studied by Haque [100]. The spikelet sterility induced by low temperature was decreased with the increase of $K^+$ supply and the increase of the K/N ratio in the rice leaves. Increasing plant frost resistance by the addition of K is associated with the increase in phospholipids, membrane permeability and improvement in the biophysical and biochemical properties of cell [101].

In brief, higher K tissue concentrations reduced chilling damage and increased cold resistance, ultimately increasing yield production [8,21]. Frost damage was inversely related to K concentration and was significantly reduced by K fertilization. Figure 4 summarizes the role of K in the plant under low temperature stress.

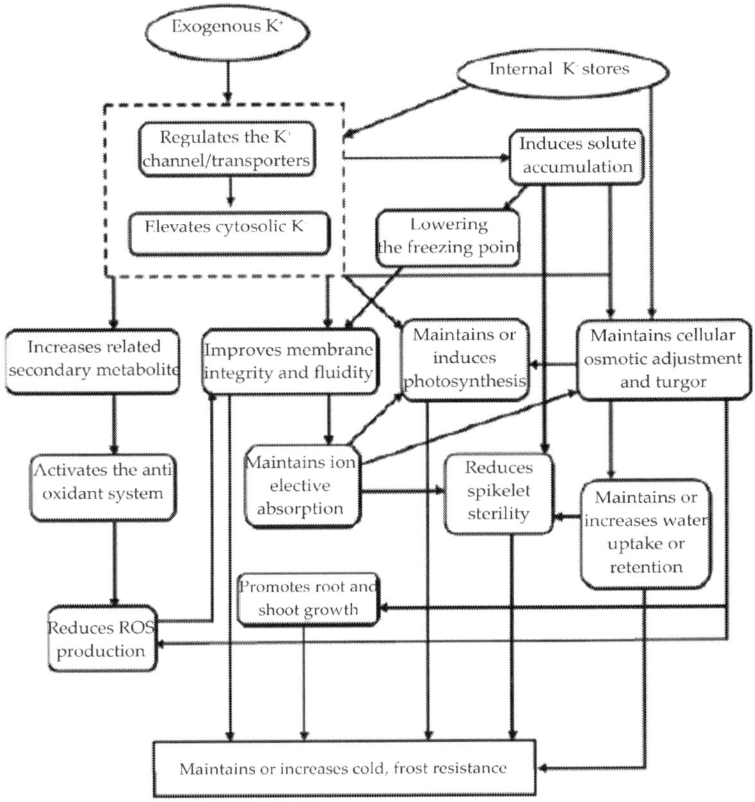


**Figure 4:** Role of K under cold and frost stress.

## Potassium and Waterlogging Stress

Waterlogging affects approximately 10% of the global land area [102] and is a serious impediment for sustainable agriculture development. Yield losses due to waterlogging may vary between 15% and 80%, depending on the crop species and growth stage, soil type and duration of the stress [103], resulting in severe economic penalties in some area. The important biological consequence of waterlogging is that the respiration of roots and micro-organisms depletes the residual oxygen and the environment becomes hypoxic (*i.e.*, oxygen levels limit mitochondrial respiration) and, later, anoxic (*i.e.*, respiration is completely inhibited) [104,105]. The low energy status under oxygen deficient conditions results in a substantial depolarization of plasma membrane potential

[106], subsequent impairment of ion transport processes through voltage-gated uptake channels and a decrease of the uptake of most essential cations (e.g., $K^+$, $NH_4^+$ or $Mg^{2+}$) [107,108]. Pang *et al.*[109] reported that hypoxia-induced $K^+$ flux responses are mediated by both inwardly rectifying potassium (KIR) and NSCC channels in the elongation zone, while KOR channels in the mature zone are likely to play a critical role. Avoiding $K^+$ loss during hypoxia or anoxia stress is the key mechanism responsible for waterlogging resistance in plants [109–112].

Furthermore, as flooding time increases, potentially toxic compounds, such as sulfides, soluble iron and manganese, ethanol, $CO_2$, ethylene, lactic acid, acetaldehyde and acetic and formic acid, were accumulated as the result of the reduced soil redox potential [106,113]. Those compounds acted on cellular membranes, leading to phospholipid oxidation and a subsequent change in membrane integrity and membrane transport [109,114]. Rapid changes in net $K^+$ were measured in response to the application of secondary metabolites (various monocarboxylic acids and phenolic acids) produced by waterlogged soils [115]. Shabala [106] assumed that organic acid uptake across the plasma membrane results in a net $H^+$influx and causes a substantial membrane depolarization. Such a depolarization will significantly affect intracellular $K^+$homeostasis by reducing $K^+$ uptake via KIR, as well as enhancing $K^+$ efflux via KOR.

Waterlogging is known to block the oxygen supply to the roots, thus inhibiting root respiration, resulting in a severe decline in energy status of root cells, affecting important metabolic processes of plants. Under waterlogged conditions, the stomata conductance, photosynthesis rate and root hydraulic conductivity of plant were hampered [116]. The oxidative damage induced by the generation of reactive oxygen species affects the integrity of membranes and induces damage to the efficiency of photosystem II, thereby, causing a considerable decrease in net photosynthetic rates [117]. Exogenous application of K could effectively ameliorate the adverse effects of waterlogging on plants. K supplement under waterlogging not only increased plant growth, photosynthetic pigments and photosynthetic capacity, but also improved plant nutrient uptake as a result of higher $K^+$, $Ca^{2+}$, N, $Mn^{2+}$ and $Fe^{2+}$ accumulation [118]. Ashraf *et al.*[118] also reported that exogenous application of K in soil and as foliar spray alleviated the adverse effects of waterlogging on cotton plants.

## IMPLICATIONS

The population of the world will exceed 9 billion by the year 2050. It is, therefore, of vital importance to improve crop yield to match the requirement for food. However, as the environment was becoming worse, the quantity and quality of crop production were significantly decreased by a variety of biotic and abiotic stresses. The practice of intensive fertilization to support massive food production for an increasing global population is a must. However, consumption of excess N fertilization and K deficiency cause a reduction in crop yields and quality in many regions. Therefore, to enable closing yield gaps and allow for a much higher productivity in many regions, a significant increase in K fertilization application is required. K is an essential plant nutrient that impacts a number of physiological and biochemical processes that are involved in plant resistance to biotic and abiotic stresses, as summarized in Figure 5.

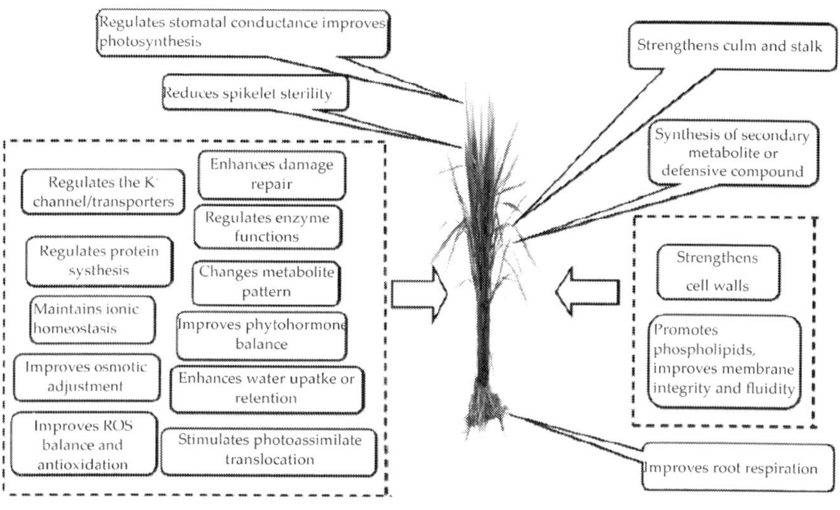


**Figure 5:** Roles of K in resisting all stresses.

Maintaining an optimum K nutritional status is essential for plant resistance to biotic and abiotic stresses. Balanced fertilization and efficient K usage in combination with other nutrients not only contribute to sustainable crop's growth, yield and quality, but also influence plant health and reduce the environmental risks. However, our understanding about the role of K in whole-plant stress response mechanisms is limited. In this paper, suggested future needs and prospects for research about the role of K in agriculture include:

- Investigating more details about the molecular mechanisms of K in plant stress resistance.
- Examining the role of K on plant resistance to biotic and abiotic stresses in differentiated cells, tissues and organs and connecting the data relevantly.
- Identifying the common or specific response of K to distinct stress and the role of K on long-term plant responses under multiple stress conditions in nature.
- Understanding the relationship between K and other nutrients in relation to plant adaptation to stresses in different agroecological systems.
- Developing models for better K recommendations based on soil, plant and environmental factors.
- Investigating more researcher on the importance of K on crop production, nutritional quality and human and animal health.

## Acknowledgements

This work was financially supported by the National Basic Research Program of China (2013CB127403) and the Nature Science Foundation of China (31172020, 31272236).

## REFERENCES

1. United States Census Bureau. International Data Base—Total Midyear Population for the World: 1950–2050, Available online: http://www.census.gov/population/international/data/idb/worldpoptotal.php (accessed on 6 June 2012).
2. Oerke, E.C. Crop losses to pests. J. Agri. Sci 2006, 144, 31–43.
3. Bray, E.A.; Bailey-Serres, J.; Weretilnyk, E. Responses to Abiotic Stresses. In Biochemistry and Molecular Biology of Plants; Buchanan, B., Gruissem, W., Jones, R., Eds.; American Society of Plant Physiologists: Rockville, MD, USA, 2000; pp. 1158–1203.
4. Marschner, P. Marschner's Mineral Nutrition of Higher Plants, 3rd ed; Academic Press: London, UK, 2012; pp. 178–189.
5. Amtmann, A.; Troufflard, S.; Armengaud, P. The effect of potassium nutrition on pest and disease resistance in plants. Physiol. Plantarum

2008, 133, 682–691.

6. Romheld, V.; Kirkby, E.A. Research on potassium in agriculture: Needs and prospects. Plant Soil 2010, 335, 155–180.
7. Cakmak, I. The role of potassium in alleviating detrimental effects of abiotic stresses in plants. J. Plant Nutr. Soil Sci 2005, 168, 521–530.
8. Kant, S.; Kafkafi, U. Potassium and Abiotic Stresses in Plants. In Potassium for Sustainable Crop Production; Pasricha, N.S., Bansal, S.K., Eds.; Potash Institute of India: Gurgaon, India, 2002; pp. 233–251.
9. Pettigrew, W.T. Potassium influences on yield and quality production for maize, wheat, soybean and cotton. Physiol. Plantarum 2008, 133, 670–681.
10. Dong, H.; Kong, X.; Li, W.; Tang, W.; Zhang, D. Effects of plant density and nitrogen and potassium fertilization on cotton yield and uptake of major nutrients in two fields with varying fertility. Field Crop Res 2010, 119, 106–113.
11. Shabala, S.; Pottosin, I.I. Potassium and potassium-permeable channels in plant salt tolerance. Signal. Commun. Plants 2010, 87–110.
12. White, P.; Karley, A. Potassium. In Cell Biology of Metals and Nutrients; Hell, R., Mendel, R.R., Eds.; Springer: Berlin/Heidelberg, Germany, 2010; pp. 199–224.
13. Oerke, E.C.; Dehne, H.W. Safeguarding production-losses in major crops and the role of crop protection. Crop Prot 2004, 23, 275–285.
14. Sarwar, M. Effects of potassium fertilization on population build up of rice stem borers (lepidopteron pests) and rice (Oryza sativa L.) yield. J. Cereals Oilseeds 2012, 3, 6–9.
15. Holzmueller, E.J.; Jose, S.; Jenkins, M.A. Influence of calcium, potassium, and magnesium on Cornus florida L. density and resistance to dogwood anthracnose. Plant Soil 2007, 290, 189–199.
16. Williams, J.; Smith, S.G. Correcting potassium deficiency can reduce rice stem diseases. Better Crops 2001, 85, 7–9.
17. Perrenoud, S. Potassium and Plant Health, 2nd ed; International Potash Institute: Bern, Switzerland, 1990; pp. 8–10.

18. Prabhu, A.S.; Fageria, N.K.; Huber, D.M. Potassium Nutrition and Plant Diseases. In Mineral Nutrition and Plant Disease; Datnoff, L.E., Elmer, W.H., Huber, D.M., Eds.; American Phytopathological Society: Saint Paul, MN, USA, 2007; pp. 57–78.
19. Nam, M.H.; Jeong, S.K.; Lee, Y.S.; Choi, J.M.; Kim, H.G. Effects of nitrogen, phosphorus, potassium and calcium nutrition on strawberry anthracnose. Plant Pathol 2006, 55, 246–249.
20. Ashley, M.K.; Grant, M.; Grabov, A. Plant responses to potassium deficiencies: A role for potassium transport proteins. J. Exp. Bot 2006, 57, 425–436.
21. Mengel, K. Principles of Plant Nutrition, 5th ed; Kluwer Academic Publishers: Dordrecht, the Netherlands, 2001; pp. 481–509.
22. DeDatta, J.G.; Mikkelson, D.S. Potassium Nutrition in Rice; American Society of Agronomy: Madison, WI, USA, 1985; pp. 665–699.
23. Hardter, R. Potassium and Biotic Stress of Plants. In Feed the Soil to Feed the People: The Role of Potash in Sustainable Agriculture; Johnston, A.E., Ed.; International Potash Institute: Basel, Switzerland, 2003; pp. 345–362.
24. Pervez, H.; Ashraf, M.; Makhdum, M.I.; Mahmood, T. Potassium nutrition of cotton (Gossypium hirsutum L.) in relation to cotton leaf curl virus disease in aridisols. Pak. J. Bot 2007, 39, 529–539.
25. Prasad, D.; Singh, R.; Singh, A. Management of sheath blight of rice with integrated nutrients. Indian Phytopathol 2010, 63, 11–15.
26. Foyer, C.H.; Vanacker, H.; Gomez, L.D.; Harbinson, J. Regulation of photosynthesis and antioxidant metabolism in maize leaves at optimal and chilling temperatures: Review. Plant Physiol. Biochem 2002, 40, 659–668.
27. Kirkby, E.A.; LeBot, J.; Adamowicz, S.; Römheld, V. Nitrogen in Physiology—An Agronomic Perspective and Implications for the Use of Different Nitrogen Forms; International Fertiliser Society: Cambridge, York, UK, 2009.
28. Egilla, J.N.; Davies, F.T.; Drew, M.C. Effect of potassium on drought resistance of Hibiscus rosa-sinensis cv. Leprechaun: Plant growth, leaf macro- and micronutrient content and root longevity. Plant Soil 2001, 229, 213–224.

29. Lindhauer, M.G. Influence of K nutrition and drought on water relations and growth of sunflower (Helianthus-annuus L.). J. Plant Nutr. Soil Sci 1985, 148, 654–669.
30. Bajji, M.; Kinet, J.M.; Lutts, S. The use of the electrolyte leakage method for assessing cell membrane stability as a water stress tolerance test in durum wheat. Plant Growth Regul 2002, 36, 61–70.
31. Wang, Z.L.; Huang, B.R. Physiological recovery of kentucky bluegrass from simultaneous drought and heat stress. Crop Sci 2004, 44, 1729–1736.
32. Premachandra, G.S.; Saneoka, H.; Ogata, S. Cell membrane stability and leaf water relations as affected by potassium nutrition of water-stressed maize. J. Exp. Bot 1991, 42, 739–745.
33. Heinen, R.B.; Ye, Q.; Chaumont, F. Role of aquaporins in leaf physiology. J. Exp. Bot 2009, 60, 2971–2985.
34. Maurel, C.; Chrispeels, M.J. Aquaporins: A molecular entry into plant water relations. Plant Physiol 2001, 125, 135–138.
35. Lian, H.L.; Yu, X.; Ye, Q.; Ding, X.S.; Kitagawa, Y.; Kwak, S.S.; Su, W.A.; Tang, Z.C. The role of aquaporin RWC3 in drought avoidance in rice. Plant Cell Physiol 2004, 45, 481–489.
36. Tyerman, S.D.; Niemietz, C.M.; Bramley, H. Plant aquaporins: Multifunctional water and solute channels with expanding roles. Plant Cell Environ 2002, 25, 173–194.
37. Galmes, J.; Pou, A.; Alsina, M.M.; Tomas, M.; Medrano, H.; Flexas, J. Aquaporin expression in response to different water stress intensities and recovery in Richter-110 (Vitis sp.): Relationship with ecophysiological status. Planta 2007, 226, 671–681.
38. Kaldenhoff, R.; Ribas-Carbo, M.; Flexas, J.; Lovisolo, C.; Heckwolf, M.; Uehlein, N. Aquaporins and plant water balance. Plant Cell Environ 2008, 31, 658–666.
39. Smart, L.B.; Moskal, W.A.; Cameron, K.D.; Bennett, A.B. MIP genes are down-regulated under drought stress in Nicotiana glauca. Plant Cell Physiol 2001, 42, 686–693.
40. Alexandersson, E.; Fraysse, L.; Sjovall-Larsen, S.; Gustavsson, S.; Fellert, M.; Karlsson, M.; Johanson, U.; Kjellbom, P. Whole gene family expression and drought stress regulation of aquaporins. Plant

Mol. Biol 2005, 59, 469–484.

41. Liu, H.Y.; Sun, W.N.; Su, W.A.; Tang, Z.C. Co-regulation of water channels and potassium channels in rice. Physiol. Plantarum 2006, 128, 58–69.
42. Cuéllar, T.; Pascaud, F.; Verdeil, J.L.; Torregrosa, L.; Adam-Blondon, A.F.; Thibaud, J.B.; Sentenac, H.; Gaillard, I. A grapevine Shaker inward K+ channel activated by the calcineurin B-like calcium sensor 1-protein kinase CIPK23 network is expressed in grape berries under drought stress conditions. Plant J 2010, 61, 58–69.
43. Kanai, S.; Moghaieb, R.E.; El-Shemy, H.A.; Panigrahi, R.; Mohapatra, P.K.; Ito, J.; Nguyen, N.T.; Saneoka, H.; Fujita, K. Potassium deficiency affects water status and photosynthetic rate of the vegetative sink in green house tomato prior to its effects on source activity. Plant Sci 2011, 180, 368–374.
44. Sahr, T.; Voigt, G.; Paretzke, H.G.; Schramel, P.; Ernst, D. Caesium-affected gene expression in Arabidopsis thaliana. New Phytol 2005, 165, 747–754.
45. Tazawa, M.; Sutou, E.; Shibasaka, M. Onion root water transport sensitive to water channel and K+ channel Inhibitors. Plant Cell Physiol 2001, 42, 28–36.
46. Guo, S.W.; Shen, Q.R.; Brueck, H. Effects of local nitrogen supply on water uptake of bean plants in a split root system. J. Integr. Plant Biol 2007, 49, 472–480.
47. Oddo, E.; Inzerillo, S.; La Bella, F.; Grisafi, F.; Salleo, S.; Nardini, A. Short-term effects of potassium fertilization on the hydraulic conductance of laurus nobilis L. Tree Physiol 2011, 31, 131–138.
48. Zwieniecki, M.A.; Melcher, P.J.; Holbrook, N.M. Hydrogel control of xylem hydraulic resistance in plants. Science 2001, 291, 1059–1062.
49. DaCosta, M.; Huang, B.R. Osmotic adjustment associated with variation in bentgrass tolerance to drought stress. J. Am. Soc. Hortic. Sci 2006, 131, 338–344.
50. Shabala, S.N.; Lew, R.R. Turgor regulation in osmotically stressed Arabidopsis epidermal root cells. Direct support for the role of inorganic ion uptake as revealed by concurrent flux and cell turgor

measurements. Plant Physiol 2002, 129, 290–299.

51. Egilla, J.N.; Davies, F.T.; Boutton, T.W. Drought stress influences leaf water content, photosynthesis, and water-use efficiency of hibiscus rosa-sinensis at three potassium concentrations. Photosynthetica 2005, 43, 135–140.
52. Jin, S.H.; Huang, J.Q.; Li, X.Q.; Zheng, B.S.; Wu, J.S.; Wang, Z.J.; Liu, G.H.; Chen, M. Effects of potassium supply on limitations of photosynthesis by mesophyll diffusion conductance in Carya cathayensis. Tree Physiol 2011, 31, 1142–1151.
53. Tomemori, H.; Hamamura, K.; Tanabe, K. Interactive effects of sodium and potassium on the growth and photosynthesis of spinach and komatsuna. Plant Prod. Sci 2002, 5, 281–285.
54. Pervez, H.; Ashraf, M.; Makhdum, M.I. Influence of potassium nutrition on gas exchange characteristics and water relations in cotton (Gossypium hirsutum L.). Photosynthetica 2004, 42, 251–255.
55. Benlloch-Gonzalez, M.; Arquero, O.; Fournier, J.M.; Barranco, D.; Benlloch, M. K+ starvation inhibits water-stress-induced stomatal closure. J. Plant Physiol 2008, 165, 623–630.
56. Benlloch-Gonzalez, M.; Romera, J.; Cristescu, S.; Harren, F.; Fournier, J.M.; Benlloch, M. K+ starvation inhibits water-stress-induced stomatal closure via ethylene synthesis in sunflower plants. J. Exp. Bot 2010, 61, 1139–1145.
57. Tsonev, T.; Velikova, V.; Yildiz-Aktas, L.; Gurel, A.; Edreva, A. Effect of water deficit and potassium fertilization on photosynthetic activity in cotton plants. Plant Biosyst 2011, 145, 841–847.
58. Shin, R.; Schachtman, D.P. Hydrogen peroxide mediates plant root cell response to nutrient deprivation. Proc. Natl. Acad. Sci. USA 2004, 101, 8827–8832.
59. Tanaka, Y.; Sano, T.; Tamaoki, M.; Nakajima, N.; Kondo, N.; Hasezawa, S. Ethylene inhibits abscisic acid-induced stomatal closure in Arabidopsis. Plant Physiol 2005, 138, 2337–2343.
60. Tanaka, Y.; Sano, T.; Tamaoki, M.; Nakajima, N.; Kondo, N.; Hasezawa, S. Cytokinin and auxin inhibit abscisic acid-induced stomatal closure by enhancing ethylene production in Arabidopsis. J. Exp. Bot 2006, 57, 2259–2266.

61. Cruz de Carvalho, M.H. Drought stress and reactive oxygen species: Production, scavenging and signaling. Plant Signal. Behav 2008, 3, 156–165.
62. Fu, J.M.; Huang, B.R. Involvement of antioxidants and lipid peroxidation in the adaptation of two cool-season grasses to localized drought stress. Environ. Exp. Bot 2001, 45, 105–114.
63. Reddy, A.R.; Chaitanya, K.V.; Vivekanandan, M. Drought-induced responses of photosynthesis and antioxidant metabolism in higher plants. J. Plant Physiol 2004, 161, 1189–1202.
64. Dat, J.; Vandenabeele, S.; Vranova, E.; van Montagu, M.; Inze, D.; van Breusegem, F. Dual action of the active oxygen species during plant stress responses. Cell. Mol. Life Sci 2000, 57, 779–795.
65. Vranova, E.; Inze, D.; van Breusegem, F. Signal transduction during oxidative stress. J. Exp. Bot 2002, 53, 1227–1236.
66. Mittler, R. Oxidative stress, antioxidants and stress tolerance. Trends Plant Sci 2002, 7, 405–410.
67. Cakmak, I. Possible roles of zinc in protecting plant cells from damage by reactive oxygen species. New Phytol 2000, 146, 185–205.
68. Abdel Wahab, A.M.; Abd-Alla, M.H. The role of potassium fertilizer in nodulation and nitrogen fixation of faba bean (Vicia faba L.) plants under drought stress. Biol. Fert. Soils 1995, 20, 147–150.
69. Peuke, A.D.; Jeschke, W.D.; Hartung, W. Flows of elements, ions and abscisic acid in Ricinus communis and site of nitrate reduction under potassium limitation. J. Exp. Bot 2002, 53, 241–250.
70. Jiang, M.; Zhang, J. Effect of abscisic acid on active oxygen species, antioxidative defence system and oxidative damage. Plant Cell Physiol 2001, 42, 1265–1273.
71. Lin, C.C.; Kao, C.H. Abscisic acid induced changes in cell wall peroxidase activity and hydrogen peroxide level in roots of rice seedlings. Plant Sci 2001, 160, 323–329.
72. Marschner, H.; Cakmak, I. Highlight intensity enhances chlorosis and necrosis in leaves of zinc, potassium, and magnesium deficient bean (Phaseolus vulgaris) plants. J. Plant Physiol 1989, 134, 308–315.

73. Munns, R. Genes and salt tolerance: Bringing them together. New Phytol 2005, 167, 645–663.
74. Parida, A.K.; Das, A.B. Salt tolerance and salinity effects on plants: A review. Ecotox. Environ. Safe 2005, 60, 324–349.
75. Munns, R.; Tester, M. Mechanisms of salinity tolerance. Annu. Rev. Plant Biol 2008, 59, 651–681.
76. Hasegawa, P.M.; Bressan, R.A.; Zhu, J.K.; Bohnert, H.J. Plant cellular and molecular responses to high salinity. Annu. Rev. Plant Phys 2000, 51, 463–499.
77. Yang, Y.; Zheng, Q.; Liu, M.; Long, X.; Liu, Z.; Shen, Q.; Guo, S. Difference in sodium spatial distribution in the shoot of two canola cultivars under saline stress. Plant Cell Physiol 2012, 53, 1083–1092.
78. Shabala, S.; Cuin, T.A. Potassium transport and plant salt tolerance. Physiol. Plantarum 2008, 133, 651–669.
79. Botella, M.A.; Martinez, V.; Pardines, J.; Cerda, A. Salinity induced potassium deficiency in maize plants. J. Plant Physiol 1997, 150, 200–205.
80. Coskun, D.; Britto, D.T.; Kronzucker, H.J. Regulation and mechanism of potassium release from barley roots: an in planta 42K+ analysis. New Phytol 2010, 188, 1028–1038.
81. Mian, A.; Oomen, R.J.; Isayenkov, S.; Sentenac, H.; Maathuis, F.J.; Very, A.A. Over-expression of an Na+-and K+-permeable HKT transporter in barley improves salt tolerance. Plant J 2011, 68, 468–479.
82. Platten, J.D.; Cotsaftis, O.; Berthomieu, P.; Bohnert, H.; Davenport, R.J.; Fairbairn, D.J.; Horie, T.; Leigh, R.A.; Lin, H.X.; Luan, S.; et al. Nomenclature for HKT transporters, key determinants of plant salinity tolerance. Trends Plant Sci 2006, 11, 372–374.
83. Byrt, C.S.; Platten, J.D.; Spielmeyer, W.; James, R.A.; Lagudah, E.S.; Dennis, E.S.; Tester, M.; Munns, R. HKT1;5-like cation transporters linked to Na+ exclusion loci in wheat, Nax2 and Kna1. Plant Physiol 2007, 143, 1918–1928.
84. Horie, T.; Hauser, F.; Schroeder, J.I. HKT transporter-mediated salinity resistance mechanisms in Arabidopsis and monocot crop

plants. Trends Plant Sci 2009, 14, 660–668.

85. Degl'Innocenti, E.; Hafsi, C.; Guidi, L.; Navari-Izzo, F. The effect of salinity on photosynthetic activity in potassium-deficient barley species. J. Plant Physiol 2009, 166, 1968–1981.

86. Qu, C.X.; Liu, C.; Gong, X.L.; Li, C.X.; Hong, M.M.; Wang, L.; Hong, F.S. Impairment of maize seedling photosynthesis caused by a combination of potassium deficiency and salt stress. Environ. Exp. Bot 2012, 75, 134–141.

87. Qu, C.X.; Liu, C.; Ze, Y.G.; Gong, X.L.; Hong, M.M.; Wang, L.; Hong, F.S. Inhibition of nitrogen and photosynthetic carbon assimilation of maize seedlings by exposure to a combination of salt stress and potassium-deficient stress. Biol. Trace. Elem. Res 2011, 144, 1159–1174.

88. Chen, Z.H.; Zhou, M.X.; Newman, I.A.; Mendham, N.J.; Zhang, G.P.; Shabala, S. Potassium and sodium relations in salinised barley tissues as a basis of differential salt tolerance. Funct. Plant Biol 2007, 34, 150–162.

89. Walker, D.J.; Leigh, R.A.; Miller, A.J. Potassium homeostasis in vacuolate plant cells. Proc. Natl. Acad. Sci. USA 1996, 93, 10510–10514.

90. Gong, X.; Chao, L.; Zhou, M.; Hong, M.; Luo, L.; Wang, L.; Ying, W.; Jingwei, C.; Songjie, G.; Fashui, H. Oxidative damages of maize seedlings caused by exposure to a combination of potassium deficiency and salt stress. Plant Soil 2011, 340, 443–452.

91. Shabala, S. Salinity and programmed cell death: Unravelling mechanisms for ion specific signalling. J. Exp. Bot 2009, 60, 709–711.

92. Devi, B.S.R.; Kim, Y.J.; Selvi, S.K.; Gayathri, S.; Altanzul, K.; Parvin, S.; Yang, D.U.; Lee, O.R.; Lee, S.; Yang, D.C. Influence of potassium nitrate on antioxidant level and secondary metabolite genes under cold stress in Panax ginseng. Russ. J. Plant Physiol 2012, 59, 318–325.

93. Suzuki, N.; Mittler, R. Reactive oxygen species and temperature stresses: A delicate balance between signaling and destruction. Physiol. Plantarum 2006, 126, 45–51.

94. Xiong, L.M.; Schumaker, K.S.; Zhu, J.K. Cell signaling during cold,

drought, and salt stress. Plant Cell 2002, 14, S165–S183.

95. Zhu, J.K. Cell signaling under salt, water and cold stresses. Curr. Opin. Plant Biol 2001, 4, 401–406.
96. Bogdevitch, I. IPI Internal Report; International Potash Institute: Basel, Switzerland, 2000.
97. Webster, D.E.; Ebdon, J.S. Effects of nitrogen and potassium fertilization on perennial ryegrass cold tolerance during deacclimation in late winter and early spring. Hortscience 2005, 40, 842–849.
98. Wang, X.M.; Li, W.Q.; Li, M.Y.; Welti, R. Profiling lipid changes in plant response to low temperatures. Physiol. Plantarum 2006, 126, 90–96.
99. McKersie, B.D.; Leshem, Y.Y. Stress and Stress Coping in Cultivated Plants; Kluwer Academic Publishers: Dordrecht, The Netherlands, 1994; pp. 181–193.
100. Haque, M.Z. Effect of nitrogen, phosphorus and potassium on spikelet sterility induced by low temperature at the reproductive stage of rice. Plant Soil 1988, 109, 31–36.
101. Hakerlerler, H.; Oktay, M.; Eryuce, N.; Yagmur, B. Effect of Potassium Sources on the Chilling Tolerance of Some Vegetable Seedlings Grown in Hotbeds. In Food Security in the WANA Region, the Essential Need for Balanced Fertilization; Johnston, A.E., Ed.; International Potash Institute: Basel, Switzerland, 1997; pp. 353–359.
102. Setter, T.L.; Waters, I. Review of prospects for germplasm improvement for waterlogging tolerance in wheat, barley and oats. Plant Soil 2003, 253, 1–34.
103. Zhou, M. Improvement of Plant Waterlogging Tolerance. In Waterlogging Signalling and Tolerance in Plants; Mancuso, S., Shabala, S., Eds.; Springer: Berlin/Heidelberg, Germany, 2010; pp. 267–285.
104. Bailey-Serres, J.; Voesenek, L.A.C.J. Flooding stress: Acclimations and genetic diversity. Annu. Rev. Plant Biol 2008, 59, 313–339.
105. Wegner, L. Oxygen Transport in Waterlogged Plants. In Waterlogging Signalling and Tolerance in Plants; Mancuso, S., Shabala, S., Eds.;

Springer: Berlin/Heidelberg, Germany, 2010; pp. 3–22.

106. Shabala, S. Physiological and cellular aspects of phytotoxicity tolerance in plants: The role of membrane transporters and implications for crop breeding for waterlogging tolerance. New Phytol 2011, 190, 289–298.

107. Colmer, T.D.; Greenway, H. Ion transport in seminal and adventitious roots of cereals during O2 deficiency. J. Exp. Bot 2011, 62, 39–57.

108. Kirmizi, S.; Bell, R.W. Responses of barley to hypoxia and salinity during seed germination, nutrient uptake, and early plant growth in solution culture. J. Plant Nutr. Soil Sci 2012, 175, 630–640.

109. Pang, J.Y.; Newman, I.; Mendham, N.; Zhou, M.; Shabala, S. Microelectrode ion and O2 fluxes measurements reveal differential sensitivity of barley root tissues to hypoxia. Plant Cell Environ 2006, 29, 1107–1121.

110. Teakle, N.L.; Bazihizina, N.; Shabala, S.; Colmer, T.D.; Barrett-Lennard, E.G.; Rodrigo-Moreno, A.; Läuchli, A.E. Differential tolerance to combined salinity and O2 deficiency in the halophytic grasses Puccinellia ciliata and Thinopyrum ponticum: The importance of K+ retention in roots. Environ. Exp. Bot 2013, 87, 69–78.

111. Mancuso, S.; Marras, A.M. Adaptative response of Vitis root to anoxia. Plant Cell Physiol 2006, 47, 401–409.

112. Mugnai, S.; Marras, A.M.; Mancuso, S. Effect of hypoxic acclimation on anoxia tolerance in Vitis roots: Response of metabolic activity and K+ fluxes. Plant Cell Physiol 2011, 52, 1107–1116.

113. Fiedler, S.; Vepraskas, M.J.; Richardson, J.L. Soil redox potential: Importance, field measurements, and observations. Adv. Agron 2007, 94, 1–54.

114. Erlejman, A.G.; Verstraeten, S.V.; Fraga, C.G.; Oteiza, P.I. The interaction of flavonoids with membranes: Potential determinant of flavonoid antioxidant effects. Free Radical Res 2004, 38, 1311–1320.

115. Pang, J.Y.; Cuin, T.; Shabala, L.; Zhou, M.X.; Mendham, N.; Shabala, S. Effect of secondary metabolites associated with anaerobic soil conditions on ion fluxes and electrophysiology in barley roots. Plant Physiol 2007, 145, 266–276.

116. Else, M.A.; Coupland, D.; Dutton, L.; Jackson, M.B. Decreased root hydraulic conductivity reduces leaf water potential, initiates stomatal closure and slows leaf expansion in flooded plants of castor oil (Ricinus communis) despite diminished delivery of ABA from the roots to shoots in xylem sap. Physiol. Plantarum 2001, 111, 46–54.

117. Ashraf, M.A. Waterlogging stress in plants: A review. Afr. J. Agri. Res 2012, 7, 1976–1981.

118. Ashraf, M.A.; Ahmad, M.S.A.; Ashraf, M.; Al-Qurainy, F.; Ashraf, M.Y. Alleviation of waterlogging stress in upland cotton (Gossypium hirsutum L.) by exogenous application of potassium in soil and as a foliar spray. Crop Pasture Sci 2011, 62, 25–38.

CHAPTER

# 4

# SODIUM TRANSPORT SYSTEM IN PLANT CELLS

Toshio Yamaguchi[1], Shin Hamamoto[2] and Nobuyuki Uozumi[2]

[1]Department of Microbiology, Faculty of Pharmacy, Niigata University of Pharmacy and Applied Life Sciences, Niigata, Japan

[2]Department of Biomolecular Engineering, Graduate School of Engineering, Tohoku University, Sendai, Japan

Since sodium, Na, is a non-essential element for the plant growth, the molecular mechanism of $Na^+$ transport system in plants has remained elusive for the last two decades. The accumulation of $Na^+$ in soil through irrigation for sustainable agricultural crop production, particularly in arid land, and by changes in environmental and climate conditions leads to the buildup of toxic level of salts in the soil. Since the latter half of the twentieth century, extensive molecular research has identified several classes of $Na^+$ transporters that play major roles in the alleviation of ionic stress by excluding toxic $Na^+$ from the cytosol or preventing $Na^+$ transport to the photosynthetic organs, and also in osmotic stress by modulating intra/extracellular osmotic balance. In this review, we summarize the current knowledge of three major $Na^+$ transporters, namely NHX, SOS1, and HKT transporters, including recently revealed characteristics of these transporters.

**Keywords**: salt tolerance, $Na^+$ transporter, plant, NHX transporters, SOS1 transporters, HKT transporters

## INTRODUCTION

Water availability and its quality greatly affect the growth of most of plant species, and keeping a sufficient amount of water in the soil is of paramount importance for agricultural crop production. Constant irrigation often leads to the accumulation of undesirable salts due to the use of poor quality water, particularly in arid lands, and it has been estimated that about 20% of irrigated land is affected by soil salinity (Yeo, 1999). Salinity stress to most plant species mainly accounts for the increase in cytoplasmic osmotic stress and ion (mainly $Na^+$)-specific toxicity (Blumwald, 2000; Blumwald et al., 2000 Munns and Tester, 2008). Overaccumulation of $Na^+$ in the cytosol causes inhibitions of protein synthesis, many enzymatic reactions (Hall and Flowers, 1973; Flowers and Läuchli, 1983; Murguía et al., 1995), and photosynthetic processes (Tsugane et al., 1999; Tsunekawa et al., 2009). Exclusion of $Na^+$ from photosynthetic organs is therefore crucial for the maintenance of an adequate metabolism and efficient carbon fixation.

To date, several classes of $Na^+$-transporters have been shown to play central roles in $Na^+$ homeostasis during salinity stress (Apse and Blumwald, 2007; Hauser and Horie, 2010; Uozumi and Schroeder, 2010). Among them, the NHX-, SOS1-, and the Class I HKT-type transporter have drawn particular attention because of their capability of transporting $Na^+$ across cellular membranes (Figure 1). Many studies have demonstrated significant involvements of these $Na^+$ transporters in $Na^+$ sequestration in plant vacuoles, $Na^+$ extrusion from cells, and $Na^+$ circulation for the alleviation of sodium stress under saline conditions. Moreover, some of these $Na^+$ transporters have been shown to improve salinity tolerance of crop plants when their encoding gene is overexpressed (Yamaguchi and Blumwald, 2005; Agarwal et al., 2013).

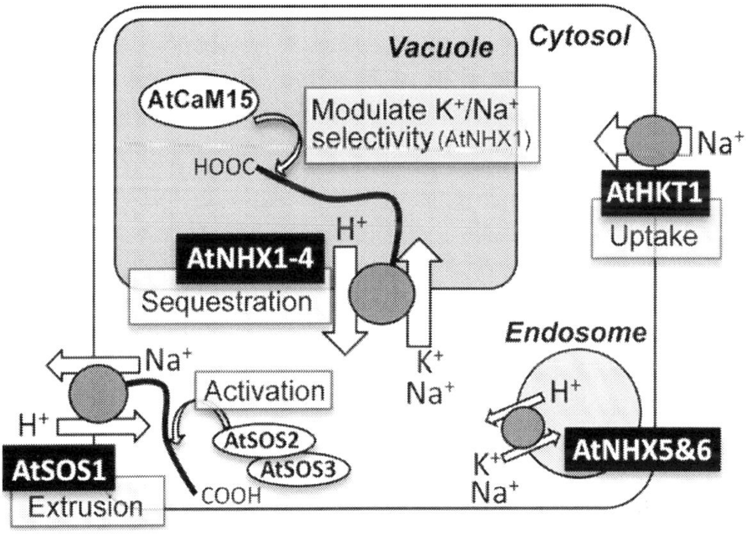


**Figure 1**: Schematic representation of subcellular localizations, functions, and regulations of NHX-, SOS1-, and HKT1-type Na+ transporters in *Arabidopsis*.

In this review, we will provide brief summaries on the properties and physiological roles of three major $Na^+$ transporters that are involved in the plant $Na^+$ tolerance, namely NHX, SOS1, and HKT transporters, from well-established to recently revealed characteristics of these transporters.

## NHX TRANSPORTERS

One of the strategies that plant cells employ for the alleviation of excess cytosolic $Na^+$ is to compartmentalize $Na^+$ into the vacuoles, and NHX-type intracellular $Na^+/H^+$ exchangers have appeared to be the best-characterized proteins involved in this process so far (Blumwald et al., 2000). Most of NHX proteins studied to date mediate electroneutral $Na^+/H^+$ and $K^+/H^+$ antiport across the membrane by utilizing the $H^+$ gradient as a driving force and are considered to be responsible for $Na^+$ sequestration into intracellular compartments under salinity stress (Bassil et al., 2012).

Plant genomes often contain multiple intracellular NHX isoforms (Bassil et al., 2012). Intracellular NHX transporters are classified into the IC-NHE/NHX family (Pardo et al., 2006), which is a part of the large cation/proton antiporter 1 (CPA1) family (Säier et al., 2006). These are further subdivided into vacuolar (Class I) or endosomal (Class II) NHXs, according to their sequence similarity and subcellular localizations of representative members (Rodríguez-Rosales et al., 2009; Bassil et al., 2012; Figure 1). Most of the plant species sequenced to date contain both types of NHXs (Rodríguez-Rosales et al., 2009; Bassil et al., 2012), and functional redundancies of vacuolar or endosomal NHXs had been recently reported in *Arabidopsis*(Bassil et al., 2011a,b; Barragan et al., 2012).

Currently, two distinct topology models have been proposed for AtNHX1. The first model was proposed based on the protease protection analysis of isolated yeast vacuoles expressing full-length AtNHX1 with epitope tags inserted into each hydrophilic loop (Yamaguchi et al., 2003). This model comprised nine transmembrane segments with a C-terminal hydrophilic tail facing toward the vacuolar lumen. Another topology model was proposed on the basis of *in vitro* translation experiments using truncated *AtNHX1*genes, which suggested that several transmembrane segments of AtNHX1 retain similar topogenic properties as human NHE1 (Sato and Sakaguchi, 2005).

The cation selectivity of AtNHX1 appears to be regulated by its C-terminal tail through the binding to a calmodulin-like protein AtCaM15, which interacts with the C-terminus of AtNHX1 in a $Ca^{2+}$ and pH dependent manner from inside of the vacuole (Yamaguchi et al., 2005; Figure 1). Interaction with AtCaM15 decreases the $Na^+$ transport activity of AtNHX1, while keeping $K^+$ transport activity almost unchanged. Given that the plant vacuole usually keeps a high $Ca^{2+}$ concentration and an acidic luminal pH, AtNHX1 would favor a $K^+/H^+$ antiport mode under normal physiological conditions due to the binding with AtCaM15. However, under salinity stress, which often causes vacuolar alkalization (Okazaki et al., 1996; Gruwel et al., 2001), disassociation of AtCaM15 would result in increased $Na^+/H^+$ antiport activity of AtNHX1 and promote compartmentation of $Na^+$ into the vacuoles (Yamaguchi et al., 2005).

Overexpression of genes encoding vacuolar NHXs has conferred salt tolerance to a range of plant species with a concomitant increase in tissue $Na^+$ (Apse et al., 1999; Zhang and Blumwald, 2001; Zhang et al., 2001; Agarwal et al., 2013). These observations, together with the fact that the

*Arabidopsis atnhx1* mutant exhibited $Na^+$ sensitivity and significantly less vacuolar $Na^+/H^+$ antiport activity (Apse et al., 2003), strongly supported the role of vacuolar NHXs in $Na^+$ compartmentation under salinity stress. However, there has been a case where the overexpression of vacuolar NHX led to accumulations of $K^+$ but not $Na^+$(Leidi et al., 2010). Involvement of endosomal NHXs in salt tolerance has also been demonstrated in tomato and *Arabidopsis*. *LeNHX2* knockdown tomato and *atnhx5/atnhx6* double knockout *Arabidopsis*plants exhibited salt sensitivities (Rodríguez-Rosales et al., 2008; Bassil et al., 2011a). The introduction of the AtNHX5 gene increased salt tolerance and $Na^+$-accumulation in Torenia (Shi et al., 2008). Tomato plants overexpressing LeNHX2 also showed increased salt tolerance. However, an increase of tissue $K^+$instead of $Na^+$ was observed in this case, suggesting that increased salt tolerance was achieved by improving cellular $K^+$-homeostasis (Huertas et al., 2013).

Besides significant roles in salt tolerance, involvement in pH regulation has been demonstrated for vacuolar NHXs (Fukada-Tanaka et al., 2000; Yamaguchi et al., 2001). Furthermore, vacuolar NHXs appeared to be involved in organ developments. The *Arabidopsis atnhx1/atnhx2* double knockout plant exhibited severe defects on the plant growth, development of reproductive organ, and leaf cell expansion under normal conditions (Bassil et al., 2011b; Barragan et al., 2012). These developmental defects seem to be attributed to the role of AtNHXs in $K^+$- rather than $Na^+$-homeostasis, since the *nhx1-1/nhx2-1* plant accumulated a lower level of $K^+$ and exhibited less tonoplast $K^+/H^+$ antiport activity than the wild type plant. Severe growth defects have also been observed in endosomal *NHX* knockout/knockdown plants. The *LeNHX2* knockdown tomato plant showed severely retarded growth as compared to the wild type (Rodríguez-Rosales et al., 2008). Similarly, *Arabidopsis atnhx5/atnhx6* double knockout lines exhibited smaller cell size, slower development of flower organs, and root growth (Bassil et al., 2011a). These results may indicate significant contributions of endosomal NHXs on growth and development, possibly via maintaining appropriate pH and/or cation balance in plant cells. Roles in endosomal trafficking have also been documented (Sottosanto et al., 2004; Bassil et al., 2011a), which have been recently summarized in excellent review articles (Bassil et al., 2012; Qiu, 2012) and therefore will not be discussed further.

## SOS1 TRANSPORTERS

Exclusion of $Na^+$ from the cytosol across the plasma membrane is also an important mechanism for the alleviation of cellular $Na^+$ toxicity in plants (Blumwald, 2000). To date, the SOS1 (salt overly sensitive 1 = AtNHX7 in *Arabidopsis*) transporters are the best-characterized class of transporters attributed with this function (Figure 1). The *SOS1* gene has been identified in several plants including *Arabidopsis* (Wu et al., 1996), rice (Martínez-Atienza et al., 2007), wheat (Xu et al., 2008), tomato (Olías et al., 2009), and *Thellungiella salsuginea* (Vera-Estrella et al., 2005). The *SOS1* gene encodes a large membrane protein (127 kDa, in the case of *Arabidopsis* AtSOS1) that is categorized into the CPA1 family along with NHXs, although it is more closely related to bacterial NhaP antiporters (Wu et al., 1996; Shi et al., 2000; Brett et al., 2005). AtSOS1 protein is localized in the plasma membrane, and appears to mediate electroneutral $Na^+/H^+$ antiport (Shi et al., 2002; Qiu et al., 2002,2003). Hydropathy prediction has indicated that the AtSOS1 protein consists of 12 putative transmembrane segments and a long C-terminal hydrophilic tail (approximately 700 aa; Shi et al., 2000). Although detailed structural properties of AtSOS1 remain elusive, a recent biochemical analysis revealed that AtSOS1 forms a homodimer (Núñez-Ramírez et al., 2012). In addition, *Arabidopsis* has another *SOS1*-like gene, termed *AtNHX8*, which seems to encode a C-terminal truncated isoform of AtSOS1. Despite its high similarity to AtSOS1 (~88% for hydrophobic region), AtNHX8 seemed to prefer $Li^+$ over $Na^+$ as its substrate (An et al., 2007).

It appears that the AtSOS1 C-terminus contains an activation and an auto-inhibitory domain that are physically associated to each other, and this interaction seems to suppress AtSOS1 activity at the resting state (Quintero et al., 2011). The activation of AtSOS1 occurs via the direct phosphorylation of the auto-inhibitory domain by the serine/threonine kinase AtSOS2, which requires a calcium binding protein, AtSOS3, or SCaBP8/CBL10, for its kinase activity (Figure 1; Qiu et al., 2002; Quan et al., 2007; Quintero et al., 2011). In addition, AtSOS2 and AtSOS3 have also shown to be required for the $Na^+$-dependent induction of *AtSOS1* gene (Shi et al., 2000).

The physiological significance of SOS1 transporters in saline environments has been well established in *Arabidopsis*, tomato, and *T. salsuginea*, as significant increases in salt sensitivity has been observed for the knockout/knockdown plants of corresponding *SOS1* genes (Wu et al., 1996; Shi et al., 2000, 2002; Oh et al., 2009; Olías et al., 2009). All of these knockout/knockdown plants tend to accumulate higher amounts of $Na^+$ than wild type plants under severe salt stress, revealing critical roles of SOS1 transporters in limiting $Na^+$ entry by promoting $Na^+$ efflux at the plasma membrane of the root cells (Shi et al., 2002; Oh et al., 2009; Olías et al., 2009). This is well in agreement with the observation that the *AtSOS1*transcript is predominantly expressed in epidermal cells of the root tip (Shi et al., 2002), and that the overexpression of *AtSOS1*, *T. salsuginea SOS1*, and rice *OsSOS1* in *Arabidopsis* successfully conferred resistance to salinity stress (Shi et al., 2003; Martínez-Atienza et al., 2007; Oh et al., 2009; Yang et al., 2009). Furthermore, SOS1 appears to play a role in protecting the root plasma membrane $K^+$ uptake under salinity stress (Qi and Spalding, 2004). Involvement in xylem loading of $Na^+$ for root to shoot transport has also been suggested for AtSOS1 under moderate salt stress and low transpiration conditions (Ding and Zhu, 1997; Shi et al., 2002). This is based on the preferential expression of *AtSOS1* in xylem parenchyma cells and decreased shoot $Na^+$ contents in the *atsos1* mutant than in wild type under hydroponic or high humidity conditions (Shi et al., 2002). On the other hand, *SlSOS1*-silenced tomato (*Solanum lycopersicum*) exhibited increased $Na^+$ content in roots and leaves as compared to wild type under salt stress, while stems of silenced plants accumulated less $Na^+$ than the wild type (Olías et al., 2009). These results prompted a suggestion that SlSOS1 possibly functioned in $Na^+$ extrusion from the leaf cells back to the xylem stream in order to protect photosynthetic organs (Olías et al., 2009). SOS1 also appears to have several other cellular functions that are not directly related to $Na^+$-homeostasis. A recent study using fluorescent probes reported that the loss of the functional *AtSOS1* gene resulted in altered pH homeostasis, both in the cytosol and the vacuole of root cells, probably due to the alteration in plasma membrane $H^+$flux (Shabala et al., 2005; Oh et al., 2010). In addition, the fragmentation of vacuoles and deficiencies in endocytosis have also been reported in *atsos1* mutants under salinity stress (Oh et al., 2010).

## HKT TRANSPORTERS

The gene encoding high-affinity $K^+$ transporter, HKT1 was first identified in the wheat genome by functional complementation screening for genes that rescue the deficiency of TRK1 and TRK2-mediated $K^+$ uptake activity in *S. cerevisiae* mutants (Schachtman and Schroeder, 1994). HKT1 belongs to the Trk/Ktr/HKT-type $K^+$ transporter superfamily found in microorganisms, plant cells, and parasites (Gaber et al., 1988; Ko and Gaber, 1988; Dosch et al., 1991; Schlösser et al., 1991; Schachtman and Schroeder, 1994; Nakamura et al., 1998; Mosimann et al., 2010). The further study on HKT1 revealed that the wheat HKT1 (TaHKT2;1) showed $Na^+/K^+$ co-transport activity at high $Na^+$ as compared to $K^+$ in extracellular compartment (Rubio et al., 1995; Gassmann et al., 1996). The $Na^+$-dependent activation of $K^+$ uptake was later found in KtrAB in *Vibrio alginolyticus* (Nakamura et al., 1998), and KtrABE in cyanobacteria *Synechosytis* PCC 6803 as well (Matsuda et al., 2004). TbHKT1 in *Trypanosoma brucei* (Mosimann et al., 2010) and TsHKT1;2 in *T. salsuginea* (Ali et al., 2012) showed higher selectivity for $K^+$ over $Na^+$. On the other hand, a strong $Na^+$ selective transport activity was found in AtHKT1;1 isolated from *Arabidopsis* (Uozumi et al., 2000). $Na^+$ selectivity was also seen in HKTs from other species including *Eucalyptus camaldulensis* and rice (Fairbairn et al., 2000; Horie et al., 2001).

An array of biochemical approaches including bacterial alkaline phosphatase-fusion analysis, glycosylation scanning, and immunofluorescence detection, revealed that AtHKT1 contains eight transmembrane domains that correspond to fourfold bacterial simple $K^+$ channels consisting of transmembrane-pore-transmembrane motifs, and N- and C-termini that face the cytoplasm (Kato et al., 2001). Trk/Ktr/HKT-type transporters appear to have evolutional correlation with the ion-conducting regions of $K^+$ channels, $Na^+$ channels, $Ca^{2+}$ channels, and Kdp-type $K^+$ pumps (Durell et al., 1999; Durell and Guy, 1999; Uozumi and Dreyer, 2012). The glycine of the $K^+$ ion selective filter in the four pore domains of Trk/Ktr/HKT-type transporters is analogous to the first glycine in the GYG signature sequence (Mäser et al., 2002b; Tholema et al., 2005; Uozumi and Dreyer, 2012). In AtHKT1, the serine in the first pore region in AtHKT1 is substituted for glycine, while the following three glycine residues are conserved. Replacement of the serine by glycine in AtHKT1 led to the increase of $K^+$ transport activity, which strongly suggested that AtHKT1 functions as a $Na^+$ selective transporter rather than a $K^+$ transport system. Based on the relationship of Gly/Ser and $Na^+/K^+$ selectivity in Trk/Ktr/HKT-type transporters in combination with sequence alignments, HKTs were categorized into two major groups (Platten et al., 2006). The conserved arginine in the middle of the last transmembrane domain of Trk/Ktr/HKT-type transporters, which is not conserved in $K^+$ channels, is one of the determinants of "transporter" activity but not "channel" activity (Kato et al., 2007). The gating of the ion transport is controlled by the sixth transmembrane in KtrB of *V. alginolyticus*(Hänelt et al., 2010a,b). This structure–function relationship is supported by the crystal structure of a TrkH from *Vibrio parahaemolyticus* (Cao et al., 2011).

A long-term exposure of a plant to $Na^+$ in soil will translocate $Na^+$ from the root to above ground tissue (Mäser et al., 2002a). AtHKT1 may participate in the $Na^+$ uptake and $Na^+$ circulation in plant body. The analysis of *AtHKT1* gene expression by the use of the GUS reporter system showed that AtHKT1 was expressed in vascular tubes (Mäser et al., 2002a; Berthomieu et al., 2003). This detailed observation, in combination with immunoelectron microscopic detection, showed that AtHKT1;1 resides at the plasma membrane of xylem parenchyma cells and phloem tissues (Berthomieu et al., 2003; Sunarpi et al., 2005). Likewise, one of the rice HKTs was present in xylem parenchyma cells (Ren et al., 2005). The *sas2*(*s*odium over-*a*ccumulation in *s*hoots) mutant exhibited increased $Na^+$ accumulation in aerial organs and decreased it in roots (Berthomieu et al., 2003). Taken together, the hypothetical model proposed here to explain the physiological role of AtHKT1 is that AtHKT1 works by unloading $Na^+$ from shoot xylem sap of *Arabidopsis* in the presence of high salinity and for recirculation of shoot to root (Figure 1; Mäser et al., 2002a; Berthomieu et al., 2003; Sunarpi et al., 2005).

Some transcriptional regulatory elements have been identified in the *AtHKT1* promoter. The tandem repeat regions found far from the *AtHKT1* initiation codon (approximately -3.9 to -5.3 kb) were responsible for the expression in roots (Rus et al., 2006). The second-half of the repeat sequence, which was located in the 3′ region of two repeat sequences, acted as enhancer elements for *AtHKT1* expression (Baek et al., 2011). Reduction of the expression in the root by the inactivation of the repeat sequence induced overaccumulation of $Na^+$ in the shoot (Rus et al., 2006). In accordance with this result, the engineered root stele-specific *AtHKT1*-overexpressor reduced salt content in the shoot, whereas *AtHKT1* expression driven by the *35S* promoter in whole plant resulted in salt hypersensitivity (Møller et al., 2009). The putative highly methylated sequence (250 bp) was found 2.6 kb upstream from the initiation codon of *AtHKT1*. It is believed that this region is a potential small RNA target site and the methylation in this region may control the expression of *AtHKT1* at a low level (Baek et al., 2011).

## CONCLUDING REMARKS

Almost two decades of extensive molecular studies have clearly established the involvement of NHX, SOS1, and HKT transporters in plant salt tolerance. The genes encoding some of these transporters have since been successfully utilized as genetic tools for enhancing salt tolerance of model and crop plants. Recent advancements have also provided evidence on the involvement of these transporters in other vital cellular processes. This indicates the possibility that further enhancement of salt tolerance by overexpression of any of the above-mentioned genes might reach its limit due to excessive perturbation of related cellular and physiological processes. A much broader and profound understanding of the molecular properties, regulations, and cellular/physiological functions of $Na^+$-transporters, may provide further insight into the development of efficient strategies for the engineering of "super salt-tolerant" crop plants and therefore progress in this research field is highly anticipated.

## ACKNOWLEDGMENTS

We thank Mario Fon for the proofreading and editing the manuscript. This work was supported by Grants-in-Aid for Scientific Research from Japan Society for the Promotion of Science (24102504, 24580135, and 25292055).

## REFERENCES

1. Agarwal, P. K., Shukla, P. S., Gupta, K., and Jha, B. (2013). Bioengineering for salinity tolerance in plants: state of art. *Mol. Biotechnol.*54, 102–123. doi: 10.1007/s12033-012-9538-3

2. Ali, Z., Park, H. C., Ali, A., Oh, D. H., Aman, R., Kropornicka, A., et al. (2012). TsHKT1;2, a HKT1 homolog from the extremophile *Arabidopsis* relative *Thellungiella salsuginea*, shows $K^+$ specificity in the presence of NaCl. *Plant Physiol.* 158, 1463–1474. doi: 10.1104/pp.111.193110

3. An, R., Chen, Q. J., Chai, M. F., Lu, P. L., Su, Z., Qin, Z. X., et al. (2007). AtNHX8, a member of the monovalent cation: proton antiporter-1 family in *Arabidopsis thaliana*, encodes a putative Li/H antiporter. *Plant J.* 49, 718–728. doi: 10.1111/j.1365-313X.2006.02990.x

4. Apse, M. P., Aharon, G. S., Snedden, W. A., and Blumwald, E. (1999). Salt tolerance conferred by overexpression of a vacuolar $Na^+/H^+$antiport in *Arabidopsis*. *Science* 285, 1256–1258. doi: 10.1126/science.285.5431.1256

5. Apse, M. P., and Blumwald, E. (2007). $Na^+$ transport in plants. *FEBS Lett.* 581, 2247–2254. doi: 10.1016/j.febslet.2007.04.014

6. Apse, M. P., Sottosanto, J. B., and Blumwald, E. (2003). Vacuolar cation/$H^+$ exchange, ion homeostasis, and leaf development are altered in a T-DNA insertional mutant of AtNHX1, the *Arabidopsis* vacuolar $Na^+/H^+$ antiporter. *Plant J.* 36, 229–239. doi: 10.1046/j.1365-313X.2003.01871.x

7. Barragan, V., Leidi, E. O., Andres, Z., Rubio, L., Luca, A. D., Fernandez, J. A., et al. (2012). Ion exchangers NHX1 and NHX2 mediate active potassium uptake into vacuoles to regulate cell turgor and stomatal function in *Arabidopsis*. *Plant Cell* 24, 1127–1142. doi: 10.1105/tpc.111.095273

8. Bassil, E., Coku, A., and Blumwald, E. (2012). Cellular ion homeostasis: emerging roles of intracellular NHX $Na^+/H^+$ antiporters in plant growth and development. *J. Exp. Bot.* 63, 5727–5740. doi: 10.1093/jxb/ers250

9. Bassil, E., Ohto, M. A., Esumi, T., Tajima, H., Zhu, Z., Cagnac, O., et al. (2011a). The *Arabidopsis* intracellular $Na^+/H^+$ antiporters NHX5 and NHX6 are endosome associated and necessary for plant

growth and development. *Plant Cell* 23, 224–239. doi: 10.1105/tpc.110.079426

10. Bassil, E., Tajima, H., Liang, Y. C., Ohto, M. A., Ushijima, K., Nakano, R., et al. (2011b). The *Arabidopsis* $Na^+/H^+$ antiporters NHX1 and NHX2 control vacuolar pH and $K^+$ homeostasis to regulate growth, flower development, and reproduction. *Plant Cell* 23, 3482–3492. doi: 10.1105/tpc.111.089581
11. Baek, D., Jiang, J., Chung, J. S., Wang, B., Chen, J., Xin, Z., et al. (2011). Regulated AtHKT1 gene expression by a distal enhancer element and DNA methylation in the promoter plays an important role in salt tolerance. *Plant Cell Physiol.* 52, 149–161. doi: 10.1093/pcp/pcq182
12. Berthomieu, P., Conéjéro, G., Nublat, A., Brackenbury, W. J., Lambert, C., Savio, C., et al. (2003). Functional analysis of AtHKT1 in *Arabidopsis* shows that $Na^+$ recirculation by the phloem is crucial for salt tolerance. *EMBO J.* 22, 2004–2014. doi: 10.1093/emboj/cdg207
13. Blumwald, E. (2000). Sodium transport and salt tolerance in plants. *Curr. Opin. Cell Biol.* 12, 431–434. doi: 10.1016/S0955-0674(00)00112-5
14. Blumwald, E., Aharon, G. S., and Apse, M. P. (2000). Sodium transport in plant cells. *Biochim. Biophys. Acta* 1465, 140–151. doi: 10.1016/S0005-2736(00)00135-8
15. Brett, C. L., Donowitz, M., and Rao, R. (2005). Evolutionary origins of eukaryotic sodium/proton exchangers. *Am. J. Physiol. Cell Physiol.* 288, C223–C239. doi: 10.1152/ajpcell.00360.2004
16. Cao, Y., Jin, X., Huang, H., Derebe, M. G., Levin, E. J., Kabaleeswaran, V., et al. (2011). Crystal structure of a potassium ion transporter, TrkH. *Nature* 471, 336–340. doi: 10.1038/nature09731
17. Ding, L., and Zhu, J. K. (1997). Reduced $Na^+$ uptake in the NaCl-hypersensitive sos1 mutant of *Arabidopsis thaliana*. *Plant Physiol.* 113, 795–799. doi: 10.1104/pp.113.3.795
18. Dosch, D. C., Helmer, G. L., Sutton, S. H., Salvacion, F. F., and Epstein, W. (1991). Genetic analysis of potassium transport loci in

*Escherichia coli*: evidence for three constitutive systems mediating uptake potassium. *J. Bacteriol.* 173, 687–696.

19. Durell, S. R., and Guy, H. R. (1999). Structural models of the KtrB, TrkH, and Trk1,2 symporters based on the structure of the KcsA $K^+$channel. *Biophys. J.* 77, 789–807. doi: 10.1016/S0006-3495(99)76932-8Durell, S. R., Hao, Y., Nakamura, T., Bakker, E. P., and Guy, H. R. (1999). Evolutionary relationship between $K^+$ channels and symporters. *Biophys. J.* 77, 775–788. doi: 10.1016/S0006-3495(99)76931-6
20. Fairbairn, D. J., Liu, W., Schachtman, D. P., Gomez-Gallego, S., Day, S. R., and Teasdale, R. D. (2000). Characterisation of two distinct HKT1-like potassium transporters from *Eucalyptus camaldulensis*. *Plant Mol. Biol.* 43, 515–525. doi: 10.1023/A:1006496402463
21. Flowers, T. J., and Läuchli, A. (1983). "Sodium versus potassium: substitution and compartmentation," in *Encyclopedia of Plant Physiology 15B, New Series, Inorganic Plant Nutrition*, eds A. Läuchli and R. A. Bieleski (Berlin: Springer), 651–681.
22. Fukada-Tanaka, S., Inagaki, Y., Yamaguchi, T., Saito, N., and Iida, S. (2000). Colour-enhancing protein in blue petals. *Nature* 407, 581. doi: 10.1038/35036683
23. Gaber, R. F., Styles, C. A., and Fink, G. R. (1988). TRK1 encodes a plasma membrane protein required for high-affinity potassium transport in *Saccharomyces cerevisiae*. *Mol. Cell. Biol.* 8, 2848–2859.
24. Gassmann, W., Rubio, F., and Schroeder, J. I. (1996). Alkali cation selectivity of the wheat root high-affinity potassium transporter HKT1. *Plant J.* 10, 869–882. doi: 10.1046/j.1365-313X.1996.10050869.x
25. Gruwel, M. L. H., Rauw, V. L., Loewen, M., and Abrams, S. R. (2001). Effects of sodium chloride on plant cells: a 31P and 23Na NMR system to study salt tolerance. *Plant Sci.* 160, 785–794. doi: 10.1016/S0168-9452(00)00424-6
26. Hall, J. L., and Flowers, T. J. (1973). The effect of salt on protein synthesis in the halophyte *Suaeda maritima*. *Planta* 110, 361–368. doi: 10.1007/BF00387064

27. Hänelt, I., Löchte, S., Sundermann, L., Elbers, K., Vor der Brüggen, M., and Bakker, E. P. (2010a). Gain of function mutations in membrane region M2C2 of KtrB open a gate controlling $K^+$ transport by the KtrAB system from *Vibrio alginolyticus*. *J. Biol. Chem.* 285, 10318–10327. doi: 10.1074/jbc.M109.089870

28. Hänelt, I., Wunnicke, D., Müller-Trimbusch, M., Vor der Brüggen, M., Kraus, I., Bakker, E. P., et al. (2010b). Membrane region M2C2 in subunit KtrB of the $K^+$ uptake system KtrAB from *Vibrio alginolyticus* forms a flexible gate controlling $K^+$ flux: an electron paramagnetic resonance study. *J. Biol. Chem.* 285, 28210–28219. doi: 10.1074/jbc.M110.139311

29. Hauser, F., and Horie, T. (2010). A conserved primary salt tolerance mechanism mediated by HKT transporters: a mechanism for sodium exclusion and maintenance of high $K^+/Na^+$ ratio in leaves during salinity stress. *Plant Cell Environ.* 33, 552–565. doi: 10.1111/j.1365-3040.2009.02056.x

30. Horie, T., Yoshida, K., Nakayama, H., Yamada, K., Oiki, S., and Shinmyo, A. (2001). Two types of HKT transporters with different properties of $Na^+$ and $K^+$ transport in *Oryza sativa*. *Plant J.* 27, 129–138. doi: 10.1046/j.1365-313x.2001.01077.x

31. Huertas, R., Rubio, L., Cagnac, O., García-Sánchez, M. J., Alché, J. D., Venema, K., et al. (2013). The $K^+/H^+$ antiporter LeNHX2 increases salt tolerance by improving $K^+$ homeostasis in transgenic tomato. *Plant Cell Environ.* doi: 10.1111/pce.12109 [Epub ahead of print].

32. Kato, N., Akai, M., Zulkifli, L., Matsuda, N., Kato, Y., Goshima, S., et al. (2007). Role of positively charged amino acids in the M2D transmembrane helix of Ktr/Trk/HKT-type cation transporters. *Channels* 1, 161–171.

33. Kato, Y., Sakaguchi, M., Mori, Y., Saito, K., Nakamura, T., Bakker, E. P., et al. (2001). Evidence in support of a four transmembrane-pore-transmembrane topology model for the *Arabidopsis thaliana* $Na^+/K^+$ translocating AtHKT1 protein, a member of the superfamily of $K^+$transporters. *Proc. Natl. Acad. Sci. USA*. 98, 6488–6493. doi: 10.1073/pnas.101556598

34. Ko, C. H., and Gaber, R. F. (1988). TRK1 and TRK2 encode structurally related $K^+$ transporters in *Saccharomyces cerevisiae. Mol. Cell.Biol.* 11, 4266–4273.

35. Leidi, E. O., Barragán, V., Rubio, L., El-Hamdaoui, A., Ruiz, M. T., Cubero, B., et al. (2010). The AtNHX1 exchanger mediates potassium compartmentation in vacuoles of transgenic tomato. *Plant J.* 61, 495–506. doi: 10.1111/j.1365-313X.2009.04073.x

36. Martínez-Atienza, J., Jiang, X., Garciadeblas, B., Mendoza, I., Zhu, J. K., Pardo, J. M., et al. (2007). Conservation of the salt overly sensitive pathway in rice. *Plant Physiol.* 143, 1001–1012. doi: 10.1104/pp.106.092635

37. Mäser, P., Eckelman, B., Vaidyanathan, R., Fairbain, D. J., Kubo, M., Yamagami, M., et al. (2002a). Altered shoot/root $Na^+$ distribution and bifurcating salt sensitivity in *Arabidopsis* by genetic disruption of the $Na^+$ transporter AtHKT1. *FEBS Lett.* 531, 157–161. doi: 10.1016/S0014-5793(02)03488-9

38. Mäser, P., Hosoo, Y., Goshima, S., Horie, T., Eckelman B., Yamada, K., et al. (2002b). Glycine residues in potassium channel-like selectivity filters determine potassium selectivity in four-loop-per-subunit HKT transporters from plants. *Proc. Natl. Acad. Sci. USA.* 99, 6428–6433. doi: 10.1073/pnas.082123799

39. Matsuda, N., Kobayashi, H., Katoh, H., Ogawa, T., Futatsugi, L., Nakamura, T., et al. (2004). $Na^+$-dependent $K^+$ uptake Ktr system from the cyanobacterium *Synechocystis* sp. *PCC* 6803 and its role in the early phases of cell adaptation to hyperosmotic shock. *J. Biol. Chem.*279, 54952–54962. doi: 10.1074/jbc.M407268200

40. Møller, I. S., Gilliham, M., Jha, D., Mayo, G. M., Roy, S. J., Coates, J. C., et al. (2009). Shoot $Na^+$ exclusion and increased salinity tolerance engineered by cell type-specific alteration of $Na^+$ transport in *Arabidopsis*. *Plant Cell* 21, 2163–2178. doi: 10.1105/tpc.108.064568

41. Mosimann, M., Goshima, S., Wenzler, T., Lüscher, A., Uozumi, N., and Mäser, P. (2010). A Trk/HKT-type $K^+$ transporter from *Trypanosoma brucei*. *Eukaryot. Cell* 9, 539–546. doi: 10.1128/EC.00314-09

42. Munns, R., and Tester, M. (2008). Mechanisms of salinity tolerance. *Annu. Rev. Plant Biol.* 59, 651–681. doi: 10.1146/annurev.arplant.59.032607.092911

43. Murguía, J. R., Bellés, J. M., and Serrano, R. (1995). A salt-sensitive 3′(2′),5′-bisphosphate nucleotidase involved in sulfate activation. *Science* 267, 232–234. doi: 10.1126/science.7809627

44. Nakamura, T., Yamamuro, N., Stumpe, S., Tsutomu, U., and Bakker, E. P. (1998). Cloning of the trkAH gene cluster and characterization of the Trk $K^+$-uptake system of *Vibrio alginolyticus*. *Microbiology* 144, 2281–2289. doi: 10.1099/00221287-144-8-2281

45. Núñez-Ramírez, R., Sánchez-Barrena, M. J., Villalta, I., Vega, J. F., Pardo, J. M., Quintero, F. J., et al. (2012). Structural insights on the plant salt-overly-sensitive 1 (SOS1) $Na^+/H^+$ antiporter. *J. Mol. Biol.* 424, 283–294. doi: 10.1016/j.jmb.2012.09.015

46. Oh, D. H., Leidi, E., Zhang, Q., Hwang, S. M., Li, Y., Quintero, F. J., et al. (2009). Loss of halophytism by interference with SOS1 expression. *Plant Physiol.* 151, 210–222. doi: 10.1104/pp.109.137802

47. Oh, D. H., Lee, S. Y., Bressan, R. A., Yun, D. J., and Bohnert, H. J. (2010). Intracellular consequences of SOS1 deficiency during salt stress. *J. Exp. Bot.* 61, 1205–1213. doi: 10.1093/jxb/erp391

48. Okazaki, Y., Kikuyama, M., Hiramoto, Y., and Iwasaki, N. (1996). Short-term regulation of cytosolic $Ca^{2+}$, cytosolic pH and vacuolar pH under NaCl stress in the charophyte alga *Nitellopsis obtusa*. *Plant Cell Environ.* 19, 569–576. doi: 10.1111/j.1365-3040.1996.tb00390.x

49. Olías, R., Eljakaoui, Z., Li, J., De Morales, P. A., Marín-Manzano, M. C., Pardo, J. M., et al. (2009). The plasma membrane $Na^+$/$H^+$antiporter SOS1 is essential for salt tolerance in tomato and affects the partitioning of $Na^+$ between plant organs. *Plant Cell Environ.* 32, 904–916. doi: 10.1111/j.1365-3040.2009.01971.x

50. Pardo, J. M., Cubero, B., Leidi, E. O., and Quintero, F. J. (2006). Alkali cation exchangers: roles in cellular homeostasis and stress tolerance. *J. Exp. Bot.* 57, 1181–1199. doi: 10.1093/jxb/erj114

51. Platten, J. D., Cotsaftis, O., Berthomieu, P., Bohnert, H., Davenport, R. J., Fairbairn, D. J., et al. (2006). Nomenclature for HKT

transporters, key determinants of plant salinity tolerance. *Trends Plant Sci.* 11, 372–374. doi: 10.1016/j.tplants.2006.06.001

52. Qi, Z., and Spalding, E. P. (2004). Protection of plasma membrane $K^+$ transport by the salt overly sensitive 1 $Na^+$–$H^+$ antiporter during salinity stress. *Plant Physiol.* 136, 2548 2555. doi: 10.1104/pp.104.049213
53. Qiu, Q. S. (2012). Plant and yeast NHX antiporters: roles in membrane trafficking. *J. Integr. Plant Biol.* 54, 66–72. doi: 10.1111/j.1744-7909.2012.01097.x
54. Qiu, Q. S., Barkla, B. J., Vera-Estrella, R., Zhu, J. K., and Schumaker, K. S. (2003). $Na^+/H^+$ exchange activity in the plasma membrane of *Arabidopsis. Plant Physiol.* 132, 1041–1052. doi: 10.1104/pp.102.010421
55. Qiu, Q. S., Guo, Y., Dietrich, M. A., Schumaker, K. S., and Zhu, J. K. (2002). Regulation of SOS1, a plasma membrane $Na^+/H^+$ exchanger in *Arabidopsis thaliana*, by SOS2 and SOS3. *Proc. Natl. Acad. Sci. USA.* 99, 8436–8441. doi: 10.1073/pnas.122224699
56. Quan, R., Lin, H., Mendoza, I., Zhang, Y., Cao, W., Yang, Y., et al. (2007). SCABP8/CBL10, a putative calcium sensor, interacts with the protein kinase SOS2 to protect *Arabidopsis* shoots from salt stress. *Plant Cell* 19, 1415–1431. doi: 10.1105/tpc.106.042291
57. Quintero, F. J., Martinez-Atienza, J., Villalta, I., Jiang, X., Kim, W. Y., Ali, Z., et al. (2011). Activation of the plasma membrane Na/H antiporter salt-overly-sensitive 1 (SOS1) by phosphorylation of an auto-inhibitory C-terminal domain. *Proc. Natl. Acad. Sci. USA.* 108, 2611–2616. doi: 10.1073/pnas.1018921108
58. Ren, Z. H., Gao, J. P., Li, L. G., Cai, X. L., Huang, W., Chao, D. Y., et al. (2005). A rice quantitative trait locus for salt tolerance encodes a sodium transporter. *Nat. Genet.* 37, 1141–1146. doi: 10.1038/ng1643
59. Rodríguez-Rosales, M. P., Gálvez, F. J., Huertas, R., Aranda, M. N., Baghour, M., Cagnac, O., et al. (2009). Plant NHX cation/proton antiporters. *Plant Signal. Behav.* 4, 265 276. doi: 10.4161/psb.4.4.7919
60. Rodríguez-Rosales, M. P., Jiang, X. J., Gálvez, F. J., Aranda, M. N., Cubero, B., and Venema, K. (2008). Overexpression of the tomato

$K^+/H^+$ antiporter LeNHX2 confers salt tolerance by improving potassium compartmentalization. *New Phytol.* 179, 366–377. doi: 10.1111/j.1469-8137.2008.02461.x

61. Rubio, F., Gassmann, W., and Schroeder, J. I. (1995). Sodium-driven potassium uptake by the plant potassium transporter HKT1 and mutations conferring salt tolerance. *Science* 270, 1660–1663. doi: 10.1126/science.270.5242.1660
62. Rus, A., Baxter, I., Muthukumar, B., Gustin, J., Lahner, B., Yakubova, E., et al. (2006). Natural variants of AtHKT1 enhance $Na^+$accumulation in two wild populations of *Arabidopsis*. *PLoS Genet.* 2:e210. doi: 10.1371/journal.pgen.0020210
63. Säier, M. H. Jr., Tran, C. V., and Barabote, R. D. (2006). TCDB: the transporter classification database for membrane transport protein analyses and information. *Nucleic Acids Res.* 34, D181–D186. doi: 10.1093/nar/gkj001
64. Sato, Y., and Sakaguchi, M. (2005). Topogenic properties of transmembrane segments of *Arabidopsis thaliana* NHX1 reveal a common topology model of the $Na^+/H^+$ exchanger family. *J. Biochem. (Tokyo)* 138, 425–431. doi: 10.1093/jb/mvi132
65. Schachtman, D. P., and Schroeder, J. I. (1994). Structure and transport mechanism of a high-affinity potassium uptake transporter from higher plants. *Nature* 370, 655–658. doi: 10.1038/370655a0
66. Schlösser, A., Kluttig, S., Hamann, A., and Bakker, E. P. (1991). Subcloning, nucleotide sequence, and expression of trkG, a gene that encodes an integral membrane protein involved in potassium uptake via the Trk system of *Escherichia coli*. *J. Bacteriol.* 173, 3170–3176.
67. Shabala, L., Cui, T. A., Newman, I. A., and Shabala, S. (2005). Salinity-induced ion flux patterns from the excised roots of *Arabidopsis*sos mutants. *Planta* 222, 1041–1050. doi: 10.1007/s00425-005-0074-2
68. Shi, H., Ishitani, M., Kim, C., and Zhu, J. K. (2000). The *Arabidopsis thaliana* salt tolerance gene SOS1 encodes a putative $Na^+/H^+$antiporter. *Proc. Natl. Acad. Sci. USA*. 97, 6896–6901. doi: 10.1073/pnas.120170197
69. Shi, H., Lee, B. H., Wu, S. J., and Zhu, J. K. (2003). Overexpression

of a plasma membrane $Na^+/H^+$ antiporter gene improves salt tolerance in *Arabidopsis thaliana. Nat. Biotechnol.* 21, 81–85. doi: 10.1038/nbt766

70. Shi, L. Y., Li, H. Q., Pan, X. P., Wu, G. J., and Li, M. R. (2008). Improvement of *Torenia fournieri* salinity tolerance by expression of *Arabidopsis* AtNHX5. *Funct. Plant Biol.* 35, 185–192 doi: 10.1071/FP07269
71. Shi, H., Quintero, F. J., Pardo, J. M., and Zhu, J. K. (2002). The putative plasma membrane $Na^+/H^+$ antiporter SOS1 controls long-distance $Na^+$ transport in plants. *Plant Cell* 14, 465–477. doi: 10.1105/tpc.010371
72. Sottosanto, J. B., Gelli, A., and Blumwald, E. (2004). DNA array analyses of *Arabidopsis thaliana* lacking a vacuolar $Na^+/H^+$ antiporter: impact of AtNHX1 on gene expression. *Plant J.* 40, 752–771. doi: 10.1111/j.1365-313X.2004.02253.x
73. Sunarpi, H. T., Motoda, J., Kubo, M., Yang, H., Yoda, K., Horie, R., et al. (2005). Enhanced salt tolerance mediated by AtHKT1 transporter-induced Na unloading from xylem vessels to xylem parenchyma cells. *Plant J.* 44, 928–938. doi: 10.1111/j.1365-313X.2005.02595.x
74. Tholema, N., Vor der Brüggen, M., Mäser, P., Nakamura, T., Schroeder, J. I., Kobayashi, H., et al. (2005). All four putative selectivity filter glycine residues in KtrB are essential for high affinity and selective $K^+$ uptake by the KtrAB system from *Vibrio alginolyticus*. *J.Biol. Chem.* 280, 41146–41154. doi: 10.1074/jbc.M507647200
75. Tsugane, K., Kobayashi, K., Niwa, Y., Ohba, Y., Wada, K., and Kobayashi, H. (1999). A recessive *Arabidopsis* mutant that grows photoautotrophically under salt stress shows enhanced active oxygen detoxification. *Plant Cell* 11, 1195–1206.
76. Tsunekawa, K, Shijuku, T., Hayashimoto, M., Kojima, Y., Onai, K., Morishita, M., Ishiura, M., et al. (2009). Identification and characterization of the $Na^+/H^+$ antiporter Nhas3 from the thylakoid membrane of *Synechocystis* sp. *PCC* 6803. *J. Biol. Chem.* 284, 16513–16521. doi: 10.1074/jbc.M109.001875

77. Uozumi, N., and Dreyer, I. (2012). "Structure–function correlates in plant ion channels," in *Comprehensive Biophysics*, Vol. 6, *Channel Proteins*, ed. E. Egelman (Amsterdam: Elsevier BV), 234–245.

78. Uozumi, N., Kim, E. J., Rubio, F., Yamaguchi, T., Muto, S., Tsuboi, A., et al. (2000). The *Arabidopsis* HKT1 gene homolog mediates inward $Na^+$ currents in *Xenopus laevis* oocytes and $Na^+$ uptake in *Saccharomyces cerevisiae. Plant Physiol.* 122, 1249–1259. doi: 10.1104/pp.122.4.1249

79. Uozumi, N., and Schroeder, J. I. (2010). "Ion channels and plant stress: past, present and future," in *Ion Channels and Plant Stress Responses*, eds V. Demidchik and F. Maathuis (Berlin: Springer), 1–22.

80. Vera-Estrella, R., Barkla, B. J., García-Ramírez, L., and Pantoja, O. (2005). Salt stress in *Thellungiella halophila* activates $Na^+$ transport mechanisms required for salinity tolerance. *Plant Physiol.* 139, 1507–1517. doi: 10.1104/pp.105.067850

81. Wu, S. J., Ding, L., and Zhu, J. K. (1996). SOS1, a genetic locus essential for salt tolerance and potassium acquisition. *Plant Cell* 8, 617–627.

82. Xu, H., Jiang, X., Zhan, K., Cheng, X., Chen, X., Pardo, J. M., et al. (2008). Functional characterization of a wheat plasma membrane $Na^+/H^+$ antiporter in yeast. *Arch. Biochem. Biophys.* 473, 8–15. doi: 10.1016/j.abb.2008.02.018

83. Yamaguchi, T., Aharon, G. S., Sottosanto, J. B., and Blumwald, E. (2005). Vacuolar $Na^+/H^+$ antiporter cation selectivity is regulated by calmodulin from within the vacuole in $Ca^{2+}$- and pH-dependent manner. *Proc. Natl. Acad. Sci. U.S.A.* 102, 16107–16112. doi: 10.1073/pnas.0504437102

84. Yamaguchi, T., Apse, M. P., Shi, H., and Blumwald, E. (2003). Topological analysis of a plant vacuolar $Na^+/H^+$ antiporter reveals a luminal C terminus that regulates antiporter cation selectivity. *Proc. Natl. Acad. Sci. U.S.A.* 100, 12510–12515. doi: 10.1073/pnas.2034966100

85. Yamaguchi, T., and Blumwald, E. (2005). Developing salt tolerant crop plants: challenges and opportunities. *Trends Plant Sci.* 10, 615–620. doi: 10.1016/j.tplants.2005.10.002

86. Yamaguchi, T., Fukada-Tanaka, S., Inagaki, Y., Saito, N., Yonekura-Sakakibara, K., Tanaka, Y., et al. (2001). Genes encoding the vacuolar $Na^{+}H^{+}$ exchanger and flower coloration. *Plant Cell Physiol.* 42, 451–461. doi: 10.1093/pcp/pce080

87. Yang, Q., Chen, Z. Z., Zhou, X. F., Yin, H. B., Li, X. F., Xin, X. F., et al. (2009). Overexpression of SOS (salt overly sensitive) genes increases salt tolerance in transgenic *Arabidopsis*. *Mol. Plant* 2, 22–31. doi: 10.1093/mp/ssn058

88. Yeo, A. R. (1999). Predicting the interaction between the effects of salinity and climate change on crop plants. *Sci. Hortic. (Amsterdam)*78, 159–174. doi: 10.1016/S0304-4238(98)00193-9

89. Zhang, H. H., Hodson, J. N., Williams, J. P., and Blumwald, E. (2001). Engineering salt-tolerant *Brassica* plants: characterization of yield and seed oil quality in transgenic plants with increased vacuolar sodium accumulation. *Proc. Natl. Acad. Sci. U.S.A.* 98, 12832–12836. doi: 10.1073/pnas.231476498

90. Zhang, H. X., and Blumwald, E. (2001). Transgenic salt-tolerant tomato plants accumulate salt in foliage but not in fruit. *Nat. Biotechnol.* 19, 765–768. doi: 10.1038/90824

CHAPTER

# 5

# THE F130S POINT MUTATION IN THE ARABIDOPSIS HIGH-AFFINITY K+ TRANSPORTER ATHAK5 INCREASES K+ OVER NA+ AND CS+ SELECTIVITY AND CONFERS NA+ AND CS+ TOLERANCE TO YEAST UNDER HETEROLOGOUS EXPRESSION

Fernando Alemán†, Fernando Caballero, Reyes Ródenas, Rosa M. Rivero, Vicente Martínez and Francisco Rubio*

Centro de Edafología y Biología Aplicada del Segura-CSIC, Murcia, Spain

**Keywords**: Potassium, sodium, cesium, selectivity, HAK, Arabidopsis, random mutagenesis, point mutation

Potassium ($K^+$) is an essential macronutrient required for plant growth, development and high yield production of crops. Members of group I of the KT/HAK/KUP family of transporters, such as HAK5, are key components for $K^+$ acquisition by plant roots at low external

$K^+$concentrations. Certain abiotic stress conditions such as salinity or $Cs^+$-polluted soils may jeopardize plant $K^+$ nutrition because HAK5-mediated $K^+$ transport is inhibited by $Na^+$ and $Cs^+$. Here, by screening in yeast a randomly-mutated collection of AtHAK5 transporters, a new mutation in AtHAK5 sequence is identified that greatly increases $Na^+$ tolerance. The single point mutation F130S, affecting an amino acid residue conserved in HAK5 transporters from several species, confers high salt tolerance, as well as $Cs^+$ tolerance. This mutation increases more than 100-fold the affinity of AtHAK5 for $K^+$ and reduces the $K_i$ values for $Na^+$ and $Cs^+$, suggesting that the F130 residue may contribute to the structure of the pore region involved in $K^+$ binding. In addition, this mutation increases the $V_{max}$ for $K^+$. All this changes occur without increasing the amount of the AtHAK5 protein in yeast and support the idea that this residue is contributing to shape the selectivity filter of the AtHAK5 transporter.

## INTRODUCTION

Potassium ($K^+$) is an essential nutrient for plants, required for plant growth and development (Marschner, 2012). It fulfills important functions related to enzyme activation, protein synthesis, maintenance of cytoplasmic pH and transmembrane voltage gradients and neutralization of negative charges (Maathuis, 2009). It is the most abundant cationic component, accounting for more than a 10% of the plant dry weight (White and Karley, 2010). As it occurs with other nutrients, roots take up $K^+$ from the soil solution through specific transport systems at the plasma membrane of epidermal and cortical cells (Marschner, 2012). Physiological and molecular studies in the model plant *Arabidopsis thaliana* have described two systems, AtHAK5 and AKT1, a high-affinity $K^+$ transporter and an inward rectifying $K^+$ channel respectively, which are the major contributors to root $K^+$ uptake (Alemán et al., 2011; Nieves-Cordones et al., 2014). The relative contribution of these two systems to $K^+$ nutrition depends on the external $K^+$concentration and several other factors such as other external ions or root cell membrane potential. At high $K^+$ concentrations (around 1 mM), within the low-affinity system described by Epstein (Epstein, 1966), AKT1 is the only system mediating $K^+$ uptake. Below 200 μM $K^+$, corresponding to the high-affinity range of concentrations, both transport systems can contribute. However, at very low $K^+$ concentrations (<10 μM) the only system involved in $K^+$ acquisition is AtHAK5 (Rubio et al., 2008, 2010; Pyo et al., 2010). This model may be extended to other plant species although the relative contribution of each of the two systems over the range of external $K^+$ concentration may vary among them. For example, in tomato or pepper plants, the AtHAK5 homologs LeHAK5 or CaHAK1 respectively, dominate $K^+$ uptake over the AKT1 homologs at low external concentrations (Martínez-Cordero et al., 2004, 2005; Nieves-Cordones et al., 2007).

Operation of the HAK5-type of transporters may be essential for $K^+$ nutrition in $K^+$ depleted soils. In addition, $K^+$ uptake under abiotic stress conditions such as salinity or $Cs^+$-polluted soils may also largely depend on HAK5-type transporters (Hampton et al., 2004; Qi et al., 2008; Alemán et al., 2009). Under salinity, uptake and accumulation of $Na^+$ in roots cells depolarize their plasma membrane (Volkov and Amtmann, 2006; Chen et al., 2007; Nieves-Cordones et al., 2008) and reduce the driving force for $K^+$uptake. As a result, AKT1 function may be importantly impaired but, $K^+$ uptake through the $K^+$-$H^+$symporter mechanism proposed for the HAK5-type transporters, may be still possible (Rodríguez-Navarro, 2000). Supporting this idea it has been reported that the Arabidopsis AtHAK5 is required for $K^+$uptake and plant growth at low $K^+$ in the presence of salinity, even though *AtHAK5* expression is reduced under high $Na^+$ concentrations (Nieves-Cordones et al., 2010). In $Cs^+$-polluted soils, $K^+$ uptake through AKT1 may be importantly inhibited because $Cs^+$ is a potent blocker of AKT1 channels (Bertl et al., 1997; White and Broadley, 2000).

Although HAK5-type transporters may be the major contributors to $K^+$ uptake under $Na^+$- or $Cs^+$-affected soils, they are inhibited by these cations. Both $Na^+$ and $Cs^+$ competitively inhibit $K^+$ uptake through HAK5-type transporters (Santa-María et al., 1997; Rubio et al., 2000). This leads to two important effects. On one hand, $K^+$ supply is reduced and in fact, it has been described that high external $Na^+$ or $Cs^+$ may lead to $K^+$deficiency (Botella et al., 1997; Carden et al., 2003; Hampton et al., 2004; Shabala et al., 2006; Adams et al., 2013). On the other hand, $Na^+$ and $Cs^+$ may accumulate into the cell by entering through HAK5 transporters (Santa-María et al., 1997; Rubio et al., 2000; Qi et al., 2008). Overall, pivotal parameters related to $Na^+$ or $Cs^+$ tolerance such as the cytoplasmic $K^+/Na^+$ (Maathuis and Amtmann, 1999; Shabala and Cuin,

2008) or $K^+/Cs^+$ (Hampton et al., 2004; Adams et al., 2013) ratios are reduced and plant growth is inhibited.

These three stress conditions, $K^+$-depleted, salt-affected or $Cs^+$-polluted soils, that may make $K^+$ nutrition largely dependent on HAK5 operation, affect agriculture worldwide. Many agricultural lands are intrinsically deficient in $K^+$ (Rengel and Damon, 2008; Römheld and Kirkby, 2010) or become $K^+$ deficient because of an inadequate fertilization that prioritize N and P over K fertilization (Römheld and Kirkby, 2010). Salinity affects more than 20% of irrigated lands worldwide and, because most crop species are glycophytes, it reduces crop productivity (Munns and Tester, 2008). Finally, although the natural concentration of $Cs^+$ in soils is low, certain areas may contain $Cs^+$ concentrations that reduce plant growth because $Cs^+$ is toxic to plants. In addition, agricultural lands affected by discharges of nuclear power plants result in polluted soils with radiocesium, which readily enters the food chain (White and Broadley, 2000).

Obtaining plants with more selective HAK5 transporters may be a solution to these agricultural problems (Niu et al., 1995). This requires function-structure relationship studies, which are scarce in literature. Some studies reported the isolation of mutated plant transporters of the HAK5-type that when expressed in yeast showed altered kinetics of $K^+$ uptake (Rubio et al., 2000; Senn et al., 2001; Garciadeblas et al., 2007; Mangano et al., 2008; Nieves-Cordones et al., 2008). Here we have focused on the identification of amino acid residues that are important for $K^+$ over $Na^+$ or $Cs^+$ selectivity of the Arabidopsis AtHAK5 transporter. Mutagenic PCR followed by selection in yeast allowed us for the identification of a highly conserved amino acid residue in the family of HAK transporters, F130, that is probably involved in the selectivity of the transporter for $K^+$.

## MATERIALS AND METHODS

### Yeast and Bacterial Strains and Growth Media

*Saccharomyces cerevisiae* yeast strain 9.3 (*MAT***a**, *ena1D:HIS3::ena4D, leu2, ura3–1, trp1–1, ade2–1, trk1D, trk2::pCK64*) (Bañuelos et al., 1995) with deleted TRK1 and TRK2 $K^+$ uptake systems and ENA1-ENA4 $Na^+$ extrusion $Na^+$-ATPases was used throughout the study.

The *Escherichia coli* strain DH5α (Hanahan, 1985) was used for plasmid DNA amplification. For yeast growth YPD (1% yeast extract, 2% peptone, 2% glucose), SD (Sherman, 1991) and Arginine Phosphate medium (AP) (Rodríguez-Navarro and Ramos, 1984) supplemented with KCl, NaCl and requirements as indicated were used. For bacterial growth, LB media supplemented with ampicillin as required was used. Yeast transformation was carried out by the high-efficiency protocol of the lithium acetate/single-stranded carrier DNA/PEG method of transformation of *S. cerevisiae* (Gietz and Schiestl, 2007). Standard procedures were used for *E. coli*growth and transformation, and DNA manipulations (Sambrook and Russcll, 2001).

For determining doubling times of yeast cells, liquid cultures were used. The optical density at 550 nm of the yeast cultures was determined in a spectrophotometer at different time points during 72 h. The optical density vs. time was plotted and the doubling time (the time required to duplicate the optical density) calculated during the exponential growth phase.

## Random AtHAK5 Mutagenesis, Mutant Selection and Site-Directed Mutagenesis

The AtHAK5 cDNA was mutagenized randomly by error prone PCR (epPCR) as described by Wong et al. (2007) with minor modifications. In short, two epPCRs biased toward transitions or transversions were carried out to amplify AtHAK5 cDNA to get a broader range of possible amino acid substitutions (Table S1). The primers used to amplify the AtHAK5 cDNA were Forward: 5′-AAAAAATATACCCCAGCCTCG-3′ and Reverse 5′-GGCGAAGAAGTCCAAAGCTGG-3, which can bind up- and downstream regions respectively of the NotI and BamHI pDR195 cloning sites (Rentsch et al., 1995), where the cDNA of AtHAK5 was inserted.

The amplified cDNA fragments and a NotI-BamHI digested pDR195 plasmid were cotransformed into yeast to allow homologous recombination (ratio 10:1 epPCR:vector) as previously described (Rubio et al., 1999). The amplified cDNA and the plasmid shared a region of more than 50 base pairs on each end to allow proper homologous recombination (Muhlrad et al., 1992). Transformants were first selected on SD medium lacking uracil and supplemented with 100 mM KCl and then replica-plated to AP medium supplemented with 0.1 mM KCl plus different concentrations of $Na^+$. The best growing colonies were selected, and their plasmids were isolated and reintroduced into yeast. Yeast retransformants that confirmed growth on high $Na^+$ were selected, their plasmids isolated and the AtHAK5 cDNA sequenced. To obtain single point mutations, PCR-based site directed mutagenesis was followed as described elsewhere (Cadwell and Joyce, 1992).

## Internal Ionic Content Determination in Yeast

Internal $K^+$, $Na^+$, and $Cs^+$ concentrations in yeast were determined after 24 h of growth in liquid AP media supplemented with $K^+$, $Na^+$, and $Cs^+$ as indicated. In short, samples of yeast cultures were filtered through a 0.8-μm-pore nitrocellulose membrane filter (Millipore, Bedford, MA) and washed with 20 mM $MgCl_2$. Ionic contents were extracted after 24 h incubation in 0.1 M HCl. The $K^+$, $Na^+$, or $Cs^+$ concentrations were determined by atomic emission spectrophotometry of the acid extracts from the cells (Rodríguez-Navarro and Ramos, 1984).

## $Rb^+$ Uptake Experiments

For kinetic characterization of $K^+$ uptake, $Rb^+$ was used in uptake experiments (Rodríguez-Navarro, 2000). Yeast cells of the 9.3 strain transformed with plasmids harboring WT AtHAK5 and the different mutants were grown overnight in AP liquid media supplemented with 3 mM $K^+$ at 28°C with shaking. After overnight incubation, cell cultures reached an optical density (550) of 0.2–0.3. Then, cells were collected by centrifugation and suspended in fresh AP media without $K^+$, and incubated for 6 h. After this 6 h period of $K^+$ starvation cells were collected by centrifugation and suspended in uptake buffer that contained 10 mM MES, 0.1 mM $MgCl_2$ and 2% glucose brought to pH 6,0 with $Ca(OH)_2$. Cells were placed at 28°C in a shaker bath and at time zero $Rb^+$ was added. Samples were taken at different time points during 10 min and collected on nitrocellulose 0.8 μm pore filters that were incubated overnight in 5 mL of 0.1 M HCl to extract their internal $Rb^+$, which was determined by atomic emission spectrophotometry. The internal $Rb^+$ concentrations per unit of cell dry weight were plotted against time and the initial rates of $Rb^+$ uptake were calculated on a dry weight basis and per unit of time (Rodríguez-Navarro and Ramos, 1984). The rates of $Rb^+$ uptake were plotted at the different external concentrations of $Rb^+$ employed and fitted to Michaelis-Menten equations to calculate the kinetic parameters. These experiments were carried out with cultures from several independent transformants producing the same results.

### $K^+$ and $Cs^+$ Depletion Experiments

For determining the $V_{max}$ of $K^+$ and $Cs^+$ uptake, depletion experiments were performed. Yeast cells were grown and starved of $K^+$ as described for the $Rb^+$ uptake experiments and then transferred to AP media to reach an optical density of 2. At zero time, $K^+$ or $Cs^+$ was added to reach a 50 μM concentration. Samples of the cell suspensions were taken at different time points and centrifuged to spin down cells. $K^+$ or $Cs^+$concentrations were determined in the supernatant by atomic emission spectrophotometry. The $K^+$ or $Cs^+$uptake rates were calculated from the difference in the external cation concentration between two time points per unit of dry weight and unit of time.

### Protein Extraction and Western Blot

Yeast cells expressing the different AtHAK5 versions as well as the empty plasmid were grown overnight in YPD media supplemented with 100 mM $K^+$. Total membranes from yeast were isolated as described elsewhere (Serrano, 1988). 25 μg of proteins from total membranes of each yeast strain were subjected to SDS/PAGE and immunoblotting following standard procedures (Sambrook and Russell, 2001). AtHAK5 protein of a predicted mass of 87.85 kDa was immunodetected with a polyclonal antibody raised in rabbit against the YGYKEDIEEPDEFE peptide present in AtHAK5 amino acid sequence, synthesized by GenScript (Piscataway, NJ USA). After immunological detection, the PVDF membrane was washed and stained with comassie blue.

## RESULTS

### Selection of AtHAK5 Mutants that Increase Tolerance to NaCl

To isolate novel HAK5 mutants, the $Na^+$ sensitive yeast strain 9.3, lacking functional TRK1 and TKR2 $K^+$uptake systems as well as ENA1-

ENA4 $Na^+$ extruding ATPases, was used for heterologous expression (Bañuelos et al., 1995). Plasmids containing putative mutants of AtHAK5 were obtained by allowing homologous recombination in yeast of mutated cDNA and plasmid fragments that rescued the $ura^-$mutation of the 9.3 yeast strain, as described in Materials and Methods Section. Colonies growing in SD-URA selective media supplemented with 100 mM $K^+$ were replica plated to different selective minimal AP media. These selective media contained 0.1 mM $K^+$ with no added $Na^+$ or with 400 mM $Na^+$. Thirty five colonies that showed a better growth in the presence of $Na^+$ were selected for further analysis. To ensure that the best performance of the selected yeast colonies was promoted by the plasmid and not by a reversion of the yeast genetic background, the plasmids were recovered from yeast cells and retransformed into the 9.3 yeast strain. The capacity of the retransformants to grow in the presence of high $Na^+$ was checked. The 9.3 strain transformed with the WT AtHAK5 was used as a control. However, as previously shown (Rubio et al., 2000), the low $V_{max}$ of $K^+$ uptake mediated by WT AtHAK5 in yeast prevented its functional complementation of *trk1, trk2* yeast mutants. Therefore, an additional control that consisted of yeast cells expressing the previously characterized AtHAK5mutant with the L776H mutation was used. This mutation increased the $V_{max}$ of $K^+$ uptake and allowed growth of *trk1, trk2* yeasts at low $K^+$ (Rubio et al., 2000). None of the yeast cells harboring the putative AtHAK5 mutants grew better than those expressing the L776H mutant (not shown).

In order to obtain AtHAK5 mutations that enhanced $Na^+$ tolerance beyond that promoted by the L776H mutant, a similar approach was undertaken, but using the cDNA of the L776H mutant as a starting material. The selective pressure was also increased by using higher $Na^+$ concentrations in the selective media. In this case, 17 $Na^+$-tolerant colonies were selected on media containing 0.1 mM $K^+$ and 600 mM $Na^+$. After retransformation and selection for growth, one colony, that was named Mut11, was selected because it showed a better growth than the L776H mutant on plates with high $Na^+$. The plasmid containing the Mut11 mutant was isolated and the AtHAK5 cDNA sequenced. It was observed that in addition to the L776H mutation, the Mut11 mutant contained the F130S mutation. To study individual effects of each of these mutations, a cDNA that encoded the single mutant F130S was generated by site directed mutagenic PCR, cloned into the pDR195 plasmid and transformed into the 9.3 strain. Growth of yeast cells expressing double and single mutants as well as WT AtHAK5 and the empty plasmid was studied by drop tests on minimal media (Figure 1).

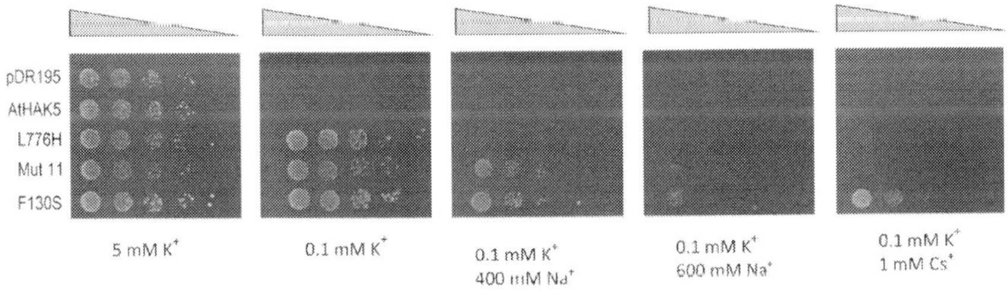


**Figure 1**: Growth of 9.3 yeast cells expressing WT and mutated AtHAK5 $K^+$transporters. Cells of the yeast strain 9.3 (trk1, trk2, ena1-4) expressing the empty plasmid pDR195, WT AtHAK5 and the mutants L776H, Mut11 (L776H, F130S) and F130S were used for growth assays on drop tests on solid media. Serial dilutions of cell suspensions were plated on minimal AP media supplemented with the indicated $K^+$, $Na^+$ and $Cs^+$ concentrations.

In the presence of 5 mM $K^+$, all yeast strains grew well. When the $K^+$ concentration was reduced to 0.1 mM $K^+$, yeast cells transformed with the empty plasmid (pDR195) or with WT AtHAK5 were unable to grow. Yeast cells expressing any of the mutated versions of AtHAK5 could grow. When 400 mM $Na^+$ was added to the 0.1 mM $K^+$ medium, only the double Mut11 mutant and the single F130S mutant were able to grow. Increasing $Na^+$ to 600 mM completely arrested growth of Mut11. However, at this high $Na^+$ concentration, the cells expressing the single mutant F130S were able to grow, although at low rates.

Tolerance to $Cs^+$ was also studied. The presence of 1 mM $Cs^+$ in the 0.1 mM $K^+$ medium completely inhibited growth of all yeast strains with the exception of those expressing the F130S mutation, that showed an outstanding growth capacity.

For a detailed study of growth rates and tolerance to $Na^+$ and $Cs^+$, liquid cultures were used. The doubling times of the different yeast strains growing in the presence of different levels of $K^+$, $Na^+$, and $Cs^+$ were determined (Table 1). At high $K^+$ (3 mM $K^+$) all strains grew although mutants did so faster than WT AtHAK5. At low $K^+$ (0.1 mM $K^+$) only the mutants grew with no significant differences among them. Including high $Na^+$ concentrations to the media gave rise to differences in growth rates among the different mutants. In the presence of 100 mM $Na^+$ new mutants, Mut11 and F130S, grew better than the L776H mutant. Further increase of $Na^+$ to 300 mM inhibited growth of L776H and reduced that of the new mutants. Importantly, the F130S showed a significant higher rate of growth than the double Mut11 mutant. When 1 mM $Cs^+$ was present in the growth media, a dramatic reduction of the growth rates was observed for all mutants, with the exception of the F130S that retained a higher rate of growth.

**Table 1**: Doubling time (h) of 9.3 yeast strain expressing the indicated version of AtHAK5 under different external concentrations of $K^+$, $Na^+$, and $Cs^+$.

| | Treatment | | | | |
|---|---|---|---|---|---|
| **Mutation** | **3 mM $K^+$** | **0.1 mM $K^+$** | **0.1 mM $K^+$ 100 mM $Na^+$** | **0.1 mM $K^+$ 300 mM $Na^+$** | **0.1 mM $K^+$ 1 mM $Cs^+$** |
| WT | $6.6 \pm 0.2^b$ | – | – | – | – |
| L776H | $3.6 \pm 0.2^a$ | $4.3 \pm 0.2^a$ | $9.5 \pm 0.6^b$ | – | $30.3 \pm 0.9^b$ |
| Mut11 | $3.9 \pm 0.3^a$ | $3.7 \pm 0.2^a$ | $5.5 \pm 0.2^a$ | $12.4 \pm 2^b$ | $32.7 \pm 1.5^b$ |
| F130S | $3.4 \pm 0.1^a$ | $3.5 \pm 0.1^a$ | $5.9 \pm 0.4^a$ | $7.3 \pm 1.8^a$ | $6.8 \pm 0.4^a$ |

*Shown are mean values of at least three replicates ± standard error. Different letters indicate significant differences within each column according to Tukey test ($p < 0.01$).*

## Internal $K^+$ and $Na^+$ Concentrations of Yeast Expressing the AtHAK5 Mutants Under Salt Stress

Internal concentrations of $K^+$ and $Na^+$ after 24 h of yeast growth in media with different external concentrations of $K^+$ and $Na^+$ were determined (Figure 2). Under control conditions (3 mM $K^+$), all strains exhibited internal $K^+$ concentrations above 200 nmol $mg^{-1}$ (Figure 2A). At low external $K^+$ (0.1 mM $K^+$), the $K^+$ concentration of yeast expressing WT AtHAK5 dropped to 52.9 ± 17.4 nmol $mg^{-1}$, whereas yeast expressing the mutants retained a high internal $K^+$ (292.4 ± 3.7, 304.8 ± 4.8, and 264.2 ± 3.4 nmol $mg^{-1}$for L776H, Mut11, and F130S respectively). The presence of 100 mM $Na^+$ reduced the internal $K^+$concentrations for WT AtHAK5 to levels below 1 nmol $mg^{-1}$, and no differences among the different mutants were found in their $K^+$ concentrations (188.0.4 ± 3.3, 218.6 ± 1.5, and 146.9 ± 2.6 nmol $mg^{-1}$ for L776H, Mut11, and F130S respectively). Further increase of $Na^+$ to 300 mM gave rise to differential results. Cells expressing the F130S mutation showed the highest $K^+$ concentration (85.6 ± 0.2 nmol $mg^{-1}$) followed by Mut11 (46.4 ± 1.8 nmol $mg^{-1}$) and L776H (38.1 ± 4.0 nmol $mg^{-1}$).

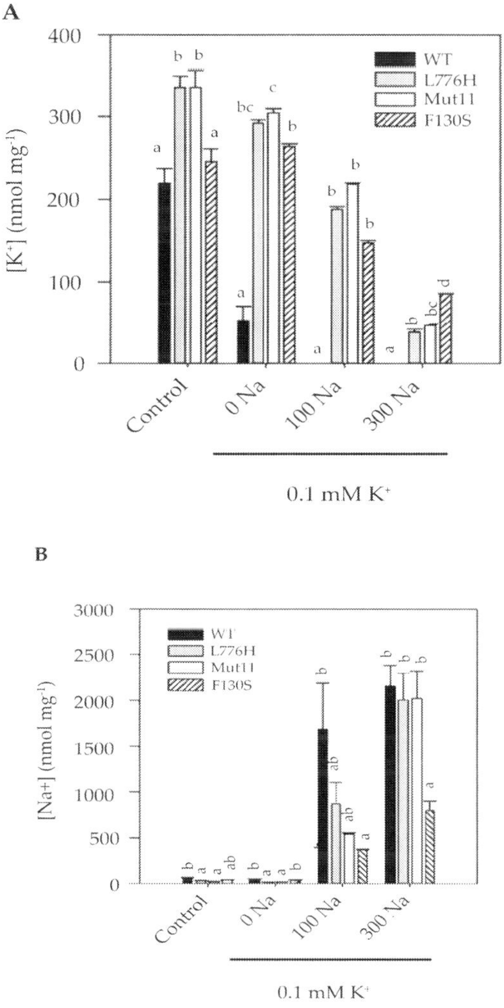


**Figure 2**: Internal $K^+$ and $Na^+$ concentrations of 9.3 yeast cells expressing WT AtHAK5 and mutated AtHAK5 $K^+$ transporters. The same strains described in

Figure 1 were grown for 24 h in minimal AP liquid media supplemented with the indicated $K^+$ and $Na^+$ concentrations. The control medium contained 3 mM $K^+$. After 24 h cells were collected by filtration and their ionic content was acid extracted. The $K^+$ and $Na^+$ concentrations in the acid extracts were determined by atomic emission spectrometry and the internal $K^+$ and Na+ concentrations calculated on a dry weight basis. Shown are averages of internal $K^+$ (A) and $Na^+$ (B) concentrations of at least 6 replicates and error bars denote standard error. Columns with different letters within each ionic treatment are significantly different according to Tukey test ($p < 0.05$).

Internal $Na^+$ concentrations were also determined (Figure 2B). In the absence of added $Na^+$, internal $Na^+$concentrations were very low. Addition of 100 mM $Na^+$ increased internal $Na^+$ in all cases, resulting the $Na^+$ concentrations of cells expressing WT AtHAK5 significantly higher than those expressing the F130S mutant. The presence of 300 mM $Na^+$ produced cells with similar high internal $Na^+$ concentrations in cells expressing WT AtHAK5, L776H, and Mut11. Importantly, cells expressing the F130S mutation showed a significant lower $Na^+$ concentration than the other strains.

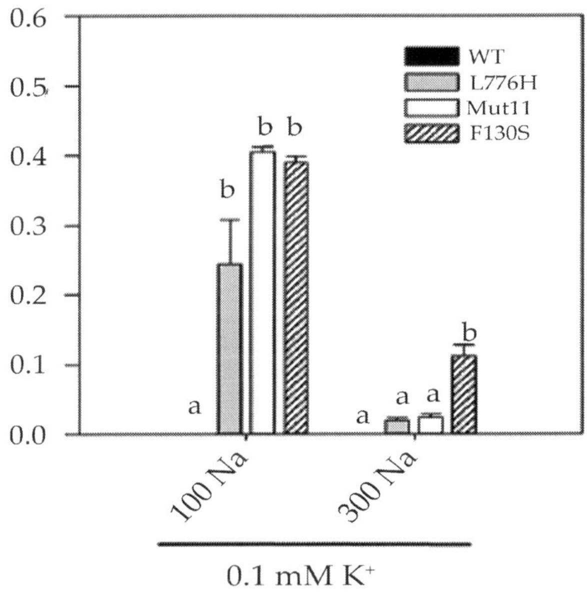


**Figure 3**: Internal $K^+/Na^+$ ratios of 9.3 yeast cells expressing WT AtHAK5 and mutated AtHAK5 $K^+$ transporters. Cells were treated and analyzed as indicated in Figure 2. Shown are the ratios of internal $K^+/Na^+$ concentrations of the cells

grown in the presence of 0.1 mM $K^+$ at 100 or 300 mM $Na^+$. Reported values are the average of at least 6 replicates and error bars denote standard error. Columns with different letters within each ionic treatment are significantly different according to Tukey test ($p < 0.05$).

The internal $K^+/Na^+$ ratios were calculated because this is one of the most important parameters for salt tolerance in yeast as well as in plants (Gaxiola et al., 1992; Maathuis and Amtmann, 1999; Gisbert et al., 2000). In the presence of 100 mM $Na^+$ (0.1 mM $K^+$), cells expressing the AtHAK5 mutants showed similarly higher $K^+/Na^+$ ratios than those expressing WT AtHAK5 (Figure 3). The presence of 300 mM $Na^+$ importantly reduced the $K^+/Na^+$ ratio in all cases. WT AtHAK5 and the L776H and Mut11 mutants showed similar low $K^+/Na^+$ ratios. The F130S mutant showed a significantly higher $K^+/Na^+$ ratio than the rest of the yeast strains.

## Internal $K^+$ and $Cs^+$ Concentrations of Yeast Expressing the Mutants Under $Cs^+$ Stress

Since the F130S mutant showed a remarkable enhanced tolerance to $Cs^+$ (Figure 1 and Table 1), we further examined the internal $K^+$ and $Cs^+$ concentrations in yeast grown for 24 h in liquid medium in the presence of at low $K^+$ (0.1 mM $K^+$) with 1 mM CsCl. The presence of 1 mM $Cs^+$ importantly reduced the internal $K^+$concentrations to 36.6 ± 1.7, 26.7 ± 1.8, 38.4 ± 3.6, and 99.3 ± 9.8 nmol $mg^{-1}$ for WT AtHAK5, L776H, Mut11, and F130S respectively (Figure 4A), far below the internal $K^+$ concentrations without $Cs^+$ (Figure 2A). However, cells expressing the F130S mutant retained a significantly higher $K^+$ concentration than the other yeast strains. When internal $Cs^+$ was determined, it was observed that cells expressing the F130S mutant accumulated a significantly lower $Cs^+$ concentration than the other strains (Figure 4B). As a result, the F130S mutation led to cells with significantly higher $K^+/Cs^+$ ratios (Figure 5), an important parameter for $Cs^+$ tolerance (Hampton et al., 2004).

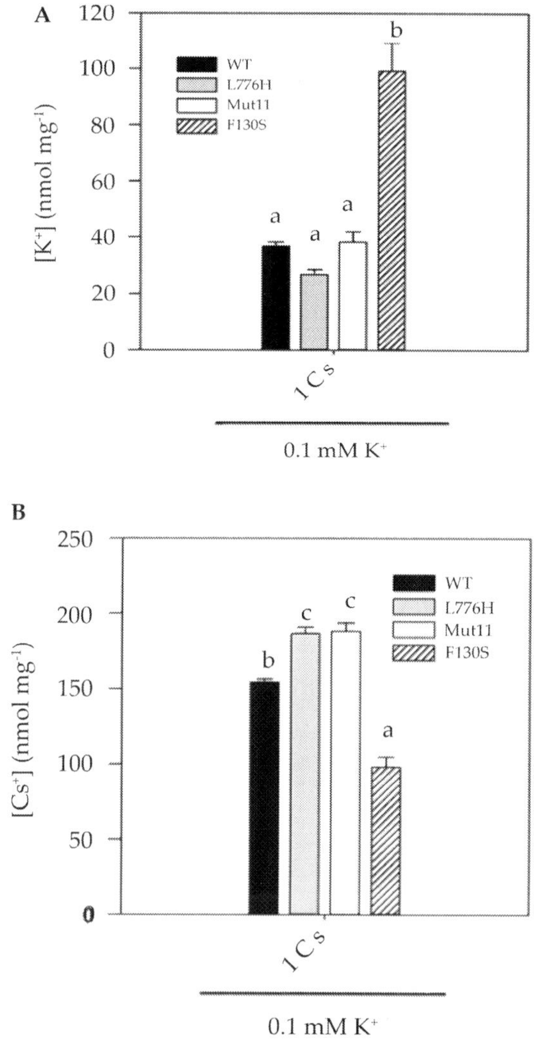


**Figure 4**: Internal $K^+$ and $Cs^+$ concentrations of 9.3 yeast cells expressing WT AtHAK5 and mutated AtHAK5 $K^+$ transporters. The same strains described in Figure 1 were grown for 24 h in minimal AP liquid media supplemented with 0.1 mM $K^+$ and 1 mM $Cs^+$. After 24 h cells were collected by filtration and their ionic content was acid extracted. The $K^+$ and $Cs^+$ concentrations in the acid extracts were determined by atomic emission spectrometry and the internal $K^+$ and Cs+concentrations calculated on a dry weight basis. Shown are averages of internal $K^+$ (A) and $Cs^+$ (B)concentrations of at least 6 replicates and error bars denote standard error. Columns with different letters within each ionic treatment are significant different according to Tukey test ($p < 0.01$).

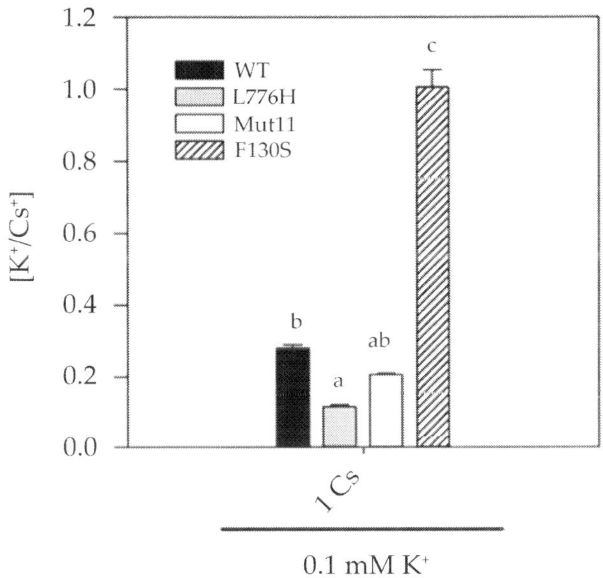


**Figure 5**: Internal $K^+/Cs^+$ ratios of 9.3 yeast cells expressing WT AtHAK5 and mutated AtHAK5 $K^+$ transporters. Cells were treated and analyzed as indicated in Figure 4. Shown are the ratios of internal $K^+/Cs^+$ concentrations of the cells grown in the presence of 0.1 mM $K^+$ and 1 mM $Cs^+$. Reported values are the average of at least 6 replicates and error bars denote standard error. Columns with different letters within each ionic treatment are significantly different according to Tukey test ($p < 0.01$).

## Kinetic Characterization of AtHAK5 Mutants

To gain further insights into the effects that the Mut11 and F130S mutations produced on AtHAK5 functionality, a kinetic characterization was performed. Yeast cells expressing the mutants were grown overnight on AP media supplemented with 3 mM $K^+$ and starved of $K^+$ for 6 h. Then, $Rb^+$ uptake experiments were performed as described in the Materials and Methods Section to calculate the initial rates of $Rb^+$ uptake at different external concentrations. The rates of $Rb^+$ uptake were plotted against the external $Rb^+$ concentration (Figure 6). Clear differences in the uptake kinetics could be observed among the mutants and WT AtHAK5. The data were fitted to Michaelis-Menten equations and $V_{max}$ and $K_m$ values determined (Table 2). All mutants increased the $V_{max}$ for $Rb^+$ with respect to WT AtHAK5. However, whereas mutants containing the L776H mutation (L776H and Mut11) showed high $V_{max}$ values around 10-fold the $V_{max}$ of WT AtHAK5, the F130S only showed a 2-fold increase in this parameter. Mut11 showed a reduction of at least 3-fold in the $K_m$ value with respect to WT AtHAK5. Importantly, the single F130S mutation led to a 100-fold increase in the affinity for $Rb^+$ in comparison to WT AtHAK5. The inhibition of $Rb^+$ uptake by $K^+$, $Na^+$ and $Cs^+$ was also determined and the $K_i$ values calculated (Table 2). As with the $Rb^+$ $K_m$ values, the $K_i$ constants were importantly reduced by the single F130S mutation. In addition, the rates of $K^+$ and $Cs^+$ depletion from a 50 μM solution were determined. It was observed that cells expressing the F130S depleted $K^+$ at higher rates than WT AtHAK5 cells (4.2 ± 0.5 vs 0.6 ± 0.3 nmol $mg^{-1}$ $min^{-1}$) and they depleted $Cs^+$ at similar rates compared to WT AtHAK5 cells (2.6 ± 0.7 nmol $mg^{-1}min^{-1}$).

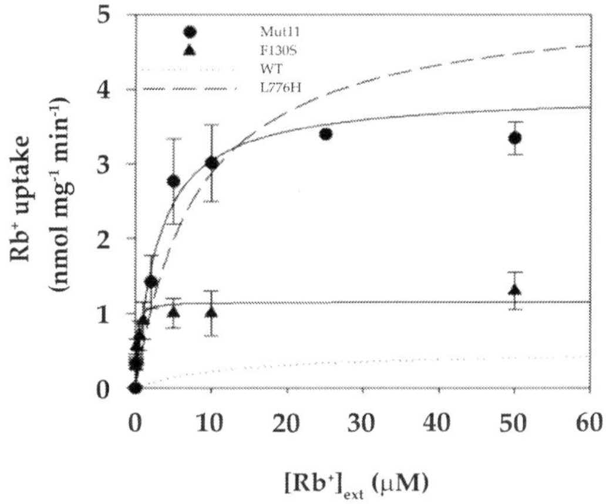


**Figure 6**: Initial rates of Rb+ uptake as a function of external $Rb^+$ in 9.3 cells expressing the AtHAK5 mutants. Cells of the 9.3 strain expressing the Mut11 and F130S mutants were grown overnight on minimal AP media supplemented with 30 mM $K^+$. After 6 h of incubation in $K^+$-free AP media cells were transferred to uptake buffer supplemented with different $Rb^+$ concentrations. The initial rates of $Rb^+$ uptake for each external $Rb^+$ concentration were determined and plotted for Mut11 (closed circles) and F130S (closed triangles). Values were fitted to Michaelis-Menten equations. The predicted curves for WT (dotted line) and the L776H mutation (dashed line) are also shown (Rubio et al., 2000).

**Table 2**: Kinetic parameters of Rb+ uptake mediated by AtHAK5 and its mutant forms in yeast.

| | $V_{max}$ (nmol mg-1 min-1) | $Rb^+$ Km (μM) | $K^+$ Ki (μM) | $Na^+$ Ki(mM) | $Cs^+$ Ki (μM) |
|---|---|---|---|---|---|
| WT* | 0.5 ± 0.3 | 12.6 ± 0.3 | 12 | 10 | 16 |
| L776H* | 5.2 ± 0.3 | 8.1 ± 1.0 | 12 | 10 | 16 |
| Mut11 | 4.1 ± 0.2 | 2.5 ± 0.7 | 5.7 ± 0.1 | 12.3 ± 2.1 | 21.3 ± 0.8 |
| F130S | 1.0 ± 0.2 | 0.1 ± 0.1 | 0.3 ± 0.2 | 0.6 ± 0.3 | 0.3 ± 0.2 |

*From (Rubio et al., 2000).

*Values at the average of at least three replicates ± standard error. $V_{max}$ (nmol $mg^{-1}$ $min^{-1}$) and $K_m$ (μM) values for $Rb^+$ uptake and $K_i$ values for inhibition of $Rb^+$ uptake by $K^+$ (μM), $Na^+$ (mM), and and $Cs^+$ (μM).*

## Detection of AtHAK5 and AtHAK5 Mutant Proteins in Yeast

The AtHAK5 protein of the different mutant versions as well as of WT AtHAK5 were immunologically detected. Proteins from total yeast membranes were extracted from liquid yeast cultures and resolved by SDS-PAGE. The AtHAK5 protein was detected by Western blot with a polyclonal antibody raised against AtHAK5. Yeast cells transformed with the empty plasmids did not produce any detectable signal (Figure 7A) whereas those expressing WT AtHAK5 and AtHAK5 mutants produced a signal band corresponding to the AtHAK5 size (predicted mass of 87.85 kDa). Differences in the amount of AtHAK5 with respect to WT AtHAK5 could be detected among the different yeast strains. L776H increased the amount of AtHAK5 protein, Mut11 showed a similar protein accumulation whereas the single F130S mutant showed a reduced amount of AtHAK5. Coomassie staining of the proteins transferred to the hybridized PVDF membrane showed that all lines contained the same amount of total protein (Figure 7B).

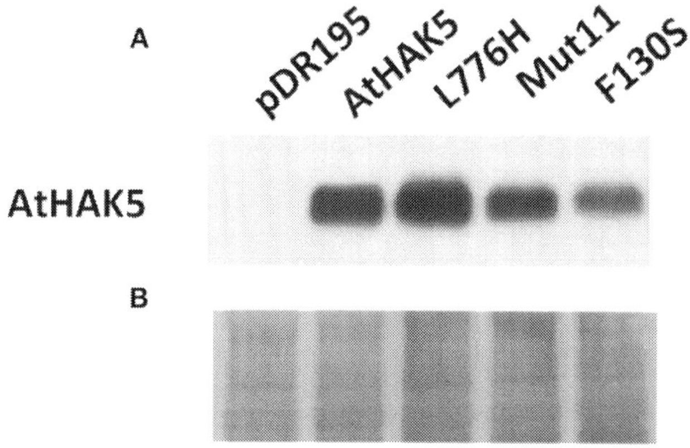


**Figure 7**: Detection of the AtHAK5 WT and mutated proteins expressed in yeast by Western blot. The yeast strains described in Figure 1 were grown overnight in YPD media supplemented with 100 mM $K^+$. Then, proteins of the total membrane fraction were isolated, resolved on polyacrylamide gel and transferred to a PVDF membrane for AtHAK5 detection by Western-bolt. A polyclonal antibody raised in rabbit against the YGYKEDIEEPDEFE peptide of AtHAK5 was used. (A) Shows the immunological detection of AtHAK5 in the different yeast strains. (B) Shows the coomasie blue staining of the hybridized PDVF membrane.

## DISCUSSION

In plants, four families of $K^+$ transport systems have been described: Shaker channels, KCO channels, HKT transporters and HAK transporters (Very and Sentenac, 2003). The members of the first three families bear putative pore forming regions derived from a $K^+$ channel ancestor subunit, region that is present in all $K^+$ channels among kingdoms (Mackinnon, 2003; Very and Sentenac, 2003; Corratgé-Faillie et al., 2010; Benito et al., 2014). This explains the high number of structure-function relationship studies on this type of transporters and the deep knowledge of their functional domains. However, this conserved pore forming region is absent in HAK transporters that, together with the lack of HAK representatives in the animal kingdom, place this family of transporters in a distinct group. This, and the scarce structure-function relationship studies on HAK transporters, make the identification of their functional domains an interesting and almost unexplored field of research. We have here undertaken the identification of mutants in the Arabidopsis AtHAK5 that help the identification of those unknown functional domains.

A previous study showed that introducing into the barley high-affinity $K^+$ transporter HvHAK1 (Santa-María et al., 1997) amino acids of the N-terminus of the barley low-affinity $K^+$ transporter HvHAK2, reduced the affinity of HvHAK1 for $K^+$ (Senn et al., 2001). Another report showed that the V336I and R591C mutants of HvHAK1 improved $K^+$ nutrition and increased $Na^+$ tolerance of yeast cells by increasing the $V_{max}$ of $K^+$ uptake (Mangano et al., 2008). In Arabidopsis, the L776H mutation of AtHAK5 increased the $V_{max}$ of $Rb^+$ uptake without affecting the $K_m$ value, improving growth of yeast cells under low $K^+$(Rubio et al., 2000). In *Physcomitrella patens*, the R443S, AQQP/VQP, L603H and G606E mutations in PpHAK1 decreased the $K_m$ and, with the exception of R443S, increased the $V_{max}$ for $Rb^+$ uptake in yeast. All these PpHAK1 mutants improved yeast growth at low $K^+$ (Garciadeblas et al., 2007). All these studies either used site-directed (Senn et al., 2001) or random mutagenic PCR (Mangano et al., 2008) or took advantage of the occurrence of spontaneous mutations in DNA (Rubio et al., 2000; Garciadeblas et al., 2007). Here, to increase the chances of isolating new mutations with higher transport selectivity, a modified mutagenic PCR approach was successfully used. According to it, different unbalanced mixtures of dNTPs were used in different reactions to favor the occurrence of transitions or transversions. To increase the selective pressure, mutants were selected on the presence of high $Na^+$ in the yeast strain 9.3, which is highly sensitive to $Na^+$ because of deletion of the endogenous $K^+$ uptake systems (TKR1, TRK2) as well the $Na^+$ extruding ATPases (ENA1-ENA4) (Bañuelos et al., 1995). In addition, the L776H mutant instead of WT AtHAK5 was used as a starting material, to generate new mutants that increase $Na^+$tolerance beyond that of the L776H mutant.

Here we show a mutation, F130S, which importantly affects the kinetics of $K^+$ uptake of a HAK5 transporter, leading to improved growth at low $K^+$ as well as enhanced tolerance to $Na^+$ and $Cs^+$ (Table 1and Figure 1). The F130S mutation in the AtHAK5 transporter increases in yeast cells 100-fold the affinity for $Rb^+$ uptake, two-fold the $V_{max}$ of $Rb^+$ uptake (Table 2 and Figure 6) and 7-fold the maximal rate of $K^+$depletion (4.2 ± 0.5 vs 0.6 ± 0.3 nmol $mg^{-1}$ $min^{-1}$ for F103S and WT respectively). In addition, important reductions in the $K_i$ values for $K^+$, $Na^+$, and $Cs^+$ are observed (Table 2). The changes in the kinetic parameters promoted by the F130S mutation lead to yeast cells that were able to maintain higher concentrations of $K^+$ and lower concentrations of $Na^+$ at high external $Na^+$ concentrations (Figure 2), as well as higher concentrations of $K^+$ and lower of $Cs^+$ in the presence of $Cs^+$ (Figure 4). Overall, yeast cells expressing the mutant showed higher $K^+/Na^+$ (Figure 3) and $K^+/Cs^+$ (Figure 5) ratios, crucial parameters for salt and $Cs^+$ tolerances in yeast as well as in plants (Gaxiola et al., 1992; Maathuis and Amtmann, 1999; Gisbert et al., 2000; Hampton et al., 2004). Importantly, the changes mediated by the F130S amino acid substitution occurred without increasing the amount of transporter protein in yeast membranes (Figure 7).

The results summarized above suggest that the F130 residue of the AtHAK5 transporter may affect a domain that is involved in $K^+$ binding and/or transport from the external solution to the cytoplasm. $K^+$, $Na^+$, and $Cs^+$ competitively inhibit HAK5-mediated $Rb^+$ uptake and they are transported through HAK5 transporters (Santa-María et al., 1997; Rubio et al., 2000; Qi et al., 2008). Therefore, the domain involved in $K^+$ binding of this type of transporters is also probably involved in $Rb^+$, $Na^+$, and $Cs^+$ binding. Then, a mutation in the $K^+$ binding site that increases the affinity for $K^+$ is expected to produce a similar effect on the affinities for $Rb^+$, $Na^+$, and $Cs^+$, which is the effect that can be deduced from the data presented here (Table 2). The idea that the F130S mutation is affecting a domain involved in the binding of $K^+$, and therefore the binding of $Na^+$ and $Cs^+$, is further supported by the lower accumulation of $Na^+$ (Figure 2B) and $Cs^+$ (Figure 4B) observed in yeast expressing this mutant in comparison with those expressing WT AtHAK5.

Recently, the topology of a bacterial member of the HAK family of transporters, the *Escherichia coli* Kup system, has been determined (Sato et al., 2014). According to this study, this type of transporters is composed of 12 transmembrane domains with the N- and C-termini facing the cytoplasmic side. The reported structure confirms the initial model proposed for AtHAK5 (Rubio et al., 2000), which was based on algorithmic prediction of transmembrane domains. This structure may be very likely extended to all members of the HAK family (Rodríguez-Navarro, 2000). Although all HAK transporters may have a similar structure they do not show extensive conserved regions. However, a study that included plant, fungal and bacterial representatives found 13 conserved glycine residues (Rodríguez-Navarro, 2000), which have been found later in many other HAK transporters. Three out of these 13 glycine residues are located in the GEGGT**F**ALY domain, highly conserved among HAK transporters and that contains the F130 residue proposed here as important for $K^+$ affinity (Figure 8). Importantly, the selectivity pore of $K^+$channels contains the glycine-rich motif GYG (Heginbotham et al., 1992) and related motifs are found in the proposed pore regions of HKT-TRK transporters (Durell and Guy, 1999; Mäser et al., 2002). Interestingly enough, mutations that affect glycine residues in several $K^+$ transport systems that belong to different families, greatly affect their capacity to mediate $K^+$ transport. The paradigmatic example is the removal of the GYG motif of a $K^+$ channel that modifies its selectivity (Heginbotham et al., 1992). Other examples are the G91S mutation in TaHKT2.1 [former TaHKT1 (Platten et al., 2006)], that abrogates $K^+$permeability (Mäser et al., 2002) or the G61S mutation in HvHAK1 that reduces the affinity and the $V_{max}$ for $K^+$ uptake (Senn et al., 2001). In conclusion, it is tempting to speculate that the GEGGT**F**ALY domain and other glycine-rich regions found along HAK proteins (Rodríguez-Navarro, 2000), are involved in shaping the pore region for $K^+$ binding, a very attractive hypothesis that needs further investigation. Modeling studies together with ion current measurements in heterologous systems should help to further characterize the structure-function relationships of this type of transporters. It is worth noting that the F130 residue studied here is conserved among almost all members of the HAK family identified so far in plants (Nieves-Cordones, personal communication). Importantly, this residue is present in all HAK transporters homologous to the high-affinity HAK5-type transporters identified in different plant species whose sequence has been deposited in the Genebank (Figure S1).

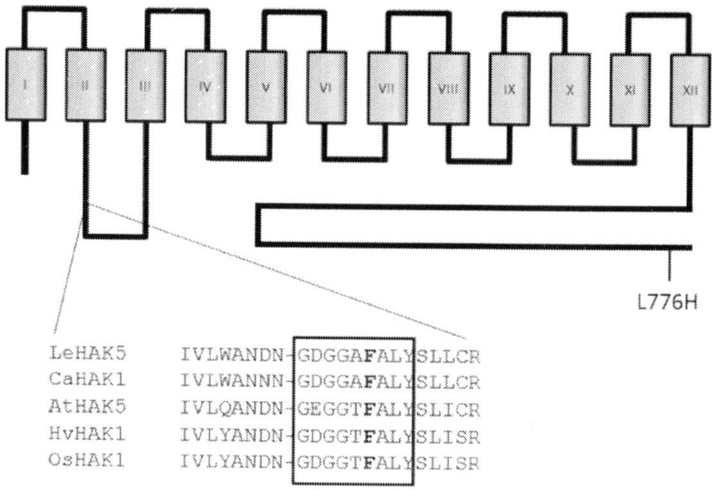


**Figure 8**: Model of AtHAK5 topology and location of mutations. The proposed model for AtHAK5 protein is shown with 12 transmembrane domains and a long cytoplasmic C-terminus. The F130 residue is located between the second and the third transmembrane domains, in a highly glycine-rich conserved region among HAK transporters. Shown is the alignment of this conserved region in representative high-affinity HAK K+ transporters of different species. The location of the L776H is also shown.

Interestingly, HAK5-type transporters can also mediate low-affinity $Na^+$ uptake (Santa-María et al., 1997) and high-affinity $Cs^+$ uptake (Rubio et al., 2000) and therefore, they may be involved in the accumulation of toxic $Na^+$ or $Cs^+$. The contribution of HAK5-type transporters to low-affinity $Na^+$ uptake and $Na^+$accumulation may be low for two main reasons. Firstly, salinity reduces the number of HAK5 transporters in the plasma membrane because high external $Na^+$ reduces the expression of the encoding genes (Nieves-Cordones et al., 2008, 2010; Alemán et al., 2009). Secondly, although yet unidentified at the molecular level, it is proposed that the main pathways for low-affinity $Na^+$ uptake are non-selective cation channels (Amtmann and Sanders, 1999; Tester and Davenport, 2003). Thus, although HAK5 transporters may be crucial to sustain $K^+$ nutrition under $K^+$-deficiency and salinity (Nieves-Cordones et al., 2010), they may not be relevant in $Na^+$ accumulation.

However, HAK5 transporters may play an important role in the accumulation of toxic $Cs^+$. Two main pathways have been proposed for $Cs^+$ uptake in plants. One is a $Ca^{2+}$-sensitive pathway that operates in $K^+$-sufficient plants, and the other is insensitive to $Ca^{2+}$ and it is only present in $K^+$-deprived plants (Heredia et al., 2002; Hampton et al., 2004; Caballero et al., 2012). This $Ca^{2+}$-insensitive pathway is very likely mediated by HAK5 transporters. Importantly, high $Cs^+$ concentrations may lead to $K^+$ starvation and induction of HAK5 genes, and HAK5 transporters may mediate $Cs^+$ uptake even in $K^+$-sufficient plants (Adams et al., 2013). Thus, furnishing plants with mutated versions of HAK5 that reduce the accumulation of $Cs^+$ could contribute to tolerance to this toxic cation. Since $Cs^+$ is a non-essential and rare cation in nature (White and Broadley, 2000), natural selection of HAK5 transporters with a high $K^+$/$Cs^+$discrimination may have not been occurred. Thus, a biotechnological approach based on random mutagenesis and selection by function, should be a promising alternative.

The selection of HAK5 mutants in yeast, as shown here, results instrumental to identify functional domains of the transporter. However, a mutant may promote higher rates of $K^+$ uptake in yeast but not in plants. This can occur for example if the mutation produces a codon which is preferred in yeast and that leads to an increased amount of the protein in the fungus (Sharp et al., 1986; Campbell and Gowri, 1990). In this sense, mutants that show an increase in the affinity (Table 2) without importantly increasing the $V_{max}$ nor the amount of protein in yeast (Figure 7) as the F130S, may provide more promising tools for biotechnological purposes.

Mutated versions of HAK5-type transporters, as the one described here, may provide biotechnological tools to improve $K^+$ nutrition of plants under $K^+$ deficiency or under certain abiotic stress conditions (Niu et al., 1995). Mutated HAK5 transporters can be expressed in plants to increase $K^+$ acquisition from low concentrations. In addition, the expression of these mutants can also provide a means of acquiring $K^+$undcr salinity or $Cs^+$-polluted soils, conditions that are known to induced $K^+$ deficiency (Botella et al., 1997; Carden et al., 2003; Hampton et al., 2004; Shabala et al., 2006) and to reduce HAK-mediated accumulation of toxic $Cs^+$.

## ACKNOWLEDGMENT

This work has been funded by grant AGL2012-33504 from Ministerio de Economía y Competitividad, Spain.

## REFERENCES

1. Adams, E., Abdollahi, P., and Shin, R. (2013). Cesium inhibits plant growth through Jasmonate signaling in *Arabidopsis thaliana. Int. J. Mol. Sci.* 14, 4545–4559. doi: 10.3390/ijms14034545
2. Alemán, F., Nieves-Cordones, M., Martínez, V., and Rubio, F. (2009). Differential regulation of the HAK5 genes encoding the high-affinity $K^+$ transporters of *Thellungiella* halophila and *Arabidopsis thaliana. Environ. Exp. Bot.* 65, 263–269. doi: 10.1016/j.envexpbot.2008.09.011
3. Alemán, F., Nieves-Cordones, M., Martínez, V., and Rubio, F. (2011). Root $K^+$ acquisition in plants: the *Arabidopsis thaliana* model. *Plant Cell Physiol.* 52, 1603–1612. doi: 10.1093/pcp/pcr096
4. Amtmann, A., and Sanders, D. (1999). Mechanisms of $Na^+$ uptake by plant cells. *Adv. Bot. Res.* 29, 75–112. doi: 10.1016/S0065-2296(08)60310-9
5. Bañuelos, M. A., Klein, R. D., Alexander, S. J., and Rodriguez-Navarro, A. (1995). A potassium transporter of the yeast *Schwanniomyces*occidentalis homologous to the Kup system of *Escherichia* coli has a high concentrative capacity. *EMBO J.* 14, 3021–3027.
6. Benito, B., Haro, R., Amtmann, A., Cuin, T. A., and Dreyer, I. (2014). The twins $K^+$ and $Na^+$ in plants. *J. Plant Physiol.* 171, 723–731. doi: 10.1016/j.jplph.2013.10.014
7. Bertl, A., Reid, J. D., Sentenac, H., and Slayman, C. L. (1997). Functional comparison of plant inward-rectifier channels expressed in yeast. *J. Exp. Bot.* 48, 405–413. doi: 10.1093/jxb/48.Special_Issue.405
8. Botella, M. A., Martínez, V., Pardines, J., and Cerdá, A. (1997). Salinity induced potassium deficiency in maize plants. *J. Plant Physiol.* 150, 200–205. doi: 10.1016/S0176-1617(97)80203-9
9. Caballero, F., Botella, M. A., Rubio, L., Fernández, J. A., Martínez, V.,

and Rubio, F. (2012). A $Ca^{2+}$-sensitive system mediates low-affinity $K^+$ uptake in the absence of AKT1 in *Arabidopsis* plants. *Plant Cell Physiol.* 53, 2047–2059. doi: 10.1093/pcp/pcs140

10. Cadwell, R. C., and Joyce, G. F. (1992). Randomization of genes by PCR mutagenesis. *PCR Methods Appl.* 2, 28–33. doi: 10.1101/gr.2.1.28
11. Campbell, W. H., and Gowri, G. (1990). Codon usage in higher plants, green algae, and *Cyanobacteria*. *Plant Physiol.* 92, 1–11. doi: 10.1104/pp.92.1.1
12. Carden, D. E., Walker, D. J., Flowers, T. J., and Miller, A. J. (2003). Single-cell measurements of the contributions of cytosolic $Na^+$ and $K^+$ to salt tolerance. *Plant Physiol.* 131, 676–683. doi: 10.1104/pp.011445
13. Chen, Z., Pottosin, I. I., Cuin, T. A., Fuglsang, A. T., Tester, M., Jha, D., et al. (2007). Root plasma membrane transporters controlling $K^+/Na^+$ homeostasis in salt-stressed barley. *Plant Physiol.* 145, 1714–1725. doi: 10.1104/pp.107.110262
14. Corratgé-Faillie, C., Jabnoune, M., Zimmermann, S., Véry, A. A., Fizames, C., and Sentenac, H. (2010). Potassium and sodium transport in non-animal cells: the Trk/Ktr/HKT transporter family. *Cell. Mol. Life Sci.* 67, 2511–2532. doi: 10.1007/s00018-010-0317-7
15. Durell, S. R., and Guy, H. R. (1999). Structural models of the KtrB, TrkH, and Trk1,2 symporters based on the structure of the KcsA $K^+$channel. *Biophys. J.* 77, 789–807. doi: 10.1016/S0006-3495(99)76932-8
16. Epstein, E. (1966). Dual pattern of ion absorption by plant cells and by plants. *Nature* 212, 1324–1327. doi: 10.1038/2121324a0
17. Garciadeblas, B., Barrero-Gil, J., Benito, B., and Rodriguez-Navarro, A. (2007). Potassium transport systems in the moss physcomitrella patens: pphak1 plants reveal the complexity of potassium uptake. *Plant J.* 52, 1080–1093. doi: 10.1111/j.1365-313X.2007.03297.x
18. Gaxiola, R., De Larrinoa, I. F., Villalba, J. M., and Serrano, R. (1992). A novel and conserved salt-induced protein is an important determinant of salt tolerance in yeast. *EMBO J.* 11, 3157–3164.

19. Gietz, R. D., and Schiestl, R. H. (2007). High-efficiency yeast transformation using the LiAc/SS carrier DNA/PEG method. *Nat. Protoc.* 2, 31–34. doi: 10.1038/nprot.2007.13
20. Gisbert, C., Rus, A. M., Bolarin, M. C., Lopez-Coronado, J. M., Arrillaga, I., Montesinos, C., et al. (2000). The yeast HAL1 gene improves salt tolerance of transgenic tomato. *Plant Physiol.* 123, 393–402. doi: 10.1104/pp.123.1.393
21. Hampton, C. R., Bowen, H. C., Broadley, M. R., Hammond, J. P., Mead, A., Payne, K. A., et al. (2004). Cesium toxicity in *Arabidopsis*. *Plant Physiol.* 136, 3824–3837. doi: 10.1104/pp.104.046672
22. Hanahan, D. (1985). "Techniques for transformation of *E. coli*," in *DNA Cloning: A Practical Approach*, Vol. 1, ed D. M. Glover (McLean, VA: IRL Press), 109.
23. Heginbotham, L., Abramson, T., and Mackinnon, R. (1992). A functional connection between the pores of distantly related ion channels as revealed by mutant $K^+$ channels. *Science* 258, 1152–1155. doi: 10.1126/science.1279807
24. Heredia, M. A., Zapico, R., Garcìa-Sánchez, M. A. J., and Fernández, J. A. (2002). Effect of calcium, sodium and pH on uptake and accumulation of radiocesium by riccia fluitans. *Aquatic Bot.* 74, 245–256. doi: 10.1016/S0304-3770(02)00107-9
25. Maathuis, F. J. M. (2009). Physiological functions of mineral macronutrients. *Curr. Opin. Plant Biol.* 12, 250–258. doi: 10.1016/j.pbi.2009.04.003
26. Maathuis, F. J. M., and Amtmann, A. (1999). $K^+$ Nutrition and $Na^+$ Toxicity: the basis of cellular $K^+/Na^+$ ratios. *Ann. Bot.* 84, 123–133. doi: 10.1006/anbo.1999.0912
27. Mackinnon, R. (2003). Potassium channels. *FEBS Lett.* 555, 62–65. doi: 10.1016/S0014-5793(03)01104-9
28. Mangano, S., Silberstein, S., and Santa-Maria, G. E. (2008). Point mutations in the barley HvHAK1 potassium transporter lead to improved $K^+$-nutrition and enhanced resistance to salt stress. *FEBS Lett.* 582, 3922–3928. doi: 10.1016/j.febslet.2008.10.036
29. Marschner, P. (2012). *Marschner's Mineral Nutrition of Higher Plants, 3rd Edn.* San Diego, CA: Academic Press.

30. Martínez-Cordero, M. A., Martinez, V., and Rubio, F. (2005). High-affinity $K^+$ uptake in pepper plants. *J. Exp. Bot.* 56, 1553–1562. doi: 10.1093/jxb/eri150

31. Martínez-Cordero, M. A., Martínez, V., and Rubio, F. (2004). Cloning and functional characterization of the high-affinity $K^+$ transporter HAK1 of pepper. *Plant Mol. Biol.* 56, 413–421. doi: 10.1007/s11103-004-3845-4

32. Mäser, P., Hosoo, Y., Goshima, S., Horie, T., Eckelman, B., Yamada, K., et al. (2002). Glycine residues in potassium channel-like selectivity filters determine potassium selectivity in four-loop-per-subunit HKT transporters from plants. *Proc. Natl. Acad. Sci. U.S.A.* 99, 6428–6433. doi: 10.1073/pnas.082123799

33. Muhlrad, D., Hunter, R., and Parker, R. (1992). A rapid method for localized mutagenesis of yeast. *Yeast* 8, 79–82. doi: 10.1002/yea.320080202

34. Munns, R., and Tester, M. (2008). Mechanisms of salinity tolerance. *Ann. Rev. Plant Biol.* 59, 651–681. doi: 10.1146/annurev.arplant.59.032607.092911

35. Nieves-Cordones, M., Aleman, F., Martinez, V., and Rubio, F. (2010). The *Arabidopsis thaliana* HAK5 $K^+$ transporter is required for plant growth and K+ acquisition from low K+ solutions under saline conditions. *Mol. Plant* 3, 326–333. doi: 10.1093/mp/ssp102

36. Nieves-Cordones, M., Alemán, F., Martínez, V., and Rubio, F. (2014). $K^+$ uptake in plant roots. the systems involved, their regulation and parallels in other organisms. *J. Plant Physiol.* 171, 688–695. doi: 10.1016/j.jplph.2013.09.021

37. Nieves-Cordones, M., Martinez-Cordero, M. A., Martinez, V., and Rubio, F. (2007). An $NH^+_4$-sensitive component dominates high-affinity $K^+$ uptake in tomato plants. *Plant Sci.* 172, 273–280. doi: 10.1016/j.plantsci.2006.09.003

38. Nieves-Cordones, M., Miller, A., Alemán, F., Martínez, V., and Rubio, F. (2008). A putative role for the plasma membrane potential in the control of the expression of the gene encoding the tomato high-affinity potassium transporter HAK5. *Plant Mol. Biol.* 68, 521–532. doi: 10.1007/s11103-008-9388-3

39. Niu, X., Bressan, R. A., Hasegawa, P. M., and Pardo, J. M. (1995).

Ion homeostasis in NaCl stress environments. *Plant Physiol.* 109, 735–742.

40. Platten, J. D., Cotsaftis, O., Berthomieu, P., Bohnert, H., Davenport, R. J., Fairbairn, D. J., et al. (2006). Nomenclature for HKT transporters, key determinants of plant salinity tolerance. *Trends Plant Sci.* 11, 372–374. doi: 10.1016/j.tplants.2006.06.001
41. Pyo, Y. J., Gierth, M., Schroeder, J. I., and Cho, M. H. (2010). High-Affinity $K^+$ transport in *Arabidopsis*: AtHAK5 and AKT1 are vital for seedling establishment and postgermination growth under low-potassium conditions. *Plant Physiol.* 153, 863–875. doi: 10.1104/pp.110.154369
42. Qi, Z., Hampton, C. R., Shin, R., Barkla, B. J., White, P. J., and Schachtman, D. P. (2008). The high affinity K+ transporter AtHAK5 plays a physiological role in planta at very low K+ concentrations and provides a caesium uptake pathway in *Arabidopsis*. *J. Exp. Bot.* 59, 595–607. doi: 10.1093/jxb/erm330
43. Rengel, Z., and Damon, P. M. (2008). Crops and genotypes differ in efficiency of potassium uptake and use. *Physiol. Plant.* 133, 624–636. doi: 10.1111/j.1399-3054.2008.01079.x
44. Rentsch, D., Laloi, M., Rouhara, I., Schmelzer, E., Delrot, S., and Frommer, W. B. (1995). NTr1 encodes a high affinity oligopeptide transporter in *Arabidopsis*. *FEBS Lett.* 370, 264–268. doi: 10.1016/0014-5793(95)00853-2
45. Rodríguez-Navarro, A. (2000). Potassium transport in fungi and plants. *Biochim. Biophys. Acta* 1469, 1–30. doi: 10.1016/S0304-4157(99)00013-1
46. Rodríguez-Navarro, A., and Ramos, J. (1984). Dual system for potassium transport in *Saccharomyces cerevisiae*. *J. Bacteriol.* 159, 940–945.
47. Römheld, V., and Kirkby, E. (2010). Research on potassium in agriculture: needs and prospects. *Plant Soil* 335, 155–180. doi: 10.1007/s11104-010-0520-1
48. Rubio, F., Alemán, F., Nieves-Cordones, M., and Martínez, V. (2010). Studies on *Arabidopsis athak5*, *atakt1* double mutants disclose the range of concentrations at which AtHAK5, AtAKT1 and unknown systems mediate $K^+$ uptake. *Physiol. Plant.* 139, 220–

228. doi: 10.1111/j.1399-3054.2010.01354.x

49. Rubio, F., Nieves-Cordones, M., Alemán, F., and Martinez, V. (2008). Relative contribution of AtHAK5 and AtAKT1 to $K^+$ uptake in the high-affinity range of concentrations. *Physiol. Plant.* 134, 598–608. doi: 10.1111/j.1399-3054.2008.01168.x

50. Rubio, F., Santa-Maria, G. E., and Rodríguez-Navarro, A. (2000). Cloning of *Arabidopsis* and barley cDNAs encoding HAK potassium transporters in root and shoot cells. *Physiol. Plant.* 109, 34–43. doi: 10.1034/j.1399-3054.2000.100106.x

51. Rubio, F., Schwarz, M., Gassmann, W., and Schroeder, J. I. (1999). Genetic selection of mutations in the high affinity $K^+$ transporter HKT1 that define functions of a loop site for reduced $Na^+$ permeability and increased $Na^+$ tolerance. *J. Biol. Chem.* 274, 6839–6847. doi: 10.1074/jbc.274.11.6839

52. Sambrook, J., and Russell, D. W. (2001). *Molecular Cloning: A Laboratory Manual*. New York, NY: Cold Spring Harbor Laboratory Press.

53. Santa-María, G. E., Rubio, F., Dubcovsky, J., and Rodriguez-Navarro, A. (1997). The HAK1 gene of barley is a member of a large gene family and encodes a high-affinity potassium transporter. *Plant Cell* 9, 2281–2289. doi: 10.1105/tpc.9.12.2281

54. Sato, Y., Nanatani, K., Hamamoto, S., Shimizu, M., Takahashi, M., Tabuchi-Kobayashi, M., et al. (2014). Defining membrane spanning domains and crucial membrane-localized acidic amino acid residues for K+ transport of a Kup/HAK/KT-type escherichia coli potassium transporter. *J. Biochem.* 155, 315–323. doi: 10.1093/jb/mvu007

55. Senn, M. E., Rubio, F., Bañuelos, M. A., and Rodriguez-Navarro, A. (2001). Comparative functional features of plant potassium HvHAK1 and HvHAK2 transporters. *J. Biol. Chem.* 276, 44563–44569. doi: 10.1074/jbc.M108129200

56. Serrano, R. (1988). $H^+$-ATPase from plasma membranes of *Saccharomyces cerevisiae* and avena sativa roots: purification and reconstitution. *Methods Enzymol.* 157, 533–544. doi: 10.1016/0076-6879(88)57102-1

57. Shabala, S., and Cuin, T. A. (2008). Potassium transport and plant

salt tolerance. *Physiol. Plant.* 133, 651–669. doi: 10.1111/j.1399-3054.2007.01008.x

58. Shabala, S., Demidchik, V., Shabala, L., Cuin, T. A., Smith, S. J., Miller, A. J., et al. (2006). Extracellular $Ca^{2+}$ ameliorates NaCl-induced K+ loss from *Arabidopsis* root and leaf cells by controlling plasma membrane K+-permeable channels. *Plant Physiol.* 141, 1653–1665. doi: 10.1104/pp.106.082388
59. Sharp, P. M., Tuohy, T. M. F., and Mosurski, K. R. (1986). Codon usage in yeast: cluster analysis clearly differentiates highly and lowly expressed genes. *Nucleic Acids Res.* 14, 5125–5143. doi: 10.1093/nar/14.13.5125
60. Sherman, F. (1991). Getting started with yeast. *Methods Enzymol.* 194, 3–21. doi: 10.1016/0076-6879(91)94004-V
61. Tester, M., and Davenport, R. J. (2003). $Na^+$ tolerance and $Na^+$ transport in higher plants. *Ann. Bot.* 91, 503–527. doi: 10.1093/aob/mcg058
62. Very, A. A., and Sentenac, H. (2003). Molecular mechanisms and regulation of $K^+$ transport in higher plants *Ann. Rev. Plant Biol.* 54, 575–603. doi: 10.1146/annurev.arplant.54.031902.134831
63. Volkov, V., and Amtmann, A. (2006). *Thellungiella halophila*, a salt-tolerant relative of *Arabidopsis thaliana*, has specific root ion-channel features supporting K+/Na+ homeostasis under salinity stress. *Plant J.* 48, 342–353. doi: 10.1111/j.1365-313X.2006.02876.x
64. White, P. J., and Broadley, M. R. (2000). Mechanisms of caesium uptake by plants. *New Phytol.* 147, 241–256. doi: 10.1046/j.1469-8137.2000.00704.x
65. White, P. J., and Karley, A. J. (2010). "Potassium," in *Cell Biology of Metals and Nutrients*, ed R. M. R. R. Hell (Heidelberg: Springer), 199–224.
66. Wong, T. S., Roccatano, D., and Schwaneberg, U. (2007). Are transversion mutations better? a mutagenesis assistant program analysis on P450 BM-3 heme domain. *Biotechnol. J.* 2, 133–142. doi: 10.1002/biot.200600201

CHAPTER

# 6

# THE ROLE OF K+ CHANNELS IN UPTAKE AND REDISTRIBUTION OF POTASSIUM IN THE MODEL PLANT ARABIDOPSIS THALIANA

Tripti Sharma[1,2], Ingo Dreyer[3] and Janin Riedelsberger[1,2]

[1]Molecular Biology, Institute for Biochemistry and Biology, University of Potsdam, Potsdam, Germany

[2]IMPRS-PMPG, Max-Planck Institute of Molecular Plant Physiology, Potsdam, Germany

[3]Centro de Biotecnología y Genómica de Plantas, Universidad Politécnica de Madrid, Madrid, Spain

Potassium ($K^+$) is inevitable for plant growth and development. It plays a crucial role in the regulation of enzyme activities, in adjusting the electrical membrane potential and the cellular turgor, in regulating cellular homeostasis and in the stabilization of protein synthesis. Uptake of $K^+$ from the soil and its transport to growing organs is essential for a healthy plant development. Uptake and allocation of $K^+$ are performed by $K^+$ channels and transporters belonging to different protein families. In this review we summarize the knowledge on the versatile physiological roles of plant $K^+$ channels and their behavior under stress conditions in the model plant *Arabidopsis thaliana*.

## INTRODUCTION

Potassium ($K^+$) is essential for growth and development of an organism. It is involved in various important cellular processes, like stabilization of protein synthesis, activation of enzymes, neutralization of negative charges on proteins and many more. In addition to the above mentioned tasks, in plants it is a key player in osmotic processes contributing to cellular turgor, cell elongation, translocation of photosynthates, maintenance of cytosolic pH homeostasis, and the setting of the membrane potential along with the proton motive force (Maathuis, 2009; Marschner, 2012). All these functions justify it being the most abundant inorganic cation in plants, contributing to up to 10% of their dry mass (Leigh and Wyn Jones, 1984).

Potassium is a major factor in resistance to drought, salinity, and fungal diseases (Amtmann et al., 2008). This explains why it is of crucial importance in agriculture affecting crop yield. For performing the tasks explained above, plants require potassium concentrations ranging between 100–200 mM in the cytoplasm (Wyn Jones and Pollard, 1983). In contrast, concentration of potassium in soil (10–100 μM) is 3–4 orders of magnitude lower (Schroeder et al., 1994). Therefore, a plant has to invest energy for the uptake of $K^+$and its distribution throughout the plant.

The transport of potassium is accomplished by a variety of transporter proteins. In the plant model organism *Arabidopsis thaliana* a total of 71 $K^+$ channels and transporters have already been identified (Mäser et al., 2001; Véry and Sentenac, 2003; Amtmann et al., 2004; Wang and Wu, 2010). They have been categorized into six different gene families, comprising of three channel families and three transporter families (KUP/HAK/KT, HKT, and CPA families; Gierth and Mäser, 2007; Chanroj et al., 2012; Gomez-Porras et al., 2012).

The three identified families of $K^+$ channels are *Shaker*, Tandem-Pore $K^+$ (TPK) and $K_{ir}$-like channels. Recent phylogenetic data, however, evidenced that $K_{ir}$-like channels in fact belong to the TPK family and originated by evolutionarily recent gene duplication and partial deletion events (Marcel et al., 2010; Voelker et al., 2010; Gomez-Porras et al., 2012). We therefore do not consider $K_{ir}$-like channels as a separate family anymore. $K^+$ channels are active as multimeric proteins composed of two or four α-subunits, which are characterized by the presence of either one or two pore (P) domains. In the functional multimeric protein, four P domains are associated to form part of the conduction pathway, including its selectivity filter. $K^+$ selective channels have the hallmark motif TXGYGD/E in their P domains (Lebaudy et al., 2007; Table 1, Figure 1).

**Table 1**: Overview of location, function and regulation of plant $K^+$ channels.

| | Channel | Other names | Voltage dependence | Location | Function | Direct and indirect effectors | | Heteromerization with | References |
|---|---|---|---|---|---|---|---|---|---|
| | | | | | | positive | negative | | |
| Shaker-like | KAT1 | – | $K_{in}$ | PM; GC (leaves, petioles) | involved in stomatal opening | $pH_{ext}$, ATP, 14-3-3, Auxin, | cGMP, CDPK and/or other kinase, ABA | KAT2, AtKC1 | Anderson et al., 1992; Schachtman et al., 1992; Hedrich et al., 1995; Hoshi, 1995; Nakamura et al., 1995; Véry et al., 1995; Ichida et al., 1997; Li et al., 1998; Hoth and Hedrich, 1999; Berkowitz et al., 2000; Kwak et al., 2001; Pilot et al., 2001; Szyroki et al., 2001; Philippar et al., 2004; Sutter et al., 2006, 2007; Sottocornola et al., 2006, 2008; Lebaudy et al., 2010; Reuff et al., 2010; Sato et al., 2010; Gonzalez et al., 2012 |
| Shaker-like | KAT2 | – | $K_{in}$ | PM; GC (leaves, petioles, stem); phloem (aerial parts) | involved in stomatal opening | $pH_{ext}$, Auxin | – | KAT1, AKT2, AtKC1 | Pilot et al., 2001; Szyroki et al., 2001; Philippar et al., 2004; Ivashikina et al., 2005; Xicluna et al., 2007; Lebaudy et al., 2008a, 2010 |
| Shaker-like | AKT1 | KT1 | $K_{in}$ | PM, root, GC, hydathodes | $K^+$ uptake from soil, stomatal opening? | CBL-CIPK | PP2C, CBL10, AtKC1 (SNARE) | AtKC1 | Sentenac et al., 1992; Basset et al., 1995; Lagarde et al., 1996; Hirsch et al., 1998; Spalding et al., 1999; Dennison et al., 2001; Szyroki et al., 2001; Xu et al., 2006; Li et al., 2006; Lee et al., 2007; Rubio et al., 2008, 2010; Honsbein et al., 2009; Geiger et al., 2009; Pyo et al., 2010; Wang and Wu, 2010; Lan et al., 2011; Ren et al., 2013 |
| Shaker-like | AKT5 | KT5 | $K_{in}$ | pollen | – | – | – | – | Lacombe et al., 2000b; Mouline et al., 2002 |
| Shaker-like | SPIK | AKT6 | $K_{in}$ | PM, pollen | pollen tube development | $pH_{ext}$ | $Ca^{2+}_{cyt}$ and CDPK11/24 | – | Mouline et al., 2002; Zhao et al., 2013 |
| Shaker-like | AKT2 | AKT2/3 AKT3 KT2/3 | $K_{weak}$ | PM; phloem; GC | $K^+$ battery (ensures phloem loading under energy limiting conditions); stomatal movement | ABA, kinase? | $pH_{cyt/ext}$, $Ca^{2+1}_{ext}$ | KAT2, AtKC1 | Marten et al., 1999; Deeken et al., 2000, 2002; Lacombe et al., 2000b; Dreyer et al., 2001; Szyroki et al., 2001; Chérel et al., 2002; Ivashikina et al., 2005; Michard et al., 2005a,b; Xicluna et al., 2007; Gajdanowicz et al., 2011; Held et al., 2011; Sandmann et al., 2011 |
| Shaker-like | AtKC1 | KAT3 AtLKT1 KC1 | regulatory subunit | PM; root; GC, hydathodes, trichomes | regulation of AKT1, and other $K_{in}$ channels? | – | – | AKT1, AKT2, KAT1, KAT2 | Szyroki et al., 2001; Reintanz et al., 2002; Pilot et al., 2003; Duby et al., 2008; Geiger et al., 2009; Honsbein et al., 2009; Grefen et al., 2010; Wang and Wu, 2010; Jeanguenin et al., 2011 |
| Shaker-like | SKOR | – | $K_{out}$ | PM; pericycle and xylem parenchyma; pollen | $K^+$ loading to xylem | $H_2O_2$ | external $K^+$, $pH_{cyt/ext}$, ABA | GORK | Gaymard et al., 1998; Lacombe et al., 2000a; Mouline et al., 2002; Pilot et al., 2003; Dreyer et al., 2004; Johansson et al., 2006; Garcia-Mata et al., 2010 |
| Shaker-like | GORK | – | $K_{out}$ | PM; root, GC | involved in stomatal closure | $H_2O_2$ | external $K^+$, $pH_{cyt/ext}$, ABA | SKOR | Ache et al., 2000; Szyroki et al., 2001; Ivashikina et al., 2001; Hosy et al., 2003; Becker et al., 2003; Dreyer et al., 2004 |
| TPK | TPK1 | KCO1 | Voltage independent | VM; [illegible] flower | $K^+$ homeostasis; osmoregulation; stomatal movement; germination | $pH_{cyt}$, GRF6 (14-3-3), $Ca^{2+}_{cyt}$ | – | ? | Deeken et al., 2003; Philippar et al., 2003; Gobert et al., 2007 |
| TPK | TPK2 | KCO2 | – | VM; root; leaves; flower | – | – | – | ? | Deeken et al., 2003; Philippar et al., 2003; Voelker et al., 2006 |
| TPK | TPK3 | KCO6 | – | VM; root, flower; seeds; senescent leaves | – | – | – | ? | Deeken et al., 2003; Philippar et al., 2003; Voelker et al., 2006 |
| TPK | TPK4 | KCO4 | Voltage independent | PM; pollen | potassium homeostasis; membrane voltage control of the growing pollen tube | heat | $pH_{cyt}$, $Ca^{2+}_{ext}$ | ? | Becker et al., 2004 |
| TPK | TPK5 | KCO5 | – | VM; senescent leaves; flower | – | – | – | ? | Deeken et al., 2003; Philippar et al., 2003; Voelker et al., 2006 |
| TPK | KCO3[2] | – | – | VM; leaves; roots; flower; stem; hydathodes | – | – | – | – | Deeken et al., 2003; Philippar et al., 2003; Voelker et al., 2006; Rocchetti et al., 2012 |

*All pH data refer to pH acidification. Abbreviations: $K_{in}$, $K_{weak}$, $K_{out}$, inwardly, weakly, outwardly rectifying potassium channel; PM, plasma membrane; VM, vacuolar membrane; GC, guard cells; CDPK, $Ca^{2+}$ dependent protein kinase; ABA, abscisic acid; CBL, calcineurin B like calcium sensors; CIPK, CBL interacting protein kinase; PP2C, 2C-type protein phosphatase; cyt, cytosolic; ext, extracellular.*

*[1] only inward currents are affected.*

*[2] KCO3 very likely evolved from TPK2 by gene duplication and a subsequent deletion of one of the two pore regions*

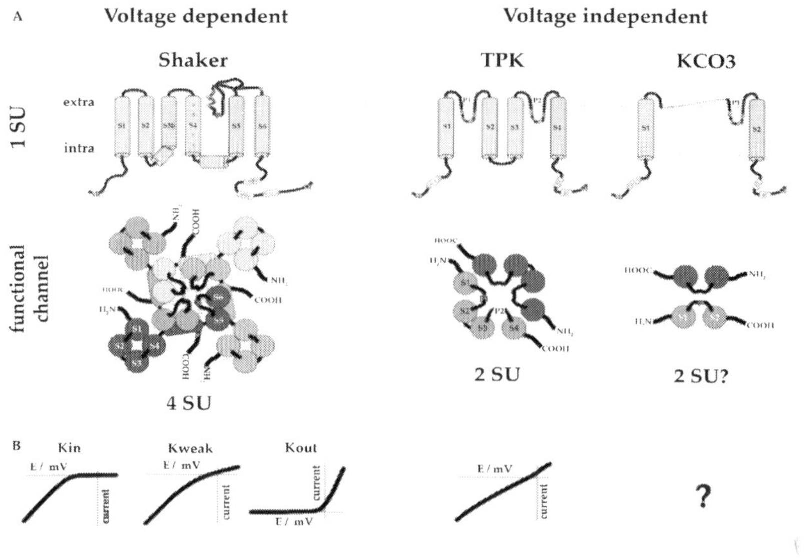


**Figure 1**: Structure and function of $K^+$ channel families in plants. The two plant K+ channel families vary in (A) structure and (B) function. Shaker channels form the most versatile family among plant K+ channels. Nine members segregate into inwardly, outwardly and weakly rectifying channels. Functional channels are tetramers and operate in a voltage dependent manner. One subunit consists of six transmembrane domains (S1–S6) and one pore domain (P). The fourth transmembrane region S4 is rich in positively charged amino acids and acts together with S1, S2, and S3 as voltage sensor. Five TPK channels have been identified. One subunit contains two pore domains (P1 and P2) and two subunits are sufficient to form a functional channel. TPKs act in a largely voltage independent manner and exhibit leak like currents. KCO3 was initially classified as a Kir-like channel showing two transmembrane regions and one pore domain. In fact, "plant Kir-like channels" originate from TPKs by partial deletion of one selectivity-filter and two transmembrane domains. In line with this notion, only stable dimers have been detected. A K+ transport function has not been shown for these truncated channels. Abbreviations: extra, extracellular side; intra, intracellular side; SU, subunit; +, positively charged amino acids; cNBD, cyclic nucleotide binding domain; anky, ankyrin repeat domain; K(T)/HA, acidic domain; EF, EF hand domain.

In 1992, AKT1 and KAT1, two inward rectifying channels from *Arabidopsis* were identified by functional complementation of yeast strains defective in potassium uptake. These two members of the *Shaker*-like channel family were the first cloned plant potassium channels

(Anderson et al., 1992; Schachtman et al., 1992; Sentenac et al., 1992).

In 1997, a first member of the TPK channel family was identified by *in silico* approaches, utilizing the *Arabidopsis* gene sequencing program. TPK channels are the plant counterparts of animal Tandem Pore (TWIK-like) channels (Czempinski et al., 1997). While searching for TPK1-related sequences in genome sequence database, KCO3 was identified and was thought to be structurally similar to animal potassium inward rectifying channels (Czempinski et al., 1999) leading to its initial classification into a separate family of plant $K_{ir}$-like channels.

# VOLTAGE INDEPENDENT K+ CHANNELS IN ARABIDOPSIS

## Tandem Pore Potassium Channels

The Tandem-Pore $K^+$ (TPK) channel family comprises six members (TPK1-TPK5 and KCO3, see also below for this special case) in the model plant *Arabidopsis thaliana*. TPK homologues were identified in higher plants and green algae (Voelker et al., 2010; Gomez-Porras et al., 2012). A phylogenetic analysis has shown that plant TPK channels are divided into two subfamilies: TPK1 belongs to one and TPK2, TPK3, TPK4, and TPK5 to the second subfamily. This sub division in families indicates a common ancestral origin of the channels TPK2, TPK3, TPK4, and TPK5; a hypothesis that was further supported by the analysis of chromosome segment duplication in the *Arabidopsis* genome (Marcel et al., 2010; Voelker et al., 2010; Gomez-Porras et al., 2012).

The first TPK channel (AtTPK1) was cloned via an *A. thaliana* EST database search for the conserved $K^+$channel pore domain motif TXGYGD (Czempinski et al., 1997). TPKs show a TM-P-TM-TM-P-TM structure with a duplicated transmembrane-pore-transmembrane module (Figure 1A). In general, these channels contain one or two $Ca^{2+}$-binding EF hands in the cytosolic C-terminal part and binding sites for 14-3-3 proteins in the cytosolic N-terminal part, as well as a putative N-glycosylation site in the luminal loop between the pore domain and the second transmembrane domain.

Functional TPK channels are built of two of such subunits and exist as dimers (Maitrejean et al., 2011). They show a high $Ca^{2+}$ dependency, which might be important for channel regulation (Latz et al., 2007a). TPK channels have been localized in the vacuolar membrane (Czempinski et al., 2002; Schönknecht et al., 2002). One exception is TPK4, which has been reported to localize in the plasma membrane (Becker et al., 2004; Dunkel et al., 2008). TPK4 shares 85% similarity (53% identity) with TPK5 but lacks the regulatory domains and the 14-3-3 protein interaction motif. It might thus be speculated that TPK4 evolved from TPK5 and subsequently underwent truncation events. Another exception from exclusive vacuolar localization might be TPK3. In Western blots TPK3 was also identified in thylakoid membranes (Zanetti et al., 2010) raising the question whether TPKs may have multiple subcellular locations.

Expression analysis of TPKs through quantitative real-time PCR experiments evidenced their presence in different plant tissues like roots, leaves and flowers (Deeken et al., 2003; Voelker et al., 2010). Among all TPKs, *TPK1* showed the highest expression levels in all tissues analysed, followed by *TPK3* and *TPK5*. Expression levels of *TPK2* and *TPK4* were very low. Elevated levels of *TPK2* transcripts were detected in stamen and pollen. *TPK3* transcript levels were more abundant in petals, stamen, seeds and senescent leaves.

## Assembly Status of Tandem-pore $K^+$ Channels

Promoter-reporter gene studies and qRT-PCR experiments revealed overlapping expression patterns for members of the TPK/KCO3 channel family (Czempinski et al., 2002; Deeken et al., 2003; Philippar et al., 2003; Becker et al., 2004; Voelker et al., 2006). Expression of *TPK1* overlaps with that of *TPK3* in root tips and with that of *TPK5* and *KCO3* in vascular tissues. Additionally, *TPK1, TPK2, TPK3*, and *TPK4* express in pollen. The overlapping expression patterns and their common localization in the tonoplast propose that heteromeric channel subunit combination might occur under different developmental stages or physiological conditions (Latz et al., 2007a).

Dimerization of TPK channels has been shown experimentally by using velocity sucrose gradient centrifugation of leaf homogenates expressing TPK1-GFP. This confirms the contribution of four pore domains to the $K^+$ selectivity filter of the TPK1 channel (Maitrejean et al., 2011). Using the same technique, AtKCO3 and AtKCO3-GFP have been observed to exist as dimers, too. These channels would thus have only two pore domains in a dimerized state (Figure 1A), which is not considered to be sufficient for an active, $K^+$-selective channel (Rocchetti et al., 2012).

With the aim of studying the assembly status of TPK/KCO family members, various experiments have been performed employing techniques like FRET and BiFC (split-YFP). Results from these approaches indicated the existence of homomeric TPK/KCO3 channels, as e.g., in the case of TPK1 or TPK5 (Voelker et al., 2006). However, so far no evidence for heteromeric channel formation could be provided. Nevertheless, there are neither convincing data ruling out this possibility. Thus it cannot be excluded that *in vivo* heteromeric channel formation might occur under different developmental and physiological conditions.

## *Localization of Tandem-pore $K^+$ Channels*

In a first approach to detect the subcellular localization, the TPK1 channel has been stably over-expressed in tobacco BY-2 cells. After protein fractionation with a sucrose gradient, this $K^+$ channel was found to co-fractionate with tonoplast markers, giving a first clue of its localization on the vacuolar membrane (Czempinski et al., 2002). Further localization studies were performed by creating GFP fusion constructs followed by their transient expression in *A. thaliana* protoplasts. Such experiments demonstrated vacuolar localization of TPK1, TPK2, TPK3, and TPK5 (Voelker et al., 2006). In contrast, when a TPK4:GFP fusion construct was expressed in onion epidermal cells, it was found to localize partially in the plasma membrane. A major fraction, however, was detected in the ER (Becker et al., 2004; Dunkel et al., 2008). This might be either due to ER-retention or may indicate that besides TPK3, been found in the tonoplast and in the thylakoid membrane, also TPK4 may exhibit at least a dual localization profile.

Unfortunately till now, no general targeting sequence is known that "guides" TPK channels to the appropriate membrane (Vitale and Hinz, 2005; Dunkel et al., 2008). With the purpose of identifying the sorting signal of vacuolar TPK channels, various chimeras were generated between TPK4 (plasma membrane protein) and TPK1 (tonoplast protein). It is not handed down why this particular pair has been chosen and not the "twins" TPK5 and TPK4; TPK4 sharing 85% of similar amino acids with TPK5. Nevertheless, the chimeras showed that complete replacement of the cytosolic C-terminus of TPK1 results in ER retention. Further detailed analysis indicated that the terminal 25 amino acids are not important for the trafficking process. An analysis of amino acids 292–308 in the C-terminus of TPK1 could identify three diacidic motifs. Out of these three motifs, mutations in (D296G/E298G) resulted in ER-stuck TPK1 proteins, suggesting that this diacidic motif is crucial for the export of TPK1 from the ER (Dunkel et al., 2008; Voelker et al., 2010). A related study on rice TPKs identified amino acids in the cytosolic C-terminal domain that determine differential targeting of TPKs to the endomembranes of the large central lytic vacuole or of protein storage vacuoles (Isayenkov et al., 2011a) indicating a general role of certain regions in the cytosolic C-terminus for channel targeting.

Retention of TPK1 channel protein in the ER also occurred when plant leaves were treated with Brefeldin A, a fungal toxin which causes redistribution of Golgi membranes. From this observation it was inferred that the transport of TPK channel proteins to the vacuolar membrane is through a Golgi-dependent pathway and that the Golgi apparatus is the first compartment crossed by the protein after it leaves the ER (Dunkel et al., 2008). Experiments in rice indicated a more complex situation of TPK targeting. TPKs targeted to the lytic vacuole indeed cross the Golgi apparatus. However, TPKs targeted to protein storage vacuoles apparently reach the endomembrane in a Golgi-independent way (Isayenkov et al., 2011a).

## *Regulation and Function of Tandem-pore $K^+$ Channels*

At present the knowledge on function and regulation of plant TPKs is limited. Research is often fuelled by comparison with related channels from other kingdoms. Animal two-pore channel activity has been shown to be regulated by interacting 14-3-3 proteins (Rajan et al., 2002). Also in plants down-regulation of $K^+$channel activity in the tonoplast has been observed to be caused by interaction with 14-3-3 proteins. In TPKs, the cytosolic N-terminus comprises a classical binding motif for 14-3-3 proteins (RSXpS/pTXP)[1]. Phosphorylation of these serine or threonine residues is crucial for the interaction with 14-3-3 proteins (Latz et al., 2007a).

TPK channels are proposed to be involved in the $K^+$ homeostasis of plant cells by allowing the controlled intracellular $K^+$ transport from and into organelles. Recent experiments employing the patch clamp technique have demonstrated a mechano-sensitive nature of TPK channels suggesting especially a role in osmoregulation. This concept was further supported by protoplast disruption assays (Maathuis, 2011) and seedling germination tests (Gobert et al., 2007).

AtTPK1 is ubiquitously expressed in *A. thaliana*. Using promoter-reporter gene (GUS) fusion, *TPK1*promoter activity was observed in root cortex, vascular tissue, mesophyll cells, guard cells and pollen grains (Czempinski et al., 2002). When expressed in yeast, TPK1 has characteristics of $K^+$-selective channels from *Vicia faba* (VK channels) previously characterized *in vivo* with strong selectivity for $K^+$ over $Na^+$ (Bihler et al., 2005; Gobert et al., 2007; Latz et al., 2007b). The activity of TPK1 is independent of the membrane voltage but was shown to be dependent on the cytosolic pH with a maximum open probability at pH 6.7, decreasing 20–30% at physiological pH 7.5–7.8. It is activated by cytosolic $Ca^{2+}$, remarkably exhibiting the highest affinity for calcium ions among the proteins tested including calmodulin. Interaction of TPK1 with the 14-3-3 protein GRF6 (General Release Factor 6) increases the channel activity in a dose dependent manner. This interaction does not play any role in targeting of the protein to the tonoplast (Latz et al., 2007a). All these data indicate that TPK1 is tightly controlled by cellular signals. TPK1 has been reported to participate in vacuolar $K^+$ release during stomatal closure and also during seed germination and radicle growth (Gobert et al., 2007).

AtTPK4 is an instantaneously activating, $K^+$ selective channel that is also found in the plasma membrane when expressed in *Xenopus* oocytes and yeast. *In planta, TPK4* exhibits low transcript abundance. It is predominantly expressed in pollen, as observed by promoter-GUS fusion analysis. TPK4 is blocked by extracellular $Ca^{2+}$ and is insensitive toward changes in extracellular pH, but it is efficiently blocked by cytosolic acidification. Activation of TPK4 by heat has also been reported (Becker et al., 2004). TPK4 is proposed to contribute to the $K^+$ conductance of the pollen tube plasma membrane, where it operates as a so called "open rectifier" with saturating current at depolarizing membrane potentials.

AtTPK5 is targeted to the tonoplast. At the mRNA level, *TPK5* shows higher abundance in senescent leaves and petals (Voelker et al., 2010). Promoter GUS studies of *TPK5* have shown expression in the vascular tissues of leaves, roots, hydathodes, floral tissues and stems. *TPK5* transcript level is increased or decreased in response to external factors.

Recently AtTPK1, AtTPK2, and AtTPK5 were functionally characterized in *Escherichia coli*. The three isoforms were able to complement the $K^+$ uptake deficient *E. coli* mutant LB2003 on low $K^+$ medium (Isayenkov and Maathuis, 2013). Interestingly, in the same experiments AtTPK3 could not complement LB2003. This may indicate that this channel might be active in a different membrane environment, as for instance the thylakoid membrane (Zanetti et al., 2010).

### *Different Isoforms of Tandem-pore $K^+$ Channels*

Tandem-pore $K^+$ channels have also been identified and characterized in plant species other than *A. thaliana*, for example *Hordeum vulgare, Nicotiana tabacum, Solanum tuberosum, Oryza sativa*(Czempinski et al., 1999; Hamamoto et al., 2008a,b; Isayenkov et al., 2011a,b). It is fascinating to see that NtTPK1 from tobacco exhibits properties different from other plant TPK channels, since it is active even in the absence of $Ca^{2+}$. Nevertheless, increase in cytosolic $Ca^{2+}$ resulted in an up to two fold increase in the $K^+$ current amplitude (Hamamoto et al., 2008a). Its current profile shows an instantaneous and a time-dependent component (Hamamoto et al., 2008b). The most interesting distinguishing feature is that two of the four identified isoforms in *N. tabacum* do not contain the conserved TXGYGD motif in the second pore domain. Instead, NtTPKb and NtTPKc possess VHG or GHG, respectively.

## Plant $K_{ir}$-Like Channels

Plant $K_{ir}$-like channels were initially classified as an own group although they are similar to TPK channels. To date, they have been found only in the genus *Arabidopsis, (A. thaliana* and *A. lyrata*; Gomez-Porras et al., 2012). Thus, they apparently emerged just recently in evolution. Phylogenetic analyses indicated them to have originated from gene duplication of an TPK channel gene followed by a partial deletion event that resulted in the loss of one pore domain (Figure 1A; Marcel et al., 2010; Voelker et al., 2010). As a consequence, a plant $K_{ir}$-like channel subunit contains only two TM and one P region. Based on that structural feature it was speculated that plant $K_{ir}$-like channels are tetramers. This concept, however, is rather questionable. The genome of *A. thaliana* contains only one gene (called KCO3) coding for a $K_{ir}$-like subunit. Recently, KCO3 could be detected only as stable dimer at the biochemical level (Rocchetti et al., 2012) pointing further to its origin from TPK channels. Very low transcript abundance has been observed for *KCO3*. Promoter-GUS fusion constructs for *KCO3* show expression in vascular tissue of leaves, roots, flower tissue and stem and also in hydathodes as seen also for *TPK5*. KCO3 might play a role in osmoregulation, as the knock-out plant for the *KCO3* gene shows reduced growth under osmotic stress condition. However, this change in the plant phenotype can be complemented by expressing a mutant *KCO3* gene with an inactive pore region. These results indicate that the function of KCO3 under osmotic stress conditions is independent of its ability to transport potassium ions (Rocchetti et al., 2012). In conclusion, based on the current knowledge, plant $K_{ir}$-like channels should be re-integrated into the TPK family, instead of being considered as a separate channel family. Their occurrence in *Arabidopsis*, only, may suggest that they just are "a freak of nature" without fundamental physiological importance outside this genus.

## VOLTAGE DEPENDENT $K^+$ CHANNELS IN ARABIDOPSIS

The so-called plant *Shaker*-family is a group of voltage gated $K^+$ channels. In *A. thaliana* it comprises nine members. This group can be divided into three subfamilies regarding their response to the membrane voltage (Lebaudy et al., 2007; Dreyer and Blatt, 2009). Six members activate upon membrane hyperpolarization and are closed when the driving force for potassium is outwardly directed. As a consequence they elicit only inward $K^+$ currents ($K_{in}$). Two members activate upon membrane depolarization. They are closed when the driving force for potassium is inwardly directed. Thus, they elicit only outward $K^+$ currents ($K_{out}$). And one member exhibits weak voltage dependence and can mediate both, $K^+$ efflux and $K^+$ influx ($K_{weak}$; Figure 1B).

Functional plant *Shaker* channels are built of four α-subunits. Each α-subunit contains six transmembrane domains and one pore domain between the fifth and the sixth transmembrane domain. The C-terminus contains various regulatory elements, like the cyclic nucleotide binding domain, an ankyrin repeat domain, the acidic domain $K_{HA}$ and in $K_{in}$ channels the $K_T$ domain (Sentenac et al., 1992; Ehrhardt et al., 1997;

Gaymard et al., 1998; Dreyer et al., 2004). Besides being functional as homotetramers, the formation of heterotetramers is common and proven to occur in plants (Dreyer et al., 1997; Lebaudy et al., 2008a).

Versatile physiological roles of plant *Shaker* channels were identified in numerous experiments. Knock-out and overexpressing mutant plants, as well as heterologous expression systems like *Saccharomyces cerevisiae, Xenopus laevis* oocytes, HEK293, COS, or Sf9 cells were used to study the functionality of $K^+$channels (Dreyer et al., 1999). The physiological roles and impacts of plant *Shaker* channels on the plant are described in the following sections.

## $K^+$ Uptake into Roots via AKT1

### *Various Conditions Necessitate Different Uptake Systems*

$K^+$ uptake from soil is performed by a well-organized system of transport proteins each contributing in its own manner (Alemán et al., 2011). All uptake systems together operate on a broad range of $K^+$concentrations and are part of an extensive regulatory network (Figure 2). As main $K^+$ uptake systems in *Arabidopsis* roots the *Shaker*-like $K^+$ channel AKT1 and the $K^+$ transporter AtHAK5 have been identified (Hirsch et al., 1998; Gierth et al., 2005; Rubio et al., 2008). At external $K^+$ concentrations below 0.01 mM the proton-driven $H^+/K^+$ co-transporter AtHAK5 is the only system responsible for $K^+$ uptake from the soil. At $K^+$ concentrations between 0.01 mM and 0.05 mM AtHAK5 and AKT1 together contribute to $K^+$uptake. At higher external $K^+$ concentrations, AKT1 together with other unknown low affinity $K^+$ uptake systems are responsible for $K^+$ uptake from the soil (Rubio et al., 2010; Pyo et al., 2010; Caballero et al., 2012).

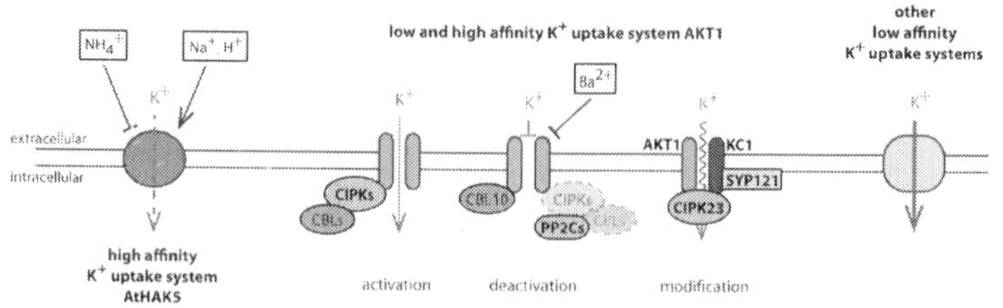


**Figure 2**: $K^+$ uptake into Arabidopsis roots and its regulation. Depending on the actual $K^+$concentration in the soil different low or high affinity $K^+$ uptake

systems are active. At $K^+$concentrations below 0.01 mM only the high affinity transporter AtHAK5 is active. It is blocked by extracellular $NH^+4$ and stimulated by extracellular $Na^+$ and $H^+$. The Shaker-like $K^+$ channel AKT1 is involved in high and low affinity $K^+$ uptake. It is a target of an extensive regulatory network that includes calcium sensors (CBLs), kinases (CIPKs), phosphatases (PP2Cs), and the ability to form heterotetramers with AtKC1. In the presence of CBL1 or CBL9, CIPK23 phosphorylates and activates AKT1. The interaction of CIPK 6, 16, and 23 each with CBL1, 2, 3, and 9 and its effect on AKT1 were shown in yeast two-hybrid assays and Xenopus laevis oocytes (Lee et al., 2007). AKT1 is deactivated by a direct interaction with CBL10, external $Ba2^+$, or dephosphorylation via PP2C phosphatases. Phosphatases act directly on AKT1 or on the CIPK-CBL machinery to inactivate AKT1 (Lan et al., 2011). Furthermore, AKT1 is able to form heterotetramers with AtKC1. The heteromeric channel exhibits changed gating and permeation properties that block efficiently potential K+ release under low external $K^+$ concentrations (Geiger et al., 2009). In addition, an interaction of CIPK23 with the heteromeric AKT1-AtKC1 was suggested and the contribution of SYP121 to the native characteristics of AKT1-AtKC1 was described (Honsbein et al., 2009).

AKT1 and AtHAK5 are affected by different environmental conditions. Both transport proteins work at different $K^+$ concentration spectra and exhibit individual sensitivity toward other ions. For instance, AtHAK5 is sensitive to ammonium ($NH^+_4$) whereas AKT1 remains unaffected in the presence of $NH^+_4$. On the contrary, $Ba^{2+}$ blocks AKT1 while AtHAK5 remains unaffected, and $Na^+$ and $H^+$ stimulate activity of AtHAK5 (Hirsch et al., 1998; Spalding et al., 1999; Rubio et al., 2008). Therefore, the different $K^+$ uptake systems complement one another and even permit $K^+$ uptake when one uptake system is disabled. AKT1 provides hence an alternative $K^+$ uptake system to the $NH^+_4$ sensitive AtHAK5 under low $K^+$ conditions.

## *Regulation of AKT1*

AKT1 itself contributes to high and low affinity $K^+$ uptake and is target of a regulatory network. Xu et al. and Li et al. showed in 2006 that CIPK23 and CBL1 or CBL9 are required to activate AKT1. The two calcineurin B-like calcium sensors CBL1 and CBL9 bind to the CBL-interacting protein kinase CIPK23, which then in turn phosphorylates AKT1. All three components (AKT1-CIPK23-CBL1/9) are essential for a functional expression of AKT1 in oocytes of *X. laevis*.

Shortly after, further components of this highly complex and flexible regulatory network were discovered. Besides several CIP kinases a 2C-type protein phosphatase (PP2C), AIP1, was shown to bind and inactivate AKT1 (Lee et al., 2007). Subsequent studies detected more interrelations between several CBLs and CIPKs with AKT1 (Lee et al., 2007; Lan et al., 2011; Ren et al., 2013). Lan et al. suggested that PP2C phosphatases also interact with the CIPK-CBL complex to inhibit the phosphorylation activity of the kinase and to dephosphorylate AKT1. And Ren et al. (2013) demonstrated that CBL10 directly binds AKT1 and diminishes its activity in a concentration-dependent and CIPK-independent manner.

Many different associations of AKT1 with CBLs, CIPKs, and PP2Cs have been reported. Grefen and Blatt (2012) argue that the method used to investigate interactions between proteins, positioning of tags and the way of analysis have a decisive impact on detectable interactions. Nevertheless, CBLs, CIPKs and PP2Cs provide a comprehensive system to regulate the $K^+$ uptake mediated by AKT1. Especially, as different CBLs are involved in different signaling pathways this phosphorylation-dephosphorylation system provides a powerful regulatory network for the plant to respond to a broad range of environmental changes (for review see Kudla et al., 2010).

## *Internal Regulation via Heteromerization*

Besides the regulation by kinases and phosphatases another member of the *Shaker*-like family alters the functionality of AKT1: AtKC1. AtKC1 is known as regulatory or silent α-subunit of $K_{in}$ *Shaker*-like channels as it shows no currents in *Xenopus* oocytes when expressed alone and affects only $K_{in}$ channels (Dreyer et al., 1997; Jeanguenin et al., 2011). Nevertheless, its participation in $K^+$ uptake and its connection to AKT1 has been recognized since long (Reintanz et al., 2002; Pilot et al., 2003). Duby et al. (2008) demonstrated AtKC1's impact on AKT1. They described that AtKC1 shifts the activation threshold of AKT1 toward more negative values. This in turn would avert $K^+$ efflux through AKT1 under unfavorable conditions. The reduction of potential outward currents prevents the plant from $K^+$ loss under low $K^+$ concentrations. However, the cost of such "a valve" is a reduced channel activity that in turn implies decreased $K^+$ influx under more favourable conditions.

Geiger et al. (2009) supported, further broadened and fine-tuned this "valve" hypothesis. They showed in electrophysiological experiments the effect of AtKC1 on AKT1 inward and outward currents under varying $K^+$ concentrations. Besides affecting the activation threshold, also the $K^+$ dependent stability of the pore has been altered in AKT1-AtKC1 heteromers. When the external $K^+$concentration drops, the permeation pathway of $K^+$ channels gets instable and collapses (Zhou et al., 2001). The threshold concentration, below which this happens, appears to be a characteristic feature of each channel. Geiger et al. (2009) found that the pore of AKT1-AtKC1 heteromers collapses at higher $K^+$concentrations than that of AKT1 homomers. Thus, heteromers comprise a more efficient block of the $K^+$passage in the unfavorable outward direction.

On top of that, the association of CIPK23 with the heteromeric AKT1-AtKC1 channel has been suggested from interaction analyses in yeast (Grefen and Blatt, 2012) along with an impact of the membrane vesicle trafficking SNARE protein SYP121 (Honsbein et al., 2009). In contrast to CIPK23, SYP121 binds only to the AtKC1 α-subunit but not to the AKT1 α-subunit. SNARE proteins are involved in vesicle targeting and fusion. Thus, $K^+$ transport is not only regulated via the channel activity but also by membrane trafficking processes. Interestingly, the transcript level of AKT1 is constant under different environmental conditions (Lagarde et al., 1996; Pilot et al., 2003). But, the expression levels of its regulators change according to environmental stimuli (Pilot et al., 2003, review: Batistic and Kudla, 2004; Tripathi et al., 2009).

## Outward Rectifiers in Roots

### *GORK in Root Hairs*

Alongside the inward rectifying $K^+$ channel AKT1, the outward rectifying $K^+$ channel GORK is expressed in root epidermal cells (Ivashikina et al., 2001). GORK activates upon membrane depolarization and its gating depends on the extracellular $K^+$ concentration. Environmental changes in the surrounding of root hairs can appear rapidly and in response the membrane depolarizes (Càrdenas et al., 2000). GORK activates under these conditions and is considered to initiate the repolarization of the membrane. By controlling the membrane potential and the turgor in root hairs, the plant can react on environmental changes, like absence or abundance of water that cause changes in solute concentrations and affect the mechanical stability and the hydration status of the root. Furthermore, the ability of GORK to sense the extracellular $K^+$ concentration is supposed to enable the root hair to sense and flexibly react on the $K^+$content in the soil.

## *Root to Shoot Communication via SKOR*

$K^+$ is transported from roots to the upper parts of the plant via the xylem. The outward rectifying *Shaker*-like channel SKOR is expressed in the pericycle and the xylem parenchyma in roots. SKOR was identified as transport protein responsible for loading $K^+$ to the xylem based on the finding that its disruption strongly reduced the $K^+$ content in the shoot while the $K^+$ content in roots remained unaffected (Gaymard et al., 1998).

In addition to the membrane voltage, SKOR is modulated by the external $K^+$concentration. In the presence of ample external $K^+$, the channel needs a higher membrane voltage to open and thus minimizes the risk to serve as an undesirable $K^+$-influx pathway. Such behavior is achieved by a complex interplay between the pore region and the last transmembrane domain of the channel that is responsible for final channel opening and closure. When the external $K^+$ concentration is high, the pore region is quite rigid and strongly interacts with the last transmembrane domain of the channel. As a consequence the channel is stabilized in a closed state. Under low external $K^+$ conditions the pore region is less occupied by $K^+$ ions. As a consequence, the pore is more flexible and does not interact with the surrounding transmembrane domains anymore. Opening of the channel is possible with less energy input, i.e., at less positive membrane voltages. If the last transmembrane domains rearrange and unclench the conduction pathway, intracellular $K^+$ ions can re-enter the pore, stabilize it in a permeable conformation and thus enable a $K^+$outward current (the $K^+$-sensing mechanisms has been animated in the supplementary material of Johansson et al., 2006).

$K^+$ distribution is also influenced by factors that are involved in stress signaling. SKOR expression is inhibited by abscisic acid (ABA). It was proposed that the reduced $K^+$ release to the xylem in response to ABA could be a possibility to adjust osmotic conditions by roots in stress situations (Gaymard et al., 1998). Besides, intra- and extracellular acidification negatively affects the SKOR currents. As the regulation via ABA appears on the transcriptional level, the pH sensitivity might be a complementary process to prevent $K^+$ loss from roots toward the shoot tissue (Lacombe et al., 2000a).

Hydrogen peroxide ($H_2O_2$) exhibits a contrary effect on SKOR currents. Reactive oxygen species function as signal and regulator in plant development and in responses to environmental stress situations (Torres and Dangl, 2005; Gapper and Dolan, 2006). Treatment with $H_2O_2$ leads to an increase in SKOR outward currents and a decrease in its half activation time (Garcia-Mata et al., 2010). This finding points to a relation between reactive oxygen species and $K^+$ partitioning during developmental processes and stress responses.

## Phloem-Allocation and Retrieval

Once loaded into the xylem, $K^+$ circulates within the whole plant. There, other $K^+$ channels contribute to the further distribution. The *Shaker*-like potassium channel AKT2[2] is mainly expressed in the vascular tissue of aerial parts and in guard cells of plants. However, it is not expressed until the plant is widely independent of carbohydrates provided by the seed (Marten et al., 1999; Deeken et al., 2000; Lacombe et al., 2000b; Szyroki et al., 2001; Ivashikina et al., 2005).

### *Charging and Using the Potassium Battery*

As the only member of the *Shaker*-like channels in plants, AKT2 features a unique channel property and can mediate both, inward and outward $K^+$ currents. AKT2 is in fact a specialized inward rectifying channel that can be changed into a non-rectifying channel. It exhibits two phosphorylation status-dependent gating modes that are inter-convertible (Dreyer et al., 2001; Chèrel et al., 2002; Michard et al., 2005a,b). The non-phosphorylated AKT2 (mode 1) is lacking its outward component and behaves like an inward rectifying channel. In contrast, the phosphorylated AKT2 (mode 2) is permanently open and able to conduct $K^+$ in the inward and in the outward direction. Two serine residues located near the intracellular side of the channel are identified as targets for phosphorylation (Michard et al., 2005a). Nevertheless, it is proposed that the two phospho-serine residues alone are not sufficient to completely convert AKT2 between its modes. Sandmann et al. (2011) proposed rather a transition via a cascade of posttranslational (so far unknown) modifications. This hypothesis is fuelled by experimental observations. A lysine within the voltage sensor enables AKT2 to sense its phosphorylation status and to change between the two modes. Replacement of the lysine by serine or arginine keeps AKT2 in its inward rectifying mode 1 (Michard et al., 2005b; Sandmann et al., 2011).

Summing up, AKT2 can modulate the membrane voltage by switching between its modes of an inward or a non-rectifying channel, respectively, and phosphorylation acts as a tool for fine tuning (Deeken et al., 2002; Michard et al., 2005a,b). Gajdanowicz et al. (2011) embedded AKT2 as a central player in a "potassium battery" model in which $K^+$ serves as mobile energy source in vascular tissues. In source tissues, the plant invests energy to load $K^+$ into the phloem sieve element companion cell complexes. The loaded potassium is then circulating with the phloem stream. Under energy limiting conditions, the AKT2 channel can be switched from its inward-rectifying to its non-rectifying mode and thus enables a passage for $K^+$ efflux. This in turn enables the use of the $K^+$ gradient between the phloem and the apoplast for the reloading of photoassimilates into the phloem. This "potassium battery" concept is illustrated in the supplementary material of Gajdanowicz et al. (2011). Limiting conditions occur for example under ATP shortage or when the $H^+$-ATPase is down-regulated by cellular signals. The normally used $H^+$ gradient is then complemented by the $K^+$ gradient. Besides tapping the "battery," the AKT2 channel is also proposed to charge it depending on its actual gating mode (Michard et al., 2005b).

### *Further Effects on AKT2*

In addition to the gating mode modulations, AKT2 was also demonstrated to act on diverse signals involved in stress responses. The expression level of AKT2 increases in the presence of ABA, light and $CO_2$assimilates (Deeken et al., 2000; Lacombe et al., 2000b). Primarily, the influences of the last two factors led to the view that AKT2 plays a role in phloem transport.

Macroscopic $K^+$currents mediated by AKT2 are modulated by changes in internal and external pH and external $Ca^{2+}$ (Marten et al., 1999; Lacombe et al., 2000b). While external $Ca^{2+}$ blocks inward currents at negative voltages in a voltage-dependent manner, acidification on both sides of the membrane diminishes AKT2 currents in the whole voltage range. Changes in pH and $Ca^{2+}$ do not affect the gating mode of the channel indicating that $H^+$ and $Ca^{2+}$ affect only the permeation pathway of AKT2. The sensitivity of AKT2 toward $Ca^{2+}$ was investigated in guard cells. Ivashikina et al. (2005) showed in experiments on guard cell protoplasts that the $Ca^{2+}$ sensitivity of $K^+$ uptake channels correlates with the presence of AKT2 subunits.

Recently, Held et al. (2011) demonstrated the association of AKT2 with CIPK6 and CBL4 and the effect of this assembly on macroscopic AKT2 currents. In contrast to the AKT1-CIPK-CBL complexes, no phosphorylation events could be detected *in vitro* so far. Held et al. therefore proposed for these findings a $Ca^{2+}$ dependent targeting of AKT2 to the plasma membrane that depends solely on the physical interaction of AKT2 with CIPK6/CBL4 rather than a regulation of the channel via phosphorylation.

## Guard Cells and its K⁺ Channel Population

Two third of the *Shaker* channel family members are expressed in guard cells. Besides AKT2, also KAT1, KAT2, AKT1, AtKC1, and GORK are detectable there and have important impacts on stomatal opening and closure (Szyroki et al., 2001; Ivashikina et al., 2005; Lebaudy et al., 2008b). Vast signal transduction pathways coordinate stomatal movement. In case of stomatal opening, they result in the activation of $K_{in}$ channels and the uptake of $K^+$ and anions, which finally leads to an increase of guard cell turgor. In case of stomatal closure, $K_{out}$ channels are activated, $K^+$ is released together with anions, water is passively flowing out and guard cell turgor decreases. Recent comprehensive reviews of signal transduction pathways that affect stomatal opening and closure were published by Pandey et al. (2007) or Kim et al. (2010). Figure 3 shows an overview of the regulation of $K^+$ channels involved in stomatal movements.

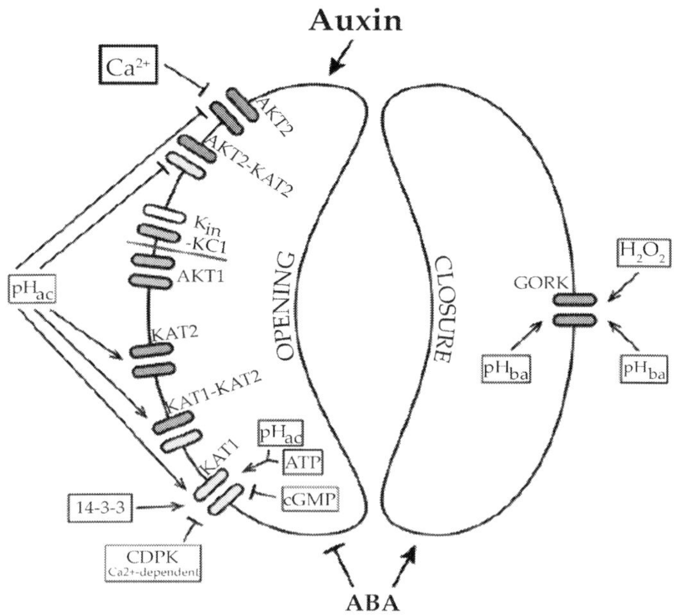


**Figure 3**: $K^+$ channels in Arabidopsis guard cells and their effectors. Changes in membrane potential lead to stomatal opening or closure, respectively. Membrane hyperpolarisation in response to $H^+$-ATPase activity activates inward-rectifying $K^+$ channels. Transcripts for $KAT_1$, $KAT_2$, $AKT_1$, $AKT_2$, and $AtKC_1$ are detectable in guard cells (Szyroki et al., 2001). All these $K^+$ channel subunits form heterotetrameric channels like $KAT_1$-$KAT_2$, $AKT_2$-$KAT_2$, and $AKT_1$-$AtKC_1$. The impact of $AKT_1$ and Kin-$AtKC_1$ on stomatal movement has not been investigated in detail. Acidification affects directly the currents through $K^+$ channels in guard cells. Kin channels are activated upon extracellular acidification, while currents through Kweak channels decrease. Besides, the Kweak channel $AKT_2$ is negatively affected by extracellular $Ca^{2+}$. $KAT_1$ currents are furthermore modulated by intracellular pH, ATP, and cGMP. ATP and cGMP have antagonistic effects. Moreover, stomatal $K^+$ channels are affected by signals via signal transduction cascades. Effects of 14-3-3 proteins, $Ca^{2+}$ and kinases have been reported. Membrane depolarization, on the other hand, caused by the inactivation of $H^+$-ATPases and activation of anion channels activates the Kout channel GORK. GORK currents are positively influenced by extra- and intra-cellular alkalinisation. Furthermore, the current enhancing effect of $H_2O_2$ is under investigation. Both, stomatal opening and closure are affected by phytohormones. While Auxin evokes stomatal opening, ABA inhibits its opening but evokes closure of stomata. Abbreviations: pHac, acidification; pHba, alkalinisation.

## Channel Variability in Guard Cells

Although $KAT_1$ represents the dominant $K_{in}$ channel in guard cells, it is not essential for stomatal opening (Ichida et al., 1997; Kwak et al., 2001; Szyroki et al., 2001). The coevally expressed $K_{in}$ channel subunits $AKT_1$, $AKT_2$, and $KAT_2$ are able to compensate for the loss of $KAT_1$. The expression pattern of all guard cell $K_{in}$ channel subunits is not exactly identical as exemplified by $KAT_1$ and $KAT_2$. Both subunits are expressed in guard cells. But, $KAT_1$ is only expressed in guard cells of leaves and petioles, while KAT2 is additionally expressed in guard cells of the stem (Nakamura et al., 1995; Pilot et al., 2001). That points to different available sets of $K^+$ channels dependent on the guard cell location. Furthermore, $K^+$ channel subunits are able to form heteromeric channels in plants (Dreyer et al., 1997; Lebaudy et al., 2008a). For $KAT_1$-$KAT_2$ heterotetramers it has been shown that their basic properties are similar to properties observed for the homotetrameric $KAT_1$ and $KAT_2$ channels (Pilot et al., 2001; Lebaudy et al., 2010). In contrast, the $AKT_2$-$KAT_2$ heterotetramer combines different properties of its parental channels and forms a new functional type of a $K^+$ channel (Xicluna et al., 2007). The gating properties of the heterotetramer are inherited from $AKT_2$, a weak-rectifying $K^+$ channel described above. The sensitivities to $Ca^{2+}$ and $H^+$are inherited from $KAT_2$. Thus, $K^+$ channel heteromers notably contribute to an increase in channel variability and enhance the regulatory possibilities of $K^+$ channels.

## $K_{in}$ Channels Contribute to Stomatal Opening

For activation of $K_{in}$ channels the membrane potential needs to be hyperpolarized. Hyperpolarization is achieved through the activity of $H^+$-ATPases that transport protons under ATP consumption out of the cell. The membrane voltage is sensed by the intrinsic voltage sensor that is formed by the transmembrane regions S1–S4. An important role is played especially by the positive charges in S4 (Figure 1A). The four voltage sensors of the channel induce conformational changes in the protein that then result in an opening of the permeation pathway. This voltage-sensitivity is modulated by other factors that interact with the channel protein. Indeed, many experiments show the sensitivity of guard cell $K_{in}$ channels to changes in pH (Hedrich et al., 1995; Hoshi, 1995; Marten et al., 1999; Pilot et al., 2001; Xicluna et al., 2007). KAT1, KAT2 and the heteromeric KAT1-KAT2 are activated by extracellular and intracellular acidification due to a shift of the voltage dependence of the channels to more positive values. A histidine residue conserved among plant $K_{in}$ channels that is located in the pore was suggested to sense pH changes in $K_{in}$ channels (Hoth et al., 1997; Hoth and Hedrich, 1999). For KST1, a $K_{in}$ channel from potato guard cells, it has been shown that this histidine is part of the pH sensor. Surprisingly, mutations of this histidine in KAT1 did not affect its pH dependence. Further investigation revealed that KAT1 senses pH changes via a sensory cloud rather than a single residue (Gonzalez et al., 2012). Besides, KAT1 is also modulated by ATP and cGMP. While cGMP reduces KAT1 currents, ATP affects KAT1 positively. Thus, ATP and cGMP show antagonistic effects (Hoshi, 1995).

Another regulator of guard cell $K_{in}$ currents might be extracellular $Ca^{2+}$. Here, AKT2 is the only channel affected directly by external $Ca^{2+}$ (Marten et al., 1999; Latz et al., 2007b). While AKT2 is blocked by $Ca^{2+}$, KAT1, KAT2 and AKT1 do not show any response (Szyroki et al., 2001; Ivashikina et al., 2005; Brüggemann et al., 1999). It is therefore proposed that experimentally observed sensitivity of guard cell $K_{in}$ channels to extracellular $Ca^{2+}$ is conferred by AKT2 subunits (Ivashikina et al., 2005).

Furthermore, effects of regulatory proteins on KAT1 have been shown. For instance, KAT1 is phosphorylated in a $Ca^{2+}$-dependent manner in the presence of CDPK-a $Ca^{2+}$-dependent protein kinase with a calmodulin-like domain (Li et al., 1998). This study used a recombinant CDPK from the bean *Vicia faba* and did not show whether KAT1 is phosphorylated directly by CDPK or rather other proteins are affected by the kinase. Berkowitz et al. (2000) showed in electrophysiological experiments that the recombinant CDPK has a negative effect on KAT1 currents. $Ca^{2+}$-dependent phosphorylation of KAT1 is further supported by a study that manipulated a protein kinase C (PKC) present in *X. laevis* oocytes (Sato et al., 2010). Upon activation of PKC that has similar target sites as plant $Ca^{2+}$-dependent kinases, KAT1 currents decline. In addition, recombinant 14-3-3 proteins from maize stimulated KAT1 currents (Sottocornola et al., 2006, 2008). These studies provide a first glimpse on the broad range of feasible effectors of $K^+$ channels in guard cells.

Additionally it was found that the channel population within the membrane undergoes regulation as well (Mikosch et al., 2006, 2009;

Sutter et al., 2006, 2007; Sieben et al., 2008; Reuff et al., 2010). It has been shown that KAT1 interacts with SNARE proteins (see above), and ABA triggers endocytosis of KAT1 from the plasma membrane. Furthermore, the ER export motif of KAT1 subunits is important for proper channel trafficking. It has been shown that efficient ER export of KAT1 depends on an acidic motif in the C-terminus. Therefore, endo- and exo-cytosis, as well as the ER export of $K^+$ channels might be another level for regulating channel densities and $K^+$ currents across the membrane.

## $K_{out}$ Channels during Stomatal Closure

$K_{out}$ channels are activated upon depolarization. Such a membrane voltage change is achieved by inhibition of the $H^+$-ATPase and activation of anion channels. GORK is the only $K_{out}$ channel identified in guard cells and is responsible for stomatal closure (Szyroki et al., 2001; Hosy et al., 2003). In contrast to inward rectifying channels, GORK currents are reduced with decreasing internal and external pH (Blatt, 1992; Ache et al., 2000). GORK also senses the external $K^+$ concentration, so that at higher external $K^+$ it requires more positive voltage for its activation (Ache et al., 2000). A similar analogy to SKOR might also hold for the direct interaction of GORK with the stress signaling molecule $H_2O_2$. GORK and SKOR share the cysteine residue that has been shown to be responsible for the activation effect in SKOR (Garcia-Mata et al., 2010). Nevertheless, the impact of $H_2O_2$ on GORK and the role of this presumed regulation in guard cell physiology still need to be investigated. Earlier reports have shown that $H_2O_2$ is an important player in stomatal signaling (reviewed by Wang and Song, 2008).

Alongside the activation of $K_{out}$ channels during stomatal closure, $K_{in}$ channels are deactivated (Blatt, 1990; Thiel et al., 1992). A knock-out mutant of the $K_{in}$ channel AKT1 has been shown to be more resistant toward water stress than wild type plants (Nieves-Cordones et al., 2012). Transpiration was reduced and stomata closure was more efficient in knock-out plants treated with ABA. Thus, the inactivation of $K_{in}$ channels favors stomatal closure but it is not essential for the process of closure itself (MacRobbie, 1998). Furthermore, very similar phenotypes of *akt1* and *cipk23* knock-out plants could be observed suggesting a regulation of AKT1 by CIPK23 also in guard cells as it has been shown already for AKT1 in roots.

## Influences of Phytohormones

The phytohormones auxin and ABA cause opposing effects on stomata. Auxin is involved in plant developmental processes and promotes stomatal opening, whereas, ABA is involved in various stress responses. It prevents the opening and promotes the closure of stomata (Gehring et al., 1990). The direct influence of ABA on guard cell $K^+$ currents was shown by Blatt and Armstrong (1993). ABA treatment leads to inactivation of $K_{in}$ and activation of $K_{out}$ channels in guard cells. Although the phytohormon affects the transcription level of the $K_{out}$ channel GORK in roots and shoots, the transcript level in guard cells remains unaffected (Becker et al., 2003). Besides, electrophysiological analyses exclude the direct effect of ABA on outward currents in guard cells. Therefore, ABA seems to affect guard cell $K_{out}$ currents indirectly. As ABA signals from roots come along with alkalinisation of the guard cell cytoplasm (Blatt and Armstrong, 1993), the pH sensitive GORK can be activated and affected by ABA by this long distance signaling pathway (Blatt, 1992; Ache et al., 2000; Becker et al., 2003).

Auxin, on the other hand, stimulates the transcription of *KAT1* and *KAT2* (Philippar et al., 2004). It is not clear, however, whether this stimulation is tissue specific as in the case of GORK or whether it is a general feature of these $K_{in}$ channel genes in different parts of the plant.

## Influence on Pollen Tube Development

The *Shaker* channel SPIK is the main $K_{in}$ channel in pollen and exclusively expressed there (Mouline et al., 2002; Zhao et al., 2013). Its disruption affects negatively pollen tube growth. The activity of SPIK is enhanced by decreasing external pH and negatively affected by the $Ca^{2+}$-dependent protein kinases CDPK11 and CDPK24. $Ca^{2+}$ affects $K_{in}$ currents only in the pollen tube but not in pollen grain protoplasts. It has been shown that the effect of $Ca^{2+}$ is dependent on the presence of both kinases. In the absence of one of the kinases $Ca^{2+}$ cannot block pollen $K_{in}$ currents. Zhao and colleagues propose that $Ca^{2+}$ acts negatively on SPIK via a kinase cascade, in which CDPK11 phosphorylates CDPK24.

## CONCLUSIONS

$K^+$ channels are important for $K^+$ uptake from the soil, its distribution within the plant and processes to maintain and support plant growth. The past two decades revealed crucial information especially for plant *Shaker* like channels regarding the structure, physiological role and-to a minor extent-regarding their regulation. In contrast, our knowledge on TPK channels is far more rudimentary. The challenge of the future of plant $K^+$ channel research will be to identify the complex regulatory networks that regulate their activity and to understand the dynamics of these networks.

## ACKNOWLEDGMENTS

This work was supported by grants from the Spanish Ministerio de Economía y Competitividad (BFU2011-28815), a Marie-Curie Career Integration Grant (FP7-PEOPLE-2011-CIG No. 303674—Regopoc), and by grants of the Deutsche Forschungsgemeinschaft to Ingo Dreyer (DFG grants DR430/8-1 and DR430/9-1). Tripti Sharma and Janin Riedelsberger were recipients of doctoral fellowships from the International Max-Planck Research School "Primary Metabolism and Plant Growth". The authors are grateful to the reviewers for their helpful comments.

## REFERENCES

1. Ache, P., Becker, D., Ivashikina, N., Dietrich, P., Roelfsema, M., and Hedrich, R. (2000). GORK, a delayed outward rectifier expressed in guard cells of *Arabidopsis thaliana*, is a $K^+$-selective, $K^+$-sensing ion channel. *FEBS Lett*. 486, 93–98. doi: 10.1016/S0014-5793(00)02248-1
2. Alemán, F., Nieves-Cordones, M., Martínez, V., and Rubio, F. (2011). Root $K^+$ acquisition in plants: the *Arabidopsis thaliana* model. *Plant Cell Physiol*. 52, 1603–1612. doi: 10.1093/pcp/pcr096
3. Amtmann, A., Armengaud, P., and Volkov, V. (2004). "Potassium nutrition and salt stress," in *Membrane Transport in Plants*, ed M. R. Blatt (Oxford: Blackwell), 316–348.
4. Amtmann, A., Troufflard, S., and Armengaud, P. (2008). The effect of potassium nutrition on pest and disease resistance in plants. *Physiol. Plant*. 133, 682–691. doi: 10.1111/j.1399-3054.2008.01075.x

5. Anderson, J. A., Huprikar, S. S., Kochian, L. V., Lucas, W. J., and Gaber, R. F. (1992). Functional expression of a probable *Arabidopsis thaliana* potassium channel in *Saccharomyces cerevisiae*. *Proc. Natl. Acad. Sci. U.S.A.* 89, 3736–3740.
6. Basset, M., Conejero, G., Lepetit, M., Fourcroy, P., and Sentenac, H. (1995). Organization and expression of the gene coding for the potassium transport system AKT1 of *Arabidopsis thaliana*. *Plant Mol. Biol.* 29, 947–958.
7. Batistic, O., and Kudla, J. (2004). Integration and channeling of calcium signaling through the CBL calcium sensor/CIPK protein kinase network. *Planta* 219, 915–924. doi: 10.1007/s00425-004-1333-3
8. Becker, D., Geiger, D., Dunkel, M., and Roller, A. (2004). AtTPK4, an *Arabidopsis* tandem-pore $K^+$ channel, poised to control the pollen membrane voltage in a pH- and $Ca^{2+}$-dependent manner. *Proc. Natl. Acad. Sci. U.S.A.* 101, 15621–15626.
9. Becker, D., Hoth, S., Ache, P., Wenkel, S., Roelfsema, M. R. G., Meyerhoff, O., et al. (2003). Regulation of the ABA-sensitive *Arabidopsis*potassium channel gene GORK in response to water stress. *FEBS Lett.* 554, 119–126. doi: 10.1016/S0014-5793(03)01118-9
10. Berkowitz, G., Zhang, X., Mercier, R., Leng, Q., and Lawton, M. (2000). Co- expression of calcium-dependent protein kinase with the inward rectified guard cell $K^+$ channel KAT1 alters current parameters in *Xenopus laevis* oocytes. *Plant Cell Physiol.* 41, 785–790. doi: 10.1093/pcp/41.6.785
11. Bihler, H., Eing, C., Hebeisen, S., Roller, A., Czempinski, K., and Bertl, A. (2005). TPK1 is a vacuolar ion channel different from the slow-vacuolar cation channel. *Plant Physiol.* 139, 417–424. doi: 10.1104/pp.105.065599
12. Blatt, M. R. (1990). Potassium channel currents in intact stomatal guard cells: rapid enhancement by abscisic acid. *Planta* 180, 445–455. doi: 10.1007/BF01160403
13. Blatt, M. R. (1992). $K^+$ channels of stomatal guard cells. Characteristics of the inward rectifier and its control by pH. *J. Gen. Physiol.* 99, 615–644.

14. Blatt, M. R., and Armstrong, F. (1993). $K^+$ channels of stomatal guard cells: abscisic- acid-evoked control of the outward rectifier mediated by cytoplasmic pH. *Planta* 191, 330–341. doi: 10.1007/BF00195690
15. Brüggemann, L., Dietrich, P., Dreyer, I., and Hedrich, R. (1999). Pronounced differences between the native $K^+$ channels and KAT1 and KST1 alpha-subunit homomers of guard cells. *Planta* 207, 370–376. doi: 10.1007/s004250050494
16. Caballero, F., Botella, M. A., Rubio, L., Fernández, J. A., Martínez, V., and Rubio, F. (2012). A $Ca^{2+}$-sensitive system mediates low-affinity $K^+$ uptake in the absence of AKT1 in *Arabidopsis* plants. *Plant Cell Physiol.* 53, 2047–2059. doi: 10.1093/pcp/pcs140
17. Càrdenas, L., Holdaway-Clarke, T. L., Saìnchez, F., Quinto, C., Feijoì, J. A., Kunkel, J. G., et al. (2000). Ion changes in legume root hairs responding to Nod factors. *Plant Physiol.* 123, 443–452. doi: 10.1104/pp.123.2.443
18. Chèrel, I., Michard, E., Platet, N., Mouline, K., Alcon, C., Sentenac, H., et al. (2002). Physical and functional interaction of the *Arabidopsis* $K^+$ channel AKT2 and phosphatase AtPP2CA. *Plant Cell* 14, 1133–1146. doi: 10.1105/tpc.000943
19. Chanroj, S., Wang, G., Venema, K., Zhang, M. W., Delwiche, C. F., and Sze, H. (2012). Conserved and diversified gene families of monovalent cation/$H^+$ antiporters from algae to flowering plants. *Front. Plant Sci.* 3:25. doi: 10.3389/fpls.2012.00025
20. Czempinski, K., Frachisse, J.-M., Maurel, C., Barbier-Brygoo, H., and Mueller-Roeber, B. (2002). Vacuolar membrane localization of the *Arabidopsis* 'two- pore' $K^+$ channel KCO1. *Plant J.* 29, 809–820. doi: 10.1046/j.1365-313X.2002.01260.x
21. Czempinski, K., Gaedeke, N., Zimmermann, S., and Mulller-Rolber, B. (1999). Molecular mechanisms and regulation of plant ion channels. *J. Exp. Bot.* 50, 955–966.
22. Czempinski, K., Zimmermann, S., Ehrhardt, T., and Mulller-Rolber, B. (1997). New structure and function in plant $K^+$ channels: KCO1, an outward rectifier with a steep $Ca^{2+}$ dependency. *EMBO J.* 16, 2565–2575. doi: 10.1093/emboj/16.10.2565
23. Deeken, R., Geiger, D., Fromm, J., Koroleva, O., Ache, P.,

Langenfeld-Heyser, R., et al. (2002). Loss of the AKT2/3 potassium channel affects sugar loading into the phloem of *Arabidopsis*. *Planta* 216, 334–344. doi: 10.1007/s00425-002-0895-1

24. Deeken, R., Ivashikina, N., Czirjak, T., Philippar, K., Becker, D., Ache, P., et al. (2003). Tumour development in *Arabidopsis thaliana*involves the *Shaker*-like $K^+$ channels AKT1 and AKT2/3. *Plant J.* 34, 778–787. doi: 10.1046/j.1365-313X.2003.01766.x
25. Deeken, R., Sanders, C., Ache, P., and Hedrich, R. (2000). Developmental and light-dependent regulation of a phloem-localised $K^+$channel of *Arabidopsis thaliana*. *Plant J.* 23, 285–290. doi: 10.1046/j.1365-313x.2000.00791.x
26. Dennison, K. L., Robertson, W. R., Lewis, B. D., Hirsch, R. E., Sussman, M. R., and Spalding, E. P. (2001). Functions of AKT1 and AKT2 potassium channels determined by studies of single and double mutants of *Arabidopsis*. *Plant Physiol.* 127, 1012–1019. doi: 10.1104/pp.010193
27. Dreyer, I., Antunes, S., Hoshi, T., Mulller-Rolber, B., Palme, K., Pongs, O., et al. (1997). Plant $K^+$ channel alpha-subunits assemble indiscriminately. *Biophys. J.* 72, 2143–2150.
28. Dreyer, I., and Blatt, M. R. (2009). What makes a gate? The ins and outs of Kv-like $K^+$ channels in plants. *Trends Plant Sci.* 14, 383–390. doi: 10.1016/j.tplants.2009.04.001
29. Dreyer, I., Horeau, C., Lemaillet, G., Zimmermann, S., Bush, D. R., Rodríguez-Navarro, A., et al. (1999). Identification and characterization of plant transporters. *J. Exp. Bot.* 50, 1073–1087.
30. Dreyer, I., Michard, E., Lacombe, B., and Thibaud, J. B. (2001). A plant *Shaker*-like $K^+$ channel switches between two distinct gating modes resulting in either inward-rectifying or "leak" current. *FEBS Lett.* 505, 233–239. doi: 10.1016/S0014-5793(01)02832-0
31. Dreyer, I., Poree, F., Schneider, A., Mittelstadt, J., Bertl, A., Sentenac, H., et al. (2004). Assembly of plant *Shaker*-like $K_{out}$ channels requires two distinct sites of the channel alpha-subunit. *Biophys. J.* 87, 858–872.
32. Duby, G., Hosy, E., Fizames, C., Alcon, C., Costa, A., Sentenac, H., et al. (2008). *At*KC1, a conditionally targeted *Shaker*-type subunit, regulates the activity of plant $K^+$ channels. *Plant J.* 53, 115–123.

doi: 10.1111/j.1365-313X.2007.03324.x

33. Dunkel, M., Latz, A., Schumacher, K., Mulller, T., Becker, D., and Hedrich, R. (2008). Targeting of vacuolar membrane localized members of the TPK channel family. *Mol. Plant* 1, 938–949.

34. Ehrhardt, T., Zimmermann, S., and Müller-Röber, B. (1997). Association of plant $K^{+}_{in}$ channels is mediated by conserved C-termini and does not affect subunit assembly. *FEBS Lett.* 409, 166–170. doi: 10.1016/S0014-5793(97)00502-4

35. Gajdanowicz, P., Michard, E., Sandmann, M., Rocha, M., Corre^a, L. G. G., Ramiìrez-Aguilar, S. J., et al. (2011). Potassium ($K^+$) gradients serve as a mobile energy source in plant vascular tissues. *Proc. Natl. Acad. Sci. U.S.A*. 108, 864–869.

36. Gapper, C., and Dolan, L. (2006). Control of plant development by reactive oxygen species. *Plant Physiol.* 141, 341–345. doi: 10.1104/pp.106.079079

37. Garcia-Mata, C., Wang, J., Gajdanowicz, P., Gonzalez, W., Hills, A., Donald, N., et al. (2010). A minimal cysteine motif required to activate the SKOR $K^+$ channel of *Arabidopsis* by the reactive oxygen species $H_2O_2$. *J. Biol. Chem.* 285, 29286–29294.

38. Gaymard, F., Pilot, G., Lacombe, B., Bouchez, D., Bruneau, D., Boucherez, J., et al. (1998). Identification and disruption of a plant *Shaker*-like outward channel involved in $K^+$ release into the xylem sap. *Cell* 94, 647–655. doi: 10.1016/S0092-8674(00)81606-2

39. Gehring, C. A., Irving, H. R., and Parish, R. W. (1990). Effects of auxin and abscisic acid on cytosolic calcium and pH in plant cells. *Proc. Natl. Acad. Sci. U.S.A*. 87, 9645–9649.

40. Geiger, D., Becker, D., Vosloh, D., Gambale, F., Palme, K., Rehers, M., et al. (2009). Heteromeric *At*KC1-AKT1 channels in *Arabidopsis*roots facilitate growth under $K^+$-limiting conditions. *J. Biol. Chem.* 284, 21288–21295.

41. Gierth, M., Maser, P., and Schroeder, J. I. (2005). The potassium transporter *At*HAK5 functions in $K^+$ deprivation-induced high-affinity $K^+$ uptake and AKT1 $K^+$ channel contribution to $K^+$ uptake kinetics in *Arabidopsis* roots. *Plant Physiol.* 137, 1105–1114. doi: 10.1104/pp.104.057216

42. Gierth, M., and Mäser, P. (2007). Potassium transporters in plants–involvement in $K^+$ acquisition, redistribution and homeostasis. *FEBS Lett.* 581, 2348–2356. doi: 10.1016/j.febslet.2007.03.035
43. Gobert, A., Isayenkov, S., Voelker, C., Czempinski, K., and Maathuis, F. J. M. (2007). The two-pore channel TPK1 gene encodes the vacuolar $K^+$ conductance and plays a role in $K^+$ homeostasis. *Proc. Natl. Acad. Sci. U.S.A.* 104, 10726–10731.
44. Gomez-Porras, J. L., Riaño-Pachón, D. M., Benito, B., Haro, R., Sklodowski, K., Rodriguez-Navarro, A., et al. (2012). Phylogenetic analysis of $K^+$ transporters in bryophytes, lycophytes, and flowering plants indicates a specialization of vascular plants. *Front. Plant Sci.* 3:167. doi: 10.3389/fpls.2012.00167
45. Gonzalez, W., Riedelsberger, J., Morales-Navarro, S. E., Caballero, J., Alzate-Morales, J. H., Gonzaìlez-Nilo, F. D., et al. (2012). The pH-sensor of the plant $K^+$ uptake channel KAT1 is built of a sensory cloud rather than of single key amino acids. *Biochem. J.* 442, 57–63.
46. Grefen, C., and Blatt, M. R. (2012). Do calcineurin B-like proteins interact independently of the serine threonine kinase CIPK23 with the $K^+$ channel AKT1? Lessons learned from a meìnage aÌ trois. *Plant Physiol.* 159, 915–919. doi: 10.1104/pp.112.198051
47. Grefen, C., Chen, Z., Honsbein, A., Donald, N., Hills, A., and Blatt, M. R. (2010). A novel motif essential for SNARE interaction with the $K^+$ channel KC1 and channel gating in *Arabidopsis*. *Plant Cell* 22, 3076–3092. doi: 10.1105/tpc.110.077768
48. Hamamoto, S., Marui, J., Matsuoka, K., Higashi, K., Igarashi, K., Nakagawa, T., et al. (2008a). Characterization of a tobacco TPK-type $K^+$channel as a novel tonoplast $K^+$ channel using yeast tonoplasts. *J. Biol. Chem.* 283, 1911–1920.
49. Hamamoto, S., Yabe, I., and Uozumi, N. (2008b). Electrophysiological properties of NtTPK1 expressed in yeast tonoplast. *Biosci. Biotechnol. Biochem.* 72, 2785–2787.
50. Hedrich, R., Moran, O., Conti, F., Busch, H., Becker, D., Gambale, F., et al. (1995). Inward rectifier potassium channels in plants differ from their animal counterparts in response to voltage and channel modulators. *Eur. Biophys. J.* 24, 107–115.
51. Held, K., Pascaud, F., Eckert, C., Gajdanowicz, P., Hashimoto, K.,

Corratgeì-Faillie, C., et al. (2011). Calcium-dependent modulation and plasma membrane targeting of the AKT2 potassium channel by the CBL4/CIPK6 calcium sensor/protein kinase complex. *Cell Res.* 21, 1116–1130. doi: 10.1038/cr.2011.50

52. Hirsch, R. E., Lewis, B. D., Spalding, E. P., and Sussman, M. R. (1998). A role for the AKT1 potassium channel in plant nutrition. *Science*280, 918–921. doi: 10.1126/science.280.5365.918
53. Honsbein, A., Sokolovski, S., Grefen, C., Campanoni, P., Pratelli, R., Paneque, M., Chen, Z., et al. (2009). A tripartite SNARE- $K^+$ channel complex mediates in channel-dependent $K^+$ nutrition in *Arabidopsis. Plant Cell* 21, 2859–2877. doi: 10.1105/tpc.109.066118
54. Hoshi, T. (1995). Regulation of voltage dependence of the KAT1 channel by intracellular factors. *J. Gen. Physiol.* 105, 309–328.
55. Hosy, E., Vavasseur, A., Mouline, K., Dreyer, I., Gaymard, F., Poree, F., et al. (2003). The *Arabidopsis* outward $K^+$ channel GORK is involved in regulation of stomatal movements and plant transpiration. *Proc. Natl. Acad. Sci. U.S.A.* 100, 5549–5554.
56. Hoth, S., Dreyer, I., Dietrich, P., Becker, D., Müller-Röber, B., and Hedrich, R. (1997). Molecular basis of plant-specific acid activation of $K^+$ uptake channels. *Proc. Natl. Acad. Sci. U.S.A.* 94, 4806–4810.
57. Hoth, S., and Hedrich, R. (1999). Distinct molecular bases for pH sensitivity of the guard cell $K^+$ channels KST1 and KAT1. *J. Biol. Chem.* 274, 11599–11603.
58. Ichida, A. M., Pei, Z. M., Baizabal-Aguirre, V. M., Turner, K. J., and Schroeder, J. I. (1997). Expression of a $Cs^+$-resistant guard cell $K^+$channel confers $Cs^+$- resistant, light-induced stomatal opening in transgenic *Arabidopsis. Plant Cell* 9, 1843–1857.
59. Isayenkov, S., Isner, J. C., and Maathuis, F. J. (2011a). Rice two-pore $K^+$ channels are expressed in different types of vacuoles. *Plant Cell*23, 756–768.
60. Isayenkov, S., Isner, J. C., and Maathuis, F. J. (2011b). Membrane localisation diversity of TPK channels and their physiological role. *Plant Signal Behav*. 6, 1201–1204.
61. Isayenkov, S., and Maathuis, F. J. (2013). *Arabidopsis thaliana* vacuolar TPK channels form functional $K^+$ uptake pathways in *Escherichia coli. Plant Signal. Behav*. 8, e24665.

62. Ivashikina, N., Becker, D., Ache, P., Meyerhoff, O., Felle, H. H., and Hedrich, R. (2001). $K^+$ channel profile and electrical properties of *Arabidopsis* root hairs. *FEBS Lett.* 508, 463–469. doi: 10.1016/S0014-5793(01)03114-3
63. Ivashikina, N., Deeken, R., Fischer, S., Ache, P., and Hedrich, R. (2005). AKT2/3 subunits render guard cell $K^+$ channels $Ca^{2+}$ sensitive. *J. Gen. Physiol.* 125, 483–492.
64. Jeanguenin, L., Alcon, C., Duby, G., Boeglin, M., Chèrel, I., Gaillard, I., et al. (2011). *At*KC1 is a general modulator of *Arabidopsis* inward *Shaker* channel activity. *Plant J.* 67, 570–582. doi: 10.1111/j.1365-313X.2011.04617.x
65. Johansson, I., Wulfetange, K., Poree, F., Michard, E., Gajdanowicz, P., Lacombe, B., et al. (2006). External $K^+$ modulates the activity of the *Arabidopsis* potassium channel SKOR via an unusual mechanism. *Plant J.* 46, 269–281. doi: 10.1111/j.1365-313X.2006.02690.x
66. Kim, T. H., Böhmer, M., Hu, H., Nishimura, N., and Schroeder, J. I. (2010). Guard cell signal transduction network: advances in understanding abscisic acid, $CO_2$, and $Ca^{2+}$ signaling. *Annu. Rev. Plant Biol.* 61, 561–591. doi: 10.1146/annurev-arplant-042809-112226
67. Kudla, J., Batistic, O., and Hashimoto, K. (2010). Calcium signals: the lead currency of plant information processing. *Plant Cell* 22, 541–563. doi: 10.1105/tpc.109.072686
68. Kwak, J. M., Murata, Y., Baizabal-Aguirre, V. M., Merrill, J., Wang, M., Kemper, A., et al. (2001). Dominant negative guard cell $K^+$channel mutants reduce inward-rectifying $K^+$ currents and light-induced stomatal opening in *Arabidopsis*. *Plant Physiol.* 127, 473–485. doi: 10.1104/pp.010428
69. Lacombe, B., Pilot, G., Gaymard, F., Sentenac, H., and Thibaud, J. B. (2000a). PH control of the plant outwardly-rectifying potassium channel SKOR. *FEBS Lett.* 466, 351–354.
70. Lacombe, B., Pilot, G., Michard, E., Gaymard, F., Sentenac, H., and Thibaud, J. B. (2000b). A *Shaker*-like $K^+$ channel with weak rectification is expressed in both source and sink phloem tissues of *Arabidopsis*. *Plant Cell* 12, 837–851.
71. Lagarde, D., Basset, M., Lepetit, M., Conejero, G., Gaymard, F., Astruc, S., et al. (1996). Tissue-specific expression of *Arabidopsis*

AKT1 gene is consistent with a role in $K^+$ nutrition. *Plant J.* 9, 195–203. doi: 10.1046/j.1365-313X.1996.09020195.x

72. Lan, W.-Z., Lee, S.-C., Che, Y.-F., Jiang, Y.-Q., and Luan, S. (2011). Mechanistic analysis of AKT1 regulation by the CBL-CIPK-PP2CA interactions. *Mol. Plant* 4, 527–536.
73. Latz, A., Becker, D., Hekman, M., Muller, T., Beyhl, D., Marten, I., et al. (2007a). TPK1, a $Ca^{2+}$-regulated *Arabidopsis* vacuole two-pore $K^+$ channel is activated by 14–3-3 proteins. *Plant J.* 52, 449–459.
74. Latz, A., Ivashikina, N., Fischer, S., Ache, P., Sano, T., Becker, D., et al. (2007b). In planta AKT2 subunits constitute a pH- and $Ca^{2+}$-sensitive inward rectifying $K^+$ channel. *Planta* 225, 1179–1191.
75. Lebaudy, A., Hosy, E., Simonneau, T., Sentenac, H., Thibaud, J., and Dreyer, I. (2008a). Heteromeric $K^+$ channels in plants. *Plant J.* 54, 1076–1082.
76. Lebaudy, A., Vavasseur, A., Hosy, E., Dreyer, I., Leonhardt, N., Thibaud, J.-B., et al. (2008b). Plant adaptation to fluctuating environment and biomass production are strongly dependent on guard cell potassium channels. *Proc. Natl. Acad. Sci. U.S.A.* 105, 5271–5276.
77. Lebaudy, A., Pascaud, F., Veìry, A.-A., Alcon, C., Dreyer, I., Thibaud, J.-B., et al. (2010). Preferential KAT1-KAT2 heteromerization determines inward $K^+$ current properties in *Arabidopsis* guard cells. *J. Biol. Chem.* 285, 6265–6274.
78. Lebaudy, A., Véry, A., and Sentenac, H. (2007). $K^+$ channel activity in plants: genes, regulations and functions. *FEBS Lett.* 581, 2357–2366. doi: 10.1016/j.febslet.2007.03.058
79. Lee, S. C., Lan, W.-Z., Kim, B.-G., Li, L., Cheong, Y. H., Pandey, G. K., et al. (2007). A protein phosphorylation/dephosphorylation network regulates a plant potassium channel. *Proc. Natl. Acad. Sci. U.S.A.* 104, 15959–15964.
80. Leigh, R. A., and Wyn Jones, R. G. (1984). A hypothesis relating critical potassium concentrations for growth to the distribution and functions of this ion in the plant cell. *New Phytol.* 97, 1–13. doi: 10.1111/j.1469-8137.1984.tb04103.x
81. Li, J., Lee, Y. J., and Assmann, S. M. (1998). Guard cells possess a calcium- dependent protein kinase that phosphorylates the KAT1

potassium channel. *Plant Physiol.* 116, 785–795. doi: 10.1104/pp.116.2.785

82. Li, L., Kim, B.-G., Cheong, Y. H., Pandey, G. K., and Luan, S. (2006). A $Ca^{2+}$ signaling pathway regulates a $K^+$ channel for low-K response in *Arabidopsis*. *Proc. Natl. Acad. Sci. U.S.A.* 103, 12625–12630.
83. Maathuis, F. J. M. (2009). Physiological functions of mineral macronutrients. *Curr. Opin. Plant Biol.* 12, 250–258.
84. Maathuis, F. J. M. (2011). Vacuolar two-pore $K^+$ channels act as vacuolar osmosensors. *New Phytol.* 191, 84–91. doi: 10.1111/j.1469-8137.2011.03664.x
85. MacRobbie, E. A. C. (1998). Signal transduction and ion channels in guard cells. *Philos. Trans. R. Soc. Lond. B.* 353, 1475–1488.
86. Maitrejean, M., Wudick, M. M., Voelker, C., Prinsi, B., Mueller-Roeber, B., Czempinski, K., et al. (2011). Assembly and sorting of the tonoplast potassium channel *At*TPK1 and its turnover by internalization into the vacuole. *Plant Physiol.* 156, 1783–1796. doi: 10.1104/pp.111.177816
87. Marcel, D., Müller, T., Hedrich, R., and Geiger, D. (2010). K+ transport characteristics of the plasma membrane tandem-pore channel TPK4 and pore chimeras with its vacuolar homologs. *FEBS letters* 584, 2433–2439. doi: 10.1016/j.febslet.2010.04.038
88. Marschner, H. (2012). *Mineral Nutrition of Higher Plants*. London: Academic Press.
89. Marten, I., Hoth, S., Deeken, R., Ache, P., Ketchum, K. A., Hoshi, T., et al. (1999). AKT3, a phloem-localized $K^+$ channel, is blocked by protons. *Proc. Natl. Acad. Sci. U.S.A.* 96, 7581–7586.
90. Mäser, P., Thomine, S., Schroeder, J., Ward, J., Hirschi, K., Sze, H., et al. (2001). Phylogenetic relationships within cation transporter families of *Arabidopsis*. *Plant Physiol.* 126, 1646–1667. doi: 10.1104/pp.126.4.1646
91. Michard, E., Dreyer, I., Lacombe, B., Sentenac, H., and Thibaud, J.-B. (2005a). Inward rectification of the AKT2 channel abolished by voltage-dependent phosphorylation. *Plant J.* 44, 783–797.
92. Michard, E., Lacombe, B., Poreìe, F., Mueller-Roeber, B., Sentenac,

H., Thibaud, J.-B., et al. (2005b). A unique voltage sensor sensitizes the potassium channel AKT2 to phosphoregulation. *J. Gen. Physiol.* 126, 605–617.

93. Mikosch, M., Hurst, A. C., Hertel, B., and Homann, U. (2006). Diacidic motif is required for efficient transport of the $K^+$ channel KAT1 to the plasma membrane. *Plant Physiol.* 142, 923–930. doi: 10.1104/pp.106.087064
94. Mikosch, M., Käberich, K., and Homann, U. (2009). ER export of KAT1 is correlated to the number of acidic residues within a triacidic motif. *Traffic* 10, 1481–1487. doi: 10.1111/j.1600-0854.2009.00962.x
95. Mouline, K., Veìry, A.-A., Gaymard, F., Boucherez, J., Pilot, G., Devic, M., et al. (2002). Pollen tube development and competitive ability are impaired by disruption of a *Shaker* $K^+$ channel in *Arabidopsis*. *Genes Dev.* 16, 339–350. doi: 10.1101/gad.213902
96. Nakamura, R. L., McKendree, W. L., Hirsch, R. E., Sedbrook, J. C., Gaber, R. F., and Sussman, M. R. (1995). Expression of an *Arabidopsis* potassium channel gene in guard cells. *Plant Physiol.* 109, 371–374. doi: 10.1104/pp.109.2.371
97. Nieves-Cordones, M., Caballero, F., Martínez, V., and Rubio, F. (2012). Disruption of the *Arabidopsis thaliana* inward-rectifier $K^+$channel AKT1 improves plant responses to water stress. *Plant Cell Physiol.* 53, 423–432. doi: 10.1093/pcp/pcr194
98. Pandey, S., Zhang, W., and Assmann, S. M. (2007). Roles of ion channels and transporters in guard cell signal transduction. *FEBS Lett.* 581, 2325–2336. doi: 10.1016/j.febslet.2007.04.008
99. Philippar, K., Büchsenschutz, K., Abshagen, M., Fuchs, I., Geiger, D., Lacombe, B., et al. (2003). The $K^+$ channel KZM1 mediates potassium uptake into the phloem and guard cells of the C4 grass *Zea mays*. *J. Biol. Chem.* 278, 16973–16981.
100. Philippar, K., Ivashikina, N., Ache, P., Christian, M., Lulthen, H., Palme, K., et al. (2004). Auxin activates KAT1 and KAT2, two $K^+$-channel genes expressed in seedlings of *Arabidopsis thaliana*. *Plant J.* 37, 815–827. doi: 10.1111/j.1365-313X.2003.02006.x
101. Pilot, G., Gaymard, F., Mouline, K., Chèrel, I., and Sentenac, H. (2003). Regulated expression of *Arabidopsis Shaker* $K^+$ channel

genes involved in $K^+$ uptake and distribution in the plant. *Plant Mol. Biol.* 51, 773–787.

102. Pilot, G., Lacombe, B., Gaymard, F., Cherel, I., Boucherez, J., Thibaud, J. B., et al. (2001). Guard cell inward $K^+$ channel activity in *Arabidopsis* involves expression of the twin channel subunits KAT1 and KAT2. *J. Biol. Chem.* 276, 3215–3221.
103. Pyo, Y. J., Gierth, M., Schroeder, J. I., and Cho, M. H. (2010). High-affinity $K^+$ transport in *Arabidopsis*: *At*HAK5 and AKT1 are vital for seedling establishment and postgermination growth under low-potassium conditions. *Plant Physiol.* 153, 863–875. doi: 10.1104/pp.110.154369
104. Rajan, S., Preisig-Muller, R., Wischmeyer, E., Nehring, R., Hanley, P. J., Renigunta, V., et al. (2002). Interaction with 14-3-3 proteins promotes functional expression of the potassium channels TASK-1 and TASK-3. *J. Physiol.* 545, 13–26. doi: 10.1113/jphysiol.2002.027052
105. Reintanz, B., Szyroki, A., Ivashikina, N., Ache, P., Godde, M., Becker, D., et al. (2002). *At*KC1, a silent *Arabidopsis* potassium channelalpha-subunit modulates root hair $K^+$ influx. *Proc. Natl. Acad. Sci. U.S.A.* 99, 4079–4084.
106. Ren, X.-L., Qi, G.-N., Feng, H.-Q., Zhao, S., Zhao, S.-S., Wang, Y., et al. (2013). Calcineurin B-like protein CBL10 directly interacts with AKT1 and modulates $K^+$ homeostasis in *Arabidopsis*. *Plant J.* 74, 258–266. doi: 10.1111/tpj.12123
107. Reuff, M., Mikosch, M., and Homann, U. (2010). Trafficking, lateral mobility and segregation of the plant $K^+$ channel KAT1*. *Plant Biol.* 12, 99–104. doi: 10.1111/j.1438-8677.2010.00355.x
108. Rocchetti, A., Sharma, T., Wulfetange, C., Scholz-Starke, J., Grippa, A., Carpaneto, A., et al. (2012). The putative $K^+$ channel subunit *At*KCO3 forms stable dimers in *Arabidopsis*. *Front. Plant Sci.* 3:251. doi: 10.3389/fpls.2012.00251.
109. Rubio, F., Alemaìn, F., Nieves-Cordones, M., and Martiìnez, V. (2010). Studies on *ArabidopsisAt*HAK5, *At*AKT1 double mutants disclose the range of concentrations at which *At*HAK5, *At*AKT1 and unknown systems mediate K uptake. *Physiol. Plant.* 139, 220–228. doi: 10.1111/j.1399-3054.2010.01354.x

110. Rubio, F., Nieves-Cordones, M., Alemaìn, F., and Martiìnez, V. (2008). Relative contribution of *At*HAK5 and *At*AKT1 to $K^+$ uptake in the high-affinity range of concentrations. *Physiol. Plant* 134, 598–608. doi: 10.1111/j.1399-3054.2008.01168.x

111. Sandmann, M., Skłodowski, K., Gajdanowicz, P., Michard, E., Rocha, M., Gomez-Porras, J. L., et al. (2011). The $K^+$ battery-regulating *Arabidopsis* $K^+$ channel AKT2 is under the control of multiple post-translational steps. *Plant Signal. Behav.* 6, 558–562.

112. Sato, A., Gambale, F., Dreyer, I., and Uozumi, N. (2010). Modulation of the *Arabidopsis* KAT1 channel by an activator of protein kinase C in *Xenopus laevis* oocytes. *FEBS J.* 277, 2318–2328. doi: 10.1111/j.1742-4658.2010.07647.x

113. Schachtman, D. P., Schroeder, J. I., Lucas, W. J., Anderson, J. A., and Gaber, R. F. (1992). Expression of an inward-rectifying potassium channel by the *Arabidopsis* KAT1 cDNA. *Science* 258, 1654–1658. doi: 10.1126/science.8966547

114. Schönknecht, G., Spoormaker, P., Steinmeyer, R., Brüggemann, L., Ache, P., Dutta, R., et al. (2002). KCO1 is a component of the slow-vacuolar (SV) ion channel. *FEBS Lett.* 511, 28–32. doi: 10.1016/S0014-5793(01)03273-2

115. Schroeder, J. I., Ward, J. M., and Gassmann, W. (1994). Perspectives on the physiology and structure of inward-rectifying $K^+$channels in higher plants: biophysical implications for $K^+$ uptake. *Annu. Rev. Biophys. Biomol. Struc.* 23, 441–471.

116. Sentenac, H., Bonneaud, N., Minet, M., Lacroute, F., Salmon, J. M., Gaymard, F., et al. (1992). Cloning and expression in yeast of a plant potassium ion transport system. *Science* 256, 663–665. doi: 10.1126/science.1585180

117. Sieben, C., Mikosch, M., Brandizzi, F., and Homann, U. (2008). Interaction of the $K^+$-channel KAT1 with the coat protein complex II coat component Sec24 depends on a di-acidic endoplasmic reticulum export motif. *Plant J.* 56, 997–1006. doi: 10.1111/j.1365-313X.2008.03658.x

118. Sottocornola, B., Gazzarrini, S., Olivari, C., Romani, G., Valbuzzi, P., Thiel, G., et al. (2008). 14-3-3 proteins regulate the potassium

channel KAT1 by dual modes. *Plant Biol.* 10, 231–236. doi: 10.1111/j.1438-8677.2007.00028.x

119. Sottocornola, B., Visconti, S., Orsi, S., Gazzarrini, S., Giacometti, S., Olivari, C., et al. (2006). The potassium channel KAT1 is activated by plant and animal 14-3-3 proteins. *J. Biol. Chem.* 281, 35735–35741.
120. Spalding, E. P., Hirsch, R. E., Lewis, D. R., Qi, Z., Sussman, M. R., and Lewis, B. D. (1999). Potassium uptake supporting plant growth in the absence of AKT1 channel activity: inhibition by ammonium and stimulation by sodium. *J. Gen. Physiol.* 113, 909–918.
121. Sutter, J.-U., Campanoni, P., Tyrrell, M., and Blatt, M. R. (2006). Selective mobility and sensitivity to SNAREs is exhibited by the *Arabidopsis* KAT1 $K^+$ channel at the plasma membrane. *Plant Cell* 18, 935–954. doi: 10.1105/tpc.105.038950
122. Sutter, J.-U., Sieben, C., Hartel, A., Eisenach, C., Thiel, G., and Blatt, M. R. (2007). Abscisic acid triggers the endocytosis of the *Arabidopsis* KAT1 $K^+$ channel and its recycling to the plasma membrane. *Curr. Biol.* 17, 1396–1402.
123. Szyroki, A., Ivashikina, N., Dietrich, P., Roelfsema, M. R., Ache, P., Reintanz, B., et al. (2001). KAT1 is not essential for stomatal opening. *Proc. Natl. Acad. Sci. U.S.A.* 98, 2917–2921.
124. Thiel, G., MacRobbie, E. A. C., and Blatt, M. B. (1992). Membrane transport in stomatal guard cells: the importance of voltage control. *J. Membr. Biol.* 126, 1–18.
125. Torres, M. A., and Dangl, J. L. (2005). Functions of the respiratory burst oxidase in biotic interactions, abiotic stress and development. *Curr. Opin. Plant Biol.* 8, 397–403.
126. Tripathi, V., Parasuraman, B., Laxmi, A., and Chattopadhyay, D. (2009). CIPK6, a CBL-interacting protein kinase is required for development and salt tolerance in plants. *Plant J.* 58, 778–790. doi: 10.1111/j.1365-313X.2009.03812.x
127. Véry, A., Gaymard, F., Bosseux, C., Sentenac, H., and Thibaud, J. (1995). Expression of a cloned plant $K^+$ channel in *Xenopus* oocytes: analysis of macroscopic currents. *Plant J.* 7, 321–332. doi: 10.1046/j.1365-313X.1995.7020321.x
128. Véry, A., and Sentenac, H. (2003). Molecular mechanisms and

regulation of $K^+$ transport in higher plants. *Annu. Rev. Plant Biol.* 54, 575–603. doi: 10.1146/annurev.arplant.54.031902.134831

129. Vitale, A., and Hinz, G. (2005). Sorting of proteins to storage vacuoles: how many mechanisms? *Trends Plant Sci.* 10, 316–323.
130. Voelker, C., Gomez-Porras, J. L., Becker, D., Hamamoto, S., Uozumi, N., Gambale, F., et al. (2010). Roles of tandem-pore $K^+$ channels in plants - a puzzle still to be solved. *Plant Biol.* 12(Suppl. 1), 56–63. doi: 10.1111/j.1438-8677.2010.00353.x
131. Voelker, C., Schmidt, D., Mueller-Roeber, B., and Czempinski, K. (2006). Members of the *Arabidopsis* AtTPK/KCO family form homomeric vacuolar channels in planta. *Plant J.* 48, 296–306. doi: 10.1111/j.1365-313X.2006.02868.x
132. Wang, P., and Song, C. P. (2008). Guard-cell signalling for hydrogen peroxide and abscisic acid. *New Phytol.* 178, 703–718. doi: 10.1111/j.1469-8137.2008.02431.x
133. Wang, Y., and Wu, W.-H. (2010). Plant sensing and signaling in response to $K^+$- deficiency. *Mol. Plant* 3, 280–287.
134. Wyn Jones, R. G., and Pollard, A. (1983). "Proteins, enzymes and inorganic ions," in *Encyclopedia of Plant Physiology*, eds A. Lauchli and A. Pirson (Berlin: Springer), 528–562.
135. Xicluna, J., Lacombe, B., Dreyer, I., Alcon, C., Jeanguenin, L., Sentenac, H., et al. (2007). Increased functional diversity of plant $K^+$channels by preferential heteromerization of the *Shaker*-like subunits AKT2 and KAT2. *J. Biol. Chem.* 282, 486–494.
136. Xu, J., Li, H.-D., Chen, L.-Q., Wang, Y., Liu, L.-L., He, L., et al. (2006). A protein kinase, interacting with two calcineurin B-like proteins, regulates $K^+$ transporter AKT1 in *Arabidopsis*. *Cell* 125, 1347–1360. doi: 10.1016/j.cell.2006.06.011
137. Zanetti, M., Teardo, E., La Rocca, N., Zulkifli, L., Checchetto, V., Shijuku, T., et al. (2010). A novel potassium channel in photosynthetic cyanobacteria. *PLoS ONE* 5:e10118. doi: 10.1371/journal.pone.0010118Zhao, L.-N., Shen, L.-K., Zhang, W.-Z., Zhang, W., Wang, Y., and Wu, W.-H. (2013). $Ca^{2+}$-Dependent Protein Kinase11 and 24 Modulate the Activity of the Inward Rectifying $K^+$ Channels in Arabidopsis Pollen Tubes. *Plant Cell* 25, 649–661. doi: 10.1105/tpc.112.103184

138. Zhou, Y., Morais-Cabral, J. H., Kaufman, A., and MacKinnon, R. (2001). Chemistry of ion coordination and hydration revealed by a $K^+$channel-Fab complex at 2.0 A resolution. *Nature* 414, 43–48. doi: 10.1038/35102009

CHAPTER

# 7

# CLUSTERING OF THE K+ CHANNEL GORK OF ARABIDOPSIS PARALLELS ITS GATING BY EXTRACELLULAR K+

Cornelia Eisenach, Maria Papanatsiou, Ellin-Kristina Hillert, Michael R. Blatt

Laboratory of Plant Physiology and Biophysics, Institute of Molecular, Cell and Systems Biology, University of Glasgow, Bower Building, Glasgow G12 8QQ, UK

## SUMMARY

GORK is the only outward-rectifying Kv-like $K^+$ channel expressed in guard cells. Its activity is tightly regulated to facilitate $K^+$ efflux for stomatal closure and is elevated in ABA in parallel with suppression of the activity of the inward-rectifying $K^+$ channel KAT1. Whereas the population of KAT1 is subject to regulated traffic to and from the plasma membrane, nothing is known about GORK, its distribution and traffic *in vivo*. We have used transformations with fluorescently-tagged GORK to explore its characteristics in tobacco epidermis and

Arabidopsis guard cells. These studies showed that GORK assembles in puncta that reversibly dissociated as a function of the external $K^+$ concentration. Puncta dissociation parallelled the gating dependence of GORK, the speed of response consistent with the rapidity of channel gating response to changes in the external ionic conditions. Dissociation was also suppressed by the $K^+$ channel blocker $Ba^{2+}$. By contrast, confocal and protein biochemical analysis failed to uncover substantial exo- and endocytotic traffic of the channel. Gating of GORK is displaced to more positive voltages with external $K^+$, a characteristic that ensures the channel facilitates only $K^+$efflux regardless of the external cation concentration. GORK conductance is also enhanced by external $K^+$ above 1 mm. We suggest that GORK clustering in puncta is related to its gating and conductance, and reflects associated conformational changes and (de)stabilisation of the channel protein, possibly as a platform for transmission and coordination of channel gating in response to external $K^+$.

## INTRODUCTION

Control of membrane transport is central to solute flux for stomatal regulation. Membrane transport of the predominant, osmotically-active ion, $K^+$, is of particular importance and is mediated largely by voltage-dependent $K^+$ channels. Inward-rectifying $K^+$ channels mediate $K^+$ uptake during stomatal opening, while outward-rectifying $K^+$ channels mediate its loss during stomatal closing (Blatt and Thiel, 1993; Blatt, 2000; Lebaudy *et al.*, 2007; Dreyer and Blatt, 2009). In Arabidopsis, the former are dominated by the inward-rectifying $K^+$ channel KAT1, which is highly-expressed in guard cells, and to a lesser extent the related channel KAT2. $K^+$ efflux across the plasma membrane is facilitated by GORK, which is the only outward rectifying $K^+$ -channel known to be expressed in the guard cells (Ache *et al.*, 2000; Hosy *et al.*, 2003; Dreyer and Blatt, 2009).

KAT1 and GORK belong to the Kv superfamily of voltage-gated ion channels that are found also in archaea, insects and mammals. These channels open and close, a process referred to as 'gating', in response to membrane voltage. Outward-rectifying $K^+$ channels of plants, including GORK, also gate in response to the extracellular $K^+$ concentration. Increasing the extracellular $K^+$ concentration displaces the voltage-dependence of gating to more positive voltages in parallel with the equilibrium voltage for $K^+$ ($E_K$) and, at concentrations above 1 mM, increases the maximum ensemble conductance (Blatt, 1988; Blatt and Gradmann, 1997; Ache *et al.*, 2000; Johansson *et al.*, 2006). Potassium is the most abundant ionic species in plants; its concentration outside, and its equilibrium, often dominates the membrane voltage in plants. Gating by extracellular $K^+$ therefore ensures that these channels activate at a voltage relative to $E_K$ and adjusts $K^+$ efflux through the channels in a compensatory manner across a wide range of $K^+$ concentrations (Blatt, 1988; Blatt and Gradmann, 1997).

The activities of guard cell $K^+$ channels are subject to a number of regulatory stimuli, including hormones such as abscisic acid (ABA) (Blatt, 2000; Blatt *et al.*, 2007; Kim *et al.*, 2010; Roelfsema and Hedrich, 2010), auxin (Thiel *et al.*, 1993; Blatt and Thiel, 1994; Bauly *et al.*, 2000), metabolites (Hedrich *et al.*, 2001; Wang and Blatt, 2011), and pathogen elicitors (Blatt *et al.*, 1999). Many of these stimuli act through intermediates of cytosolic-free [$Ca^{2+}$] ([$Ca^{2+}$]$_i$), pH and protein (de-) phosphorylation (Blatt *et al.*, 1990; Blatt and Grabov, 1999; Blatt, 2000; Kim *et al.*, 2010). Channel responses to these intermediates are usually very rapid and are underpinned by changes in channel gating, but over longer time periods can extend to membrane vesicle traffic and changes in channel population at the membrane. Much less is known of the scope of channel regulation by vesicle traffic or its regulation. To date, only the KAT1 $K^+$ channel has been studied in any detail (Hurst *et al.*, 2004; Meckel *et al.*, 2005; Sutter *et al.*, 2006, 2007; Eisenach *et al.*, 2012). KAT1 is localised at the plasma membrane in punctate domains from which ABA and [$Ca^{2+}$]$_i$ trigger its endocytosis. Its recycling to the plasma membrane depends on the vesicle-trafficking protein SYP121 (=SYR1/PEN1) (Leyman *et al.*, 1999) and underpins the phenomenon of so-called 'programmed closure' of stomata (Eisenach *et al.*, 2012). By contrast, nothing is known of the localisation, distribution and trafficking of the outward-rectifying $K^+$ channels, including GORK. Here we show that GORK is localised in discrete puncta at the plasma membrane that are similar to, but distinct from, those reported previously for KAT1-GFP. Unexpectedly, GORK-GFP puncta proved to be insensitive to ABA or the presence of SYP121, but channel distribution was affected by increasing external $K^+$ concentrations in parallel with the dependence of channel gating on the cation. We propose that the gating of GORK with external $K^+$ concentration is associated with a re-organisation of the channel proteins in clusters within the plane of the plasma membrane.

# RESULTS

## GORK-GFP is a Functional Outward Rectifying $K^+$ Channel That Localises in Puncta

We generated GORK-GFP constructs for expression under the control of the Ubiquitin-10 gene promoter to give moderate over-expression *in vivo* (Grefen *et al.*, 2010b) and verified functionality of the recombinant protein after expression in *Xenopus* oocytes. Oocytes injected with GORK-GFP cRNA showed an outward current when clamped to voltages positive of $E_K$, with current relaxations typical of the GORK channel and block by $TEA^+$ . Relative conductances ($G/G_{max}$) were well-fitted to a Boltzmann function with a half-maximal activation voltage ($V_{1/2}$) that was displaced in parallel with $E_K$, much as has previously been shown for GORK (Ache *et al.*, 2000), the outward-rectifier of *Vicia* (Blatt, 1988; Blatt and Gradmann, 1997) and for SKOR, the close homolog of the GORK $K^+$ channel (Johansson *et al.*, 2006).

We generated stable transformants with GORK-GFP in wild-type Arabidopsis and in the *syp121* mutant that lacks the vesicle-trafficking (SNARE) protein SYP121 (=SYR1/PEN1) (Leyman *et al.*, 1999) and is known to affect $K^+$ channel traffic and function (Honsbein *et al.*, 2009; Grefen *et al.*, 2010a; Eisenach *et al.*, 2012). Leaves of $T_2$ and $T_3$ lines from three independent transformation events were selected; in every case GORK-GFP fluorescence appeared in puncta around the cell periphery (Figure 1), both in wild-type and *syp121*mutant backgrounds. Fluorescence was detected in epidermal cells and guard cells (Figure 1a), distinct from chloroplast fluorescence (Figure 1b), and was localised to the cell periphery (Figure 1c–j). Intracellular GORK-GFP signals were observed in 7% ($n$ = 72) of stomata, mostly in very young, developing tissue. Analysis of puncta size yielded diameters of of 0.63 ± 0.01 μm ($n$ = 97) that were normally distributed and well-removed from the diffraction limit near 300 nm (Figure 1k). These dimensions compare favourably with the mean diameter of 0.5 μm observed for microdomains formed by KAT1-GFP in tobacco (Sutter *et al.*, 2006). Although similar in size and distribution, the puncta formed by RFP-tagged GORK did not co-localize (Figures S3 and S4) with puncta formed by the GFP-tagged KAT1 $K^+$ channel (Sutter *et al*., 2006, 2007). When expressed in tobacco, fluorescence was occasionally observed within epidermal cells; nonetheless, after plasmolysis GORK-GFP fluorescence was observed in Hechtian strands, indicating that the punctate structures were localised to the plasma membrane.

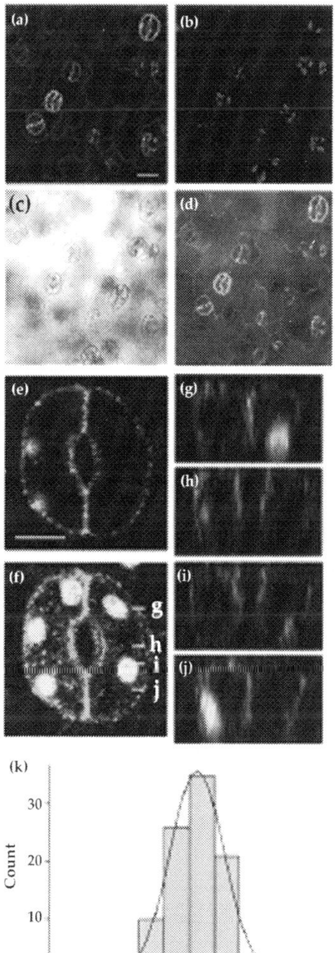


**Figure 1**: GORK localises in puncta at the periphery of Arabidopsis guard cells and epidermal cells.

(a–d) Confocal fluorescence images of GORK-GFP (green, a), chlorophyll fluorescence (red, b), bright field (c) and the corresponding overlay image (d) from the epidermis of an Arabidopsis leaf expressing *pUBQ10::GORK-GFP*. Scale bar, 20 μm.

(e) Lateral optical section through a single pair of Arabidopsis guard cells showing punctate distribution of GORK-GFP (green) and chloroplasts (yellow). Scale bar, 5 μm.

(f) 3D projection of the same guard cell pair in (e) reconstructed from a *z*-stack collected at intervals of 0.8 μm.

(g–j) *Z*-plane sections derived from (f) at the positions indicated by white lines in (f).

(k) Histogram analysis of the diameters of GORK-GFP puncta. Data were fitted to a gaussian function (solid line), indicating a normal distribution of diameters about a mean of 632 ± 1 nm.

## GORK-GFP Puncta are Sensitive to the External KCL Concentration

KAT1-GFP is internalised from the plasma membrane following application of ABA (Sutter *et al.*, 2007), and by treatments known to elevate $[Ca^{2+}]_i$ (Grabov and Blatt, 1998), studies showing its recycling by SYP121-mediated traffic underpins programmed stomatal closure (Leyman *et al.*, 1999; Eisenach *et al.*, 2012). We used the treatments to elevate $[Ca^{2+}]_i$, anticipating that GORK traffic would complement that of the KAT1 channel. Intact leaves were incubated for 2 h under light after infiltration with Depolarising Buffer (DB) containing 100 mM KCl and 0 $CaCl_2$ to open stomata, and stomatal closure was induced by exchange with Hyperpolarising Buffer (HB) containing 0.1 mM KCl and 10 mM $CaCl_2$ (Eisenach *et al.*, 2012). We found that GORK-GFP did not internalise following HB treatment, unlike KAT1 (Eisenach *et al.*, 2012). Instead, the punctate distribution of GORK-GFP was lost in DB, the fluorescence redistributing around the cell periphery. This process was fully reversible on transfer back to HB and was evident in both wild-type and *syp121*mutant Arabidopsis. However, treatments with ABA had no effect on GORK-GFP distribution (Figures S6 and S7). We reasoned that the changes in the punctate distribution observed between DB and HB were likely to be related to the ionic content of the buffer rather than to $Ca^{2+}$ or ABA signalling *per se*. We repeated the experiment with solutions of 100 mM KCl and 0.1 mM KCl alone. Again, we found, GORK-GFP puncta were clearly visible in 0.1 mM KCl (Figure 2a,b) but, in the same guard cells, the punctate distribution was lost when the external KCl concentration was raised to 100 mM (Figure 2c,d). To exclude an effect of osmotic strength, we repeated these experiments with 0.1 mM KCl including Mannitol to adjust the osmotic strength to that of the 100 mM KCl (200 ± 10 mOsmol). No difference in GORK-GFP distribution was observed between treatments with and without mannitol, thus excluding osmotic effects as the explanation for the changing GORK-GFP distribution.

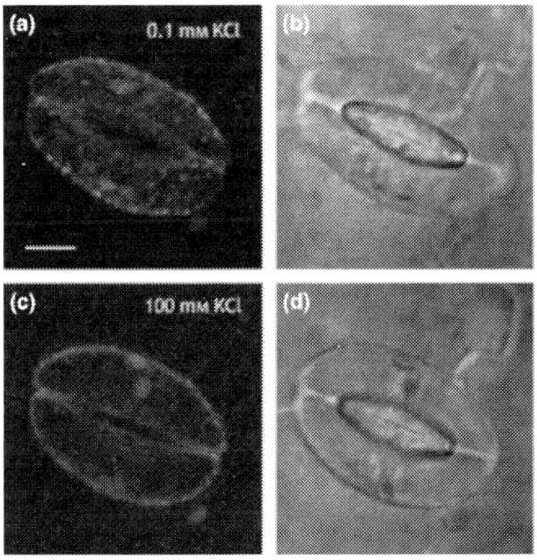


**Figure 2**: GORK-GFP fluorescence is punctate in 0.1 mMKCl, but is diffuse in 100 mM KCl.

GORK-GFP fluorescence (a, c) and brightfield images (b, d) in a pair of guard cells in 0.1 mM KCl (a, b) and after transfer to 100 mM KCl (c, d). Scale bar, 5 µm. Images in (a) and (c) are overlaid, 3D projections of GORK-GFP(green) and chloroplast (red) fluorescence from *z*-stacks taken at intervals of 0.7 µm. Equivalent results were obtained with Hyperpolarising and Depolarising Buffers (see text).

## GORK-GFP Redistribution is not Associated with Endocytic Traffic

The dissolution of KAT1 puncta is marked by a substantial increase in mobility and endocytosis of the channels (Sutter *et al.*, 2007; Eisenach *et al*., 2012). To determine whether GORK-GFP dispersal was similarly associated with endocytosis from the plasma membrane, we first examined the mobility of GORK-GFP by time-lapse and by fluorescence recovery after photobleaching (FRAP) analysis. Figure 3 illustrates one of eight independent time-lapse experiments with GORK-GFP puncta identified at the stomatal surface in tangential scans (Figure 3a,b). Kymographic analysis (Figure 3c) yielded no evidence of lateral movement. We used FRAP analysis in Arabidopsis to determine the recovery of fluorescence at the cell periphery, calculating the mobile fraction of GORK-GFP fluorescence from the signal recovery. GORK-GFP was photobleached locally at the periphery of guard cells preincubated in 0.1 and 100 mM KCl, and its recovery within the photobleached regions was monitored. Figure 3(d) summarizes the result of 12 independent experiments. FRAP analysis yielded similar results in 0.1 and 100 mM KCl: fluorescence recovered with a mean halftime of 36 ± 4 sec in 0.1 mM KCl and 28 ± 7 sec in 100 mM KCl, a difference that was not significant at the $P < 0.05$ level. Additionally, in both KCl concentrations the GORK-GFP signal was predominantly non-mobile, with a mobile fraction of 9 ± 1% in 100 mM KCl and of 14 ± 6% in 0.1 mM KCl. Again, the difference was not statistically significant at the $P < 0.05$ level.

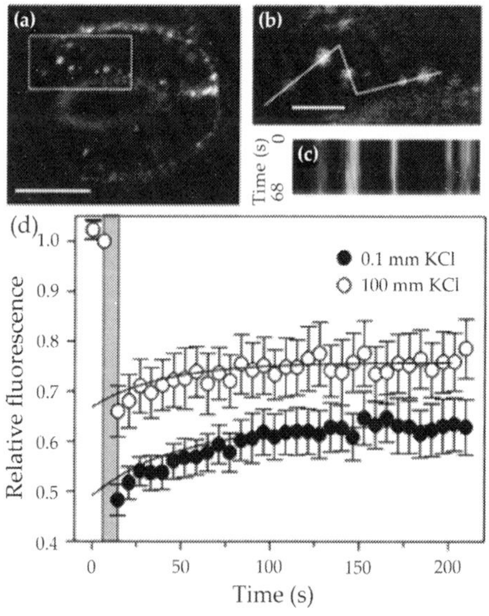


**Figure 3**: The macroscopic mobility of GORK-GFP at the cell periphery is not affected by extracellular KCl concentration.

(a) Tangential section of a guard cell pair with GORK-GFP puncta in 0.1 mM KCl. Scale bar, 5 μm.

(b) Close-up of the boxed region in (a) with the linescan for kymographic analysis as indicated. Scale bar, 2 μm.

(c) Kymograph over the linescan in (b) collected over a time interval of 68 sec. The *x*-axis corresponds to the linescan dimension and the *y*-axis corresponds to time. The analysis shows that GORK-GFP is positionally stable over this time period.

(d) Fluorescence recovery after photobleaching (FRAP) of the GORK-GFP recorded from guard cells in 0.1 mM KCl (closed circles, $n = 12$) and in 100 mM KCl (open circles, $n = 11$). Fluorescence was normalised to the signals at the start of the experiments after correcting for fluorescence decay (Sutter *et al*., 2006, 2007) and fluorescence recoveries after the photobleaching period (grey) were fitted to a single exponential function (solid lines) to derive halftimes and amplitudes for recovery.

We used aqueous, two-phase partitioning as an independent test for GORK redistribution between the plasma membrane and endomembranes, including endocytotic vesicles (Sutter *et al*., 2006, 2007; Honsbein *et al*., 2009). GORK polyclonal antibodies were raised against a synthetic peptide corresponding to a GORK-specific N-terminal amino-acid sequence. The antibodies were purified and verified for specificity in detecting GORK after immunoblotting. Figure 4(a) shows an immunoblot of soluble (S) and microsomal (M) protein fractions probed with αGORK antibody and the pre-immune serum. The antibody detected a band near the 80 kDa marker in the microsomal protein fraction and close to the 93.8 kDa predicted for GORK. Analysis of microsomal proteins extracts from wild-type plants, a transgenic GORK-GFP overexpressing line probed with αGFP antibody, and the *gork* mutant confirmed this band as the GORK protein (Figure 4b).

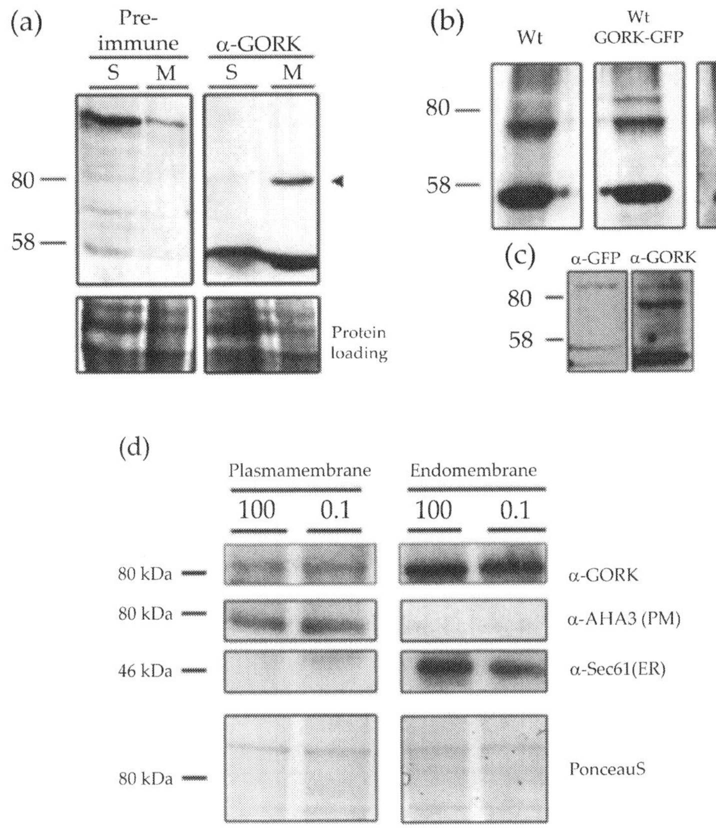


**Figure 4**: αGORK antibody raised to an N-terminal peptide shows no substantial redistribution of the $K^+$channel between plasma membrane and endomembranes in Arabidopsis.

(a) Immunoblot (top) and Ponecau S stain (bottom) of wild-type Arabidopsis soluble (S) and microsomal (M) protein fractions, probed either with pre-immune serum (*left*) and with the purified αGORK antibody (*right*). The αGORK antibody uniquely detected a single band around 80 kDa in the microsomal fraction (*arrowhead, right*).

(b) Immunoblot of Arabidopsis microsomal fractions from wild type (Wt), a transgenic line expressing GORK-GFP (GORK-GFP), and the *gork* null mutant. Note the double band in the GORK-GFP fraction corresponds to endogenous GORK and recombinant GORK-GFP. The 80 kDa band is missing from the *gork* mutant. The lower intensity of the GORK-GFP band probably is a consequence of co-suppression of the transgene.

(c) Immunoblot of Arabidopsis microsomal fractions from a transgenic line expressing GORK-GFP and probed with αGFP antibody. Note that the antibody detects a specific band above 80 kDa. An additional band is evident on stripping and reprobing with αGORK antibody, consistent with the presence of the native channel.

(d) Endogenous GORK is not affected by the external buffer or KCl concentration.

Immunoblot and PonceauS stain of plasma membrane and endomembrane protein fractions from aqueous two-phase partitioning of Arabidopsis leaves pretreated with either 0.1 or 100 mM KCl (0.1 and 100, *indicated above*). Molecular weights (*left*) and staining or antibody used for detection (*right*) are indicated. PonceauS stain of the membrane (*bottom*) shows equal protein loading within each gel. The membrane was probed with αGORK antibody (αGORK) before stripping and re-probing, first with an antibody against the plasma membrane $H^+$-ATPase AHA3 (αAHA3), and an antibody against the ER-localised translocon Sec61 (αSec61).

We separated plasma membrane and endomembrane fractions after pretreatments of leaf tissues with 0.1 and 100 mM KCl. Figure 4(c) shows an immunoblot of plasma membrane and endomembrane protein fractions from one of two independent experiments, each yielding similar results. Equal loading between samples of plasma and of endomembrane fractions was verified by Ponceau staining and purity was confirmed by probing against the plasma membrane marker AHA3 (Pardo and Serrano, 1989) and the endomembrane marker Sec61 (Yuasa *et al*., 2005). GORK bands of comparable intensity were observed in plasma membrane fractions irrespective of pretreatments with 0.1 and 100 mM KCl (Figure 4c, *left*). GORK bands were also detected in the endomembrane fractions (Figure 4c, *right*). An absolute comparison with the plasma membrane fractions is not possible, because of differences in total protein, but a ratiometric comparison between treatments showed little difference in relative distributions. The band density ratios for GORK between plasma membrane fractions in 0.1 and 100 mM KCl yielded values of 1.3 in the first experiment and 1.1 in the second. Thus, KCl treatments did not have a major effect on GORK distribution to the plasma membrane, indicating that vesicle traffic to and from the plasma membrane was unlikely to contribute appreciably to the changes observed in GORK-GFP distribution *in vivo*.

## GORK Redistribution is Fast and Mirrors Alkali Cation-Dependent Channel Gating

We analysed the time-course of puncta dispersal, expressing GORK in tobacco epidermis which showed the same reversible formation of puncta and simplified manipulations. Transformed leaf sections were infiltrated with 0.1 mM KCl and images collected before one half of the sections were re-infiltrated with 100 mM KCl. Samples were randomised and images were collected and analysed in blind assays following the second infiltration. The numbers of cells showing puntate distributions were assessed and then normalised to the measurements prior to the second infiltration. Figure 5 summarises these data with images collected before (a,c) and after (b,d) infiltration with 100 mM KCl, and shows a rapid decrease in the percentage of cells showing puncta, the count dropping to approximately 40% (Figure 5e) within 10 min of treatments. Exponential fittings yielded a halftime of 2.4 ± 0.5 min for the response. This analysis overestimates the halftime, as the experiments do not take into account the $K^+$ diffusion time across the cell wall which can slow diffusion by a factor of $10^{-2}$ to $10^{-4}$ (Canny, 1990). Assuming a $10^{-3}$ -fold decrease, for $K^+$ the effect implies a diffusion coefficient of $6.10^{-9}$ $cm^2$ $sec^{-1}$ and an increase in the root mean square time for bulk diffusion across the 2 μm of cell wall to 100–200 sec (Hille, 2001), roughly equivalent to the time for GORK-GFP puncta dispersal. These values contrast with the dispersion of KAT1-GFP puncta during endocytosis (Sutter *et al.*, 2007), which displayed a halftime of 10–12 min and was complete only after 40–60 min following ABA treatments. The comparison supports our earlier observations indicating that the redistribution of GORK-GFP is not associated with endocytosis.

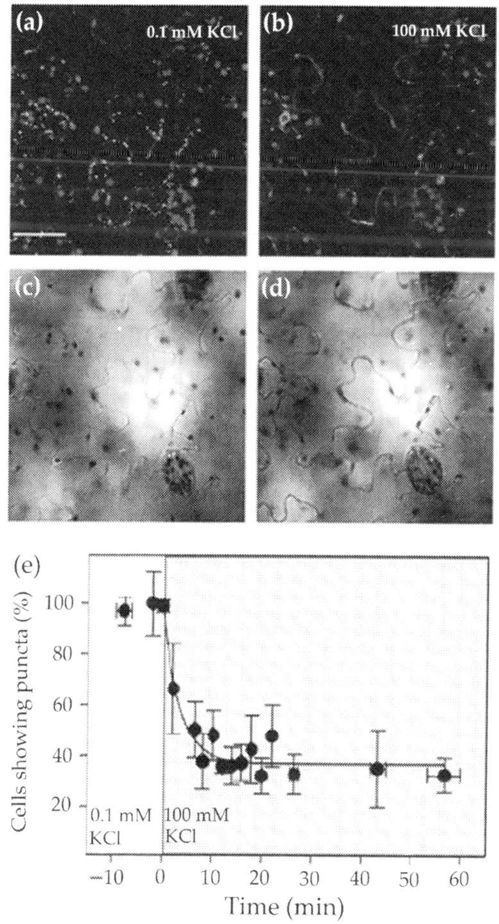


**Figure 5**: GORK-GFP puncta disperse rapidly on transfer to 100 mM KCl.

(a–d) Confocal images (a, b) of GORK-GFP (green) and chloroplast fluorescence (red) expressed in tobacco epidermal cells with corresponding bright field (c, d) images collected in 0.1 mM KCl (a, c) and 30 min after transfer to 100 mM KCl (b, d) showing conversion from punctate to diffuse distribution of the channel. Scale bar, 50 μm.

(e) The percentage of transformed tobacco cells displaying GORK-GFP puncta as a function of time on transfer from 0.1 to 100 mM KCl (grey). Data are means ±SE of $n \geq 5$ independent experiments and were fitted by non-linear least-squares to an exponential decay function of time (solid line).

The gating of GORK, and its homologs in *Vicia* and tobacco, responds over a similar time course to changes in alkali cation concentrations, including $Cs^+$ (Blatt, 1988; Armstrong *et al.*, 1995; Blatt and Gradmann, 1997; Blatt *et al.*, 1999; Bauly *et al.*, 2000). We examined whether the distribution GORK in puncta was sensitive to the $K^+$ or to the $Cl^-$ concentration, and whether it showed ionic sensitivities similar to those of gating. Experiments were carried out as before with pre-infiltrations of 0.1 mM KCl, but with either 100 mM$KNO_3$ or 50 mM $MgCl_2$ in the second infiltration. Leaf sections were examined for GORK-GFP distribution after 20 min. The results (Figure 6a) showed a similar decrease in punctate frequency to 45 ± 10% in leaf sections treated with 100 mM $KNO_3$ but not with 50 mM $MgCl_2$. To avoid the 'digitizing' effect of counting cells, we also quantified GORK distribution by analysis of the relative standard deviations (RSD) in fluorescence intensity around the periphery of the cells. This approach yielded the same pattern in $K^+$dependence. As a further test, we also replaced the KCl solution with 100 mM CsCl. Again, these substitutions gave a punctate distribution of 33 ± 12% and a corresponding decrease in RSD, similar to that 100 mM KCl. We challenged leaf sections with different $K^+$ concentrations for comparison with GORK channel gating. Leaf disks were pretreated as before, then split for blind image acquisition and the section fragments re-infiltrated with 0.1, 10, 30, 50 and 100 mM KCl. The results (Figure 6b) showed that the percentage of cells with puncta decreased progressively with increasing KCl concentrations above 10 mM, although a statistical analysis belies the trend. This pattern is similar to that of the ensemble conductance maximum ($G_{max}$) observed for the outward-rectifying $K^+$ channels in both *Vicia* and Arabidopsis guard cells (Figure 6c). Quantified by RSD, the effect of KCl concentration is evident across the entire concentration range. Finally, we tested whether the $K^+$ channel blocker $Ba^{2+}$ (Armstrong and Taylor, 1980; Roelfsema and Prins, 1997; Romano *et al.*, 1998; Hamilton *et al.*, 2000) might affect the distribution of GORK-GFP. In this case, leaf disks were pretreated with 0.1 mM KCl and split as before, one half of each disk then re-infiltrated either with 50 mM KCl or first with 30 mM $BaCl_2$ and, after 30 min, with 30 mM $BaCl_2$ plus 50 mM KCl. Figure 6c shows that infiltrating with $BaCl_2$ was sufficient to prevent the dispersal of the GORK puncta. These findings indicate a close association between GORK gating and the distribution of GORK-GFP in puncta and we return to the observations below.

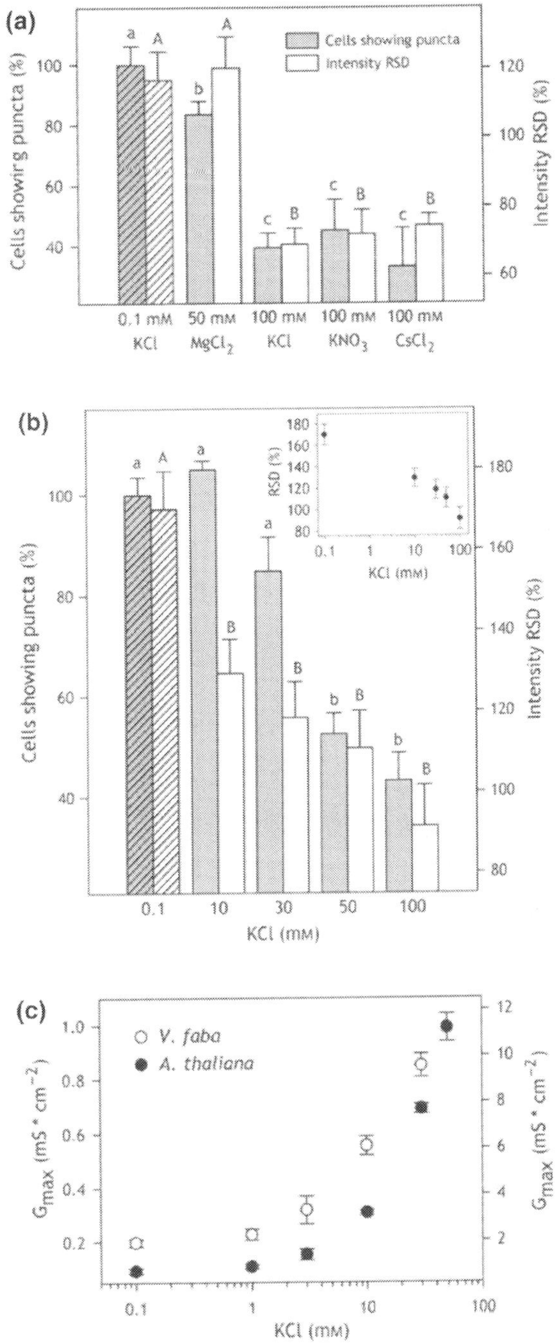
(a)
Cells showing puncta
Intensity RSD
Cells showing puncta (%)
Intensity RSD (%)
0.1 mM KCl
50 mM MgCl2
100 mM KCl
100 mM KNO3
100 mM CsCl2
(b)
Cells showing puncta (%)
Intensity RSD (%)
RSD (%)
KCl (mM)
(c)
V. faba
A. thaliana
Gmax (mS * cm−2)
KCl (mM)

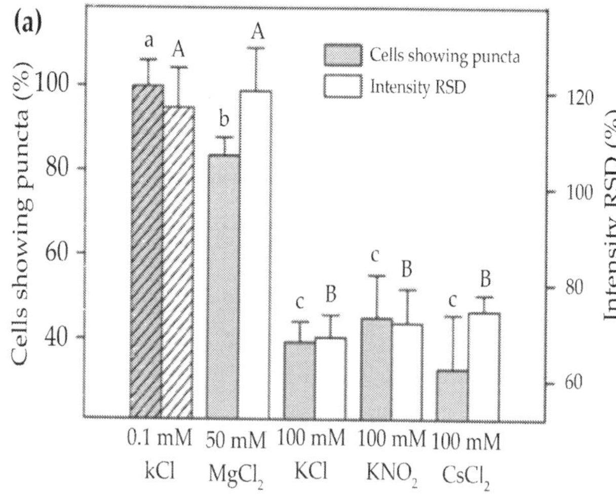


**Figure 6**: GORK-GFP dispersion is subject to $K^+$, not $Cl^-$, is equally sensitive to $K^+$ substitution with $Cs^+$, and is suppressed by pretreatment with the channel blocker $Ba^{2+}$.

(a) Quantification of GORK-GFP puncta in 0.1 mM KCl (diagonal shading) and after treatment with 100 mM KCl, 100 mM $KNO_3$, 50 mM $MgCl_2$ and 100 mM CsCl. Punctate distribution was analysed by two methods. The left ordinate and grey bars represent the percentage of cells showing puncta in any one image. The right ordinate and white bars represent the relative standard deviation (RSD) of intensities determined from a 1-pixel-wide line around the periphery of cells and normalized to the intensity means. Data are means ±SE of $n > 9$ independent experiments for each treatment.

(b) Percentage of cells showing GORK-GFP puncta (left ordinate, grey bars) and the intensity SD (right ordinate, white bars) in 0.1 mM KCl (diagonal shading) and after treatment with 10, 30, 50 and 100 mM KCl. Data are means ±SE of $n > 6$ independent for each treatment. *Inset*: Intensity RSD as a function of [KCl] outside plotted on a logarithmic scale.

(c) Maximum ensemble conductance of GORK in Arabidopsis guard cells (filled circles) and the homologous, outward-rectifying $K^+$ current in *Vicia* guard cells (open circles) as a function of $[K^+]$ outside. Data are derived and replotted as means ± SE from prior publications [see (Blatt, 1988; Thiel and Blatt, 1991; Thiel *et al.*, 1992; Blatt and Gradmann, 1997; Brearley *et al.*, 1997; Blatt *et al.*, 1999; Leyman *et al.*, 1999; Sokolovski and Blatt, 2004; Chen *et al.*, 2012; Wang *et al.*, 2012)].

(d) Percentage of cells showing GORK-GFP puncta (left ordinate, grey bars) and the normalized relative standard deviation in intensity (RSD; right ordinate, white bars) in 0.1 mM KCl (diagonal shading), after 30 min pretreatment with 30 mM of the $K^+$ channel blocker $BaCl_2$, and after addition of 50 mM KCl to the 30 mM $BaCl_2$-treated samples. Data are means ±SE of $n = 5$ independent experiments. Significance at $P < 0.05$ was analysed by ANOVA and pair-wise multiple comparison and are indicated in each case by lettering for the percentage of cells showing puncta (small letters) and RSD (capital letters).

## DISCUSSION

Only recently has attention in channel regulation turned to the traffic of ion channel proteins and to changes in the population of ion channels active at the plasma membrane. Notable among the few studies in plants is the ABA-evoked endocytosis of the KAT1 $K^+$ channel, its slower recycling to the plasma membrane (Sutter *et al.*, 2007), and a role for the vesicle-trafficking protein SYP121 in recovery from programmed closure of guard cells (Eisenach *et al.*, 2012). Until now, no information has been forthcoming relating to the traffic of GORK, the major outward-rectifying $K^+$ channel that facilitates $K^+$ efflux during stomatal closure in Arabidopsis, although changes in ensemble channel conductance and the voltage-dependence of gating are known for these and homologous outward-rectifying channels (Blatt, 1990, 1992; Thiel *et al.*, 1992). We anticipated these effects on GORK current might result from channel traffic, but the studies outlined above contradict this expectation. Here we report that the GORK $K^+$ channel, like KAT1, appears in discrete puncta at the guard cell periphery. These clusters form independently of the structures assembled by KAT1 (Meckel *et al.*, 2004; Sutter *et al.*, 2006, 2007) and are not affected by ABA, but form and disperse reversibly with external $K^+$ concentration. Three additional lines of evidence associate GORK puncta and their dispersal with short-term changes in channel gating and conductance, rather than with channel traffic. (i) The mobility of GORK-GFP was unaffected by ABA, nor was it altered between the punctate and dispersed forms of the channel. (ii) Two-phase partitioning experiments yielded no substantial difference in membrane distribution between punctate and dispersed forms of the channel, unlike KAT1 (Sutter *et al.*, 2007; Eisenach *et al.*, 2012). (iii) Dispersal of the GORK puncta was rapid and demonstrably sensitive to extracellular cations, parallelling gating and block in these channels. We suggest that conformational changes of the channel and its gate with $K^+$ are connected structurally to the clustering of GORK within the plane of the plasma membrane. Such clustering may signal changes in the external ionic environment, possibly by exploiting the changing capacity of GORK, when in puncta, to interact with other cytoplasmic or plasma membrane proteins.

## GORK Puncta are Distinct from the Trafficking Platforms of the Kat1 Channel

Clustering of the GORK $K^+$ channels within puncta at the surface of tobacco epidermal cells and Arabidopsis guard cells complements studies of the other major $K^+$ channel subgroup, represented by KAT1 (Meckel *et al.*, 2004; Sutter *et al.*, 2006, 2007; Homann *et al.*, 2007), and supports electrophysiological evidence that many functional $K^+$ channels in plants are distributed non-homogenously over the plasma membrane surface (Tester, 1990; Hille, 2001). The puncta formed by GORK are similar in size to those of KAT1, but are physically distinct as evident from co-localisation studies (Figures 1, S3 and S4). GORK puncta also differ from those of KAT1 in their mobility and sensitivity to external stimuli. Whereas the puncta of KAT1 are lost in ABA (Sutter *et al.*, 2007) and reform in a manner sensitive to the vesicle-trafficking protein SYP121 (Eisenach *et al.*, 2012), we found GORK to be insensitive to ABA and SYP121 but dependent on the extracellular $K^+$ concentration (Figures 2, 5 and 6). These findings alone distinguish the structural characteristics of the two $K^+$ channels.

Physical clustering of channel and other proteins is well known also in animal cells (Lai and Jan, 2006; O›Connell *et al.*, 2010; Zilly *et al.*, 2011; Pristera *et al.*, 2012) and, in neurons, often demarcates platforms for vesicle fusion and channel traffic (Lai and Jan, 2006; Deutsch *et al.*,

2012). Much the same conclusion was reached for the puncta formed of the KAT1 channel when expressed in tobacco (Sutter *et al.*, 2006, 2007): Our analysis of GORK-GFP recovery after photobleaching suggested that roughly 10% may be mobile within the cell (Figure 3), possibly through vesicle traffic. However, two-phase partitioning failed to uncover a substantial change in GORK distribution between plasma- and endomembranes fractions, even following transfer to 100 mM KCl to evoke a dispersal of the puncta. GORK puncta also dissociated reversibly in the *syp121* mutant that is defective in KAT1 trafficking to the plasma membrane (Eisenach *et al.*, 2012). We cannot discount a role for vesicle traffic in the mobility of GORK, but these results indicate that any such contribution to GORK clustering and dispersal is small and of a magnitude that might only be resolved by fluorescence analysis *in vivo* over extended time periods (Luu *et al.*, 2012).

We stress that plant ion channels, including GORK, normally express at levels that are too low for detection by confocal microscopy. We used GORK-GFP constructs driven by the Ubiquitin-10 promoter, *pUBQ10*, that drives expression at lower levels that the 35S CaMV promoter (Grefen *et al.*, 2010b), and we confirmed channel function of GORK-GFP. We cannot rule out that *pUBQ10*-driven expression misrepresents GORK distribution. Nonetheless, GORK-GFP generally was not observed to accumulate in the endoplasmic reticulum, and GORK puncta dissociated reversibly *in vivo*. These findings support the idea that clustering of the $K^+$ channels is physiologically relevant. Clustering of membrane proteins in lipid rafts is common and often accompanied by restricted lateral diffusion (Bhat *et al.*, 2005; Lai and Jan, 2006; Mongrand *et al.*, 2010). Such rafts have been implicated in signal transduction, polarised secretion, membrane traffic and pathogen entry. Membrane rafts are generally acknowledged to form below the resolution limit of the light microscope and, thus, are 5- to 10-fold smaller than the GORK-GFP clusters we observed. However, such small-scale rafts may aggregate in response to external stimuli (Bhat *et al.*, 2005; Mongrand *et al.*, 2010), potentially underpinning the larger clusters observed with the GORK $K^+$ channel. Our FRAP analysis (Figure 3) showed that cluster dispersion was not associated with a significant increase in mobility or altered macroscopic diffusion. Thus, it may be that GORK remains anchored with other proteins in raft-like microdomains, even in high external $K^+$.

## A Physiological role for GORK Clustering Associated with $K^+$-Dependent Gating?

The most striking aspects of the physical organisation of GORK are its parallels to the $K^+$-sensitivity of gating and ensemble conductance. The activities of the Kv-like channels GORK and SKOR in Arabidopsis, as well as outward-rectifying $K^+$ channels in other plants, show a characteristic dependence on external alkali cation concentration as well as voltage (Blatt, 1988; Armstrong *et al.*, 1995; Blatt and Gradmann, 1997; Blatt *et al.*, 1999; Ache *et al.*, 2000; Bauly *et al.*, 2000; Johansson *et al.*, 2006). These characteristics include a displacement in channel conductance in parallel with the equlibrium voltage for $K^+$, an increase in the ensemble conductance maximum, and the ability of $Cs^+$ to substitute for $K^+$ outside. Channel gating responds in a near-instantaneous manner to $K^+$ within the timescale of bulk solution changes, and changes in ensemble conductance are only marginally slower (Blatt, 1988; Blatt and Gradmann, 1997). Johansson *et al.* (2006) have shown that external cations affect the $K^+$ channel through interactions of the pore loop and pore-lining α-helices that, in turn, are likely to influence the conformational states of the outer ring of polypeptides formed by the voltage-sensor domains of the channels (Dreyer and Blatt, 2009): in effect, occupancy of the inner vestibule of the pore at high $K^+$ concentrations appears to favour a temporally-stable, 'locked-closed' state. We speculate that these conformations may affect the local packing of the $K^+$ channel proteins within the plasma membrane, which becomes evident at a macroscopic level as cluster dissolution. Barium suppression of the GORK puncta dispersal (Figure 6), and the parallels in timecourse for cluster dissolution and gating in 100 mM $K^+$, makes this hypothesis especially compelling. Barium enters the pores of many $K^+$ channels (Vergani *et al.*, 1997; Hamilton *et al.*, 2000; Jiang and MacKinnon, 2000; Hille, 2001), including GORK (Roelfsema and Prins, 1997; Ache *et al.*, 2000), and is thought to block by occluding the selectivity filter (Jiang and MacKinnon, 2000; Rowley and Roux, 2013). It follows that, with $Ba^{2+}$ present, the channel is likely to become trapped in the 'locked-closed' state with $K^+$ in the inner vestibule of the channel, isolated from the pool of $K^+$ in the cytosol and, of course, from changes in $K^+$ concentration outside. Thus, the effect on GORK cluster dispersal implies a close connection to channel conformation in gating and to channel conductance.

What role might reversible clustering of the $K^+$ channels serve? In addition to the voltage-dependence of the channel gate, at least two physiological phenomena may be connected to this behaviour. First, we suggest that clustering may serve to control the pool of GORK channels active in the plasma membrane by sequestering channels, possibly in a non-activatable state. Such behaviour might be seen to withdraw channels from the active pool at the membrane much as an iceberg constrains a volume of water within a crystalline array, preventing its movement and ability to act as a general solvent. For the channels at low external $K^+$ concentrations, the effect of such an 'iceberg' model might be seen to confer a synergy to the voltage-dependence of the channels as the $K^+$ concentration decreases while reducing the maximum $K^+$ flux capacity of the channel population (Blatt, 1988, 1990, 1992; Thiel *et al*., 1992; Armstrong *et al*., 1995; Blatt *et al*., 1999; Bauly *et al*., 2000) (see also Figure 6). Similar concepts have been proposed to explain the clustering of mammalian $K^+$ channels (Misonou and Trimmer, 2004), although correlating channel clustering and ensemble conductance has proven difficult (O›Connell *et al*., 2010).

We can propose a second role for GORK clustering that builds on this iceberg analogy. $K^+$ flux across the plasma membrane is known to be tightly linked to membrane surface area and volume control (Zonia and Munnik, 2007). Work from this laboratory has uncovered the wide-spread interaction between subsets of vesicle-trafficking (SNARE) proteins and $K^+$ channels that have profound effects on channel gating and net $K^+$uptake (Honsbein *et al*., 2009, 2011; Grefen *et al*., 2010a). These interactions are implicated in priming the SNAREs to catalyse membrane fusion (Karnik *et al*., 2013). It is plausible, therefore, that the cell might exploit the $K^+$-sensitivity of GORK clustering to regulate vesicle traffic of other membrane components, adjusting the rate of cell surface expansion with $K^+$ uptake and cell volume. For example, $K^+$ accumulation is enhanced in guard cells at high external $K^+$, which also promotes the dissolution of GORK clusters. We can imagine that 'freeing' the channels from within the clusters facilitates SNARE and other interactions to accelerate vesicle traffic for the increase in cell volume.

In conclusion, we have uncovered a new and unusual process that may help to understand how the $K^+$channels of Arabidopsis guard cells, and other plant cells, sense and respond to external $K^+$ concentration.

To date, the phenomenon of $K^+$-dependent channel clustering in plants appears unique to the outward-rectifying $K^+$ channel GORK, but we anticipate that similar characteristics will be forthcoming for other outward-rectifying $K^+$ channels in plants. For GORK, the properties of channel clustering shows many hallmarks of $K^+$-dependent gating and conductance, implicating a close connection that is now open to testing, for example through a combination of protein domain exchange with a non-clustering channel such as AKT1 (Honsbein *et al*., 2009). We suspect that the organisation of GORK and other channels in this manner is likely to contribute both as a mechanism regulating the channels and as a platform for sensing and responding to environmental cues.

# MATERIAL AND METHODS

## Molecular Cloning

The GORK coding sequence was amplified from rosette-leaf cDNA using the oligonucleotide primers 5′-attB1 TA ATG GGA CGT CTC CGG-3′ and 5′-attB2 G TGT TTG ATC AGT AGT ATC ACT G-3′, recombined into Gateway™ entry vector pDONR207 with BP clonase II (Life Technologies, http://www.lifetechnologies.com), and transformed into the *E. coli* strain *CopyCutter EPI400* (Cambio, http://www.cambio.co.uk) (Grefen *et al*., 2009). Clones were verified by sequencing (GATC, http://www.gatc-biotech.com/) and recombined into expression vectors via LR-reaction (Life Technologies). The *pUBQ10* vector system was used to generate GORK constructs C-terminally tagged with GFP and RFP (Grefen *et al*., 2010b). Constructs were transformed into *Agrobacterium tumefaciens* (strain GV3101) and were verified by rescue in *E. coli* and restriction digest analysis. A single *A. tumefaciens* clone was used for transient transformation of tobacco and stable transformation of Arabidopsis.

The coding sequence for GORK-GFP was amplified from the *pUBQ10 -DEST* expression construct for re-cloning into the oocyte expression vector *pGT-DEST* (Grefen *et al*., 2010b) using the oligonucleotide primers 5′-A GTT AAC ATG GGA CGT CTC CGG AGA C-3′ and 5′-C AGG CCT TTA TAA CTT GTA CAG CTC GTC CAT GC-3′. These primers incorporated 5′ *HpaI* and *StuI* overhangs, respectively. Blunt-end restriction of the oocyte expression vector *pGT-DEST* with *MscI* and of the PCR product with *HpaI* and *StuI* was followed by ligation using T4 DNA Ligase (Promega, http://www.promega.com/) before verification by sequencing.

## Electrophysiology

GORK-GFP was expressed in *Xenopus* oocytes using *pGT GORK-GFP* plasmid DNA (Vergani *et al.*, 1998; Grefen *et al.*, 2010a), and currents were recorded by two-electrode voltage clamp as described previously (Johansson and Blatt, 2006). Measurements were carried out using microelectrodes filled with 3 M $K^+$-acetate and with oocytes bathed in 10 mM HEPES-NaOH, pH 7.2, with 1.8 mM $MgCl_2$, 1.8 mM $CaCl_2$ and additions of 10 mM, 30 mM or 96 mM KCl balanced by NaCl to give a total salt concentration of 100 mM.

## Plant Growth, Transformation and Microscopy

*Arabidopsis thaliana* Col0, *syp121* and *gork* [SALK_082258C (Alonso *et al.*, 2003)] mutant plants were grown and transformed by floral dip (Clough and Bent, 1998) and $T_1$ transformants were selected for BASTA®resistance (1:1000 dilution; Bayer Cropscience, http://www.cropscience.bayer.com). Seeds of the $T_2$, $T_3$ and $T_4$ generations were selected in the same manner. Leaves of *Nicotiana tabacum* (tobacco) were transformed by infiltration with *Agrobacterium* carrying the GORK-GFP construct (Geelen *et al.*, 2002), and were analysed 72–120 h after transformation.

GORK-GFP fluorescence was resolved on a LSM510-META confocal microscope (Zeiss, http://www.zeiss.com) using the 488-nm line of an Argon laser and fluorescence was collected as reflectance from a 545-nm dichroic mirror after passage through a 505–530 nm interference filter. Chloroplast fluorescence was collected concurrently after passage through the 545-nm dichroic mirror and a long pass 560 nm emission filter. Fluorescence recovery after photobleaching (FRAP) was carried out after fluorophore bleaching with the 488 laser using a pixel time of 1.6 sec (2.4 mJ exposure energy). Bleaching, detection and analysis settings were standardised between experiments for ROI dimensions, laser power, magnification and detection settings and fluorescence was corrected for decay during acquisition.

Imaging was carried out using young rosette leaves of Arabidopsis mounted in a custom-build perfusion chamber that enabled continuous perfusion and solution changes during experiments. Buffers included 0.003% Silwet L77 (Lehle Seeds, http://www.lehleseeds.com) to ensure cuticle penetration. For tobacco, imaging was carried out on 1-cm diameter leaf discs after vacuum infiltration with treatment buffers before remounting. All buffers were based on 10 mM NaMES, pH6.1. The salts KCl, $KNO_3$, $CsCl_2$, $MgCl_2$ and $BaCl_2$were added as indicated and osmotic strength was adjusted using mannitol. Abscisic acid (ABA) was prepared as before (Blatt and Armstrong, 1993) and diluted in buffer to 40 μM for use. For ABA treatments, leaves were infiltrated and incubated in 5 mM $Ca^{2+}$-MES, pH 6.1, 10 mM KCl for 2 h under 100 μmol $m^{-2}$ $sec^{-1}$white light. Images were collected, the leaves reinfiltrated using the same buffer with the addition of ABA, and the leaves re-imaged after 30 and 60 min. Tobacco disc analysis was carried out in blind assays. GORK puncta were quantified both as the percentage of cells showing puncta and by relative standard deviation (RSD) after normalizing to the intensity means.

## Protein Biochemistry

GORK distribution between plasma membrane and endomembrane fractions was determined by aqueous two-phase partitioning. Arabidopsis leaf tissues were infiltrated in either 100 mM KCl or 0.1 mM KCl buffers and incubated for 2 h. All subsequent steps were carried out at 4°C as described before (Honsbein *et al.*, 2009). Proteins were separated by electrophoresis and transferred onto PVDF membranes (Honsbein *et al.*, 2009; Karnik *et al.*, 2013) and were probed overnight with primary antibody, either αGORK [1:100]; αAHA [1:5000, (Villalba *et al.*, 1991)], αSec61 [1:3000, (Yuasa *et al.*, 2005)], or αGFP [1:200; Abcam, http://www.abcam.co.uk]. Horseradish peroxidase-coupled, α-rabbit IgG secondary antibody (1:100 000) was used for detection with WestFemto SuperSignal (Pierce; Thermo Scientific, http://www.thermofischer.com). In some experiments, membranes were re-probed after stripping by incubation in 100 mM β-Mercaptoethanol, 2% SDS, 62.5 mM Tris-HCl, pH 6.7 at 70°C for 45 min.

### Data Analysis

Statistical analyses were carried out using SIGMAPLOT 11 (Systat Software, Inc., http://www.sigmaplot.com). Non-linear least-squares fittings used a Marquardt–Levenberg algorithm (Marquardt, 1963). Significance was tested using Student›s *t*-test and analysis of variance (ANOVA). Otherwise, data are reported as means ± SE of *n* observations.

## ACKNOWLEDGEMENTS

We thank Christopher Grefen and Annegret Honsbein for support in molecular cloning and electrophysiology, respectively, and Amparo Ruiz-Prado for plant maintenance. This work was supported by a postgraduate grant-in-aid from Plant Bioscience Ltd. (CE), by BBSRC grants BB/H009817/1 and BB/I024496/1 (MRB), and by a Begonia Trust scholarship (MP).

## REFERENCES

1. Ache, P., Becker, D., Ivashikina, N., Dietrich, P., Roelfsema, M.R.G. and Hedrich, R. (2000) GORK, a delayed outward rectifier expressed in guard cells of Arabidopsis thaliana, is a K+-selective, K+-sensing ion channel. FEBS Lett. 486, 93–98.
2. Alonso, J.M., Stepanova, A.N., Leisse, T.J. et al. (2003) Genome-wide insertional mutagenesis of Arabidopsis thaliana. Science, 301, 653–657.
3. Armstrong, C.M. and Taylor, S.R. (1980) Interaction of barium ions with potassium channels in squid giant axons. Biophys. J. 30, 473–488.
4. Armstrong, F., Leung, J., Grabov, A., Brearley, J., Giraudat, J. and Blatt, M.R. (1995) Sensitivity to abscisic acid of guard cell K+ channels is suppressed by abi1-1, a mutant Arabidopsis gene encoding a putative protein phosphatase. Proc. Natl Acad. Sci. USA, 92, 9520–9524.
5. Bauly, J.M., Sealy, I.M., MacDonald, H. et al. (2000) Overexpression of auxin-binding protein enhances the sensitivity of guard cells to auxin. Plant Physiol. 124, 1229–1238.

6. Bhat, R.A., Miklis, M., Schmelzer, E., Schulze-Lefert, P. and Panstruga, R. (2005) Recruitment and interaction dynamics of plant penetration resistance components in a plasma membrane microdomain. Proc. Natl Acad. Sci. USA, 102, 3135–3140.
7. Blatt, M.R. (1988) Potassium-dependent bipolar gating of potassium channels in guard cells. J. Membr. Biol. 102, 235–246.
8. Blatt, M.R. (1990) Potassium channel currents in intact stomatal guard cells: rapid enhancement by abscisic acid. Planta, 180, 445–455.
9. Blatt, M.R. (1992) K+ channels of stomatal guard cells: characteristics of the inward rectifier and its control by pH. J. Gen. Physiol. 99, 615–644.
10. Blatt, M.R. (2000) Cellular signaling and volume control in stomatal movements in plants. Annu. Rev. Cell Dev. Biol. 16, 221–241.
11. Blatt, M.R. and Armstrong, F. (1993) K+ channels of stomatal guard cells: abscisic acid-evoked control of the outward rectifier mediated by cytoplasmic pH. Planta, 191, 330–341.
12. Blatt, M.R. and Grabov, A. (1999) H+-mediated control of ion channels in guard cells of higher plants. In pH in Biological Systems (Egginton, S., Taylor, E.W. and Raven, J.A., eds). Cambridge: Cambridge University Press, pp. 155–176.
13. Blatt, M.R. and Gradmann, D. (1997) K+-sensitive gating of the K+ outward rectifier in Vicia guard cells. J. Membr. Biol. 158, 241–256.
14. Blatt, M.R. and Thiel, G. (1993) Hormonal control of ion channel gating. Annu. Rev. Plant Physiol. Mol. Biol. 44, 543–567.
15. Blatt, M.R. and Thiel, G. (1994) K+ channels of stomatal guard cells: bimodal control of the K+ inward-rectifier evoked by auxin. Plant J. 5, 55–68.
16. Blatt, M.R., Thiel, G. and Trentham, D.R. (1990) Reversible inactivation of K+ channels of Vicia stomatal guard cells following the photolysis of caged inositol 1,4,5- trisphosphate. Nature, 346, 766–769.
17. Blatt, M.R., Grabov, A., Brearley, J., HammondKosack, K. and Jones, J.D.G. (1999) K+ channels of Cf-9 transgenic tobacco guard

cells as targets for Cladosporium fulvum Avr9 elicitor-dependent signal transduction. Plant J. 19, 453–462.

18. Blatt, M.R., Garcia-Mata, C. and Sokolovski, S. (2007) Membrane transport and Ca2+ oscillations in guard cells. In Rhythms in Plants (Mancuso, S. and Shabala, S., eds). Berlin: Springer, pp. 115–134.
19. Brearley, J., Venis, M.A. and Blatt, M.R. (1997) The effect of elevated CO2 concentrations on K+ and anion channels of Vicia faba L. guard cells. Planta, 203, 145–154.
20. Canny, M.J. (1990) Rates of diffusion in wheat leaves. New Phytol. 116, 263–268.
21. Chen, Z.H., Eisenach, C., Xu, X.Q., Hills, A. and Blatt, M.R. (2012) Protocol: optimised electrophyiological analysis of intact guard cells from Arabidopsis. Plant Methods, 8, 15–25.
22. Clough, S.J. and Bent, A.F. (1998) Floral dip: a simplified method for Agrobacterium-mediated transformation of Arabidopsis thaliana. Plant J. 16, 735–743.
23. Deutsch, E., Weigel, A.V., Akin, E.J., Fox, P., Hansen, G., Haberkorn, C.J., Loftus, R., Krapf, D. and Tamkun, M.M. (2012) Kv2.1 cell surface clusters are insertion platforms for ion channel delivery to the plasma membrane. Mol. Biol. Cell, 23, 2917–2929.
24. Dreyer, I. and Blatt, M.R. (2009) What makes a gate? The ins and outs of Kv-like K+ channels in plants. Trends Plant Sci. 14, 383–390.
25. Eisenach, C., Chen, Z.H., Grefen, C. and Blatt, M.R. (2012) The trafficking protein SYP121 of Arabidopsis connects programmed stomatal closure and K+ channel acivity with vegetative growth. Plant J. 69, 241–251.
26. Geelen, D., Leyman, B., Batoko, H., Di Sansabastiano, G.P., Moore, I. and Blatt, M.R. (2002) The abscisic acid-related SNARE homolog NtSyr1 contributes to secretion and growth: evidence from competition with its cytosolic domain. Plant Cell, 14, 387–406.
27. Grabov, A. and Blatt, M.R. (1998) Membrane voltage initiates Ca2+ waves and potentiates Ca2+ increases with abscisic acid in stomatal guard cells. Proc. Natl Acad. Sci. USA, 95, 4778–4783.
28. Grefen, C., Obrdlik, P. and Harter, K. (2009) The determination of

protein-protein interactions byt the mating-based split-ubiquitin system (mbSUS). Methods Mol. Biol. 479, 1–17.

29. Grefen, C., Chen, Z.H., Honsbein, A., Donald, N., Hills, A. and Blatt, M.R. (2010a) A novel motif essential for SNARE interaction with the K+ channel KC1 and channel gating in Arabidopsis. Plant Cell, 22, 3076–3092.
30. Grefen, C., Donald, N., Hashimoto, K., Kudla, J., Schumacher, K. and Blatt, M.R. (2010b) A ubiquitin-10 promoter-based vector set for fluorescent protein tagging facilitates temporal stability and native protein distribution in transient and stable expression studies. Plant J. 64, 355–365.
31. Hamilton, D.W.A., Hills, A., Kohler, B. and Blatt, M.R. (2000) Ca2+ channels at the plasma membrane of stomatal guard cells are activated by hyperpolarization and abscisic acid. Proc. Natl Acad. Sci. USA, 97, 4967–4972.
32. Hedrich, R., Neimanis, S., Savchenko, G., Felle, H.H., Kaiser, W.M. and Heber, U. (2001) Changes in apoplastic pH and membrane potential in leaves in relation to stomatal responses to CO2, malate abscisic acid or interruption of water supply. Planta, 213, 594–601.
33. Hille, B. (2001) Ionic Channels of Excitable Membranes. Sunderland, MA: Sinauer Press.
34. Homann, U., Meckel, T., Hewing, J., Hutt, M.T. and Hurst, A.C. (2007) Distinct fluorescent pattern of KAT1: GFP in the plasma membrane of Vicia faba guard cells. Eur. J. Cell Biol. 86, 489–500.
35. Honsbein, A., Sokolovski, S., Grefen, C., Campanoni, P., Pratelli, R., Paneque, M., Chen, Z.H., Johansson, I. and Blatt, M.R. (2009) A tripartite SNARE-K+ channel complex mediates in channel-dependent K+ nutrition in Arabidopsis. Plant Cell, 21, 2859–2877.
36. Honsbein, A., Blatt, M.R. and Grefen, C. (2011) A molecular framework for coupling cellular volume and osmotic solute transport control. J. Exp. Bot. 62, 2363–2370.
37. Hosy, E., Vavasseur, A., Mouline, K. et al. (2003) The Arabidopsis outward K+ channel GORK is involved in regulation of stomatal movements and plant transpiration. Proc. Natl Acad. Sci. USA, 100, 5549–5554.

38. Hurst, A.C., Meckel, T., Tayefeh, S., Thiel, G. and Homann, U. (2004) Trafficking of the plant potassium inward rectifier KAT1 in guard cell protoplasts of Vicia faba. Plant J. 37, 391–397.

39. Jiang, Y.X. and MacKinnon, R. (2000) The barium site in a potassium channel by X-ray crystallography. J. Gen. Physiol. 115, 269–272.

40. Johansson, I. and Blatt, M.R. (2006) Interactive domains between pore loops of the yeast K+ channel TOK1 associate with extracellular K+ sensitivity. Biochem. J. 393, 645–655.

41. Johansson, I., Wulfetange, K., Poree, F. et al. (2006) External K+ modulates the activity of the Arabidopsis potassium channel SKOR via an unusual mechanism. Plant J. 46, 269–281.

42. Karnik, R., Grefen, C., Bayne, R., Honsbein, A., Koehler, T., Kioumourtzoglou, D., Williams, M., Bryant, N.J. and Blatt, M.R. (2013) Arabidopsis Sec1/Munc18 protein SEC11 is a competitive and dynamic modulator of SNARE binding and SYP121-dependent vesicle traffic. Plant Cell, 25, 1368–1382.

43. Kim, T.H., Bohmer, M., Hu, H.H., Nishimura, N. and Schroeder, J.I. (2010) Guard cell signal transduction network: advances in understanding abscisic acid, CO2, and Ca2+ signaling. Annu. Rev. Plant Biol. 61, 561–591.

44. Lai, H.C. and Jan, L.Y. (2006) The distribution and targeting of neuronal voltage-gated ion channels. Nat. Rev. Neurosci. 7, 548–562.

45. Lebaudy, A., Very, A.A. and Sentenac, H. (2007) K+ channel activity in plants: genes, regulations and functions. FEBS Lett. 581, 2357–2366.

46. Leyman, B., Geelen, D., Quintero, F.J. and Blatt, M.R. (1999) A tobacco syntaxin with a role in hormonal control of guard cell ion channels. Science, 283, 537–540.

47. Luu, D.T., Martiniere, A., Sorieul, M., Runions, J. and Maurel, C. (2012) Fluorescence recovery after photobleaching reveals high cycling dynamics of plasma membrane aquaporins in Arabidopsis roots under salt stress. Plant J. 69, 894–905.

48. Marquardt, D. (1963) An algorithm for least-squares estimation of nonlinear parameters. J. Soc. Ind. Appl. Math. 11, 431–441.

49. Meckel, T., Hurst, A.C., Thiel, G. and Homann, U. (2004) Endocytosis against high turgor: intact guard cells of Vicia faba constitutively endocytose fluorescently labelled plasma membrane and GFP-tagged K+-channel KAT1. Plant J. 39, 182–193.
50. Meckel, T., Hurst, A.C., Thiel, G. and Homann, U. (2005) Guard cells undergo constitutive and pressure-driven membrane turnover. Protoplasma, 226, 23–29.
51. Misonou, H. and Trimmer, J.S. (2004) Determinants of voltage-gated potassium channel surface expression and localization in mammalian neurons. Crit. Rev. Biochem. Mol. Biol. 39, 125–145.
52. Mongrand, S., Stanislas, T., Bayer, E.M.F., Lherminier, J. and Simon-Plas, F. (2010) Membrane rafts in plant cells. Trends Plant Sci. 15, 656–663.
53. O'Connell, K.M.S., Loftus, R. and Tamkun, M.M. (2010) Localization-dependent activity of the Kv2.1 delayed-rectifier K+ channel. Proc. Natl Acad. Sci. USA, 107, 12351–12356.
54. Pardo, J.M. and Serrano, R. (1989) Structure of a plasma membrane H+-ATPase gene from the plant Arabidopsis thaliana. J. Biol. Chem. 264, 8557–8562.
55. Pristera, A., Baker, M.D. and Okuse, K. (2012) Association between tetrodotoxin resistant channels and lipid rafts regulates sensory neuron excitability. PLoS One, 7, e40078.
56. Roelfsema, M.R. and Hedrich, R. (2010) Making sense out of Ca2+signals: their role in regulating stomatal movements. Plant, Cell Environ. 33, 305–321.
57. Roelfsema, M.G. and Prins, H.A. (1997) Ion channels in guard cells of Arabidopsis thaliana (L.) Heynh. Planta, 202, 18–27.
58. Romano, L.A., Miedema, H. and Assmann, S.M. (1998) Ca2+-permeable, outwardly-rectifying K+ channels in mesophyll cells of Arabidopsis thaliana. Plant Cell Physiol. 39, 1133–1144.
59. Rowley, C.N. and Roux, B. (2013) A computational study of barium blockades in the KcsA potassium channel based on multi-ion potential of mean-force calculations and free energy perturbation. J. Gen. Physiol. 142, 451–463.
60. Sokolovski, S. and Blatt, M.R. (2004) Nitric oxide block of

outward-rectifying K+ channels indicates direct control by protein nitrosylation in guard cells. Plant Physiol. 136, 4275–4284.

61. Sutter, J.U., Campanoni, P., Tyrrell, M. and Blatt, M.R. (2006) Selective mobility and sensitivity to SNAREs is exhibited by the Arabidopsis KAT1 K+ channel at the plasma membrane. Plant Cell, 18, 935–954.

62. Sutter, J.U., Sieben, C., Hartel, A., Eisenach, C., Thiel, G. and Blatt, M.R. (2007) Abscisic acid triggers the endocytosis of the Arabidopsis KAT1 K+ channel and its recycling to the plasma membrane. Curr. Biol. 17, 1396–1402.

63. Tester, M. (1990) Plant ion channels: whole-cell and single-channel studies. New Phytol. 114, 305–340.

64. Thiel, G. and Blatt, M.R. (1991) The mechanism of ion permeation through K+ channels of stomatal guard cells voltage-dependent block by Na+. J. Plant Physiol. 138, 326–334.

65. Thiel, G., MacRobbie, E.A.C. and Blatt, M.R. (1992) Membrane transport in stomatal guard cells: the importance of voltage control. J. Membr. Biol. 126, 1–18.

66. Thiel, G., Blatt, M.R., Fricker, M.D., White, I.R. and Millner, P.A. (1993) Modulation of K+ channels in Vicia stomatal guard cells by peptide homologs to the auxin-binding protein C-terminus. Proc. Natl Acad. Sci. USA, 90, 11493–11497.

67. Vergani, P., Miosga, T., Jarvis, S.M. and Blatt, M.R. (1997) Extracellular K+ and Ba2+ mediate voltage-dependent inactivation of the outward-rectifying K+ channel encoded by the yeast gene TOK1. FEBS Lett. 405, 337–344.

68. Vergani, P., Hamilton, D., Jarvis, S. and Blatt, M.R. (1998) Mutations in the pore regions of the yeast K+ channel YKC1 affect gating by extracellular K+. EMBO J. 17, 7190–7198.

69. Villalba, J.M., Lutzelschwab, M. and Serrano, R. (1991) Immunocytolocalization of plasma membrane H+-ATPase in maize coleoptiles and enclosed leaves. Planta, 185, 458–461.

70. Wang, Y. and Blatt, M.R. (2011) Anion channel sensitivity to cytosolic organic acids implicates a central role for oxaloacetate in integrating ion flux with metabolism in stomatal guard cells. Biochem. J. 439, 161–170.

71. Wang, Y., Papanatsiou, M., Eisenach, C., Karnik, R., Williams, M., Hills, A., Lew, V.L. and Blatt, M.R. (2012) Systems dynamic modelling of a guard cell Cl- channel mutant uncovers an emergent homeostatic network regulating stomatal transpiration. Plant Physiol. 160, 1956–1972.

72. Yuasa, K., Toyooka, K., Fukuda, H. and Matsuoka, K. (2005) Membrane-anchored prolyl hydroxylase with an export signal from the endoplasmic reticulum. Plant J. 41, 81–94.

73. Zilly, F.E., Halemani, N.D., Walrafen, D., Spitta, L., Schreiber, A., Jahn, R. and Lang, T. (2011) Ca2+ induces clustering of membrane proteins in the plasma membrane via electrostatic interactions. EMBO J. 30, 1209–1220.

74. Zonia, L. and Munnik, T. (2007) Life under pressure: hydrostatic pressure in cell growth and function. Trends Plant Sci. 12, 90–97.

CHAPTER

# 8

# SALT-DEPENDENT REGULATION OF A CNG CHANNEL SUBFAMILY IN ARABIDOPSIS

Annette Kugler[1], Barbara Köhler[2], Klaus Palme[3], Patricia Wolff[3] and Petra Dietrich[1]

[1]Molecular Plant Physiology, Department Biology, University of Erlangen, Erlangen, Germany,

[2]Institute of Biochemistry and Biology, University of Potsdam, Potsdam-Golm, Germany and

[3]Institute of Biology II/Botany, Faculty of Biology, University of Freiburg, Freiburg, Germany

## ABSTRACT

### Background

In *Arabidopsis thaliana*, the family of cyclic nucleotide-gated channels (CNGCs) is composed of 20 members. Previous studies indicate that plant CNGCs are involved in the control of growth processes and responses to abiotic and biotic stresses. According to their proposed function as cation entry pathways these channels contribute to cellular cation homeostasis, including calcium and sodium, as well as to stress-related signal transduction. Here, we studied the expression patterns and regulation of *CNGC19* and *CNGC20*, which constitute one of the five CNGC subfamilies.

## Results

GUS, GFP and luciferase reporter assays were used to study the expression of *CNGC19* and *CNGC20* genes from *Arabidopsis thaliana* in response to developmental cues and salt stress. *CNGC19* and *CNGC20* were differentially expressed in roots and shoots. The *CNGC19* gene was predominantly active in roots already at early growth stages. Major expression was observed in the phloem. *CNGC20* showed highest promoter activity in mesophyll cells surrounding the veins. Its expression increased during development and was maximal in mature and senescent leaves. Both genes were upregulated in the shoot in response to elevated NaCl but not mannitol concentrations. While in the root, *CNGC19* did not respond to changes in the salt concentration, in the shoot it was strongly upregulated in the observed time frame (6-72 hours). Salt-induction of *CNGC20* was also observed in the shoot, starting already one hour after stress treatment. It occurred with similar kinetics, irrespective of whether NaCl was applied to roots of intact plants or to the petiole of detached leaves. No differences in K and Na contents of the shoots were measured in homozygous T-DNA insertion lines for *CNGC19* and *CNGC20*, respectively, which developed a growth phenotype in the presence of up to 75 mM NaCl similar to that of the wild type.

## Conclusion

Together, the results strongly suggest that both channels are involved in the salinity response of different cell types in the shoot. Upon salinity both genes are upregulated within hours. *CNGC19* and *CNGC20* could assist the plant to cope with toxic effects caused by salt stress, probably by contributing to a re-allocation of sodium within the plant.

# BACKGROUND

Salinity has become a major constraint in crop production. Understanding the mechanisms, which enable growth under saline conditions is of high scientific and agricultural interest [1, 2]. Sodium uptake and distribution within the plant is a major determinant for the salt sensitivity of a plant. Sustained exposure to elevated salt concentrations leads to the transfer and accumulation of NaCl in the shoot tissue, where it can inhibit leaf growth. Prevention of $Na^+$ entry into the root, transport to and allocation within the leaf, and sequestration into the vacuole are strategies with which plants cope with high salt environment. Accordingly, the overexpression of the vacuolar $Na^+/H^+$ antiporter NHX1, for instance, improves salt-tolerance in Arabidopsis [3]. Within the shoot, ion allocation can vary between cell types as found in mesophyll and epidermis of barley and wheat, where differences for $K^+$ and $Cl^-$ were measured [4, 5]. $Na^+$ can either be retained in older leaves reducing transport to young organs or translocated to petioles and leaf margins to protect the lamina from excessive entry of salt as described for *Medicago citrine* and *Ricinus communis* [6, 7]. Hence, control of $Na^+$ and $K^+$ fluxes on the whole plant level guarantees the maintenance of a high cytosolic $K^+/Na^+$-ratio, which is crucial for growth in saline soils. In Arabidopsis, transporters contributing to $Na^+$homeostasis include plasma membrane (SOS1) and vacuolar $Na^+/H^+$ antiporters (e.g. NHX1), and the plasma membrane uniporter HKT1 [2].

*AtSOS1* is expressed in epidermal cells at the root tip and in xylem parenchyma cells of roots and shoots [8]. Altogether, data showed that SOS1 controls $Na^+$ extrusion out of the root and long-distance transport, limiting $Na^+$ accumulation in plant cells. The ability of tomato (*Solanum lycopersicum)* plants to retain $Na^+$ in the stems, and thus to prevent $Na^+$ from reaching the photosynthetic tissues, is largely dependent on the function of *Sl*SOS1, the functional homolog of *At*SOS1 [9].

While NHX1 and SOS1 export $Na^+$ from the cytosol on the expense of the proton gradient, $Na^+$ entry follows its electrochemical gradient. Members of two gene families, the high affinity $K^+$ transporter family HKT, and the cyclic nucleotide-gated ion channel family, CNGC, have been shown to mediate $Na^+$ uptake and regulation of long distance transport. Proteins belonging to the HKT family control $Na^+$ unloading in the xylem of Arabidopsis, rice and wheat [1], and therefore control the long-distance transport of $Na^+$ to the leaf. The Arabidopsis genome contains a single *HKT* homolog, *AtHKT1*, which belongs to the subfamily of HKT transporters that encode low affinity $Na^+$ uniporters. Loss-of-function mutations in *At*HKT1 render plants $Na^+$hypersensitive and disturb the distribution of $Na^+$ between roots and shoots.

Members of the cyclic nucleotide-gated channel (CNGC) family belong to the group of nonselective cation channels and enable the uptake of $Na^+$, $K^+$, and $Ca^{2+}$[10]. CNG channels are assumed to activate upon binding of cellular cAMP or cGMP to the ligand-binding site. Within the C-terminus of the channel, a partially overlapping binding domain for calmodulin allows $Ca^{2+}$-calmodulin binding and is proposed to destabilize the open state. Functional expression of plant CNG channels in *Xenopus* oocytes or animal cell lines has not been reproducibly successful; hence a detailed biophysical characterization of these channels including their gating and permeation characteristics still remains to be performed. The Arabidopsis CNGC gene family comprises 20 members [11]. Phenotypical analysis of loss-of-function mutants showed that members play a role in plant growth and the response to pathogen attack [10]. CNGC10 is involved in Arabidopsis› tolerance towards salt. Mature plants of *CNGC10* antisense lines were more sensitive to salt stress and contained higher $Na^+$concentrations in shoots compared with wild-type [12]. In contrast, salt-grown seedlings of the antisense lines developed longer roots compared to the wild type. Likewise, *cngc3* mutant seedlings showed slightly enhanced growth in the presence of elevated NaCl or KCl concentrations compared to wild type plants [13]. So far, members tested have been localized in the plasma membrane [13, 14, 15, 16], suggesting a direct function in cation entry into the cell.

In this study, we show that both *CNGC19* and *CNGC20* respond to salinity with increased gene activity and accumulation of transcripts in the shoot. Salt treatment of roots or cut leaves induced the shoot regulation of *CNGC20*, suggesting that NaCl itself is the root-to-shoot signal. Although the loss of either channel did not lead to a salt-related growth phenotype, the strong upregulation by NaCl underlines their role in the salinity response, which is discussed on the basis of their distinct expression pattern.

## RESULTS AND DISCUSSION

We investigated the expression pattern and regulation by salt stress of the group IVA of Arabidopsis CNG channels [11], consisting of *CNGC19* and *CNGC20*. Both genes are arranged in tandem on chromosome 3. On the amino acid level, the two proteins share 73% identity.

## Distinct Expression Patterns of CNGC19 and CNGC20

The age- and tissue-dependent expression pattern was visualized in plants expressing β-glucuronidase under the control of the CNG channel promoter. In case of *CNGC19*, blue staining was visible starting one day after the radicle's emergence from the seedling. Promoter activity was detected mainly in roots and to a lesser extent in shoots of plants of different developmental stages (Fig. 1A-E). *CNGC19* showed expression only in the vasculature, and was observed during the development of lateral roots, as soon as the stelar long-distance transport system develops. It was not detected in root meristematic tissue (Fig. 1F-H). The staining of two strands within the stele indicated that the expression is located in phloem tissue (Fig. 1I-K). This was supported by the staining of root cross sections (Fig. 1L). In the shoot, *CNGC19* promoter activity was observed in the vasculature, too (Fig. 1C, M).

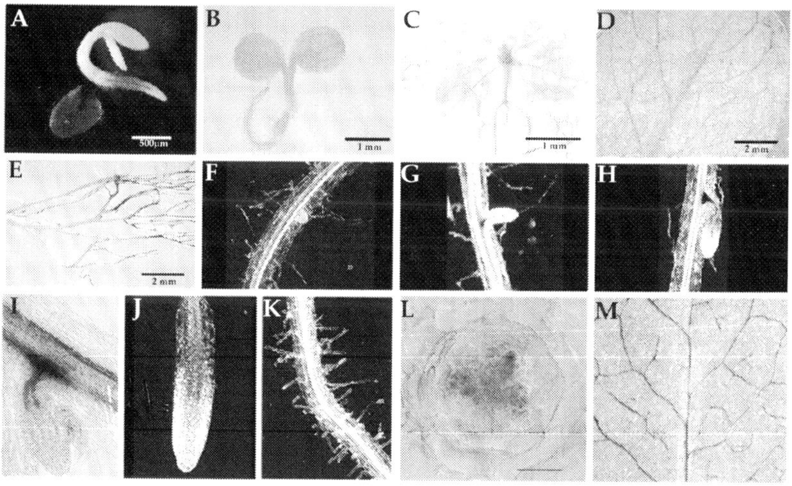


**Figure 1**: *CNGC19* expression in vascular tissues. GUS staining of plants at different ages carrying the *CNGC19p:GUS*construct: 2 (A), 4 (B), 18 (C), 39 (D, E) days after stratification. (F-H) Pictures of developing lateral roots where blue staining is not detected before vasculature formation. (I-K) GUS staining visible in the phloem strands of the root. (L) Cross section of the root with GUS staining in the phloem but not in xylem cells of the stele. The bar represents 20

µm. (M) Part of an adult leaf showing GUS staining in the vasculature. Note, that the plant shown in (M) has been grown on salt-containing agar (see Fig. 3).

GUS staining in plants expressing the β-glucuronidase gene under control of the *CNGC20* promoter was visible in the roots of young seedlings (Fig. 2A), but was more pronounced in shoot tissue of mature plants (Fig. 2B), as well as in the carpel (Fig. 2C) and crown leaves of flowers (not shown). While within the root, cortex tissue was stained, in the shoot *CNGC20* gene expression was mainly observed in the mesophyll tissue surrounding the veins and in the petioles (Fig. 2B, E). Expression in guard cells could also be observed (Fig. 2F). Staining of epidermal cells, however, was weak. To test this, tobacco leaves were transiently transformed with a *CNGC20*promoter-GFP fusion construct. Indeed, promoter activity was evident in epidermal cells (Fig. 2G).

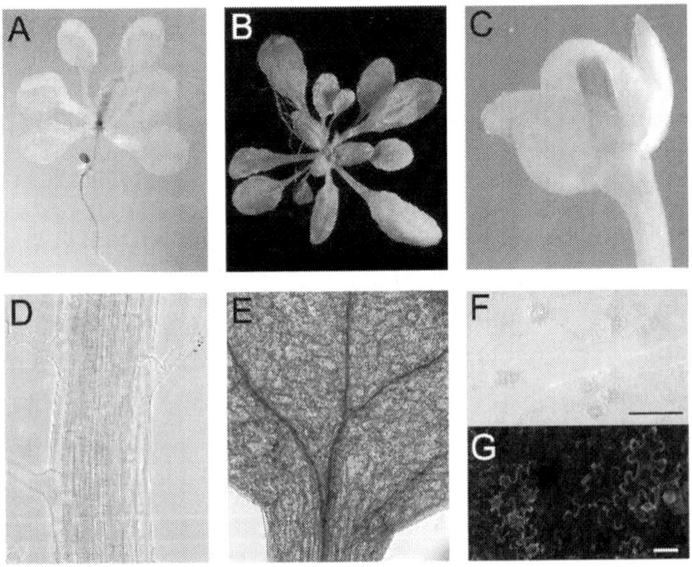


**Figure 2**: *CNGC20* expression in roots and shoots. GUS staining of plants carrying the *CNGC20p:GUS* construct was observed in roots of young seedlings (A), in petioles and tissue surrounding the veins of adult leaves (B), and carpels (C). On the cellular level, GUS staining was detected in the root cortex (D), mesophyll surrounding the veins (E), and in guard cells (F). (G) GFP fluorescence in epidermal cells after Agrobacterium-mediated transformation of *Nicotiana benthamiana* leaves with the *CNGC20p:GFP* construct. The bars represent 50 µm.

Data in this study indicate specific expression patterns for both genes. Interestingly, CNGC19 is found in the vasculature, which is surrounded by CNGC20 expressing cells. Thus, the two genes may fulfil similar functions in different but adjacent tissues.

## CNGC19 and cngc20expression is Regulated by Salinity

Previous publications reported on the participation of nonselective cation channels in the plant's response to salt stress [17, 18]. To investigate possible effects of salinity on *CNGC19* and *CNGC20*, we monitored gene activities using reporter genes and quantitative RT-PCR. When plants transformed with *CNGC19p:GUS* were grown for one week on half-strength MS-agar in the absence or presence of 150 mM NaCl, GUS-staining of seedlings and whole plants revealed strong promoter activity in root tissue under both control and salt stress conditions (Fig. 3D, E). In the aerial parts of the plant, the staining intensity was increased in 65% (13 out of 20) of the plants tested after the salt stress period compared to controls (Fig. 3A, B).

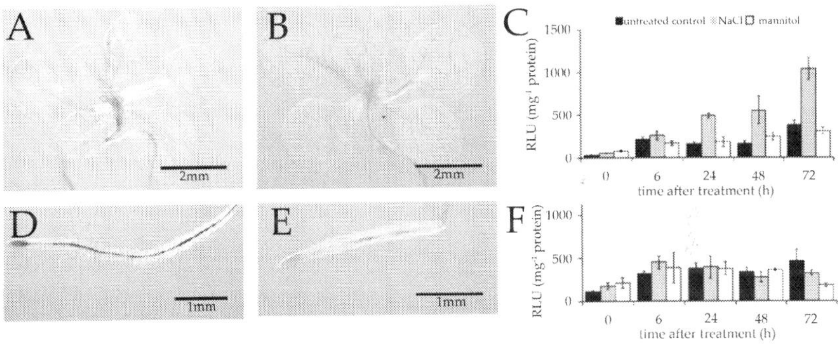


**Figure 3**: Salt regulation of *CNGC19*. (A-E) GUS staining of 18-day old plants carrying the *CNGC19p:GUS* construct, which were grown in the absence (A, D) or presence (B, E) of 150 mM NaCl for 7 days. (C, F) Luciferase activity in *CNGC19p:LUC* transformed plants. Promoter activity was determined as relative luciferase luminescence units (RLU) in shoots (C) and roots (F) normalized to the respective protein content. Roots of 12-day old plants were either untreated (black bars), or treated with 200 mM NaCl (blue bars) or 300 mM mannitol (white bars). Luciferase activities were determined after stress application at time points indicated. Data represent mean ± SEM (n = 3).

Transgenic plants expressing the luciferase gene under the control of the *CNGC19* promoter were used to determine the gene activity as the luciferase luminescence intensity normalized to the protein content of tissue extracts (relative luciferase luminescence). The relative luciferase luminescence in 12-day old plants was higher in root than in shoot tissue (Fig. 3C, F). The same approach was used to monitor gene regulation by NaCl treatment. After application of 200 mM NaCl to the root, *CNGC19* gene activity increased only in the shoot (Fig. 3C) but not in the root (Fig. 3F). The increase continued during 72 hours of salt stress, corresponding to a steady upregulation of *CNGC19* gene activity. In the presence of 300 mM mannitol, *CNGC19* was not affected in the same manner, indicating that the response was specific and mainly due to the ionic rather than the osmotic component of the stress. These results are well in agreement with whole genome array data on cDNA isolated from 13-day old plants, which show a time-dependent accumulation of CNGC19 transcripts in the shoot [19]. In the root, transcript levels rose transiently within the first hour after salt treatment, but returned to control levels within 6 hours.

Kilian and colleagues [19] did not detect a salt-dependent regulation of *CNGC20*. We therefore decided to test the salt-sensitivity at developmental stages, where *CNGC20* is predominantly expressed. 6-week old plants expressing luciferase under control of the *CNGC20* promoter reported indeed a salt-dependent increase in *CNGC20* expression (Fig. 4A, B). Application of 200 mM NaCl to roots led to a significant increase of luciferase in the leaves already within one hour, which then remained elevated during two days of constant salt stress (Fig. 4A). *CNGC20* was not upregulated in control conditions or in the presence of 300 mM mannitol. Kinetics of the salt response of *CNGC20* indicated that tissue not directly in contact with the stress responded rapidly. To test, whether this could be due to the uptake and translocation of salt to the shoot tissue, the stress was applied directly to the leaves by putting the petioles of detached leaves into the NaCl solution. As in the root treatment, *CNGC20* in the shoot responded within one hour as indicated by the increase in relative luminescence, while control treatments had no effect (Fig. 4B). Salt-dependent regulation of *CNGC20* was supported by quantitative RT-PCR experiments. Transcript abundance was assessed in detached leaves, revealing a salt-dependent increase of *CNGC20* transcript in the shoot (Fig. 4C), which was significantly different from controls.

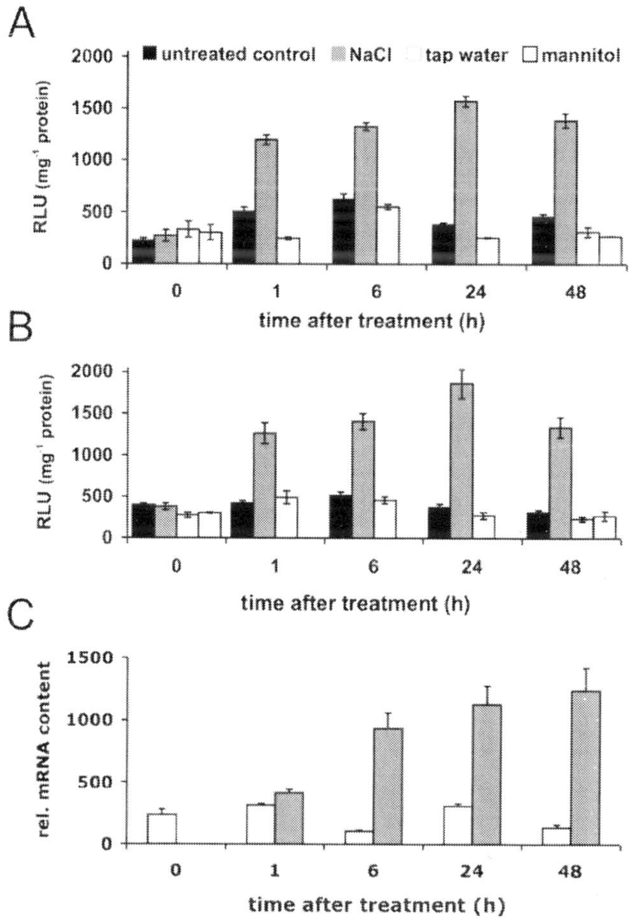


**Figure 4**: Salt regulation of *CNGC20*. CNGC20 promoter activity was monitored in 6-week old plants, using the luciferase reporter (A, B) or quantitative RT-PCR (C). (A) Luciferase activity in leaves from plants carrying the *CNGC20p:LUC*construct, at indicated time points after treatment of the roots with 200 mM NaCl (blue bars), 300 mM mannitol (white bars), tap water (yellow bars) or in untreated material (black bars). Data represent mean ± SEM (n = 3). (B) Petioles of detached leaves were put into Hoagland Medium (black bars), 200 mM NaCl-, 300 mM mannitol-, or tap water-containing solution and luciferase activity was quantified at time points indicated. Data represent mean ± SEM (n = 3). (C) *CNGC20* transcript levels in detached leaves after treatment with 200 mM NaCl or tap water. Transcript levels are expressed relative to the respective actin control. Data represent mean ± SEM (n = 7-8). Color code in (B) and (C) as in (A).

Together, promoter activities assessed by reporter genes and transcript levels determined by quantitative RT-PCR reported qualitatively similar results on the salt-dependent upregulation of *CNGC19* and *CNGC20* in leaves of mature plants. Our data point to a physiological response to the accumulation of NaCl itself rather than to an osmotic shock response. The results show that salt treatment of the root triggers a response of *CNGC20* in the leaf. Such long-distance signaling from root to shoot might be mediated by hormones, such as abscisic acid (ABA) [20, 21]. However, the response of isolated leaves demonstrated that salt perception and signal transduction can take place in the aerial parts of the plant. Thus, it appears likely that the regulation of *CNGC20* depends on the direct transfer of NaCl to the shoot. It remains unclear in which cell types the signal perception takes place and whether this is the same for both genes. Since *CNGC19* is expressed in phloem tissue and *CNGC20* in mesophyll cells nearby the phloem, it is interesting to note that the *CNGC20* induction kinetics saturates much earlier compared to that of *CNGC19*. Whether or not this is related to the time-dependent distribution of NaCl within the shoot remains to be clarified.

Salt stress, like many other abiotic stresses, can elicit a transient increase in cytosolic $Ca^{2+}$ [22]. In Arabidopsis seedlings, cGMP levels increased rapidly (<5 s) and to different degrees after salt and osmotic stress [23]. Interestingly, Donaldson and colleagues provided evidence that salt stress activates two cGMP signaling pathways - an osmotic, calcium-independent pathway and an ionic, calcium-dependent pathway. It is tempting to suggest that CNGC19 and CNGC20 might be suitable candidates taking part in these early responses, possibly linking cGMP- and $Ca^{2+}$-signaling.

Increased expression of *CNGC20* was detected quickly within one hour. *CNGC19* responded a bit slower within 24 hours. The strong induction of the expression by NaCl implies a function in the adaptation to salinity. In this respect it is interesting that salt stress affects both genes mainly in the shoot, where most dramatic changes occur [1]. Control of $Na^{+}$ accumulation in the shoot is of major importance for the adaptation to salt stress. As most sensitive plants display poor ability to sequester

$Na^+$ in leaf vacuoles, they have to rely on other mechanisms to cope with the $Na^+$ delivered to leaf cells. Both CNGC19 and CNGC20 represent possible $Na^+$ entry pathways into cells, and could participate in the $Na^+$ distribution within the leaf. For instance, CNGC19 could participate in $Na^+$ sequestration into phloem parenchyma cells and CNGC20 in $Na^+$ sequestration into the mesophyll of petioles. A translocation of $Na^+$ to petioles is known from species that tolerate salt [6, 7], but might occur to a certain extent also in Arabidopsis. Since $Na^+$ is preferentially deposited in older leaves, a role in compartmentation is supported by the fact that *CNGC20* is mainly expressed in older leaves.

Expression of *CNGC19* was detected in the phloem, strengthening the hypothesis about a function in phloem loading and unloading. CNGC19 could be involved in $Na^+$ recirculation from shoots to roots, where $Na^+$ might be extruded, or at least in $Na^+$ redistribution between tissues. In the upper parts of the roots and in the stem, a direct transfer of sodium ions from xylem to phloem tissues is thought to play a role in the control of $Na^+$translocation towards the shoot [24, 25, 26, 27]. This would require $Na^+$ uptake into the phloem. $Na^+$ assays of the phloem sap revealed high concentrations up to 80 mM in some species [28], but the physiological significance of such data was interpreted contradictorily [29, 30]. According to the expression pattern and expected ion channel characteristics, it is tempting to hypothesize that CNGC19, similar to AKT2/3 [31], might play a role in membrane potential stabilization and therewith might indirectly affect phloem (re)loading of metabolites.

## Phenotypical Analysis of CNGC19 and CNGC20Mutants

Arabidopsis T-DNA insertion lines from the Salk collection [32] were investigated with T-DNA insertions in the first (Salk_027306, *cngc19-1*) and fourth exon (Salk_129133, *cngc20-1*), respectively (Fig. 5A). PCR using cDNA isolated from wild type and mutants confirmed that *CNGC19* and *CNGC20* expression was virtually absent from the respective homozygous T-DNA mutants (Fig. 5B, C).

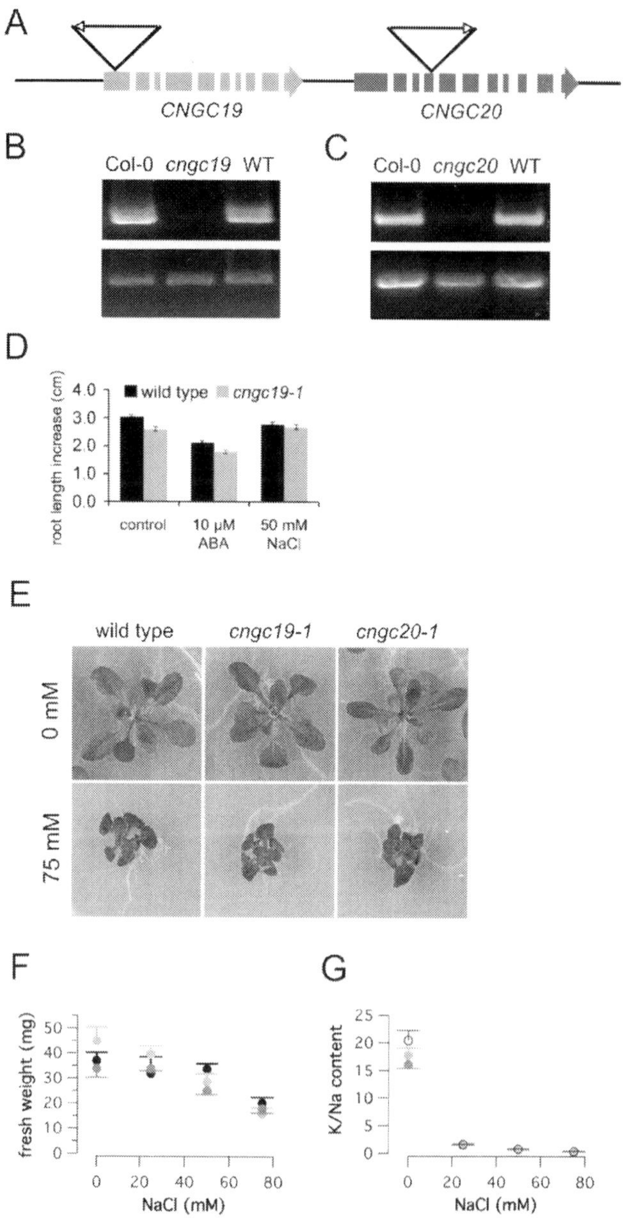


**Figure 5**: Analysis of *cngc19* and *cngc20* T-DNA insertion lines. (A) Genomic organization of *CNGC19* and *CNGC20* and the respective T-DNA insertions for *cngc19-1* (SALK line 027306) and *cngc20-1* (SALK line 129133). Exons are shown in bold. (B) Absence of *CNGC19* mRNA from *cngc19-1* plants (lane 2), but presence in Col-0 (lane 1) and a backcrossed *CNGC19* wild type (WT,

lane 3). *Upper traces*: PCR result using *CNGC19* gene-specific primers, which amplified a 348 bp fragment downstream of the T-DNA insertion, *lower traces*: *Actin2* primers were used as a control. (C) Corresponding PCR analysis of the T-DNA insertion line *cngc20-1* with cDNA from Col-0 (lane 1), *cngc20-1* (lane 2) and a backcrossed *CNGC20* wild type (WT, lane 3). *CNGC20* gene-specific primers, which span a 310 bp fragment downstream of the T-DNA insertion, were used. No PCR fragment was amplified from cDNA from homozygous *cngc20-1* (lane 2). (D) Unchanged root growth of *cngc19-1* plants. Root length increase of *cngc19-1* and wild type seedlings were measured during a 7-day growth period, starting 4 days after stratification. Mutant (orange bars) and wild type (black bars) plants grew vertically on half-strength MS agar plates in the absence (control) or presence of 10 μM ABA or 50 mM NaCl. Data represent mean ± SEM (n = 30). (E) Absence of a salt-dependent growth phenotype in *cngc19-1* and *cngc20-1*. Photographs show representative plants from wild type, *cngc19-1*, and *cngc20-1* after a 12-day growth period in the absence or presence of 75 mM NaCl. (F) Fresh weight of wild type (black circles), *cngc19-1* (orange circles), and *cngc20-1* (green circles) shoots after a 12-day growth period in the presence of 0, 25, 50, or 75 mM NaCl. Data represent mean ± SEM (n = 6). The dry weight did also not differ significantly between wild type and mutants (not shown). (G) K/Na content in shoots of plants shown in (E) and (F) as a function of the applied salt-concentration. Data represent mean ± SEM (n = 6). Color code as in (F).

We analyzed the CNG channel mutants in conditions mimicking high salinity and in the presence of abscisic acid (ABA), which plays a crucial role in root-to-shoot and cellular signaling in response to salt stress [1]. Although *CNGC19* is expressed in the root tissue of young seedlings, *cngc19-1* displayed no root growth phenotype under control or saline (50 mM NaCl) conditions. The growth was also not affected by 10 μM ABA compared to the wild type (Fig. 5D). Similarly, the germination was indistinguishable from wild type in the presence of 50 mM NaCl or 10 μM ABA (data not shown). However, root growth is usually less affected than leaf growth during $Na^+$ toxicity, and the root elongation rate recovers remarkably well after exposure to NaCl or other osmotica [30]. Regarding the whole plant, it is primarily the mature leaf where $Na^+$ toxicity is manifested. As both *CNGC19* and *CNGC20* are upregulated in shoot tissue of salt-stressed plants, we compared the shoot growth of mutants and wild type. After a 5 day growth period on half-strength MS-agar, seedlings were transferred to agar plates containing 0, 25, 50, or 75 mM NaCl and grown for another twelve days. No salt-dependent phenotype could be observed for *cngc19-1* and *cngc20-1* mutants compared to the wild type (Fig. 5E, F).

To test if $Na^+$ or $K^+$ accumulation is affected in the mutants, contents were determined with ICP. Although the plants displayed a reduction of fresh weights with increasing NaCl content in the media (Fig. 5F), the K:Na content ratio of wild type and mutant shoots of plants grown for twelve days on plates containing 0, 25, 50 or 75 mM NaCl did not differ (Fig. 5G). These findings suggest that both uptake of sodium/potassium and extrusion of $Na^+$ are unaltered in the mutants.

Phenotypic characterization of loss-of-channel mutants does not allow deducing an explicit role of CNGC19 and CNGC20 during the nonselective uptake of $Na^+$ during salt stress. Due to their expression pattern, they could be involved in salt stress-dependent signal transduction or distribution of sodium throughout the plant. In the mesophyll and the root cortex, the specific role of CNGC20 might be masked by the activity of other nonselective cation channels having partially redundant functions. *CNGC20* and *CNGC3*, for instance, are both expressed in the root cortex. *Cngc3* loss-of-function mutants grew slightly better in the presence of 40 to 80 mM NaCl [13]. This together with short and long-term $Na^+$ influx experiments led to the conclusion that CNGC3 is functioning in sodium (and potassium) influx. Only during initial stages of salt stress, CNGC3 contributed considerably to $Na^+$ influx. Its participation in ion uptake in salinity-adapted plants seems to be limited [13].

In leaves, CNGC10 was located in the plasma membrane of mesophyll, palisade parenchyma and epidermal cells [14]. Mature plants of *CNGC10* antisense lines were more sensitive towards salt stress and their shoots contained higher $Na^+$ concentrations compared with wild type [12]. GUS staining showed that *CNGC10*expression in leaves is patchy; it comprised vascular tissue and mesophyll (B. Köhler, unpublished results). Thus, at least a partially redundant function to CNGC20 appears possible.

Apart from pollen-specific CNGCs, the CNGCs investigated so far show a broad expression pattern [13, 14, 16, 33]. By comparison, expression of *CNGC19* is relatively confined. Thus, further cell-type specific functional assays are required to assess its physiological role *in planta*.

## CONCLUSION

*CNGC19* is expressed in the phloem and *CNGC20* in the epidermis and the mesophyll, mainly in petioles. Upon salinity, both genes are upregulated within hours in the shoot, where most dramatic changes happen [1]. Salt-dependent regulation of *CNGC20* occurred in the shoot, irrespective of whether NaCl was applied to the roots of intact plants or to the petioles of detached leaves. At first glance, it seems puzzling that a cell promotes the upregulation of genes encoding proteins that provide $Na^+$ entry pathways and therefore would contribute to increase the cytosolic sodium levels. However, under severe salt stress, CNG channels represent a fast and effective way to redistribute sodium throughout the whole plant. The fact that both *CNGC19* and *CNGC20* were upregulated by salt rather than by osmolytes indicates a role in salt adaptation. Therefore we propose a distributive role for CNGC19 and CNGC20 enabling the plant to cope with toxic effects caused by salt stress.

## METHODS

### Plant Material and Growth Conditions

*Arabidopsis thaliana* Col-0 ecotype and transgenic plants in Col-0 background were used. Plants were grown on soil in a growth chamber at a photoperiod of either 16 h (long day LD) or 8 h (short day SD). For sterile cultivation, seeds were sown on half-strength MS agar pH 5.8 containing 1% sucrose and 0.8% phytagar (Duchefa). For culture on sand (1-2 mm aquarium grit), nutrients were supplied by modified Hoagland medium [34], containing 1.25 mM $KNO_3$, 1.5 mM $Ca(NO_3)_2$, 0.75 mM $MgSO_4$, 0.5 mM $KH_2PO_4$, 50 μM KCl, 50 μM $HBO_3$, 10 μM $MnSO_4$, 2 μM $ZnSO_4$, 1.5 μM $CuSO_4$, 0.075 μM $(NH_4)_6Mo_7O_{24}$, 72 μM $FeSO_4$, 89.28 μM EDTA, pH 6. In all conditions, plants grew at 22°C and about 80-100 μmol/$m^2$sec light intensity.

### Generation of Transgenic Plants

For reporter gene studies, a 1.15-kb promoter region of the *CNGC20* gene was introduced into the binary vector pVKH-35S-pA1 [35], where it replaced the 35S promoter in front of the *uidA* gene, resulting in the binary vector, pVKH-*CNGC20p:GUS*. Additionally, the promoter region was inserted into the destination vector pMDC206 [36], using gateway technology (Invitrogen). In case of *CNGC19*, a 1.82-kb promoter region was amplified by PCR using the primers PC19 for (5'-CCGCTCGAGAGCAACATGACAAACTTCTTC) and PC19rev (5'-CTAGCTAGCTTTTTATTTCAGAAACCCAAAATCTAGGGC). PCR fragments and the binary vector pGPTV-HPT [37] were cut with *Xho*I and *Nhe*I, and *Sal*I and *Xba*I, respectively, and ligated. The resulting plasmid was named pGPTV-*CNGC19p:GUS*. For luciferase studies, the luciferase$^+$ gene was introduced in pMDC206, where it replaced the GFP coding region, resulting in the new destination vector pMDC206-luc. A 1.26-kb promoter fragment of the *CNGC19* gene and the *CNGC20* promoter region were inserted into pMDC206-luc via gateway cloning. *Agrobacterium tumefaciens* strain GV3101 carrying the binary plasmids was used to transform *A. thaliana* Col-0 [38]. Transgenic plants were tested for reporter gene activities. Homozygous lines were produced for each construct.

## GUS Histochemical Assay

GUS staining followed the method of Jefferson et al. [39]. The plant tissue was cleared in 70% ethanol for 1-2 days. For vibratome sectioning, the tissue was embedded directly after staining into 5% agarose in PBS (0.8% NaCl, 0.02% KCl, 0.144% $Na_2HPO_4$, 0.024% $KH_2PO_4$, pH 7.4 HCl) and then cut into 30 μm-sections with a vibratome (Model 1500; The Vibratome Company, St. Louis, USA).

## Agrobacterium-Mediated Transient Expression in Nicotiana Benthamiana

Overnight cultures of *Agrobacterium tumefaciens* strain *C58C1* transformed with the *CNGC20p:GFP* construct, and Agrobacterium strain *p19* featuring a viral-encoded suppressor of gene-silencing were used for coinfiltration into the abaxial side of *Nicotiana benthamiana* leaves [40]. Confocal images were taken 4 days after infiltration using a Leica confocal microscope (TCS SP II; Leica Microsystems, Wetzlar, Germany). Fluorescence was excited with a UV argon laser at 488 nm, and emission of GFP was detected in the range from 497-547 nm. Emission of chlorophyll was collected at 644-731 nm and transmitted light was detected at 779-840 nm.

## Luciferase Activity Assay

Seedlings of a homozygous *CNGC19p:LUC* line were grown vertically on half-strength MS agar plates in LD conditions for 12 days. For stress application, 3 ml of solution containing either 200 mM NaCl, 300 mM mannitol, or tap water was applied to the root tissue. 50 mg samples of root and shoot tissue were harvested at different time points after treatment, homogenized and suspended in CCLR buffer (Promega Corp. Madison, USA). After 30 min incubation on ice, extracts were cleared by 30 min centrifugation at 4°C at 10.000 g and the supernatant was stored at -80°C.

*CNGC20p:LUC* transformed plants were grown for 6 weeks either on sand or on soil. Sand-cultured plants were salt-treated by replacing the modified Hoagland medium with medium supplemented with 200 mM NaCl or 300 mM mannitol. For direct stress treatment of the shoot, leaves of soil-grown plants were cut and petioles placed in 200 mM NaCl or 300 mM mannitol solution. Leaf extracts were prepared after the indicated incubation times.

The frozen luciferase extracts were thawed on ice and luciferase luminescence was determined in a 50 μl aliquot after addition of 150 μl LAR buffer (Promega), using an Orion II 96 microplate luminometer (Berthold Detection Systems GmbH, Pforzheim, Germany). Protein contents of the samples were determined using the Roti-Nanoquant kit (Carl Roth GmbH, Karlsruhe, Germany). After background subtraction, the relative luminescence (RLU) was determined by normalization to the total protein content.

## Quantitative RT-PCR

Total RNA of shoots of 9-week old plants was isolated using TRIZOL reagent [41]. First strand cDNA was prepared from 7.5 μg of RNA in a total volume of 10 μl using the RevertAid H Minus first-strand cDNA synthesis kit (Fermentas, St. Leon-Rot, Germany) and diluted for RT-PCR 20-fold in water. PCR was performed in a Rotogene 2000 (Corbett, Concorde, USA) with the LightCycler-FastStart Quanti Tect SYBR Green PCR Kit (Qiagen, Hilden, Germany), using the *CNGC20* gene-specific primers (5'-CCTCGAACGCTCTTCTGTAAA and 5'-CTAGTTATAGCCTTTAGTTTGTA). Actin2 primers (5'-ATTTCAGATGCCCAGAAGTCTTGTT and 5'-GAAACATTTTCTGTGAACGATTCCT) were used to normalize the *CNGC20* mRNA level to that of actin.

## Isolation of T-DNA Insertion Lines

Seeds of T-DNA insertion lines for *CNGC19* (Salk_027306, named *cngc19-1*) and *CNGC20* (Salk_129133, named *cngc20-1*) were obtained from the SALK institute (http://signal.salk.edu/cgi-bin/tdnaexpress, [32]). Homozygous mutants were genotypically identified through PCR using a gene-specific primer (CNGC19: 5'-TGCACATCCCTAATGTCCA; CNGC20: 5'-GATGGCCGATGACTAAAGC) in combination with a T-DNA border primer (5'-CTGGCGTAATAGCGAAGACG) and a PCR using a gene-specific primer pair (CNGC19: 5'-TGCCCTAGATTTTGGGTTTC and 5'-AAATACTCTTGTGTCAGCTGCTATG; CNGC20: 5'-TCCCCTCTTCTTCTTCCTCATAAA and 5'-AACCAG TAGGAGCTCTAACGTAAC). For determination of *CNGC19*transcript levels,totalRNAisolatedfromroottissueof200*Arabidopsisthaliana*plants vertically grown for 12 days on half-strength MS medium was transcribed into cDNA as described above. *CNGC19* gene-specific primers binding downstreamoftheT-DNAinsertion(5'-GAAACTTGGAACTTTGGAGC and 5'-CTACCAAACCAAACATCATCAT) were used to verify the lack of *CNGC19* mRNA in *cngc19-1* plants. *CNGC20*transcript levels were assayed in total RNA isolated from leaves of 5-week old plants. PCR was carried out on transcribed cDNA with a *CNGC20* gene-specific primer set binding downstream of the T-DNA insertion in *cngc20-1* (5'-CCTCGAACGCTCTTCTGTAAA-3' and 5'-CTAGTTATAGCCTTTAGTTTGTA). Transcript levels in the mutants were compared to the ones in Col-0 wild type as well as in a backcrossed wild type line. Actin2-specific primers were used as controls in all reactions. No *CNGC19* and *CNGC20* transcripts were detected in *cngc19-1* and *cngc20-1* mutants, respectively.

## ICP Analysis

Shoot dry weights of wild type, *cngc19-1* and *cngc20-1* plants grown for 12 days in MS-agar containing 0, 25, 50 or 75 mM NaCl were determined after 72 h incubation at 60°C. K and Na content analysis was performed in an ICP emission spectrometer JY 70 Plus (Division d›Instruments S.A./Jobin, France) after solubilization of the plant material in 1 ml conc. $HNO_3$ for 10 h at 170°C under pressure (10 bar) followed by a dilution step (1:10) in deionized water.

# DECLARATIONS

## Acknowledgements

AK is a fellow of the Konrad Adenauer Foundation. We are grateful to Franz Klebl for help with quantitative RT-PCR experiments and acknowledge the technical assistance by Elisabeth Dunkel. We thank Georg Nagel and Elfriede Reisberg (University of Würzburg, Germany) for ICP measurements and Dirk Becker (University of Würzburg, Germany) for aid with the generation of luciferase constructs. Work in KP's laboratory was supported by the Collaborative Research Center 592, the Excellence Initiative of the German Federal and State Governments (EXC 294), Graduiertenkolleg 1305, Bundesministerium für Forschung und Technik (BMBF), Deutsches Zentrum für Luft und Raumfahrt and the European Space Agency. This work was funded by DFG grants (SPP1108, FOR964) and a grant of the Dr. Hertha and Helmut Schmauser foundation to PD.

# REFERENCES

1. Munns R, Tester M: Mechanisms of salinity tolerance. Annu Rev Plant Biol. 2008, 59: 651-681. 10.1146/annurev.arplant.59.032607.092911.
2. Zhu JK: Regulation of ion homeostasis under salt stress. Curr Opin Plant Biol. 2003, 6: 441-445. 10.1016/S1369-5266(03)00085-2.
3. Apse MP, Aharon GS, Snedden WA, Blumwald E: Salt tolerance conferred by overexpression of a vacuolar Na+/H+ antiport in Arabidopsis. Science. 1999, 285: 1256-1258. 10.1126/science.285.5431.1256.
4. James RA, Munns R, von Caemmerer S, Trejo C, Miller C, Condon TA: Photosynthetic capacity is related to the cellular and subcellular partitioning of Na+, K+ and Cl. Plant Cell Environ. 2006, 29: 2185-2197. 10.1111/j.1365-3040.2006.01592.x.

5. Karley AJ, Leigh RA, Sanders D: Differential ion accumulation and ion fluxes in the mesophyll and epidermis of barley. Plant Physiol. 2000, 122: 835-844. 10.1104/pp.122.3.835.
6. Sibole JV, Cabot C, Poschenrieder C, Barcelo J: Ion allocation in two different salt-tolerant Mediterranean Medicago species. J Plant Physiol. 2003, 160: 1361-1365. 10.1078/0176-1617-00811.
7. Jeschke JPS: Ionic interactions of petiole and lamina during the life of a leaf of castor bean (Ricinus communis L.) under moderately saline conditions. J Exp Bot. 1991, 42: 1105-1116. [http://jxb.oxfordjournals.org/cgi/content/abstract/42/8/1051]
8. Shi H, Quintero FJ, Pardo JM, Zhu JK: The putative plasma membrane Na+/H+ antiporter SOS1 controls long-distance Na+ transport in plants. Plant Cell. 2002, 14: 465-477. 10.1105/tpc.010371.
9. Olias R, Eljakaoui Z, Li J, de Morales PA, Marin-Manzano MC, Pardo JM, Belver A: The plasma membrane Na+/H+ antiporter SOS1 is essential for salt tolerance in tomato and affects the partitioning of Na+ between plant organs. Plant Cell Environ. 2009, 32: 904-916. 10.1111/j.1365-3040.2009.01971.x.
10. Kaplan B, Sherman T, Fromm H: Cyclic nucleotide-gated channels in plants. FEBS Lett. 2007, 581: 2237-2246. 10.1016/j.febslet.2007.02.017.
11. Mäser P, Thomine S, Schroeder JI, Ward JM, Hirschi K, Sze H, Talke IN, Amtmann A, Maathuis FJ, Sanders D, et al: Phylogenetic relationships within cation transporter families of Arabidopsis. Plant Physiol. 2001, 126: 1646-1667. 10.1104/pp.126.4.1646.
12. Guo KM, Babourina O, Christopher DA, Borsics T, Rengel Z: The cyclic nucleotide-gated channel, AtCNGC10, influences salt tolerance in Arabidopsis. Physiol Plant. 2008, 134: 499-507. 10.1111/j.1399-3054.2008.01157.x.
13. Gobert A, Park G, Amtmann A, Sanders D, Maathuis FJ: Arabidopsis thaliana cyclic nucleotide gated channel 3 forms a non-selective ion transporter involved in germination and cation transport. J Exp Bot. 2006, 57: 791-800. 10.1093/jxb/erj064.
14. Christopher DA, Borsics T, Yuen CY, Ullmer W, Andeme-Ondzighi C, Andres MA, Kang BH, Staehelin LA: The cyclic nucleotide gated cation channel AtCNGC10 traffics from the ER via Golgi vesicles

to the plasma membrane of Arabidopsis root and leaf cells. BMC Plant Biol. 2007, 7: 48-10.1186/1471-2229-7-48.

15. Borsics T, Webb D, Andeme-Ondzighi C, Staehelin LA, Christopher DA: The cyclic nucleotide-gated calmodulin-binding channel AtCNGC10 localizes to the plasma membrane and influences numerous growth responses and starch accumulation in Arabidopsis thaliana. Planta. 2007, 225: 563-573. 10.1007/s00425-006-0372-3.
16. Frietsch S, Wang YF, Sladek C, Poulsen LR, Romanowsky SM, Schroeder JI, Harper JF: A cyclic nucleotide-gated channel is essential for polarized tip growth of pollen. Proc Natl Acad Sci USA. 2007, 104: 14531-14536. 10.1073/pnas.0701781104.
17. Maathuis FJ: The role of monovalent cation transporters in plant responses to salinity. J Exp Bot. 2006, 57: 1137-1147. 10.1093/jxb/erj001.
18. Demidchik V, Maathuis FJ: Physiological roles of nonselective cation channels in plants: from salt stress to signalling and development. New Phytol. 2007, 175: 387-404. 10.1111/j.1469-8137.2007.02128.x.
19. Kilian J, Whitehead D, Horak J, Wanke D, Weinl S, Batistic O, D'Angelo C, Bornberg-Bauer E, Kudla J, Harter K: The AtGenExpress global stress expression data set: protocols, evaluation and model data analysis of UV-B light, drought and cold stress responses. Plant J. 2007, 50: 347-363. 10.1111/j.1365-313X.2007.03052.x.
20. Fricke W, Akhiyarova G, Veselov D, Kudoyarova G: Rapid and tissue-specific changes in ABA and in growth rate in response to salinity in barley leaves. J Exp Bot. 2004, 55: 1115-1123. 10.1093/jxb/erh117.
21. Fricke W, Akhiyarova G, Wei W, Alexandersson E, Miller A, Kjellbom PO, Richardson A, Wojciechowski T, Schreiber L, Veselov D, et al: The short-term growth response to salt of the developing barley leaf. J Exp Bot. 2006, 57: 1079-1095. 10.1093/jxb/erj095.
22. Knight H, Trewavas AJ, Knight MR: Calcium signalling in Arabidopsis thaliana responding to drought and salinity. Plant J. 1997, 12: 1067-1078. 10.1046/j.1365-313X.1997.12051067.x.
23. Donaldson L, Ludidi N, Knight MR, Gehring C, Denby K:

Salt and osmotic stress cause rapid increases in Arabidopsis thaliana cGMP levels. FEBS Lett. 2004, 569: 317-320. 10.1016/j.febslet.2004.06.016.

24. Davenport RJ, Munoz-Mayor A, Jha D, Essah PA, Rus A, Tester M: The Na+ transporter AtHKT1;1 controls retrieval of Na+ from the xylem in Arabidopsis. Plant Cell Environ. 2007, 30: 497-507. 10.1111/j.1365-3040.2007.01637.x.
25. Sunarpi , Horie T, Motoda J, Kubo M, Yang H, Yoda K, Horie R, Chan WY, Leung HY, Hattori K, et al: Enhanced salt tolerance mediated by AtHKT1 transporter-induced Na unloading from xylem vessels to xylem parenchyma cells. Plant J. 2005, 44: 928-938. 10.1111/j.1365-313X.2005.02595.x.
26. Berthomieu P, Conejero G, Nublat A, Brackenbury WJ, Lambert C, Savio C, Uozumi N, Oiki S, Yamada K, Cellier F, et al: Functional analysis of AtHKT1 in Arabidopsis shows that Na+ recirculation by the phloem is crucial for salt tolerance. Embo J. 2003, 22: 2004-2014. 10.1093/emboj/cdg207.
27. Jeschke JPS: Cation and chloride partitioning through xylem and phloem within the whole plant of Ricinus communis L. under conditions of salt stress. J Exp Bot. 1991, 42: 1105-1116. 10.1093/jxb/42.9.1105.
28. Jeschke WD, Pate JS: Temporal patterns of uptake, flow and utilization of nitrate, reduced nitrogen and carbon in a leaf of salt-treated castor bean (Ricinus communis L.). J Exp Bot. 1992, 43: 393-402. 10.1093/jxb/43.3.393.
29. Munns R, King RW: Abscisic acid is not the only stomatal inhibitor in the transpiration stream of wheat plants. Plant Physiol. 1988, 88: 703-708. 10.1104/pp.88.3.703.
30. Munns R: Comparative physiology of salt and water stress. Plant Cell Environ. 2002, 25: 239-250. 10.1046/j.0016-8025.2001.00808.x.
31. Deeken R, Geiger D, Fromm J, Koroleva O, Ache P, Langenfeld-Heyser R, Sauer N, May ST, Hedrich R: Loss of the AKT2/3 potassium channel affects sugar loading into the phloem of Arabidopsis. Planta. 2002, 216: 334-344. 10.1007/s00425-002-0895-1.

32. Alonso JM, Stepanova AN, Leisse TJ, Kim CJ, Chen H, Shinn P, Stevenson DK, Zimmerman J, Barajas P, Cheuk R, et al: Genome-wide insertional mutagenesis of Arabidopsis thaliana. Science. 2003, 301: 653-657. 10.1126/science.1086391.

33. Köhler C, Merkle T, Roby D, Neuhaus G: Developmentally regulated expression of a cyclic nucleotide-gated ion channel from Arabidopsis indicates its involvement in programmed cell death. Planta. 2001, 213: 327-332. 10.1007/s004250000510.

34. Arnon DI: Vitamin B1 in relation to the growth of green plants. Science. 1940, 92: 264-266. 10.1126/science.92.2386.264.

35. Baumann E, Lewald J, Saedler H, Schulz B, Wisman E: Successful PCR-based reverse genetic screens using an En-1-mutagenised Arabidopsis thaliana population generated via single-seed descent. Theor Appl Genet. 1998, 97: 729-734. 10.1007/s001220050949.

36. Curtis MD, Grossniklaus U: A gateway cloning vector set for high-throughput functional analysis of genes in planta. Plant Physiol. 2003, 133: 462-469. 10.1104/pp.103.027979.

37. Becker D, Kemper E, Schell J, Masterson R: New plant binary vectors with selectable markers located proximal to the left T-DNA border. Plant Mol Biol. 1992, 20: 1195-1197. 10.1007/BF00028908.

38. Clough SJ, Bent AF: Floral dip: a simplified method for Agrobacterium-mediated transformation of Arabidopsis thaliana. Plant J. 1998, 16: 735-743. 10.1046/j.1365-313x.1998.00343.x.

39. Jefferson RA, Kavanagh TA, Bevan MW: GUS fusions: beta-glucuronidase as a sensitive and versatile gene fusion marker in higher plants. Embo J. 1987, 6: 3901-3907.

40. Voinnet O, Rivas S, Mestre P, Baulcombe D: An enhanced transient expression system in plants based on suppression of gene silencing by the p19 protein of tomato bushy stunt virus. Plant J. 2003, 33: 949-956. 10.1046/j.1365-313X.2003.01676.x.

41. Chomczynski P, Sacchi N: Single-step method of RNA isolation by acid guanidinium thiocyanate-phenol-chloroform extraction. Analytical biochemistry. 1987, 162: 156-159. 10.1016/0003-2697(87)90021-2.

CHAPTER

# 9

# HETEROLOGOUS EXPRESSION AND FUNCTIONAL ANALYSIS OF RICE GLUTAMATE RECEPTOR-LIKE FAMILY INDICATES ITS ROLE IN GLUTAMATE TRIGGERED CALCIUM FLUX IN RICE ROOTS

Jun Ni[1], Zhiming Yu[1] , Guankui Du[2] , Yanyan Zhang[1] , Jemma L. Taylor[3] , Chenjia Shen[1] , Jing Xu[1] , Xunyan Liu[1] , Yifeng Wang[4] and Yunrong Wu

[1]College of Life and Environmental Sciences, Hangzhou Normal University

[2]Department of Biochemistry, Hainan Medical College

[3]School of Life Sciences, Gibbet Hill Campus, University of Warwick

[4]State Key Laboratory of Rice Biology, China National Rice Research Institute, Chinese Academy of Agricultural Sciences

[5]State Key Laboratory of Plant Physiology and Biochemistry, College of Life Science, Zhejiang University

## ABSTRACT

### Background

Tremendous progress has been made in understanding the functions of the GLUTAMATE RECEPTOR-LIKE (GLR) family in *Arabidopsis*. Still,

the functions of OsGLRs in rice, especially the ion channel activities, are largely unknown.

## Results

Using the aequorin-based luminescence imaging system, we screened the specificity of amino acids involved in the induction of $Ca^{2+}$ flux in rice roots. Of all the amino acids tested, glutamate (Glu) was the only one to trigger $Ca^{2+}$flux significantly in rice roots. Detailed analysis showed a dose response of $Ca^{2+}$ increase to different concentrations of Glu. In addition, the $Ca^{2+}$ spike response to Glu was rapid, within 20 s after the application. A desensitization assay and pharmacological tests showed that the Glu-triggered $Ca^{2+}$ flux is mediated by OsGLRs. Whole genome analysis identified 13 *OsGLR* genes in rice, and these genes have various expression patterns in different tissues. Subcellular localization studies showed that all the OsGLRs examined are likely localized to the plasma membrane. Bacteria growth assays showed that at least OsGLR2.1 and OsGLR3.2 have the potential to mediate ion uptake in bacteria. Further analysis using Fura-2-based $Ca^{2+}$ imaging revealed a Glu-triggered $Ca^{2+}$increase in OsGLR2.1-expressing human embryonic kidney (HEK) cells.

## Conclusions

Our work provides a molecular basis for investigating mechanisms of Glu-triggered $Ca^{2+}$ flux in rice.

**Keywords:** Aequorin Glu GLR Calcium HEK cells Rice

# BACKGROUND

Glutamate (Glu), as a fast excitatory neurotransmitter, has an important role in the animal nervous system (Watkins and Jane 2006). Glu activates the ionotropic glutamate receptors (iGluRs) to mediate the $Na^+$, $K^+$ and $Ca^{2+}$ entry into cells, catalyzing the signaling across the neurosynaptic gaps (Dingledine et al. 1999). Since the discovery of *GLUTAMATE RECEPTOR-LIKE* (*GLR*) genes in plants (Lacombe et al. 2001; Lam et al. 1998), which are closely related to the mammalian *iGluRs*, more and more evidence has indicated the existence of similar signaling pathways analogous to those found in animals (Forde 2014).

*Arabidopsis* has a total of 20 *GLR* genes, and phylogenetic analysis has shown that these genes can be subdivided into three clades (Chiu et al. 1999; Chiu et al. 2002; Lacombe et al. 2001). The basic domain structures of plant GLRs and animal iGluRs are conserved and phylogenetic analysis suggested that the separation of GLRs can be traced back to the last universal common ancestor of plants, metazoans and bacteria (Price et al. 2012). Recently, molecular genetics and electrophysiological approaches have made exciting advances in our understanding of various GLR functions in plants. AtGLR1.1 integrates and regulates the different aspects of carbon, nitrogen and water balance that are required for normal plant growth and development (Kang et al. 2004; Kang and Turano 2003). AtGLR1.2 mediates a plant signaling mechanism between male gametophytes and pistil tissue that is important for the pollen tube growth (Michard et al. 2011). AtGLR1.4 accounts for methionine (Met)-induced membrane depolarization in *Arabidopsis* leaves (Tapken et al. 2013). AtGLR3.3 is involved in plant defense signaling and the control of root gravitropism (Manzoor et al. 2013; Miller et al. 2010). AtGLR3.5 modulates cytosolic $Ca^{2+}$ level to counteract effect of abscisic acid in seed germination (Kong et al. 2015). AtGLRs also mediate mechanical wound signaling and elicitor/pathogen-mediated defense signaling in *Arabidopsis* (Manzoor et al. 2013; Mousavi et al. 2013). Moreover, AtGLR3.2 and AtGLR3.4 were reported to form heteromeric channels to affect lateral root development via $Ca^{2+}$ signaling in the phloem (Vincill et al. 2013).

In contrast with Glu or glycine (Gly) activated iGluRs, *Arabidopsis* GLRs have similar amino acid-gated ion channel activities, but with a broader agonist profile (Forde 2014). Six amino acids are considered to be the agonists of AtGLR3.3 dependent $Ca^{2+}$ influx in *Arabidopsis*, and a mutation of *AtGLR3.3* gene inhibits the rise of $Ca^{2+}$ triggered by these amino acids (Qi et al. 2006). Further research demonstrates that these six effective amino acids are not equivalent agonists, but grouped into hierarchical classes based on their ability to desensitize the response mechanism (Stephens et al. 2008). Human embryonic kidney (HEK) cells expressing AtGLR3.4 showed wide agonist profile with asparagine (Asn) and serine (Ser) as strong agonists and Gly less so (Vincill et al. 2012).

Interestingly, AtGLR1.4 expressing in *Xenopus* oocytes, functioned as a cation channel that responded to an even broader range of amino acids with Met being the most effective and most potent agonist (Tapken et al. 2013). An aequorin-based luminescence recording system showed that strong $Ca^{2+}$ responses were induced by all the amino acids tested, with the highest $Ca^{2+}$ amplitude for cysteine (Cys) and lowest one for Asn (Zhu et al. 2013).

Compared with the great progresses of GLRs in dicotyledonous *Arabidopsis*, less information has been obtained from the monocot, rice. The only gene characterized was *OsGLR3.1*, which positively regulates cell proliferation in the root apical meristem and is required for survival of meristematic cells (Li et al. 2006). However, the ion channel activities of OsGLRs are still completely unknown.

In our previous research, we established an aequorin-based luminescence imaging system in rice, which was able to reflect the concentration of cytosolic free $Ca^{2+}$ ($[Ca^{2+}]_i$) level in rice roots (Zhang et al. 2015). Here, using this system, we present the Glu-specific triggered $Ca^{2+}$ response in rice roots. Further analysis using Fura-2-based $Ca^{2+}$imaging revealed a Glu triggered $Ca^{2+}$ increase in OsGLR2.1-expressing HEK cells. Our results indicated an OsGLR mediated $Ca^{2+}$ changes in response to Glu in rice roots.

# RESULTS

## Glu Specifically Triggered $Ca^{2+}$ Influx in Rice Roots

To investigate the amino acid-triggered $Ca^{2+}$ influx in rice, we employed an aequorin-based luminescence imaging system, which is able to reflect the $[Ca^{2+}]_i$ level in rice roots (Zhang et al. 2015). After the reconstitution of aequorin by spraying seedlings with coelenterazine, the aequorin luminescence of rice roots was recorded using a photo-counting camera by treating plants with nine different amino acids respectively. As shown in Fig. 1, Glu induced a very strong $[Ca^{2+}]_i$ increase, while the other eight amino acids had little effect on the $[Ca^{2+}]_i$ induction compared with the control treatment. This showed that Glu is able to trigger $Ca^{2+}$ influx specifically in rice roots.

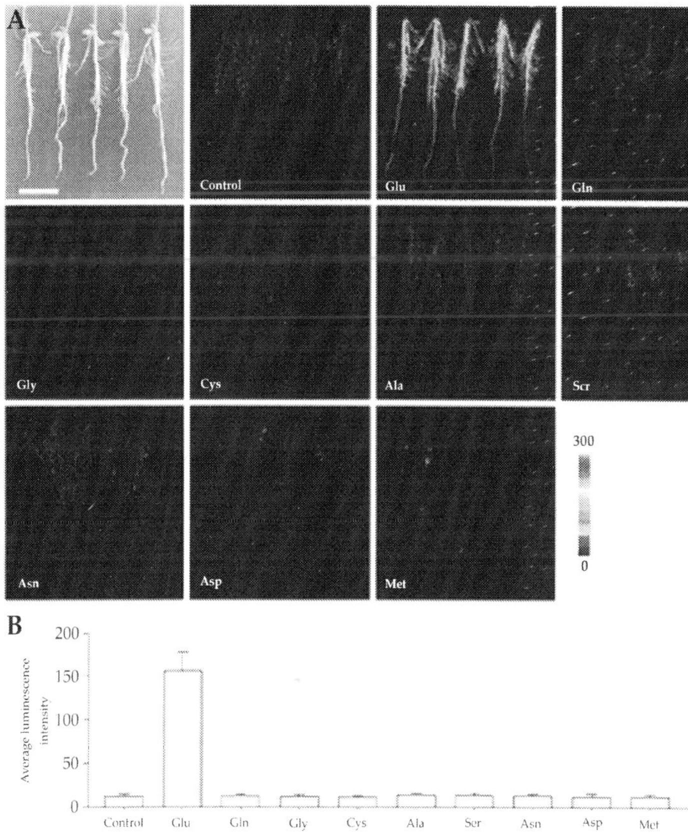


**Figure 1**: Glutamate specifically triggers $Ca^{2+}$ influx in rice roots. **a** Pseudocolor images of aequorin luminescence in roots treated with different amino acids. The relationship between luminescence intensity and the pseudocolor images are scaled by a pseudocolor bar and the numbers next to the pseudocolor bar are maximum and minimum values of luminescence intensity. The image on the left of the top is the treated roots view in white light. Bar=2 cm. Water was used instead of amino acid in the control treatment. **b** The bar chart of luminescence signal intensity of every treatment. The results were obtained from at least three independent experiments (mean ± sd; $n=10$)

## The Characteristics of $[Ca^{2+}]_i$ Response to Glu in Rice Roots

In order to investigate the $[Ca^{2+}]_i$ increase in response to Glu in rice roots, we examined the aequorin-based luminescence under various concentrations of Glu treatments. Overall, the more Glu that was applied, the stronger the luminescence signals would be (Fig. 2a). Detailed analysis showed that the concentration-dependent $[Ca^{2+}]_i$ increase was in agreement with Michaelis-Menten function with the Km value of 9.60 mM (Fig. 2b). To investigate the time courses of $[Ca^{2+}]_i$ responses induced by different concentrations of Glu in rice roots, the average luminescence intensity of continuous images with exposure time of 20 s, were analyzed and a comparison of basic parameters (amplitudes, durations and phases) of $[Ca^{2+}]_i$ responses were made. As shown in Fig. 2c, a strong luminescence signal was detected in the first image, which collected the luminescence signal within the first 20 s. However, almost no luminescence signal was detected after the first 20 s. This showed that Glu rapidly induced a sharp spike of $[Ca^{2+}]_i$ within 20 s, which quickly declined to the basal level after the spike. The amplitudes of luminescence signals varied according to different concentrations of Glu, while both the durations and phases were the same. These results were similar to the NaCl triggered $[Ca^{2+}]_i$ increase in rice and osmotic triggered $[Ca^{2+}]_i$ increase in *Arabidopsis* (Yuan et al. 2014; Zhang et al. 2015), indicating the existence of sensory channels to mediate Glu-triggered $[Ca^{2+}]_i$ increase in rice roots.

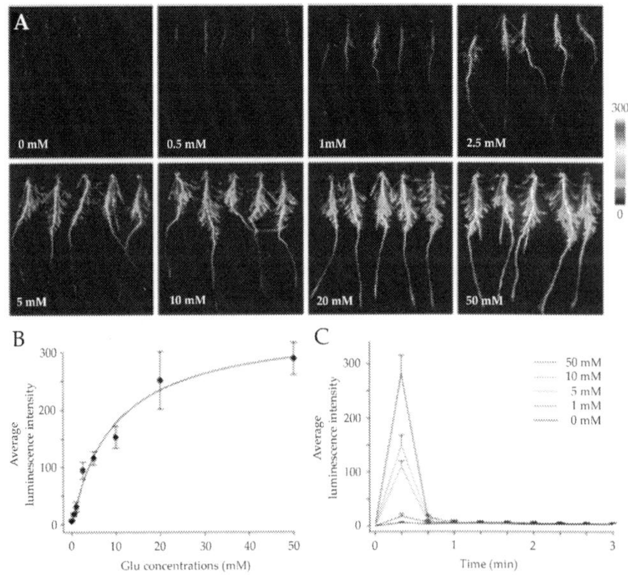


**Figure 2**: The characteristics of $[Ca^{2+}]_i$ response to Glu in rice roots. **a** Pseudocolor images of aequorin luminescence in roots treated with different con-

centrations of Glu. The relationship between luminescence intensity and the pseudocolor images are scaled by a pseudocolor bar and the numbers next to the pseudocolor bar are maximum and minimum values of luminescence intensity. **b** Concentration dependence of $[Ca^{2+}]_i$ increase in rice roots. Data were fitted by the Michaelis-Menten function. **c** The time courses of $[Ca^{2+}]_i$ changes induced by different concentrations of Glu in rice roots. The results were obtained from at least three independent experiments (mean ± sd; $n = 10$)

## Glu-Triggered $[Ca^{2+}]_i$ Increase is Mediated by OsGLRs

It is believed that GLRs are key components of a $Ca^{2+}$ influx mechanism in response to amino acids in plants (Dennison and Spalding 2000). Thus, we examined whether the Glu-triggered $[Ca^{2+}]_i$ increase in rice roots is also mediated by OsGLRs. Desensitization is an important mode of ligand-gated $Ca^{2+}$ channels at synapses in the animal nervous system (Jones and Westbrook 1996). Research in *Arabidopsis* showed that the phenomenon of desensitization also exists in the Glu activated, GLR mediated $Ca^{2+}$ entry (Meyerhoff et al. 2005; Qi et al. 2006). To determine whether the Glu- triggered $[Ca^{2+}]_i$ increase is the manifestation of the same mechanism (e.g. the activation of OsGLRs), we examined the $[Ca^{2+}]_i$ increases in response to two sequential Glu treatments. The first application of Glu triggered an obvious $[Ca^{2+}]_i$ increase, while after one hour's recovery, a second application of Glu with the same concentration, caused about 70 % reduction in the $[Ca^{2+}]_i$ increase compared with the first one (Additional file 1). To determine the dependence of *de novo* protein synthesis in the recovery of desensitization, we pretreated rice roots with cycloheximide (CHX), a well-known translation inhibitor. As a result, we observed a more serious inhibition (about 90 % reduction) in the $[Ca^{2+}]_i$ increase in response to a second application of Glu (Additional file 1). These results were very similar to that of AtGLR3.4-expressing mesophyll cells (Meyerhoff et al. 2005), indicating the possibility that OsGLRs mediate the Glu-triggered $[Ca^{2+}]_i$ increase in rice roots.

To confirm the participation of OsGLRs in the Glu-triggered $[Ca^{2+}]_i$ increase, we conducted pharmacological tests in the aequorin-based luminescence imaging system. $La^{3+}$ and $Gd^{3+}$ are agonists of $Ca^{2+}$, and they have been used as $Ca^{2+}$ channel blockers to inhibit $Ca^{2+}$ flux (Tracy et al. 2008). In our experiment, both $LaCl_3$ and $GdCl_3$ had significant inhibitory effects in Glu-triggered $[Ca^{2+}]_i$ increase. Furthermore, all the concentrations tested had similar inhibitory effects, indicating an effective inhibition function of $LaCl_3$ and $GdCl_3$ treatments (Fig. 3). 6-Cyano-7-Nitroquinoxaline-2,3-dione (CNQX) and 6,7-Dinitroquinoxaline-2,3-dione (DNQX) are GLR antagonists, and application of these drugs inhibits the channel activities of GLRs in plants (Meyerhoff et al. 2005; Michard et al. 2011; Tapken et al. 2013). In our experiment, both antagonists had significant inhibitory effects. In addition, dosage effects were observed in the inhibition of Glu-triggered $[Ca^{2+}]_i$ increase (Fig. 3). CNQX and DNQX were dissolved by Dimethyl sulfoxide (DMSO) in the stock solutions, and the highest concentration of DMSO in the working solution was 0.5 % (*v/v*). In order to investigate the effect of DMSO in the Glu-induced $[Ca^{2+}]_i$ increase, we examined the luminescence after various concentrations of DMSO pretreatments, and found that the effect 0.5 % of DMSO had no significant differences to the water control (Additional file 2). These results above showed that Glu-triggered $[Ca^{2+}]_i$ increase in rice roots is likely to be mediated by OsGLRs.

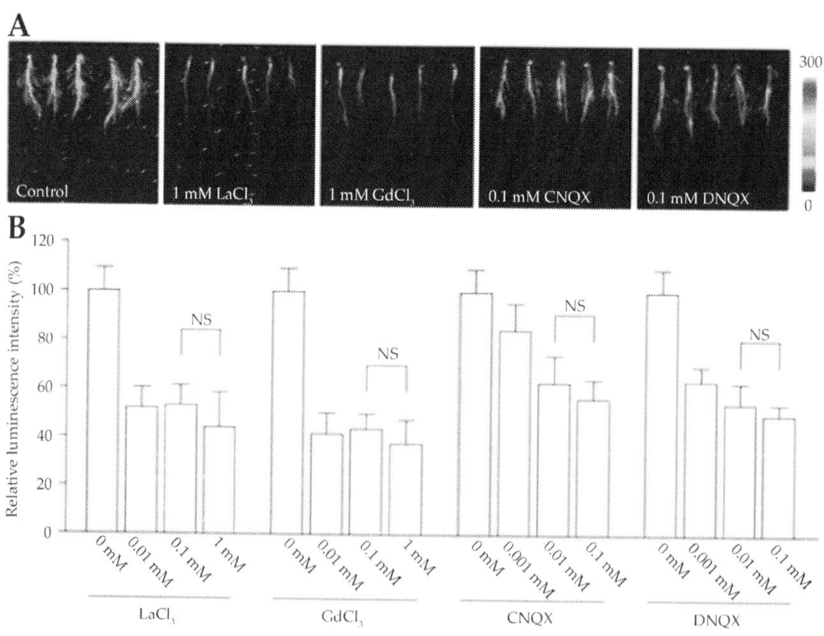


**Figure 3**: The pharmacological test of Glu-induced $[Ca^{2+}]_i$ increases. **a** Treatment of $Ca^{2+}$ agonists ($LaCl_3$ and $GdCl_3$) and GLR-specific antagonists (CNQX and DNQX) reduced Glu-induced $[Ca^{2+}]_i$ increases. The relationship between luminescence intensity and the pseudocolor images are scaled by a

pseudocolor bar and the numbers next to the pseudocolor bar are maximum and minimum values of luminescence intensity. **b** The effects of different concentrations of $Ca^{2+}$ agonists and GLR-specific antagonists in Glu-induced $[Ca^{2+}]_i$ increases. The results were obtained from at least three independent experiments (mean ± sd; $n=10$; NS, not significant $P>0.05$; Student's $t$-test)

## Characterization of *OsGLR* Gene Family in Rice

Since there is no comprehensive information about rice OsGLR family, searches of public genomic data were carried out. A total of 13 *OsGLR* genes that are most closely related to *Arabidospis* GLRs were identified by phytozome (http://www.phytozome.net/). Because there was no standard annotation assigned to these newly-identified genes, we named these genes according to their position in the phylogenetic tree (Fig. 4). The names of the *OsGLR* genes, the locus ID, the chromosome location, the isoelectric point (pI), the molecular weight (Mw) and predicted subcellular localizations are shown in Additional file 3. The decoded polypeptides were predicted to contain 3–6 transmembrane helices (TMHs), and most of them have the signature domains of animal iGluRs, including the 'three-plus-one' transmembrane domains (Lam et al. 1998) (Additional file 4). To detect the evolutionary relationships of GLRs, a phylogenetic tree was generated from alignments of 13 OsGLRs in rice and 20 AtGLRs in *Arabidopsis* (Fig. 4 and Additional file 5). We grouped the GLRs into three clades according to the previously reported work in *Arabidopsis* (Chiu et al. 2002). Detailed analysis showed that the phylogenetic relationships between *Arabidopsis* and rice are different among the three clades of GLRs. GLRs in clade I and clade II separated very early, while GLRs in clade III are evolutionary very close to each other (Fig. 4).

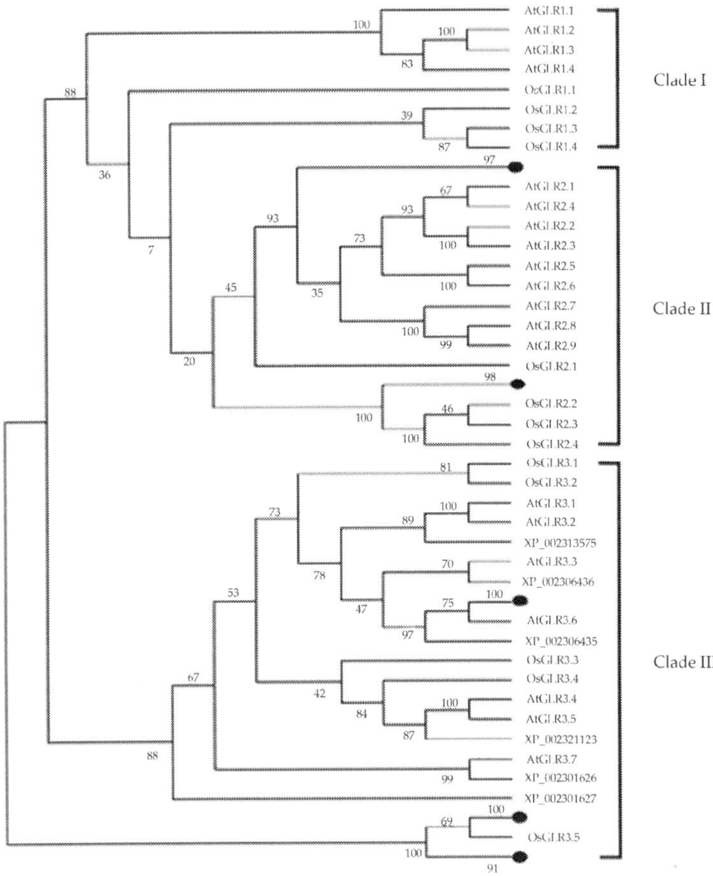


**Figure 4**: Phylogenetic relationships of GLRs in *Arabidopsis*, rice and poplar. Unrooted phylogenetic tree was constructed using MEGA6 by Maximum-Likelihood method. Three different GLR clades were marked by *lines*. *Black dots* represent compressed subtree of GLRs from poplar. [GenBank: XP_002313575, GenBank: XP_002306436, GenBank: XP_002306435, GenBank: XP_002321123, GenBank: XP_002301626 and GenBank: XP_002301627] are accession numbers of GLRs from poplar. Bootstrap values from 500 replicates are indicated at each branch

## Tissue-Specific Expression of *OsGLR* Genes

We examined the expression patterns of identified *OsGLR* genes in the five different tissues by real-time quantitative reverse transcription polymerase chain reaction (qRT-PCR). The results revealed variations

increases vital for osmosensing in Arabidopsis. Nature 514:367–371

55. Zhang S, Huganir RL (1999) Calmodulin modification of NMDA receptors. Methods Mol Biol 128:103–111
56. Zhang Y, Wang Y, Taylor JL, Jiang Z, Zhang S, Mei F, Wu Y, Wu P, Ni J (2015) Aequorin-based luminescence imaging reveals differential calcium signalling responses to salt and reactive oxygen species in rice roots. J Exp Bot 66:2535–2545
57. Zhu X, Feng Y, Liang G, Liu N, Zhu JK (2013) Aequorin-based luminescence imaging reveals stimulus- and tissue-specific Ca2+ dynamics in Arabidopsis plants. Mol Plant 6:444–455

CHAPTER

# 10

# PLANT HIGH-AFFINITY POTASSIUM (HKT) TRANSPORTERS INVOLVED IN SALINITY TOLERANCE: STRUCTURAL INSIGHTS TO PROBE DIFFERENCES IN ION SELECTIVITY

Shane Waters [1], Matthew Gilliham [2] and Maria Hrmova [1]

[1]Australian Centre for Plant Functional Genomics and Waite Research Institute, University of Adelaide, Glen Osmond, South Australia 5064, Australia

[2]School of Agriculture, Food and Wine, Waite Research Institute and ARC Centre of Excellence in Plant Energy Biology, University of Adelaide, Glen Osmond, South Australia 5064, Australia

## ABSTRACT

High-affinity Potassium Transporters (HKTs) belong to an important class of integral membrane proteins (IMPs) that facilitate cation transport across the plasma membranes of plant cells. Some members of the HKT protein family have been shown to be critical for salinity tolerance in commercially important crop species, particularly in grains, through exclusion of $Na^+$ ions from sensitive shoot tissues in plants. However, given the number of different HKT proteins expressed in plants, it is likely that different members of this protein family perform in a range of functions. Plant breeders and biotechnologists have attempted to manipulate *HKT* gene expression through genetic engineering and more conventional plant breeding methods to improve the salinity tolerance of commercially important crop plants. Successful manipulation of a biological trait is more likely to be effective after a thorough understanding of how the trait, genes and proteins are interconnected at the whole plant level. This article examines the current structural and functional knowledge relating to plant HKTs and how their structural features may explain their transport selectivity. We also highlight specific areas where new knowledge of plant HKT transporters is needed. Our goal is to present how knowledge of the structure of HKT proteins is helpful in understanding their function and how this understanding can be an invaluable experimental tool. As such, we assert that accurate structural information of plant IMPs will greatly inform functional studies and will lead to a deeper understanding of plant nutrition, signalling and stress tolerance, all of which represent factors that can be manipulated to improve agricultural productivity.

**Keywords:** Bacterial TrkH $K^+$ transporter, cation exclusion, protein structure and function, selectivity filter and pore; structural analysis

## BACKGROUND

In sequenced genomes, around 30% of all genes are predicted to encode integral membrane proteins (IMPs) [1,2] that are fully embedded in the phospholipid bilayer of a biological membrane. These proteins usually make up around 50% of the total mass of a membrane, but can account for as much as 75% [3]. IMPs are crucial to cell function, as they are a vital point of interaction between cell organelles and the cytoplasm and between the cell as a whole and the extracellular environment. Common types of IMPs include proteins involved in energy transduction (e.g., bacteriorhodopsins, [4]), cell adhesion (e.g., integrins, [5]), catalytic function (e.g., curdlan synthase, [6]), protein-protein interaction (e.g., hyaluronan receptors, [7]) and transport (e.g., aquaporins, [8]). IMPs receive and transmit signals, as well as control the movement of solutes across membranes [9,10]. Those solutes that have high molecular mass or carry charge (e.g., ions, metabolites and sugars), cannot diffuse across phospholipid bilayers, so their movement across a biological membrane is facilitated by transport proteins [9,11]. The transport rate across biological membranes of small and non-polar solutes and gases (e.g., water, urea, ammonia and $CO_2$) can also be mediated by IMPs [12]. Thus, solute transporters are known to be the key players of many well-characterised aspects of physiology in mammalian and plant systems [13–15]. A demonstration of the importance of solute transporters and channels is that they are predicted to be targets of more than half of all pharmaceuticals on the market [1,16]. Despite this importance, structural knowledge of IMPs lags behind that of soluble proteins, with less than 400 unique structures of IMP resolved as of December 2012 [16,17]. Notably, fewer than ten of these 400 unique structures are from plants [18].

The scarcity of structural knowledge about IMPs is due to the additional challenges associated with obtaining IMP structures as opposed to determining the structures of soluble proteins [2,16,18]. These include: (i) the difficulty in obtaining sufficient amounts of IMPs from native or recombinant sources; (ii) recovering IMPs from membranes without denaturing them; (iii) frequent cytotoxicity issues when over-expressing IMPs in recombinant systems and (iv) the complex crystallisation conditions of IMPs.

This article focuses on a particular class of IMP; the plant High-affinity Potassium Transporters (HKTs). HKT transporters occur only in plants, but have sequence and functional similarity to the TrkH/KtrB classes of cation transporters from bacteria and fungi [19,20] and are considered to have a similar structure to these proteins [21,22]. HKTs are segregated into two sub-groups based on their transport selectivity. Group 1 are described as $Na^+$ uniporters, while group 2 are thought to allow both $Na^+$ and $K^+$ transport (and possibly symport of these ions under specific conditions) and $Na^+$ uniport at high $Na^+$ concentrations. Individual plant species often contain multiple *HKT* genes, e.g., rice has nine, although only seven are functional (in the Nipponbare cultivar) [23]. The HKT proteins are of particular interest to plant biologists, as some members of this class have an important role in helping plants to tolerate high soil salinity, which represents a major agricultural problem worldwide [24,25].

# HKT TRANSPORTERS AND THEIR FUNCTION

## Function of HKT Transporters Is Important for Plants Tolerating Soil Salinity

Salinity has come to mean the occurrence of salts (primarily sodium chloride, NaCl) in ground water or the soil solution at levels that inhibit the growth of plants [25]. The actual concentration of salts needed to have a negative influence on plant growth varies and is dependent on many factors, including plant species, soil type and water availability [25]. From an agricultural perspective, soil is usually considered saline when the electrical conductivity (ECe) of the soil solution exceeds 4 dS/m [24,25]. Both intra- and inter-specific natural variation in salinity tolerance exists in plants [24,26]. Sensitive species, such as *Arabidopsis thaliana* and *Oryza sativa*, struggle to survive in the presence of 100 mM NaCl, whereas *Atriplex amnicola* (saltbush) can complete its lifecycle in 600 mM NaCl [24]. Intraspecific variation is common and probably best documented in cereals [27,28]. The mechanisms that result in salt tolerance and account for variation within and between species are numerous, but unsurprisingly, solute transport proteins are often critical, particularly those involved in $Na^+$ transport [24,29].

HKT proteins have been characterised predominantly as monovalent cation transporters [30], although reports of $Mg^{2+}$ and $Ca^{2+}$ permeability in some members has been proposed sporadically [31,32]. HKT proteins-mediated transport of $Na^+$ is known to be an important component of salinity tolerance in several species, including *Arabidopsis*[26,33,34], rice [22,28,35] and wheat [36–38]. *HKT* genes are frequently discovered underlying quantitative trait loci (QTL) that explain significant variation in salinity tolerance within mapping populations of wheat [37,39,40]. It has been demonstrated that a wide range of alleles for specific *HKT* genes can be present and that differences in these alleles may explain a component of the natural variation in salinity tolerance in *Arabidopsis*[26,41]. For example, Baxter and co-workers [26,41] used genome-wide association mapping, genetic complementation and gene expression studies to identify *cis*-regulatory expression level polymorphisms at the AtHKT1;1 locus. Some reports challenge the assumption that $Na^+$ exclusion leads to better salinity tolerance [26,27], with either no correlation or higher $Na^+$ concentration in the shoot and lower *HKT* expression correlating with better salt tolerance, while many other reports find the opposite [28,37]. Regardless, *HKT* alleles are an important component of salt tolerance conferred by either strategy. This paradox may be explained by simultaneous differences and natural variation in the other components that underpin salinity tolerance, such as the ability to tolerate the effects of salt accumulation in the shoot or cope with the osmotic components of salinity [24].

The important function of HKTs in assisting plants to survive high soil salinity has been confirmed in gene knockout experiments that made plants more sensitive to salt [33,34,42] and through transgenic expression of *HKT* genes in other plant species, *i.e.*, expression of the *AtHKT1;1* gene in rice allowed the rice plants to grow better under saline conditions [35].

While it is clearly very beneficial to manipulate *HKT* gene expression in crop plants to improve their salinity tolerance in some instances, as recently demonstrated by Munns *et al.*[36], it is important to properly understand function, regulation and expression patterns of the HKT transporters to achieve the desired result. For example, Møller *et al.*[43] overexpressed the *Arabidopsis HKT1;1* gene using a constitutive promoter and a promoter that was specific for stele cells of roots. Transgenic plants overexpressing *AtHKT1;1* constitutively accumulated large amounts of salt in the shoot and were relatively more salt sensitive than control plants, whilst plants over-expressing the same gene specifically in root stele cells accumulated less salt in the shoot and became more salt tolerant [43]. Evidence also exists that different HKT proteins may have different roles within a plant, so it is important to choose the correct allele, when modifying HKT expression to improve salt tolerance. TmHKT1;5-A is located in the cells around the root xylem, where the transporter moves $Na^+$ down its electrochemical gradient, which in most physiological circumstances is likely to be out of the xylem [36]. Similarly, AtHKT1;1 is thought to be present around xylem and phloem tissues in the root and shoot, and it also increases $Na^+$ retention in the root directly [44]. AtHKT1;1 has also been proposed to recirculate $Na^+$ in the phloem, although this function has been questioned [24,30,34]. The possible involvement of HKTs in $Na^+$ recirculation is still controversial and has been discussed in detail in a recent review by Hauser and Horie [45]. On the other hand, TaHKT2;1, OsHKT2;1 and OsHKT2;4 appear to be expressed in the outer part of the root, including root hairs, and may actually provide an entry point for $Na^+$ into plant roots from the soil solution. In these circumstances, downhill electrochemical gradient for $Na^+$ entry into plant cells is proposed to assist with high affinity $K^+$ entry into roots when $K^+$ is at low concentrations in the soil solution [32,46–48]. With this in mind, it is clearly important to have a comprehensive understanding of how a particular HKT protein functions, before attempting to manipulate expression of that *HKT* or the selection of that allele and its promoter to increase the productivity of a plant species or a variety grown in saline soil.

## Ion Selectivity of HKT Transporters

The discovery of plant HKT proteins was made by Schachtman and Schroeder [46], who identified a plant transport protein, which could rescue a mutant yeast strain (*Saccharomyces cerevisiae* CY162) defective in $K^+$ transport. A cDNA of the wheat-root plasmid encoding the *HKT1* gene permitted mutant yeast growing on media containing 30 μM $K^+$[46]. These authors subsequently expressed this *HKT1* gene in *Xenopus laevis* oocytes and showed that the resulting transport protein was selective for the $K^+$ ions, while allowing transport of some $Cs^+$ and $Rb^+$, but only of trace amounts of $Na^+$ or $NH_4^+$. For these reasons, this transport protein was subsequently designated as a "High-affinity Potassium Transporter" (HKT) [46]. However, this protein, now known as TaHKT2;1 [49], is now thought to be a $Na^+/K^+$symporter at micromolar concentrations of $Na^+$ and a $Na^+$ uniporter at millimolar concentrations of $Na^+$, after more detailed analysis in *Xenopus* oocytes [50,51].

There are numerous *HKT* genes in commercially important grain species [46], but there is only one in the model plant *Arabidopsis*, designated *AtHKT1;1*[52,53]. Most functional studies have indicated that HKT proteins are monovalent cation transporters, with the cation selectivity varying between individual members of the HKT family [46,54,55], although some findings have suggested divalent cation transport by HKT proteins [31,32]. Some HKT proteins are highly selective for $Na^+$, while others are more promiscuous and will move $Na^+$ and $K^+$, e.g., OsHKT2;1 and OsHKT2;2, respectively [55]. Further, some HKT proteins even change their selectivity depending on an ionic environment, e.g., EcHKT1;1 [56]. Many individual *HKT* genes have now been cloned and expressed in various heterologous systems, to evaluate and characterise transport selectivity and behaviour of the specific HKT transport protein. These includes *HvHKT2;1* expressed in yeast [57], *AtHKT1;1* in *Xenopus* oocytes [34], yeast [53] and *E. coli*[53], *EcHKT1;1* and *EcHKT1;2* in *Xenopus* oocytes and *E. coli*[56,58] and *TaHKT2;1* expressed in yeast and *Xenopus* oocytes [46,50]. These experiments in heterologous systems describe HvHKT2;1 as a $Na^+$selective uniporter [57] and OsHKT2;1 and TaHKT2;1 as $Na^+/K^+$ symporters at low $Na^+$ concentrations and a $Na^+$ uniporter at high $Na^+$ concentrations [34,46,50,53,59]. EcHKT1;1 and EcHKT1;2 have been reported to be permeable to $Ca^{2+}$ and $Mg^{2+}$, as well as being $Na^+/K^+$ symporters, even at high $Na^+$concentrations [56,58]. Several authors have observed $Na^+/K^+$ symport by the AtHKT1;1, HvHKT1;5, Ni-OsHKT1;5 and Po-OsHKT1;5 proteins in yeast, which has not been observed in *Xenopus* oocytes [46,57,60]. These observations lead to uncertainty, as to what would be the actual behaviour of HKT proteins *in planta*.

This uncertainty is compounded by paucity of direct evidence showing $Na^+/K^+$ symport in wheat [61], barley [62] and rice [63] roots, calling into question the putative role of HKT2-type proteins in plants in the uptake of $K^+$, in $K^+$ deficient conditions, using the electrochemical gradient for $Na^+$[57,64]. This means that either the results from heterologous systems are incorrect or the activity of the HKTs *in planta* has been masked in these particular experiments through the activity of other plant proteins or in the specific conditions imposed. At the same time, there is evidence of HKT activity from plants that mirrors the activity found in oocytes and yeast for instance $Na^+/K^+$ symport in Arabidopsis [65] and for OsHKT2;1 in a plant cell expression system [55]. Speculation as to the accuracy of extrapolating HKT function *in planta* from heterologous systems has been called into question by a number of groups for reasons ranging from artefacts induced by the choice of vector used for protein overexpression to the inadvertent ascribing of endogenous channel activity of heterologous systems to that of HKTs [48,66]. This extends to the observation of divalent cation transport ($Mg^{2+}$ and $Ca^{2+}$) by OsHKT2;4 when expressed in *Xenopus* oocytes [32], which could not be reproduced in a subsequent study [48]. This is something that the researchers need to be mindful of, when interpreting data from various heterologous expression systems [55,66].

While the transport selectivity of some HKT transporters is well established, the actual mechanism of transport is less clear. Whether HKT predominantly mediates $Na^+$, $K^+$, $Mg^{2+}$ or $Ca^{2+}$, there have been several proposed mechanisms as to how the HKT proteins mediate movement of these ions. It was originally suggested that $H^+$ symport may provide the gradient needed to move ions across the plasma membrane [46,67], although subsequent experimental evidence seemed to suggest that this is unlikely. Active transport by HKT proteins powered by the ATP hydrolysis has been ruled out, due to the lack of a recognisable ATP-binding domain [46]. Both *in vitro* and *in planta* experiments now indicate that the most likely mechanism of HKT proteins is to facilitate the movement of monovalent cations down their electrophysiological gradient [24,68]. Some HKTs potentially having $Na^+/K^+$ symport capacity when either the $Na^+$ or $K^+$ gradient allows the other ion to enter cells against its electrochemical gradient [50,56,59]; however, this mechanism is highly dependent on specific ionic conditions imposed on heterologous systems, which may not be relevant to *in planta* function. Before manipulating HKT-mediated cation movement *in planta* within the context of altering salinity tolerance in plants, it would be beneficial to describe the characteristics of transport mechanisms of monovalent cations by HKT proteins in general. It would also be useful to know what the structural determinants of selectivity are for specific HKT.

Expression patterns and tissue localisation varies significantly between different HKT transporters, likely reflecting *in planta* function of the individual proteins [32,34,44]. There is a consensus that all HKT proteins characterised so far are plasma membrane IMPs [24,32,34,44,45]. Tissue localisation however, varies significantly. OsHKT1;5 [22] and TaHKT1;5 [37] are expressed in the cells around the root xylem in rice and wheat, while OsHKT1;4 [22] and TaHKT1;4 [38] are expressed in the corresponding cells in the leaf sheath. AtHKT1;1, on the other hand, is expressed around the phloem and xylem tissues in *Arabidopsis*[33,34]. TaHKT2;1 and OsHKT2;1 have quite different localisation though, as these proteins appear to be expressed on the exterior of root and root hairs rather than around the vascular tissue [32,46] and assist with $K^+$ and $Na^+$ entry into roots [69].

# MOLECULAR STRUCTURE OF PLANT HKT TRANSPORTERS

Despite the importance of HKT transport proteins, there are no resolved three-dimensional (3D) structures of these proteins in the Protein Data Bank [70]. The structural prediction information for the HKT proteins is based mostly on bioinformatical analyses using bacterial homologues, such as the KcsA $K^+$ channel [71]. Recently, molecular models of the rice HKT transporters OsHKT1;5 and OsHKT1;4, based on a crystal structure of the bacterial TrkH $K^+$ transporter from *Vibrio parahaemolyticus* (VpTrkH) [21], have been constructed, and the structural data were reconciled with transcriptomic and physiological data [22]. A VpTrkH crystal structure was used here, as this protein and the rice HKT transporters share functional properties as a consequence of sequence and structural similarities [21,22].

## Molecular Modelling of Plant HKT Transporters

Protein structure determines protein function [18,72,73]. Therefore, the 3D structure of a protein provides vital information as to how a protein functions [73]. Often proteins with very different sequences can have similar structure and function, although the reverse case can occur, where proteins of very similar amino acid sequences have very different functions [60]. For instance, the plant *Equisetum arvense* (horsetail) has very high levels of silicon (Si), but when Gregoire *et al.* [8] searched the plant's genome using the sequences of known Si transporters, they were unable to find any homologues. When a family of Si transporters were eventually identified in *Equisetum arvense*, the genes were found to be similar to sequences of aquaporin genes, while the predicted structure was similar to the other Si transporters [8]. Often singular amino acid residues play critical roles in structure of proteins. For example, it has been demonstrated that a single amino acid substitution (Pro to Leu) can substantially alter the functional behaviour of a bacterial PilQ secretin, allowing it to auto-assemble into multimeric assemblies *in vitro*, a process that requires association with another protein for the native PilQ secretin [74].

Since 2006, when the current naming convention for HKT proteins was proposed [46], there have been a lot of new sequences added to public databases, and therefore, new structural bioinformatics analyses can be performed. To this end, we have performed a phylogenetic analysis of 46 amino acid sequences of known plant HKT proteins from mono- and di-cotyledonous plant species (Table 1) and constructed a phylogenetic tree (Figure 1). Protein sequences were obtained using the BLAST tool from the NCBI database [75] to search for plant sequences that relate to the EcHKT1;2 (AF176036_1), OsHKT1;5 (A2WNZ9.2) and AtHKT1;1 (Q84TI7.1) sequences (Table 1). Many of the annotations available have only been assigned from sequences and are annotated as "Predicted protein, cation transporter HKT1-like" or similar. However, all of the annotated sequences cluster as predicted [49,76], with type 1 and type 2 HKT entries in separate clades. The three proteins standing separate at the top of the tree, *i.e.*, ScTRK1 and ScTRK2 from *Saccharomyces cerevisiae* and VpTrkH from *V. parahaemolyticus* (Figure 1), represent yeast and bacterial $K^+$ transporters/symporters that have been included into the analysis. The VpTrkH protein was used as the template structure for molecular modelling shown in Figure 2.

The phylogenetic tree (Figure 1) clearly shows that all of the type 2 HKT proteins are in the clade below these three proteins, while the type 1 HKT proteins are in the bottom half of the tree. As predicted by Platten *et al.* [46], there are no dicotyledonous HKT proteins in the type 2 clade. However, it can be noted that the type 1 HKT entries have been subdivided into monocotyledonous and dicotyledonous groups (Figure 1). All the dicotyledonous HKT entries (and none of the monocotyledonous HKT entries) are marked by the bracket at the bottom of the tree. This suggests that a divergence between the monocotyledonous and dicotyledonous *HKT* genes must have occurred early during evolution of both types of plants. It is unclear as yet whether there are any major functional differences between the dicotyledonous and monocotyledonous HKT proteins; however, the eucalyptus (dicotyledonous) EcHKT1;1 and EcHKT1;2 are the only HKT proteins known to be sensitive to solution osmolarity [58]. As noted by Møller and Tester [77], care should be taken, when extrapolating transport selectivity data obtained from dicotyledonous (e.g., *Arabidopsis*) to monocotyledonous species (e.g., rice and wheat). Another recently published phylogenetic analysis [78] suggested segregation between HKTs from simple bryophytes and lycophytes and HKTs from higher plants. That analysis only included *HKT* genes from five species [78], but combined with the phylogenetic analysis included in this study, the possibility is raised that in future, as more information becomes available, HKTs may be reclassified into more subclasses then the current two.

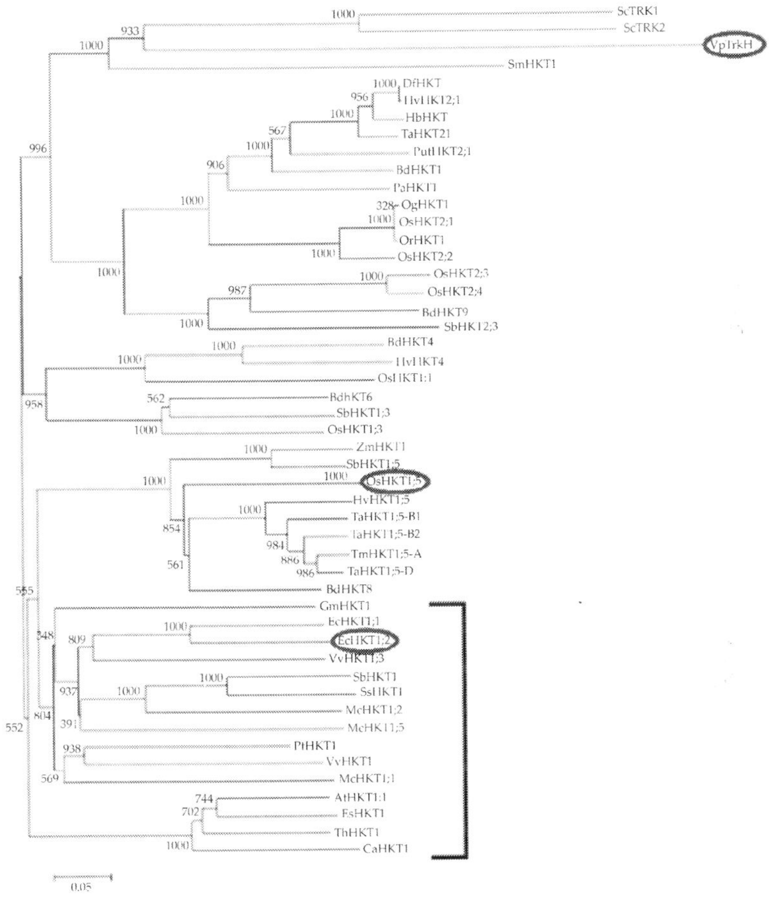


**Figure 1:** Phylogenetic analysis of plant HKT transporters. All published sequences were retrieved from the NCBI database and aligned with ClustalX2 [79]. The phylogenetic tree was constructed based on aligned protein sequences using Neighbour-Joining algorithm [80] with a Bootstrap value of 1000 using ClustalX2 [79]. The scale bar indicates the substitution rate per site. All dicotyledonous HKT entries (and none of monocotyledonous ones) are bracketed. The red circles denote proteins that are shown in Figure 2.

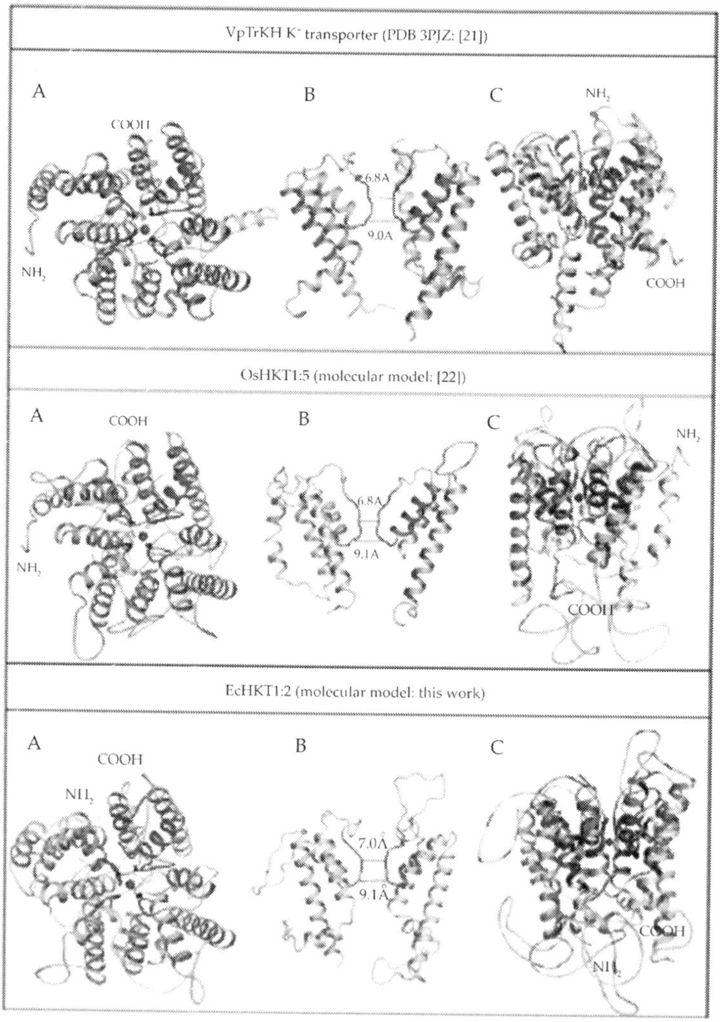


**Figure 2:** Cartoon representations of 3D structure of VpTrkH (PDB 3PJZ) (top panels) and molecular models of the *Oryza sativa*OsHKT1;5 (middle panels) and *Eucalyptus camaldulensis* EcHKT1;2 (bottom panels) transporters. (**A**) The overall folds of transporters are coloured in green (VpTrkH) orange (OsHKT1;5) and cyan (EcHKT1;2). Purple spheres indicate the $K^+$ (VpTrkH) and $Na^+$(OsHKT1;5 and EcHKT1;2) ions. The selectivity filter signatures illustrated in black sticks are: Gly-Gly-Gly-Gly for VpTrkH and Ser-Gly-Gly-Gly for OsHKT1;5 and EcHKT1;2. The residues contained in the selectivity pores are coloured in red. These residues are likely to affect pore rigidity and dispositions of residues controlling cation selectivity and transport rates; (**B**) The cut-out images show geometry of selectivity pores. Their width dimensions are

indicated; (**C**) Cartoon representations of VpTrkH, OsHKT1;5 and EcHKT1;2 colour-coded by conservation score of amino acid residues as predicted by ConSurf [81]. The views in panels C are rotated by 90° along the *x*-axis with respect to the views in panels A. The $NH_2$- and COOH-termini of the transporters are shown.

**Table 1:** List of entries used for phylogenetic analysis of plant High-affinity Potassium Transporters (HKTs) proteins (*cf.* Figure 1). Accession numbers of protein sequences were obtained with the BLAST tool from the NCBI database to search for sequences that relate to the OsHKT1;5, EcHKT1;2 and AtHKT1;1 entries.

| Protein | Species | Common name | Accession number |
|---|---|---|---|
| AtHKT1;1 | *Arabidopsis thaliana* | Thale cress | Q84TI7.1 |
| BdHKT1 | *Brachypodium distachyon* | Purple false brome | XP_003560515.1 |
| BdHKT4 | *Brachypodium distachyon* | Purple false brome | XP_003581628.1 |
| BdHKT6 | *Brachypodium distachyon* | Purple false brome | XP_003570995.1 |
| BdHKT8 | *Brachypodium distachyon* | Purple false brome | XP_003564102.1 |
| BdHKT9 | *Brachypodium distachyon* | Purple false brome | XP_003563514.1 |
| CaHKT1 | *Cochlearia anglica* | Scurvy grass | AFH37929.1 |
| DfHKT | *Diplachne fusca* | Brown beetle grass | AEM55592.1 |
| EcHKT1;1 | *Eucalyptus camaldulensis* | River redgum | AF176035_1 |
| EcHKT1;2 | *Eucalyptus camaldulensis* | River redgum | AF176036_1 |
| EsHKT1 | *Eutrema salsugineum* | Saltwater cress | AFJ23835.1 |
| GmHKT1 | *Glycine max* | Soybean | XP_003540998.1 |
| HbHKT | *Hordeum brevisubulatum* | Short-awned barley | AER42622.1 |
| HvHKT1;5 | *Hordeum vulgare* | Barley | ABK58096.1 |
| HvHKT4 | *Hordeum vulgare* | Barley | AEM44690.1 |
| HvHKT2;1 | *Hordeum vulgare* | Barley | AEM55590.1 |
| McHKT1;1 | *Mesembryanthemum crystallinum* | Ice plant | AF367366_1 |
| McHKT1;2 | *Mesembryanthemum crystallinum* | Ice plant | AAO73474.1 |
| MtHKT1;5 | *Medicago truncatula* | Alfalfa | AES77170.1 |
| OgHKT1 | *Oryza glumipatula* | Wild rice | ABD15858.1 |
| OrHKT1 | *Oryza rufipogon* | Red rice | AAY33540.1 |
| OsHKT1;1 | *Oryza sativa* | Rice | Q7XPF8.2 |
| OsHKT1;3 | *Oryza sativa* | Rice | Q6H501.1 |
| OsHKT1;5 | *Oryza sativa* | Rice | A2WNZ9.2 |
| OsHKT2;1 | *Oryza sativa* | Rice | A2YGP9.2 |
| OsHKT2;2 | *Oryza sativa* | Rice | Q93XI5.1 |
| OsHKT2;3 | *Oryza sativa* | Rice | Q8L481.1 |

| | | | |
|---|---|---|---|
| OsHKT2;4 | *Oryza sativa* | Rice | Q8L4K5.1 |
| PaHKT1 | *Phragmites australis* | Common reed | BAE44384.1 |
| PtHKT1 | *Populus trichocarpa* | Poplar | EEF03794.1 |
| PutHKT2;1 | *Puccinellia tenuiflora* | Alkali grass | ACT21087.1 |
| SbiHKT1;5 | *Sorghum bicolor* | Sorghum | EES02856.1 |
| SbiHKT1;3 | *Sorghum bicolor* | Sorghum | EES04614.1 |
| SbiHKT2;3 | *Sorghum bicolor* | Sorghum | EER90327.1 |
| SbHKT1 | *Salicornia bigelovii* | Dwarf saltwort | ADG45565.1 |
| SmHKT1 | *Selaginella moellendorffii* | Starry spike moss | EFJ18587.1 |
| SsHKT1 | *Suaeda salsa* | Seepweed | AAS20529.2 |
| TaHKT1;5-D | *Triticum aestivum* | Bread wheat | ABG33949.1 |
| TaHKT2;1 | *Triticum aestivum* | Bread wheat | AAA52749 |
| TaHKT1;5-B1 | *Triticum aestivum* | Bread wheat | ABG33947.1 |
| TaHKT1;5-B2 | *Triticum aestivum* | Bread wheat | ABG33948.1 |
| TmHKT1;5-A | *Triticum monococcum* | Einkorn wheat | ABG33946.1 |
| ThHKT1 | *Thellungiella halophila* | Salt cress | BAJ34563.1 |
| VvHKT1 | *Vitis vinifera* | Grape vine | XP_002270986.1 |
| VvHKT1;3 | *Vitis vinifera* | Grape vine | XP_002267717.1 |
| ZmHKT1 | *Zea mays* | Maize/corn | AEK27028.1 |
| ScTRK1 | *Saccharomyces cerevisiae* | Yeast | AAA34728 |
| ScTRK2 | *Saccharomyces cerevisiae* | Yeast | AAA35172 |
| VpTrkH | *Vibrio parahaemolyticus* | Not applicable | Q87TN7.1 |

Currently, there is very little experimentally derived structural knowledge about any HKT protein. There are no resolved structures of any HKT protein [70]. However, the topology of the AtHKT1;1 protein was examined by Kato *et al.*[52]. This was determined by inserting various tags (FLAG, PhoA alkaline phosphatase and *N*-glycosylation sites) into the AtHKT1;1 protein sequence and then synthesising the protein in vesicles in a cell-free system or expressing in *E. coli*. The synthesised protein was probed, based on the various tags to determine, on which side of the vesicle/cell membrane the individual tags locate. The data showed that the AtHKT1;1 protein had eight membrane-spanning helices, and the $NH_2$- and COOH-termini of AtHKT1;1 were located both on the cytoplasmic side of the membrane, when the protein was expressed in *E. coli*. Various experiments have also been done to mutate individual amino acid residues in order to determine which residues are critical to transport selectivity and function [82,83].

In the absence of experimentally determined structure, it is possible to infer structural data from comparisons with known structures or from computational predictions, such as comparative (homology) modelling [16,72]. These methods are limited by the availability of appropriate templates. For example, as there are no 3D structures of the HKT proteins available, the models of the rice OsHKT1;5 protein and the eucalyptus EcHKT1;2 (unpublished data) could provide this predictive structural information (Figure 2). OsHKT1;5 is a $Na^+$ selective transporter [28], EcHKT1;2 is a $Na^+$/$K^+$symporter and possibly transports $Mg^{2+}$ and $Ca^{2+}$, as well [56,58]. Both models were built based on the crystal structure of the bacterial $K^+$ selective transporter VpTrkH [21], as previously described [22]. Briefly, the sequence of EcHKT1;2 was aligned with that of the VpTrkH $K^+$ transporter (Protein Data Bank 3PJZ) [21], and the resultant alignments were tested for the positions of secondary structures. The manually corrected aligned sequences were used as input parameters to generate a molecular model of EcHKT2;1 in complex with $Na^+$ using Modeller 9v8 [84]. The best model of EcHKT1;2 was selected from 40 models based on the Modeller 9v8 objective function and the most favourable energy scoring parameters.

The bioinformatics and structural analyses of the plant HKT cation transporters showed that these proteins have the conserved "selectivity filter" motif of Ser-Gly-Gly-Gly for type 1 HKT proteins and Gly-Gly-Gly-Gly for type 2 HKT proteins [49]. Type 1 HKT proteins are highly selective for $Na^+$, while type 2 HKT could facilitate movement of $Na^+$ and $K^+$. Mutating the Ser residue of a type one HKT to Gly would change the transport characteristics of the protein to be similar to the type 2 HKT transporters, at least in the *Xenopus* oocyte expression system [76]. OsHKT2;1 is an exception to this rule, in that the protein is defined as a class 2 HKT, but has the Ser-Gly-Gly-Gly selectivity filter motif [55,76]. EcHKT1;2 is also something of an exception, as this protein has the Ser-Gly-Gly-Gly motif of type 1 HKT, but transports $K^+$, as well as $Na^+$, which is usually associated with the type 2 HKT proteins [49,56]. This suggests that other structural elements besides the key selectivity filter and pore residues (in $K^+$channels known as the pore helix residues [20]) may help in determining the substrate selectivity of HKT proteins. However, without a resolved 3D structure of HKT proteins, it is challenging to predict which specific motifs of the protein sequence are responsible for transport selectivity. Therefore, at the moment, it is necessary to experimentally determine transport selectivity of each HKT of interest, rather than predict the ion selectivity from its sequence.

Also displayed in Figure 2 (panels C) is the 3D structure of VpTrkH and the molecular models of OsHKT1;5 and EcHKT1;2 that are colour-coded according to the conservation values assigned by the ConSurf Server [81]. This analysis reports conservation of individual amino acid residues in each structure [81]. Cartoon representations of molecular models of OsHKT1;5 and EcHKT1;2, and the crystal structure of VpTrkH indicate that a series of the putative selectivity filter residues occur on the extracellular side of the protein [46,76]. These residues (Figure 2, highlighted in black in panel A) are located in the narrowest point in pores through each of the HKT proteins, which supports the hypothesis that these residues are the primary determinants of substrate selectivity of these transporters. Further, there is a short segment consisting of about seven amino residues (marked red in Figure 2, panels A and B) lining the narrow part of pores before they open up to a cavity. This cavity would not restrict the movement of individual monovalent ions. The conformation of this short loop is likely to be the key determinant of cation permeation into the pore helix of the HKT proteins. While the four selectivity filter residues are critical in this regard (Ser-Gly-Gly-Gly *versus* Gly-Gly-Gly-Gly), the ConSurf analysis showed that most of the remaining amino acid residues forming the pore helices are highly conserved and are, therefore, also likely to be critical for transport function. Last, but not least, the ion conducting channel is bound on the cellular interior by a highly conserved Arg residue, which is a unique feature of the TrkH transporters and almost all bacterial superfamily of $K^+$ transporters and has not been observed in any other known $K^+$ channel structures [21,85]. Remarkably, these highly conserved Arg and neighbouring Gly residues (Gly-Arg motif) with the highest conservation score of 9 as assigned by ConSurf [81], occur in both plant HKT transporters (data not shown).

Table 2 lists the amino acid residues lining the narrow pore of all three proteins displayed in Figure 2(coloured in red in panels A and B), that are colour-coded in Table 2 according to conservation score predicted by ConSurf. The pore regions are rich in the Thr and Ser residues, and while both HKT proteins contain an Arg-His pair at the beginning of the third loop of the pore, none of the proteins have any acidic residues prior to the selectivity filter. The presence of polar residues, such as Thr and Ser, in particular, their oxygen atoms along the backbone and side-chain oxygen atoms around the entrance to the pores, is likely to help to co-ordinate cations as they enter the selectivity filter. On the other hand, the Glu residues immediately after the selectivity filter in the HKT proteins present a negative charge at the narrowest point of the selectivity filter, which would prevent any anion permeability, such as, *i.e.*, $Cl^-$.

**Table 2:** The amino acid residues, given in three-letter codes, lining pores of the EcHKT1;2, OsHKT1;5 and VpTrkH transporters. The predicted selectivity filter residues are underlined.

| Protein | Accession number | Pore residues[1] | | | | | | | Residue number |
|---|---|---|---|---|---|---|---|---|---|
| EcHKT1;2 | AF176036_1 | Thr | Thr | Val | Ser | Ser | Met | Ser | 91–97 |
| | | Ala | Ser | Cys | Gly | Phe | Val | Pro | 259–265 |
| | | Arg | His | Thr | Gly | Glu | Thr | Val | 383–389 |
| | | Gly | Asn | Val | Gly | Phe | Thr | Thr | 488–494 |
| OsHKT1;5 | A2WNZ9.2 | Thr | Val | Ser | Ser | Met | Val | Ala | 73–79 |
| | | Ala | Asn | Cys | Gly | Phe | Val | Pro | 261–267 |
| | | Arg | His | Ser | Gly | Glu | Met | Val | 388–394 |
| | | Gly | Asn | Val | Gly | Phe | Ser | Thr | 492–498 |
| VpTrkH | Q87TN7.1 | Thr | Thr | Thr | Gly | Ala | Thr | Val | 110–116 |
| | | Ala | Ile | Gly | Gly | Phe | Ser | Thr | 219–225 |
| | | Thr | Thr | Ala | Gly | Phe | Thr | Thr | 319–325 |
| | | Asn | Asn | Leu | Gly | Pro | Gly | Leu | 436–442 |

[1] Residues are coloured according to conservation predictions analysed by the ConSurf server [81].

1 2 3 4 5 6 7 8 9

Variable Average Conserved

It is surprising that EcHKT1;2 transports $K^+$ in *Xenopus* oocytes, while the selectivity filter motif suggests that EcHKT1;2 is a type 1 HKT with $Na^+$ selective transport. This is likely to become more apparent if multiple HKT of each clade and of known selectivity are included in the same analysis. However, at this point, it is premature to explain this behavioural discrepancy, as another reason could exist besides the homology modelling prediction based on a $K^+$ selective transporter. Future modelling against the structures of $Na^+$ selective transporters and $Na^+/K^+$ symporters, when they become available, may reveal the underlying structural reason for an unusual transport behaviour of EcHKT1;2's. It would also be advisable to confirm the transport behaviour of EcHKT1;2 *in planta* and in other systems, as currently the transport selectivity of EcHKT1;2 has only been defined using yeast and *Xenopus* oocytes, which may not always accurately reflect true function of the EcHKT1;2 protein in its native environment.

## Why Is There So Little Structural Data for HKT Transporters?

While IMPs make up a significant proportion of the mass of cell membranes, many membrane proteins are only present on specific membranes, e.g., on a plasma membrane or on a specific organelle [3]. Consequently, individual proteins often occur in very small amounts in their native environments [11]. Limitations in space in the membrane and in co-factors needed for successful membrane insertion can result in overexpressed proteins failing to insert and forming mis-folded "inclusion bodies" in cells [86]. These inclusion bodies can sometimes be solubilised and refolded [86], but it is essential to make sure that refolded proteins are functional before attempting crystallisation, as the structure of a mis-folded and non-functional protein is of little value [18].

Soluble proteins can be expressed with a secretion signal, so the protein is secreted into the growthmedium, where it has limited effect on the survival of the cells. IMPs have to be expressed into a cell membrane to fold correctly and, consequently, may have severe effects on cell survival [31,87], requiring specific media and growth conditions to generate sufficient protein without disturbing the cell culture [11,88].

Purification of IMPs can be performed using standard protein purification methods, such as expressing with a His tag on the protein and using a nickel/cobalt-based affinity chromatography to isolate a protein effectively. However, these methods require the addition of a surfactant to solubilise IMPs. The choice of surfactant is critical, as a surfactant must be efficient enough to separate IMPs from the membrane, without denaturing the protein and must have appropriate chemical structure to protect hydrophobic surfaces of membrane proteins from solution to prevent agglutination, while at the same time, not obscuring the tag being used for purification [89,90].

Crystallisation of IMPs also requires a mixture of hydrophobic and hydrophilic environments [18,90]. This can be achieved by including phospholipids and sometimes gentle surfactants in the crystallisation mixtures [91]. The phospholipids are essential for IMPs to achieve a correct conformation, as they mediate hydrophobic matching that is required between IMP and its native membrane environment [92]. The hydrophilic surfaces of IMPs must be exposed and not obscured by a lipid or surfactant to allow protein-protein interactions, or IMPs do not form an ordered lattice and a crystal [89]. Unfortunately, lipids and surfactants in a protein crystal can interfere with X-ray diffraction; therefore, multiple combinations of additives may have to be tried to find conditions, under which a crystal diffracts well [89].

Das *et al.*[93] have noted that because the new environment, after IMP, is extracted from its native environment and can change protein structure, it is necessary to test the functionality of extracted IMPs. This is usually possible with an enzyme, but may not be practical with transport proteins. However, comparing structures obtained under different conditions, when they exist, can provide an insight into different forms that IMP can take. This is important as crystallised IMP may not be in the exact native conformation. Including substrates or co-factors can increase the likelihood of IMPs taking on a native conformation. It is also important to remember that a crystal structure is a 'freeze frame' image of a dynamic molecule, which moves between different states (*i.e.*, open and closed forms of a transporter). Thus, structures with and without an ion can give clues as to the changes that occur, as transport proteins mediate their function, *i.e.*, transport.

## CONCLUSIONS AND PERSPECTIVES

Transport selectivity of some HKT proteins is defined, but there is insufficient data comparing different alleles or mutations to allow a plant breeder to rationally select for a specific allele [37,38,94]. The expression patterns of the HKT proteins *in planta* also need to be more clearly defined. Knowing what cations a particular HKT transporter permeates across a membrane bilayer only reveals a part of the protein function. Exactly which membrane the ions are crossing, on a tissue and a cellular level, will determine the actual effect *in planta*. Expressing marker genes with the HKT promoters and probing for a presence RNA of a specific *HKT* gene (*i.e.*, *in situ* PCR) can reveal the tissues that the protein, or at least its RNA, is expressed in. However, localising the protein ultimately requires a method of detecting the actual protein, not just the corresponding RNA. For example, in the case of HKT proteins, this would undoubtedly require the development of highly specific antibodies [32].

Proteins rarely act in isolation. There are interactions with other proteins or cellular components, which activate, repress and otherwise modulate function of a protein. To date, no data have been reported that identified interaction partners for any HKT protein. This is an area which should be investigated, as it may turn out that the most effective way of modifying HKT function *in planta* is to modify their interaction partners, e.g., to increase the activity of a specific kinase or other modifying enzyme that might regulate transport function.

It would also be highly desirable to determine 3D structures of HKT transport proteins, as homology modelling can only take us so far in understanding the mode of action of HKT proteins at the molecular levels. There are also other experimental methods available to gain insights into protein structure, such as small-angle X-ray scattering and circular dichroism spectroscopy, which reveal the shapes and content of secondary structures within protein folds, respectively.

Last, but not least, performing experiments to analyse transport selectivity in clean systems, without endogenous proteins, *i.e.*, in proteoliposomes instead of yeast or *Xenopus* oocytes systems, would help to clarify the actual ion selectivity and rates of transport of individual HKT transporters. Finally, it would be worthwhile to perform more detailed mutational analysis of amino acid residues [83] besides those positioned in selectivity filters and pores, especially those that are predicted to be highly conserved across various types of HKT proteins.

## ACKNOWLEDGMENTS

Shane Waters thanks the University of Adelaide for his Postgraduate Award and the Australian Centre for Plant Functional Genomics for support. This work was supported by the DP120100900 and LP120100201 grants from the Australian Research Council awarded to MH.

## REFERENCES

1. Marassi, F.M.; Das, B.B.; Lu, G.J.; Nothnagel, H.J.; Park, S.H.; Son, W.S.; Tian, Y.; Opella, S.J. Structure determination of membrane proteins in five easy pieces. *Methods* 2011, *55*, 363–369.
2. Souda, P.; Ryan, C.M.; Cramer, W.A.; Whitelegge, J. Profiling of integral membrane proteins and their post translational modifications using high-resolution mass spectrometry. *Methods* 2011, *55*, 330–336.
3. Alberts, B.; Johnson, A.; Lewis, J.; Raff, M.; Roberts, K.; Walter, P. *Molecular Biology of the Cell*, 4th ed; Garland Science: New York, NY, USA, 2003; pp. 583–655.
4. Ito, H.; Sumii, M.; Kawanabe, A.; Fan, Y.; Furutani, Y.; Brown, L.S.;

Kandori, H. Comparative FTIR study of a new fungal rhodopsin. *J. Phys. Chem. B* 2012, *116*, 11881–11889.

5. Hu, P.; Luo, B.-H. Integrin bi-directional signaling across the plasma membrane. *J. Cell Physiol*2013, *228*, 306–312.
6. Periasamy, A.; Shadiac, N.; Amalraj, A.; Garajova, S.; Nagarajan, Y.; Waters, S.; Mertens, H.D.T.; Hrmova, M. Cell-free protein synthesis of membrane (1,3)-β-D-glucan (curdlan) synthase: Co-translational insertion in liposomes and reconstitution in nanodiscs. *Biochim. Biophys. Acta* 2013, *1828*, 743–757.
7. Bourguignon, L.Y.W.; Zhu, H.; Shao, L.; Chen, Y.-W. CD44 Interaction with c-Src kinase promotes cortactin-mediated cytoskeleton function and hyaluronic acid-dependent ovarian tumor cell migration. *J. Biol. Chem* 2001, *276*, 7327–7336.
8. Gregoire, C.; Rémus-Borel, W.; Vivancos, J.; Labbé, C.; Belzile, F.; Bélanger, R.R. Discovery of a multigene family of aquaporin silicon transporters in the primitive plant *Equisetum arvense*. *Plant J*2012, *72*, 320–330.
9. Hedrich, R. Ion channels in plants. *Physiol. Rev* 2012, *92*, 1777–1811.
10. Rosenbaum, D.M.; Rasmussen, S.G.F.; Kobilka, B.K. The structure and function of G-protein-coupled receptors. *Nature* 2009, *459*, 356–363.
11. Haferkamp, I.; Linka, N. Functional expression and characterisation of membrane transport proteins. *Plant Biol* 2012, *14*, 675–690.
12. Tyerman, S.D.; Niemietz, C.M.; Bramley, H. Plant aquaporins: multifunctional water and solute channels with expanding roles. *Plant Cell Environ* 2002, *25*, 173–194.
13. Chatelain, F.C.; Bichet, D.; Douguet, D.; Feliciangeli, S.; Bendahhou, S.; Reichold, M.; Warth, R.; Barhanin, J.; Lesage, F. TWIK1, a unique background channel with variable ion selectivity. *Proc. Natl. Acad. Sci. USA* 2012, *109*, 5499–5504.
14. Zimmerli, C.; Ribot, C.; Vavasseur, A.; Bauer, H.; Hedrich, R.; Poirier, Y. PHO1 expression in guard cells mediates the stomatal response to abscisic acid in *Arabidopsis*. *Plant J* 2012, *72*, 199–211.
15. Ikeda, M.; Beitz, E.; Kozono, D.; Guggino, W.B.; Agre, P.; Yasui, M.

Characterization of aquaporin-6 as a nitrate channel in mammalian cells. Requirement of pore-lining residue threonine 63. *J. Biol. Chem* 2002, *277*, 39873–39879.

16. Nugent, T.; Jones, D.T. Membrane protein structural bioinformatics. *J. Struct. Biol* 2012, *179*, 327–337.
17. Gabanyi, M.J.; Adams, P.D.; Arnold, K.; Bordoli, L.; Carter, L.G.; Flippen-Andersen, J.; Gifford, L.; Haas, J.; Kouranov, A.; McLaughlin, W.A.; *et al.* The Structural Biology Knowledgebase: A portal to protein structures, sequences, functions, and methods. *J. Struct. Funct. Genomics* 2011, *12*, 45–54.
18. Shadiac, N.; Nagarajan, Y.; Waters, S.; Hrmova, M. Close allies in membrane protein research: Cell-free synthesis and nanotechnology. *Mol. Membr. Biol.* 2013. [CrossRef]
19. Corratgé-Faillie, C.; Jabnoune, M.; Zimmermann, S.; Véry, A.A.; Fizames, C.; Sentenac, H. Potassium and sodium transport in non-animal cells: The Trk/Ktr/HKT transporter family. *Cell. Mol. Life Sci* 2010, *67*, 2511–2532.
20. Durell, S.R.; Guy, H.R. Structural models of the KtrB, TrkH, and Trk1,2 symporters based on the structure of the KcsA $K^+$ channel. *Biophys. J* 1999, *77*, 789–807.
21. Cao, Y.; Jin, X.; Huang, H.; Derebe, M.G.; Levin, E.J.; Kabaleeswaran, V.; Pan, Y.; Punta, M.; Love, J.; Weng, J.; *et al.* Crystal structure of a potassium ion transporter, TrkH. *Nature* 2011, *471*, 336–340.
22. Cotsaftis, O.; Plett, D.; Shirley, N.; Tester, M.; Hrmova, M. A two-stage model of $Na^+$ exclusion in rice explained by 3D modeling of HKT transporters and alternative splicing. *PLoS One* 2012, *7*, e39865.
23. Garciadeblas, B.; Senn, M.E.; Banuelos, M.A.; Rodriguez-Navarro, A. Sodium transport and HKT transporters: The rice model. *Plant J* 2003, *34*, 788–801.
24. Munns, R.; Tester, M. Mechanisms of salinity tolerance. *Annu. Rev. Plant Biol* 2008, *59*, 651–681.
25. Rengasamy, P. World salinization with emphasis on Australia. *J. Exp. Bot* 2006, *57*, 1017–1023.

26. Baxter, I.; Brazelton, J.N.; Yu, D.; Huang, Y.S.; Lahner, B.; Yakubova, E.; Li, Y.; Bergelson, J.; Borevitz, J.O.; Nordborg, M.; *et al.* A coastal cline in sodium accumulation in *Arabidopsis thaliana* is driven by natural variation of the sodium transporter AtHKT1;1. *PLoS Genet* 2010, *6*, e1001193.
27. Genc, Y.; McDonald, G.K.; Tester, M. Reassessment of tissue $Na^+$ concentration as a criterion for salinity tolerance in bread wheat. *Plant Cell Environ* 2007, *30*, 1486–1498.
28. Ren, Z.H.; Gao, J.P.; Li, L.G.; Cai, X.L.; Huang, W.; Chao, D.Y.; Zhu, M.Z.; Wang, Z.Y.; Luan, S.; Lin, H.X. A rice quantitative trait locus for salt tolerance encodes a sodium transporter. *Nat. Genet*2005, *37*, 1141–1146.
29. Plett, D.C.; Møller, I.S. $Na^+$ transport in glycophytic plants: What we know and would like to know. *Plant Cell Environ* 2010, *33*, 612–626.
30. Horie, T.; Hauser, F.; Schroeder, J.I. HKT transporter-mediated salinity resistance mechanisms in *Arabidopsis* and monocot crop plants. *Trends Plant Sci* 2009, *14*, 660–668.
31. Horie, T.; Brodsky, D.E.; Costa, A.; Kaneko, T.; Lo Schiavo, F.; Katsuhara, M.; Schroeder, J.I. $K^+$transport by the OsHKT2;4 transporter from rice with atypical $Na^+$ transport properties and competition in permeation of $K^+$ over $Mg^{2+}$ and $Ca^{2+}$ ions. *Plant Physiol* 2011, *156*, 1493–1507.
32. Lan, W.-Z.; Wang, W.; Wang, S.-M.; Li, L.-G.; Buchanan, B.B.; Lin, H.-X.; Gao, J.-P.; Luan, S. A rice high-affinity potassium transporter (HKT) conceals a calcium-permeable cation channel. *Proc. Natl. Acad. Sci. USA* 2010, *107*, 7089–7094.
33. Davenport, R.J.; Munoz-Mayor, A.; Jha, D.; Essah, P.A.; Rus, A.; Tester, M. The $Na^+$ transporter AtHKT1,1 controls retrieval of $Na^+$ from the xylem in *Arabidopsis*. *Plant Cell Environ* 2007, *30*, 497–507.
34. Berthomieu, P.; Conejero, G.; Nublat, A.; Brackenbury, W.J.; Lambert, C.; Savio, C.; Uozumi, N.; Oiki, S.; Yamada, K.; Cellier, F.; *et al.* Functional analysis of AtHKT1 in *Arabidopsis* shows that $Na^+$recirculation by the phloem is crucial for salt tolerance. *EMBO J* 2003, *22*, 2004–2014.

35. Plett, D.; Safwat, G.; Gilliham, M.; Møller, I.S.; Roy, S.; Shirley, N.; Jacobs, A.; Johnson, A.; Tester, M. Improved salinity tolerance of rice through cell type-specific expression of *AtHKT1;1*. *PLoS One* 2010, *5*, e12571.

36. Munns, R.; James, R.A.; Xu, B.; Athman, A.; Conn, S.J.; Jordans, C.; Byrt, C.S.; Hare, R.A.; Tyerman, S.D.; Tester, M.; *et al.* Wheat grain yield on saline soils is improved by an ancestral $Na^+$transporter gene. *Nat. Biotechnol* 2012, *30*, 360–364.

37. Byrt, C.S.; Platten, J.D.; Spielmeyer, W.; James, R.A.; Lagudah, E.S.; Dennis, E.S.; Tester, M.; Munns, R. HKT1;5-like cation transporters linked to $Na^+$ exclusion loci in wheat, *Nax2* and *Kna1*. *Plant Physiol* 2007, *143*, 1918–1928.

38. Huang, S.; Spielmeyer, W.; Lagudah, E.S.; James, R.A.; Platten, J.D.; Dennis, E.S.; Munns, R. A sodium transporter (HKT7) is a candidate for *Nax1*, a gene for salt tolerance in durum wheat. *Plant Physiol* 2006, *142*, 1718–1727.

39. Dubcovsky, J.; Maria, G.S.; Epstein, E.; Luo, M.C.; Dvorak, J. Mapping of the $K^+/Na^+$discrimination locus *Kna1* in wheat. *Theor. Appl. Genet* 1996, *92*, 448–454.

40. Dvořak, J.; Noaman, M.M.; Goyal, S.; Gorham, J. Enhancement of the salt tolerance of *Triticum turgidum* L. by the *Kna1* locus transferred from the *Triticum aestivum* L. chromosome 4D by homoeologous recombination. *Theor. Appl. Genet* 1994, *87*, 872–877.

41. Rus, A.; Baxter, I.; Muthukumar, B.; Gustin, J.; Lahner, B.; Yakubova, E.; Salt, D.E. Natural variants of *AtHKT1* enhance $Na^+$ accumulation in two wild populations of *Arabidopsis*. *PLoS Genet* 2006, *2*, e210.

42. Mäser, P.; Eckelman, B.; Vaidyanathan, R.; Horie, T.; Fairbairn, D.J.; Kubo, M.; Yamagami, M.; Yamaguchi, K.; Nishimura, M.; Uozumi, N.; *et al.* Altered shoot/root $Na^+$ distribution and bifurcating salt sensitivity in *Arabidopsis* by genetic disruption of the $Na^+$ transporter *AtHKT1*. *FEBS Lett* 2002, *531*, 157–161.

43. Møller, I.S.; Gilliham, M.; Jha, D.; Mayo, G.M.; Roy, S.J.; Coates, J.C.; Haseloff, J.; Tester, M. Shoot $Na^+$ exclusion and increased salinity tolerance engineered by cell type-specific alteration of

$Na^+$transport in *Arabidopsis*. *Plant Cell* 2009, *21*, 2163–2178.

44. Sunarpi; Horie, T.; Motoda, J.; Kubo, M.; Yang, H.; Yoda, K.; Horie, R.; Chan, W.Y.; Leung, H.Y.; Hattori, K.; Konomi, M.; *et al.* Enhanced salt tolerance mediated by AtHKT1 transporter-induced $Na^+$unloading from xylem vessels to xylem parenchyma cells. *Plant J.* 2005, *44*, 928–938.
45. Hauser, F.; Horie, T. A conserved primary salt tolerance mechanism mediated by HKT transporters: a mechanism for sodium exclusion and maintenance of high $K^+/Na^+$ ratio in leaves during salinity stress. *Plant Cell Environ* 2010, *33*, 552–565.
46. Schachtman, D.P.; Schroeder, J.I. Structure and transport mechanism of a high-affinity potassium uptake transporter from higher plants. *Nature* 1994, *370*, 655–658.
47. Laurie, S.; Feeney, K.A.; Maathuis, F.J.M.; Heard, P.J.; Brown, S.J.; Leigh, R.A. A role for HKT1 in sodium uptake by wheat roots. *Plant J* 2002, *32*, 139–149.
48. Sassi, A.; Mieulet, D.; Khan, I.; Moreau, B.; Gaillard, I.; Sentenac, H.; Véry, A.-A. The rice monovalent cation transporter OsHKT2;4: Revisited ionic selectivity. *Plant Physiol* 2012, *160*, 498–510.
49. Platten, J.D.; Cotsaftis, O.; Berthomieu, P.; Bohnert, H.; Davenport, R.J.; Fairbairn, D.J.; Horie, T.; Leigh, R.A.; Lin, H.X.; Luan, S.; *et al.* Nomenclature for HKT transporters, key determinants of plant salinity tolerance. *Trends Plant Sci* 2006, *11*, 372–374.
50. Gassmann, W.; Rubio, F.; Schroeder, J.I. Alkali cation selectivity of the wheat root high-affinity potassium transporter HKT1. *Plant J* 1996, *10*, 869–882.
51. Rubio, F.; Gassmann, W.; Schroeder, J.I. Sodium-driven potassium uptake by the plant potassium transporter HKT1 and mutations conferring salt tolerance. *Science* 1995, *270*, 1660–1663.
52. Kato, Y.; Sakaguchi, M.; Mori, Y.; Saito, K.; Nakamura, T.; Bakker, E.P.; Sato, Y.; Goshima, S.; Uozumi, N. Evidence in support of a four transmembrane-pore-transmembrane topology model for the *Arabidopsis thaliana* $Na^+/K^+$ translocating AtHKT1 protein, a member of the superfamily of $K^+$ transporters. *Proc. Natl. Acad. Sci. USA* 2001, *98*, 6488–6493.

53. Uozumi, N.; Kim, E.J.; Rubio, F.; Yamaguchi, T.; Muto, S.; Tsuboi, A.; Bakker, E.P.; Nakamura, T.; Schroeder, J.I. The *Arabidopsis HKT1* gene homolog mediates inward $Na^+$ currents in *Xenopus laevis* oocytes and $Na^+$ uptake in *Saccharomyces cerevisiae*. *Plant Physiol* 2000, *122*, 1249–1260.

54. Ardie, S.W.; Xie, L.N.; Takahashi, R.; Liu, S.K.; Takano, T. Cloning of a high-affinity $K^+$ transporter gene *PutHKT2;1* from *Puccinellia tenuiflora* and its functional comparison with OsHKT2;1 from rice in yeast and *Arabidopsis*. *J. Exp. Bot* 2009, *60*, 3491–3502.

55. Yao, X.; Horie, T.; Xue, S.; Leung, H.-Y.; Katsuhara, M.; Brodsky, D.E.; Wu, Y.; Schroeder, J.I. Differential sodium and potassium transport selectivities of the rice OsHKT2;1 and OsHKT2;2 transporters in plant cells. *Plant Physiol* 2010, *152*, 341–355.

56. Fairbairn, D.J.; Liu, W.H.; Schachtman, D.P.; Gomez-Gallego, S.; Day, S.R.; Teasdale, R.D. Characterisation of two distinct HKT1-like potassium transporters from *Eucalyptus camaldulensis*. *Plant Mol. Biol* 2000, *43*, 515–525.

57. Haro, R.; Banuelos, M.A.; Senn, M.A.E.; Barrero-Gil, J.; Rodriguez-Navarro, A. HKT1 mediates sodium uniport in roots. Pitfalls in the expression of HKT1 in yeast. *Plant Physiol* 2005, *139*, 1495–1506.

58. Liu, W.H.; Fairbairn, D.J.; Reid, R.J.; Schachtman, D.P. Characterization of two HKT1 homologues from *Eucalyptus camaldulensis* that display intrinsic osmosensing capability. *Plant Physiol* 2001, *127*, 283–294.

59. Oomen, R.J.F.J.; Benito, B.; Sentenac, H.; Rodriguez-Navarro, A.; Talon, M.; Very, A.-A.; Domingo, C. HKT2;2/1, a $K^+$-permeable transporter identified in a salt-tolerant rice cultivar through surveys of natural genetic polymorphism. *Plant J* 2012, *71*, 750–762.

60. Horie, T.; Yoshida, K.; Nakayama, H.; Yamada, K.; Oiki, S.; Shinmyo, A. Two types of HKT transporters with different properties of $Na^+$ and $K^+$ transport in *Oryza sativa*. *Plant J* 2001, *27*, 129–138.

61. Maathuis, F.J.M.; Verlin, D.; Smith, F.A.; Sanders, D.; Fernandez, J.A.; Walker, N.A. The physiological relevance of $Na^+$-coupled $K^+$-transport. *Plant Physiol* 1996, *112*, 1609–1616.

62. Schulze, L.M.; Britto, D.T.; Li, M.; Kronzucker, H.J. A

pharmacological analysis of high-affinity sodium transport in barley (*Hordeum vulgare* L.): A $^{24}Na^+/^{42}K^+$ study. *J. Exp. Bot* 2012, *63*, 2479–2489.

63. Horie, T.; Costa, A.; Kim, T.H.; Han, M.J.; Horie, R.; Leung, H.-Y.; Miyao, A.; Hirochika, H.; An, G.; Schroeder, J.I. Rice OsHKT2;1 transporter mediates large $Na^+$ influx component into $K^+$-starved roots for growth. *EMBO J* 2007, *26*, 3003–3014.
64. Walker, N.A.; Sanders, D.; Maathuis, F.J.M.; Rubio, F.; Gassmann, W.; Schroeder, J.I. High-affinity potassium uptake in plants. *Science* 1996, *273*, 977–979.
65. Spalding, E.P.; Hirsch, R.E.; Lewis, D.R.; Qi, Z.; Sussman, M.R.; Lewis, B.D. Potassium uptake supporting plant growth in the absence of AKT1 channel activity—Inhibition by ammonium and stimulation by sodium. *J. Gen. Physiol* 1999, *113*, 909–918.
66. Banuelos, M.A.; Haro, R.; Fraile-Escanciano, A.; Rodriguez-Navarro, A. Effects of polylinker uATGs on the function of grass HKT1 transporters expressed in yeast cells. *Plant Cell. Physiol*2008, *49*, 1128–1132.
67. Schachtman, D.; Liu, W.H. Molecular pieces to the puzzle of the interaction between potassium and sodium uptake in plants. *Trends Plant Sci* 1999, *4*, 281–287.
68. Jabnoune, M.; Espeout, S.; Mieulet, D.; Fizames, C.; Verdeil, J.-L.; Conéjéro, G.; Rodríguez-Navarro, A.; Sentenac, H.; Guiderdoni, E.; Abdelly, C.; *et al.* Diversity in expression patterns and functional properties in the rice HKT transporter family. *Plant Physiol* 2009, *150*, 1955–1971.
69. Wang, T.B.; Gassmann, W.; Rubio, F.; Schroeder, J.I.; Glass, A.D.M. Rapid up-regulation of HKT1, a high-affinity potassium transporter gene, in roots of barley and wheat following withdrawal of potassium. *Plant Physiol* 1998, *118*, 651–659.
70. Berman, H.M.; Westbrook, J.; Feng, Z.; Gilliland, G.; Bhat, T.N.; Weissig, H.; Shindyalov, I.N.; Bourne, P.E. The protein data bank. *Nucleic Acids Res* 2000, *28*, 235–242.
71. Raja, M. The potassium channel KcsA: A model protein in studying membrane protein oligomerization and stability of oligomeric assembly? *Arch. Biochem. Biophys* 2011, *510*, 1–10.

72. Hrmova, M.; Fincher, G.B. Functional Genomics and Structural Biology in the Definition of Gene Function. In *Plant Genomics*; Langridge, P., Gustafson, P., Eds.; Humana Press Inc: Totowa, NJ, USA, 2009; pp. 199–227.

73. Hunter, M.S.; Fromme, P. Toward structure determination using membrane-protein nanocrystals and microcrystals. *Methods* 2011, *55*, 387–404.

74. Nickerson, N.N.; Abby, S.S.; Rocha, E.P.C.; Chami, M.; Pugsley, A.P. A single amino acid substitution changes the self-assembly status of a type IV piliation secretin. *J. Bacteriol* 2012, *194*, 4951–4958.

75. Geer, L.Y.; Marchler-Bauer, A.; Geer, R.C.; Han, L.; He, J.; He, S.; Liu, C.; Shi, W.; Bryant, S.H. The NCBI BioSystems database. *Nucleic Acids Res* 2010, *38*, 492–496.

76. Mäser, P.; Hosoo, Y.; Goshima, S.; Horie, T.; Eckelman, B.; Yamada, K.; Yoshida, K.; Bakker, E.P.; Shinmyo, A.; Oiki, S.; *et al.* Glycine residues in potassium channel-like selectivity filters determine potassium selectivity in four-loop-per-subunit HKT transporters from plants. *Proc. Natl. Acad. Sci. USA* 2002, *99*, 6428–6433.

77. Møller, I.S.; Tester, M. Salinity tolerance of *Arabidopsis*: A good model for cereals? *Trends Plant Sci*2007, *12*, 534–540.

78. Gomez-Porras, J.L.; Riaño-Pachón, D.M.; Benito, B.; Haro, R.; Sklodowski, K.; Rodriguez-Navarro, A.; Dreyer, I. Phylogenetic analysis of $K^+$ transporters in bryophytes, lycophytes, and flowering plants indicates a specialization of vascular plants. *Front. Plant Sci* 2012, *3*, 167–167.

79. Larkin, M.A.; Blackshields, G.; Brown, N.P.; Chenna, R.; McGettigan, P.A.; McWilliam, H.; Valentin, F.; Wallace, I.M.; Wilm, A.; Lopez, R.; *et al.* Clustal W and Clustal X version 2.0. *Bioinformatics* 2007, *23*, 2947–2948.

80. Saitou, N.; Nei, M. The neighbor-joining method: A new method for reconstructing phylogenetic trees. *Mol. Biol. Evol* 1987, *4*, 406–425.

81. Landau, M.; Mayrose, I.; Rosenberg, Y.; Glaser, F.; Martz, E.; Pupko, T.; Ben-Tal, N. ConSurf 2005: The projection of evolutionary conservation scores of residues on protein structures. *Nucleic Acids Res* 2005, *33*, 299–302.

82. Rubio, F.; Schwarz, M.; Gassmann, W.; Schroeder, J.I. Genetic selection of mutations in the high affinity $K^+$ transporter HKT1 that define functions of a loop site for reduced $Na^+$ permeability and increased $Na^+$ tolerance. *J. Biol. Chem* 1999, *274*, 6839–6847.

83. Diatloff, E.; Kumar, R.; Schachtman, D.P. Site directed mutagenesis reduces the $Na^+$ affinity of HKT1, an $Na^+$ energized high affinity $K^+$ transporter. *FEBS Lett* 1998, *432*, 31–36.

84. Šali, A.; Blundell, T.L. Comparative protein modelling by satisfaction of spatial restraints. *J. Mol. Biol* 1993, *234*, 779--815.

85. Kato, N.; Akai, M.; Zulkifli, L.; Matsuda, N.; Kato, Y.; Goshima, S.; Hazama, A.; Yamagami, M.; Guy, H.R.; Uozumi, N. Role of positively charged amino acids in the M2(D) transmembrane helix of Ktr/Trk/HKT type cation transporters. *Channels* 2007, *1*, 161–171.

86. Scalise, M.; Galluccio, M.; Pochini, L.; Indiveri, C. Over-expression in *Escherichia coli*, purification and reconstitution in liposomes of the third member of the OCTN sub-family: The mouse carnitine transporter OCTN3. *Biochem. Biophys. Res. Commun* 2012, *422*, 59–63.

87. Li, T.; Hu, Y.-J.; Hao, Z.-P.; Li, H.; Wang, Y.-S.; Chen, B.-D. First cloning and characterization of two functional aquaporin genes from an arbuscular mycorrhizal fungus *Glomus intraradices*. *New Phytol* 2013, *197*, 617–630.

88. Farrokhi, N.; Hrmova, M.; Burton, R.A.; Fincher, G.B. Heterologous and Cell Free Protein Expression Systems. In *Plant Genomics*; Gustafson, J.P., Langridge, P., Somers, D.J., Eds.; Humana Press: Totowa, NJ, USA, 2009; pp. 175–198.

89. Bolla, J.R.; Su, C.-C.; Yu, E.W. Biomolecular membrane protein crystallization. *Philos. Mag* 2012, *92*, 2648–2661.

90. Luecke, H. Atomic Resolution Structures of Integral Membrane Proteins Using Cubic Lipid Phase Crystallization. In *Structural Biology of Membrane Proteins*; Grisshammer, R., Buchanan, S.K., Eds.; Royal Society of Chemistry: Cambridge, UK, 2006; pp. 173–191.

91. Walas, F.; Matsumura, H.; Luisi, B. An Introduction to Detergents and Their Use in Membrane Protein Studies. In *Structural Biology*

*of Membrane Proteins*; Grisshammer, R., Buchanan, S.K., Eds.; Royal Society of Chemistry: Cambridge, UK, 2006; pp. 72–95.

92. Andersen, O.S.; Koeppe, R.E., II. Bilayer thickness and membrane protein function: An energetic perspective. *Annu. Rev. Biophys. Biomol. Struct.* 2007, *36*, 107–130.
93. Das, B.B.; Nothnagel, H.J.; Lu, G.J.; Son, W.S.; Tian, Y.; Marassi, F.M.; Opella, S.J. Structure determination of a membrane protein in proteoliposomes. *J. Am. Chem. Soc* 2012, *134*, 2047–2056.
94. Mondini, L.; Nachit, M.; Porceddu, E.; Pagnotta, M.A. Identification of SNP mutations in DREB1, HKT1, and WRKY1 genes involved in drought and salt stress tolerance in durum wheat (*Triticum turgidum L. var durum*). *OMICS* 2012, *16*, 178–187.

CHAPTER

# 11

# PHYLOGENETIC ANALYSIS AND PROTEIN STRUCTURE MODELLING IDENTIFIES DISTINCT CA2+/CATION ANTIPORTERS AND CONSERVATION OF GENE FAMILY STRUCTURE WITHIN ARABIDOPSIS AND RICE SPECIES

Jon K. Pittman[1] and Kendal D. Hirschi[2]

[1]Faculty of Life Sciences, The University of Manches

[2]USDA/ARS Children's Nutrition Research Center, Baylor College of Medicine

## ABSTRACT

### Background

The $Ca^{2+}$/Cation Antiporter (CaCA) superfamily is an ancient and widespread family of ion-coupled cation transporters found in nearly all kingdoms of life. In animals, $K^+$-dependent and $K^+$-indendent $Na^+/Ca^{2+}$ exchangers (NCKX and NCX) are important CaCA members. Recently it was proposed that all rice and Arabidopsis CaCA proteins should be classified as NCX proteins. Here we performed phylogenetic analysis of CaCA genes and protein structure homology modelling to further characterise members of this transporter superfamily.

### Findings

Phylogenetic analysis of rice and Arabidopsis CaCAs in comparison with selected CaCA members from non-plant species demonstrated that these genes form clearly distinct families, with the $H^+$/Cation exchanger (CAX) and cation/$Ca^{2+}$ exchanger (CCX) families dominant in higher plants but the NCKX and NCX families absent. NCX-related $Mg^{2+}/H^+$ exchanger (MHX) and CAX-related $Na^+/Ca^{2+}$ exchanger-like (NCL) proteins are instead present. Analysis of genomes of ten closely-related rice species and four Arabidopsis-related species found that CaCA gene family structures are highly conserved within related plants, apart from minor variation. Protein structures were modelled for OsCAX1a and OsMHX1. Despite exhibiting broad structural conservation, there are clear structural differences observed between the different CaCA types.

### Conclusions

Members of the CaCA superfamily form clearly distinct families with different phylogenetic, structural and functional characteristics, and therefore should not be simply classified as NCX proteins, which should remain as a separate gene family.

**Keywords:** Cation transport Proton/calcium exchanger Sodium/calcium exchanger Magnesium/proton exchanger Phylogenetics Protein structure

## FINDINGS

### The $Ca^{2+}$/Cation Antiporter (CaCA) Superfamily

Control of ion concentrations is critical to cellular function. Such ion homeostasis is dependent on transporters, including ion-coupled transporters like the $Na^+/Ca^{2+}$ exchanger (NCX), $Na^+/Ca^{2+}$, $K^+$ exchanger (NCKX), cation/$Ca^{2+}$ exchanger (CCX) and $H^+$/Cation exchanger (CAX), which are members of the CaCA superfamily (Cai and Lytton 2004). CaCAs are abundant throughout life; CAX and the prokaryotic-specific

YRBG-type exchangers are abundant in bacteria, while NCX, NCKX and CCX genes are abundant in animals (Cai and Lytton 2004; Cai and Clapham 2012; Emery et al. 2012). In the land plant lineage NCKX and NCX have been lost but other types, including CAX, CCX and MHX ($Mg^{2+}/H^+$ exchanger) are present within all plants (Shigaki et al. 2006; Emery et al. 2012; Gaash et al. 2013). Many plant CaCAs have been functionally characterised and have key roles in cation transport and homeostasis (Hirschi et al. 1996; Shaul et al. 1999; Hirschi et al. 2000; Kamiya et al. 2005; Morris et al. 2008; Wang et al. 2012). Recently, Singh et al. (2015) performed phylogenetic analysis of CaCA genes from rice and Arabidopsis, and argued that all genes should be classified as members of the NCX family due to the ubiquitous presence of the so-called 'NCX domain' and tight phylogenetic relationship. Here we have performed further phylogenetic analysis of rice and Arabidopsis CaCAs, along with protein structure modelling to demonstrate that CaCA members form clearly distinct families with different structural and functional characteristics, and are not all NCXs. In addition, we have examined the genomes of closely related rice and Arabidopsis species to identify the conservation of CaCA genes within these genera.

## Phylogenetic Analysis Supports Distinct CaCA Gene Families

Clear separation of CaCA family members within a phylogenetic tree is more apparent when members from each family are included, including NCKX and YRBG genes absent in higher plants. Therefore a phylogenetic tree was generated for rice (*Oryza sativa* Japonica) and Arabidopsis (*Arabidopsis thaliana*) CaCAs and with members from chosen species throughout life.

CaCA proteins have a conserved core structure of ten transmembrane (TM) helices and two conserved α-repeat regions within TM 2–3 and 7–8, which include residues for cation binding. High variability in amino acid sequence between distinct CaCAs, particularly within tail regions, means that full length sequences generate poor alignments and potentially inaccurate phylogenetic reconstruction. Therefore three sets of trees were compared: those generated from full-length sequence; 'core domain' sequence lacking tail sequence before TM1 and after TM10; and α2-repeat region sequence, which is conserved in all sequences, including NCL ($Na^+/Ca^{2+}$ exchanger-like) sequences, which lack the α1-repeat (Emery et al. 2012; Wang et al. 2012). All trees showed clear separation between the four major CaCA families: CAX, NCX, NCKX and CCX, but YRBG genes were only clearly separated in the core domain tree (Fig. 1). MHX genes, which are unique to land plants and present instead of animal-like NCXs (Emery et al. 2012; Gaash et al. 2013), are shown as a separate clade within the NCX family cluster. NCL genes have been identified as new CaCA members very recently (Wang et al. 2012) and appear to be found only in photosynthetic organisms (Emery et al. 2012). In full length and core domain trees, NCLs arguably form a distinct family, but in the α2-repeat tree, NCL genes were tightly grouped with CAXs, agreeing with previous analysis that despite exhibiting $Na^+/Ca^{2+}$ exchange activity (Li et al. 2016), AtNCL and related NCLs are more closely related to CAXs than NCXs (Emery et al. 2012).

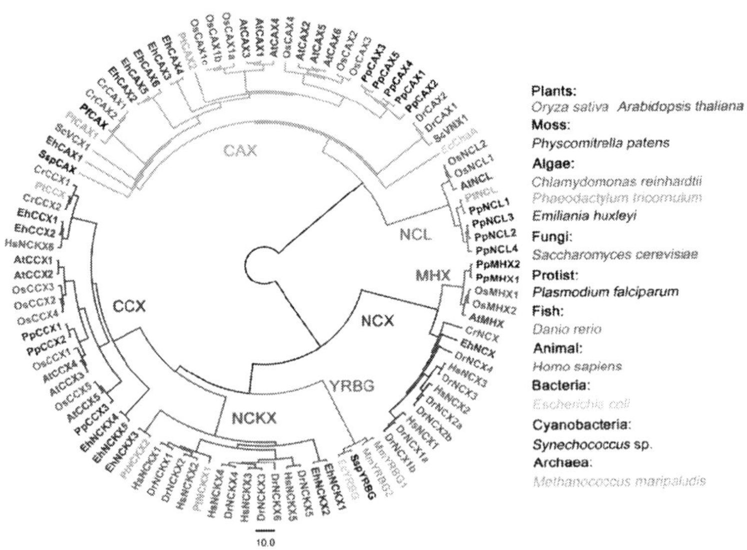


**Figure 1**: Phylogenetic analysis of the CaCA superfamily. The tree was derived from alignments of conserved 'core domain' hydrophobic region amino acid sequences obtained from the genomes of rice and Arabidopsis in comparison with selected algae, fungal, protist, animal, bacterial and archaebacterial species, and constructed using the maximum likelihood method. A consensus tree following 1000 bootstrap replications are shown. The branch length scale bar indicates the evolutionary distance of ten amino acid substitutions per site

CaCA genes are ubiquitous within plants but can vary in number; CAX gene numbers vary from 4 in *Vitis vinifera* to 8 in *Glycine max* (Emery et al. 2012) while all plants have one MHX apart from *O. sativa* (Japonica and Indica) and *Mimulus guttatus*, which have two MHXs (Gaash et al. 2013). It is unknown whether CaCA gene family structures are conserved amongst closely related species. Here we surveyed the genomes of three species closely related to *A. thaliana*, and nine additional rice species, then phylogenetic analysis using full length sequences was performed. For Arabidopsis relatives each gene family is highly stable; each species has one MHX, one NCL, five CCXs and six CAXs, except that *E. salsugineum* has just two Type 1B CAXs (*CAX2*, *CAX5*, *CAX6*) (Fig. 2a). Gene family structure was also highly conserved within *Oryza* species: there were no differences between CCX genes, and all CAX and NCL genes were conserved, except that a *CAX2* gene was undetected in *O. meridionalis* and a *NCL1*gene was undetected in *O. glaberrima* (Fig. 2b). All rice genomes possessed a *MHX1* gene but *MHX2* was only found in five of the ten species. Gaash et al. (2013) suggested that MHX gene duplication occurred before the split between the *O. sativa* Indica and Japonica subspecies.

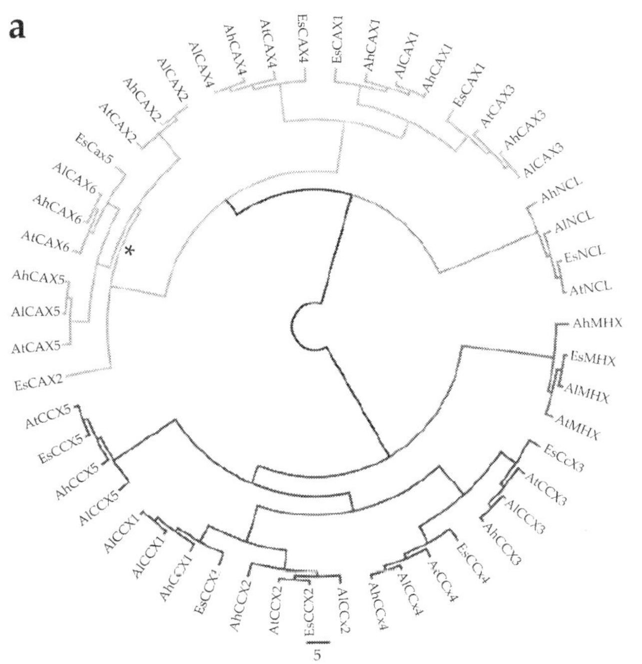

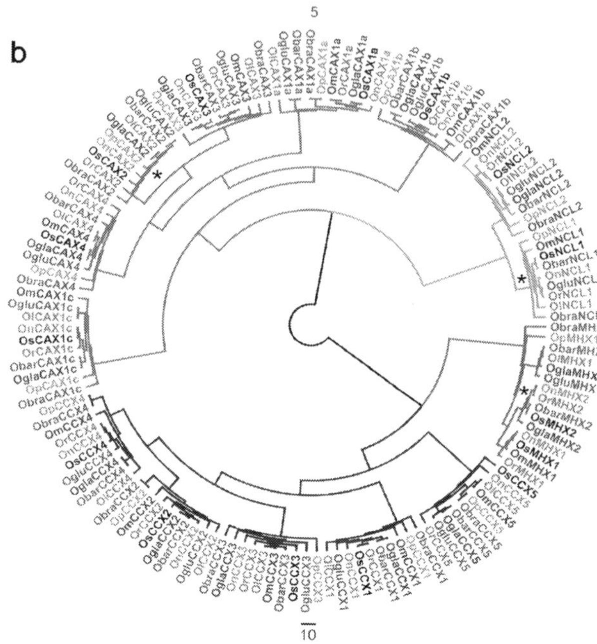


**Figure 2**: Conservation of CaCA genes in close relatives of *Arabidopsis thaliana* and *Oryza sativa*. **a** Phylogenetic analysis of CaCA proteins from *A. thaliana, A. halleri, A. lyrata* and *Eutrema salsugineum*. **b** Phylogenetic analysis of CaCA proteins from *O. sativa, Oryza barthii, O. brachyantha, O. glaberrima, O. glumipatula, O. longistaminata, O. meridionalis, O. nivara, O. punctate* and *O. rufipogon*. CaCA gene families are colour coded as shown in Fig. 1. Phylogenetic analysis was performed as described in Fig. 1. Asterisks indicate gene clades where one or more species are absent

Consistent with this conclusion, we found *MHX2* in four AA genome rice species (*O. barthii, O. glaberrima, O. nivara, O. rufipogon*), but not in the BB genome (*O. punctata*) or FF genome (*O. brachyantha*) species, suggesting that gene duplication occurred during AA genome rice evolution but *MHX2* was subsequently lost in some AA genome species like *O. glumipatula*and *O. longistaminata*.

## Protein Structure Homology Modelling Supports CAX and NCX/MHX Distinction

Recent crystallographic analysis of MjNCX from *Methanococcus jannaschii* (Liao et al. 2012) and CAX proteins from archaea, bacteria and yeast (Nishizawa et al. 2013; Waight et al. 2013; Wu et al. 2013), including *Saccharomyces cerevisiae* ScVCX1, provided the first detailed structural information of CaCAs. Although MjNCX displays $Na^+/Ca^{2+}$ exchange activity, phylogenetically it is a YRBG member (Liao et al. 2012; Gaash et al. 2013), and related to MmYRBG1 (Fig. 1). However, MjNCX has previously been used to successfully generate protein homology models of NCX proteins from nematodes (He and O›Halloran 2014), and there is strong evidence that a 10 TM structure is conserved between NCX and MjNCX proteins (John et al. 2013; Ren and Philipson 2013). As yet, there is no structural information for plant exchangers. Using ScVCX1 and MjNCX structures as templates, protein homology modelling was performed for OsCAX1a and OsMHX1. A model of OsCAX1a was generated (Fig. 3a) showing high similarity to ScVCX1. OsMHX1 was harder to model, and despite showing some alignment with MjNCX, it was not identical, highlighting the distinction between a $Na^+/Ca^{2+}$ transporting YRBG family member and MHX proteins. Previous evaluation of MHX protein structure by comparison with NCX proteins and hydropathy prediction analysis suggested that MHX proteins have 9 TM helices (Gaash et al. 2013). However, this assumption was based on earlier biochemical models of a 9 TM topology for NCX1, but more recent studies demonstrate that NCX1 is composed of 10 TM helices (Ren and Philipson 2013; Szerencsei et al. 2013). This topology of NCX1 and consensus hydropathy prediction for MHX proteins provides confidence in this 10 TM helices model of OsMHX1. Despite overall conservation in core structure between CaCA proteins, there are clear structural differences between plant CAX and MHX (Fig. 3a), and non-plant CAX and NCX proteins. For example, an additional redundant N-terminal TM helix (denoted MR) is found in most CAX proteins, and a cytosolic domain rich in acidic residues (the acidic helix) is a common feature linking the pseudo-symmetrical halves of CAX proteins (Fig. 3a). OsCAX1a also appears to be more tightly packed in the membrane than OsMHX1.

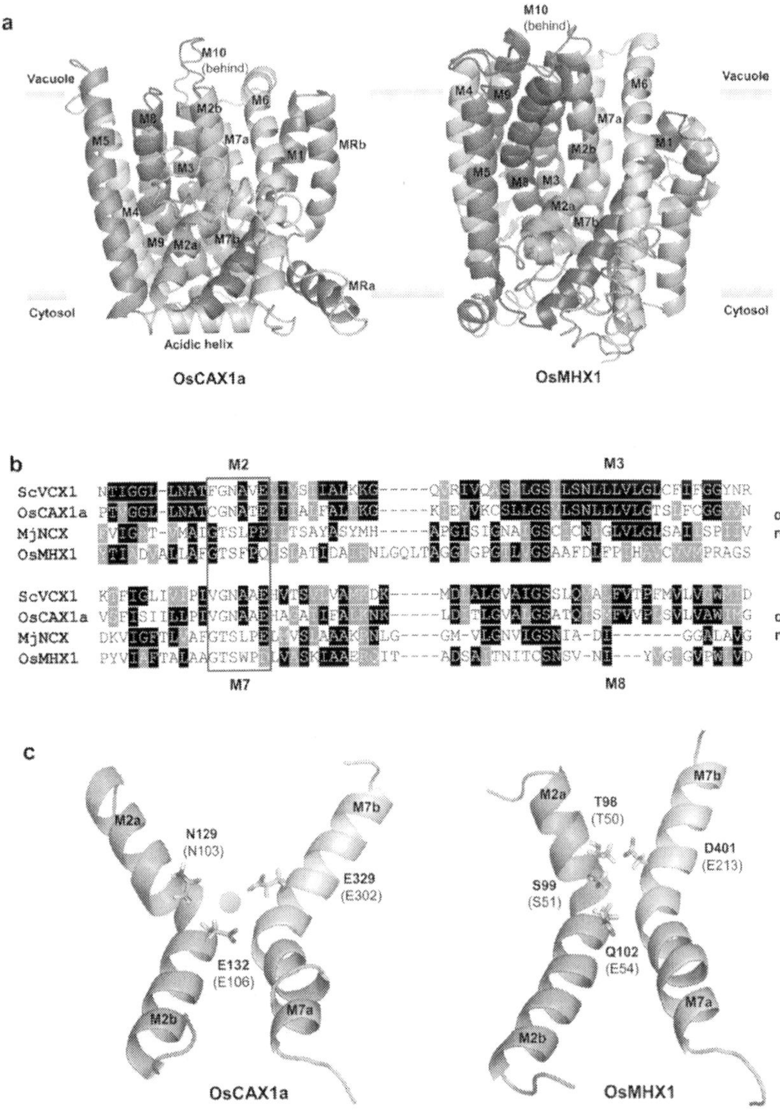


**Figure 3**: Comparative structure models of OsCAX1a and OsMHX1 determined by homology modelling. **a** Models of OsCAX1a and OsMHX1 generated using the known structures of ScVCX1 and MjNCX, respectively. The proteins are viewed from within the membrane. OsCAX1a has 11 predicted transmembrane (TM) helices with the first non-conserved TM labelled as MR. TM helices that can be clearly sub-divided (*kinked*) are referred to as 'a' and 'b'. M1 to M10 in both proteins are coloured equivalently, and the MR and acidic helix domains, and non-TM regions are shaded in grey. **b** Sequence alignment of the α-repeat

sequences within helices M2/M3 and M7/M8. The signature sequence (GNXXE for CAX; GXSXPE/D for NCX) is boxed in red. **c** Comparison of the putative cation binding pocket of OsCAX1a and OsMHX1. Key residues of M2 and M7 that bind to a $Ca^{2+}$ ion in ScVCX1 (relative to OsCAX1a) and in MjNCX (relative to OsMHX1) are indicated in parentheses while the equivalent residues in the same positions are shown. A $Ca^{2+}$ ion is indicated in the OsCAX1a pocket. Substitution of a Glu residue (E54) with a Gln residue (Q102) in OsMHX1 may explain the lack of $Ca^{2+}$ binding but $Mg^{2+}$-binding residues are unknown.

Although the α1- and α2-repeat regions are highly conserved features of CaCAs, there is clear sequence variation, as demonstrated by the phylogenetic variation within the α2-repeat tree. Comparison of primary sequence for the α1- and α2-repeat regions from two CAX proteins and two NCX/MHX proteins demonstrates the sequence similarity within CaCA classes but sequence distinction between them (Fig. 3b). CAX proteins possess an α-repeat signature sequence of GNXXE while NCX proteins have a signature sequence of GXSXPE/D (Cai and Lytton 2004). The Glu (or Asp) residues within M2 and M7 are suggested to be critical in coordinating the geometry of the $Ca^{2+}$ ion (Fig. 3c), with other residues acting to stabilise this coordination (Liao et al. 2012; Waight et al. 2013). These negatively charged Glu/Asp residues are highly conserved, with an exception being the non-charged Gln residue (Gln-102 in OsMHX1) within the MHX α1-repeat, suggested in part to explain the lack of $Ca^{2+}$ binding by MHXs (Gaash et al. 2013). The M2 and M7 TM helices are strongly kinked, and this topology yields a cation binding pocket in combination with the tilted, weakly bent M3 and M8 helices (Liao et al. 2012; Nishizawa et al. 2013). Nishizawa et al. (2013), Waight et al. (2013), and Wu et al. (2013) indicated clear differences in the kink angle of M2 between MjNCX and the CAX proteins. Furthermore, conformation of the M1/M6 'gating bundle' is opposed in CAX relative to NCX. These structural variations explain the differences between the mutually-exclusive $Ca^{2+}$ and $H^+$binding of CAX and the 1 $Ca^{2+}$/3 $Na^+$ binding of NCX (Nishizawa et al. 2013). MjNCX shows an outward-facing conformation while ScVXC1, AfCAX and YfkE have an inward-facing conformation (Nishizawa et al. 2013; Waight et al. 2013; Wu et al. 2013). OsCAX1a also displays inward facing conformation, while the closer similarity of OsMHX1 with CAX topologies compared to an NCX topology, particularly with regard to the position of M6 suggests that MHX also shows an inward-facing conformation.

## CONCLUSIONS

NCX/MHX, NCKX, YRBG, CCX and CAX/NCL proteins are a unique family of ion exchangers that share conserved structural features but also clear distinctions. While phylogenetic distinctions between NCX and MHX genes or CAX and NCL genes could be argued, there is clear phylogenetic separation between the CAX/NCL, NCX/MHX, NCKX, YRBG and CCX families (Fig. 1). Thus it is certainly not supported for these proteins to be combined into a single gene family named 'NCX' as proposed by Singh et al. (2015); rather the evidence overwhelmingly shows that NCX is a separate family within the CaCA superfamily. Nomenclature changes can lead to confusion, but sometimes it is appropriate. The CaCA genes originally named *AtCAX7 – AtCAX11* were renamed *AtCCX1 – AtCCX5* after phylogenetic evidence clearly showed these genes to be members of the CCX rather than CAX family (Shigaki et al. 2006). However, it is not appropriate to change the CaCA gene nomenclature as proposed (Singh et al. 2015). More pertinently, there are clear functional differences between these proteins; for example, exchange of $H^+/Ca^{2+}$(AtCAX1), $H^+/Mn^{2+}$ (AtCAX2), $H^+/Mg^{2+}$ (AtMHX), $H^+/K^+$ (AtCCX3), and $Na^+/Ca^{2+}$ (AtNCL) (Hirschi et al. 1996; Shaul et al. 1999; Pittman et al. 2004; Morris et al. 2008; Li et al. 2016), strongly argues against a common classification of these proteins as $Na^+/Ca^{2+}$ exchangers.

## ACKNOWLEDGEMENTS

We thank Sandip Patel for advice on homology modelling. JKP is grateful to funding from BBSRC (BB/K000942/1).

## REFERENCES

1. Cai X, Clapham DE (2012) Ancestral Ca2+ signaling machinery in early animal and fungal evolution. Mol Biol Evol 29(1):91–100
2. Cai XJ, Lytton J (2004) The cation/Ca2+ exchanger superfamily: phylogenetic analysis and structural implications. Mol Biol Evol 21(9):1692–1703
3. Emery L, Whelan S, Hirschi KD, Pittman JK (2012) Protein phylogenetic analysis of Ca2+/cation antiporters and insights into their evolution in plants. Front Plant Sci 3:1

4. Gaash R, Elazar M, Mizrahi K, Avramov-Mor M, Berezin I, Shaul O (2013) Phylogeny and a structural model of plant MHX transporters. BMC Plant Biol 13(1):75
5. He C, O'Halloran DM (2014) Analysis of the Na+/Ca2+ exchanger gene family within the Phylum nematoda. PLoS One 9(11):e112841
6. Hirschi KD, Zhen RG, Cunningham KW, Rea PA, Fink GR (1996) CAX1, an H+/Ca2+ antiporter from Arabidopsis. Proc Natl Acad Sci U S A 93(16):8782–8786
7. Hirschi KD, Korenkov VD, Wilganowski NL, Wagner GJ (2000) Expression of Arabidopsis CAX2 in tobacco. Altered metal accumulation and increased manganese tolerance. Plant Physiol 124(1):125–133
8. John SA, Liao J, Jiang Y, Ottolia M (2013) The cardiac Na+-Ca2+ exchanger has two cytoplasmic ion permeation pathways. Proc Natl Acad Sci U S A 110(18):7500–7505
9. Kamiya T, Akahori T, Maeshima M (2005) Expression profile of the genes for rice cation/H+ exchanger family and functional analysis in yeast. Plant Cell Physiol 46(10):1735–1740
10. Li P, Zhang G, Gonzales N, Guo Y, Hu H, Park S, Zhao J (2016) Ca2+-and diurnal rhythm-regulated Na+/Ca2+ exchanger AtNCL affects flowering time and auxin signaling in Arabidopsis. Plant Cell Environ. 39(2):377-392
11. Liao J, Li H, Zeng W, Sauer DB, Belmares R, Jiang Y (2012) Structural insight into the ion-exchange mechanism of the sodium/calcium exchanger. Science 335(6069):686–690
12. Morris J, Tian H, Park S, Sreevidya CS, Ward JM, Hirschi KD (2008) AtCCX3 Is an Arabidopsis endomembrane H+-dependent K+ transporter. Plant Physiol 148(3):1474–1486
13. Nishizawa T, Kita S, Maturana AD, Furuya N, Hirata K, Kasuya G, Ogasawara S, Dohmae N, Iwamoto T, Ishitani R, Nureki O (2013) Structural basis for the counter-transport mechanism of a H+/Ca2+ exchanger. Science 341(6142):168–172
14. Pittman JK, Shigaki T, Marshall JL, Morris JL, Cheng NH, Hirschi KD (2004) Functional and regulatory analysis of the Arabidopsis thaliana CAX2 cation transporter. Plant Mol Biol 56(6):959–971

15. Ren X, Philipson KD (2013) The topology of the cardiac Na+/Ca2+ exchanger, NCX1. J Mol Cell Cardiol 57:68–71
16. Shaul O, Hilgemann DW, de-Almeida-Engler J, Van Montagu M, Inze D, Galili G (1999) Cloning and characterization of a novel Mg2+/H+ exchanger. EMBO J 18(14):3973–3980
17. Shigaki T, Rees I, Nakhleh L, Hirschi KD (2006) Identification of three distinct phylogenetic groups of CAX cation/proton antiporters. J Mol Evol 63(6):815–825
18. Singh A, Kumar R, Tripathi A, Gupta B, Pareek A, Singla-Pareek S (2015) Genome-wide investigation and expression analysis of Sodium/Calcium exchanger gene family in rice and Arabidopsis. Rice 8(1):21
19. Szerencsei RT, Kinjo TG, Schnetkamp PPM (2013) The topology of the C-terminal sections of the NCX1 Na+/Ca2+ exchanger and the NCKX2 Na+/Ca2+-K+ exchanger. Channels 7(2):109–114
20. Waight AB, Pedersen BP, Schlessinger A, Bonomi M, Chau BH, Roe-Zurz Z, Risenmay AJ, Sali A, Stroud RM (2013) Structural basis for alternating access of a eukaryotic calcium/proton exchanger. Nature 499(7456):107–110
21. Wang P, Li Z, Wei J, Zhao Z, Sun D, Cui S (2012) A Na+/Ca2+ exchanger-like protein (AtNCL) involved in salt stress in Arabidopsis. J Biol Chem 287(53):44062–44070
22. Wu M, Tong S, Waltersperger S, Diederichs K, Wang M, Zheng L (2013) Crystal structure of Ca2+/H+ antiporter protein YfkE reveals the mechanisms of Ca2+ efflux and its pH regulation. Proc Natl Acad Sci U S A 110(28):11367–11372

CHAPTER

# 12

# DIFFERENT EVOLUTIONARY HISTORIES OF TWO CATION/PROTON EXCHANGER GENE FAMILIES IN PLANTS

Inês S Pires[1,2], Sónia Negrão[1], Melissa M Pentony[2], Isabel A Abreu[1], Margarida M Oliveira1 and Michael D Purugganan[2]

[1]Instituto de Tecnologia Química e Biológica, Universidade Nova de Lisboa, Av. da República, 2780-157 Oeiras, Portugal and iBET

[2]Department of Biology and Center for Genomics and Systems Biology, New York University

## ABSTRACT

### Background

Gene duplication events have been proposed to be involved in the adaptation of plants to stress conditions; precisely how is unclear. To address this question, we studied the evolution of two families of antiporters. Cation/proton exchangers are important for normal cell function and in plants, $Na^+,K^+/H^+$ antiporters have also been implicated in salt tolerance. Two well-known plant cation/proton antiporters are

NHX1 and SOS1, which perform $Na^+$ and $K^+$ compartmentalization into the vacuole and $Na^+$ efflux from the cell, respectively. However, our knowledge about the evolution of *NHX* and *SOS1* stress responsive gene families is still limited.

## Results

In this study we performed a comprehensive molecular evolutionary analysis of the NHX and SOS1 families. Using available sequences from a total of 33 plant species, we estimated gene family phylogenies and gene duplication histories, as well as examined heterogeneous selection pressure on amino acid sites. Our results show that, while the *NHX* family expanded and specialized, the *SOS1* family remained a low copy gene family that appears to have undergone neofunctionalization during its evolutionary history. Additionally, we found that both families are under purifying selection although *SOS1* is less constrained.

## Conclusions

We propose that the different evolution histories are related with the proteins' function and localization, and that the NHX and SOS1 families are examples of two different evolutionary paths through which duplication events may result in adaptive evolution of stress tolerance.

# BACKGROUND

It has been argued that gene duplications underlie mechanisms to achieve stress adaptation [1, 2]. There is little evidence, however, to support this. Additionally, it is not known if there is one or multiple ways of achieving stress adaptation through gene duplication. To address this question, we focused our attention on salt tolerance, and the evolutionary histories of two gene families involved in this trait.

Salt tolerance is a complex trait that is thought to have originated multiple times in plants [3], and it is important to understand the nature and molecular evolution of salt tolerance mechanisms throughout the history of land plants. This will be key not only to our understanding of how plants adapt against the disruptive effects of high salt concentrations in the soil, but also to point out directions for possible crop improvement. Indeed farmers are increasingly facing loss of crop production due to this abiotic stress, since currently > 20% of the world's arable land is affected by high salinity [4].

Cation/proton exchangers are essential to the normal function of the cell. Besides helping regulating internal pH, cell volume, and cytoplasmic ion homeostasis [5, 6, 7], these transporters have also been shown to be involved in vesicular trafficking and protein targeting [8, 9]. In plants, $Na^+$,$K^+$/$H^+$ antiporters are also associated with salt tolerance [10, 11], and a series of studies have targeted these cation exchangers to improve agronomically important crops [12, 13].

All $Na^+$,$K^+$/$H^+$ exchangers belong to a superfamily of monovalent cation/proton antiporters (CPA) and are divided in two families, CPA1 and CPA2 [14]. The CPA1 family is composed of two main groups, one containing plasma membrane-bound proteins and the other intracellular proteins [14]. Within these two groups are well-characterized antiporters associated with salt tolerance in plants, the salt overly sensitive 1 (SOS1) and sodium hydrogen exchanger (NHX) proteins. SOS1 proteins are localized in the plasma membrane [15, 16], and are responsible for $Na^+$ efflux from the cell. NHX proteins, on the other hand, are intracellular proteins that compartmentalize $Na^+$ and $K^+$ in the vacuole [17, 18, 19].

In *Arabidopsis thaliana*, the NHX protein family has six members that are classified into two classes [17]. The class I proteins, AtNHX1-4, are localized in the tonoplast [17] and have equal affinity for $Na^+$ and $K^+$[7, 20]. Class II proteins include AtNHX5 and 6, which are localized in the endosomal compartment of the Golgi and trans-Golgi network [18] and have a higher affinity for $K^+$ compared to $Na^+$[6].

Previous reports have shown that the Arabidopsis *NHX* genes which encode these proteins have different expression patterns and responsiveness to abiotic stresses [21, 22]. *AtNHX1* and *2* were shown to be expressed at high levels in all plant tissues, while *AtNHX3* and *4* were mainly expressed in root and flower tissues, respectively [21, 22]. *AtNHX5* was also expressed in all tissues but at lower levels [21, 22], while *AtNHX6* expression was detected only in shoots and roots by RT-PCR [22]. In addition, while *AtNHX1*-3 were shown to be responsive to both salt stress and abscisic acid (ABA) [22, 23], *AtNHX5* was only responsive to salt stress, suggesting that its response is ABA-independent [22]. Moreover, all *NHX* genes promoted the recovery of a salt sensitive yeast mutant [21, 22] and numerous studies have shown that overexpression of Arabidopsis *NHX1* and *5*, or rice *NHX1* (among others), resulted in increased salt tolerance in transgenic plants [10, 24, 25, 26, 27].

The other protein that characterizes this cation/proton exchanger family is SOS1, which belongs to the well-known salt tolerance Salt-Overly-Sensitive (SOS) pathway [28, 29, 30]. Salt stress elicits a cytosolic calcium signal, which functions as a major secondary-messenger signalling molecule. A myristoylated calcineurin B-like protein (SOS3) senses the salt-elicited calcium signal, and upon $Ca^{2+}$ binding SOS3 undergoes dimerization and enhances the serine/threonine protein kinase activity of SOS2. The SOS3/SOS2 complex is targeted to the plasma membrane and activates SOS1 [28, 29, 30].

SOS1 has been called NHX7 by several authors, and was thus thought to be part of the NHX gene family [12, 16]. Previous studies, however, regarding the phylogeny of cation/hydrogen transporters [14, 31] suggest that it is distinct from the other NHX proteins and should more appropriately retain its designation as SOS1. For this reason, we refer to the Arabidopsis proteins SOS1/NHX7 and NHX8 (highly similar to, but shorter than SOS1) as SOS1A and SOS1B, respectively. SOS1B, like SOS1A, is localized to the plasma membrane [15, 16]. While SOS1A is a $Na^+/H^+$ antiporter [32], however, SOS1B seems to only transport $Li^+$[16]. In addition, the tonoplast exchangers in Arabidopsis are also regulated by the SOS pathway, specifically by SOS2 [33], thus suggesting coordination between tonoplast and plasma membrane antiporters.

The importance of these two families of proteins in plant salt tolerance is well established. By promoting the efflux of $Na^+$, SOS1A helps protect the cell from the deleterious effects of this ion. Additionally, SOS1A seems to have an important role in long-distance $Na^+$ transport, thus helping to regulate the $Na^+/K^+$ ratio in roots and shoots [17]. On the other hand, NHX family members have been described as essential to $Na^+$compartmentalization in the vacuole, protecting the cell from the deleterious effects of this ion and maintaining cytoplasmic ion homeostasis [10, 34]. Recent studies, however, suggest instead that increased salt tolerance of *NHX* over-expressing plants result from an improved ability to retain $K^+$ after stress induction [35, 36, 37].

Despite the key roles these genes play in salt tolerance in plants, little is known about the evolutionary histories of both gene families. Understanding the evolution of $Na^+$,$K^+$/$H^+$ antiporters can help clarify the mechanisms leading to plant stress adaptation associated with gene duplication events [1, 2, 38]. In this study, we report a comprehensive molecular evolutionary analysis of both the NHX and SOS1 plant protein families. We reconstructed the phylogeny and the history of duplication events for each family, as well as determined the selection pressure on amino acid sites within these proteins. Our purpose here is not to identify novel *NHX* and *SOS1* plant genes, or to do a phylogeny analysis of the entire monovalent CPA gene family, as in others [39]. Instead, we show that the *NHX* and *SOS1* gene families have very different evolution trajectories and suggest that these divergent evolutionary histories are related to the evolution of their function and cellular localization. Finally, we suggest that the NHX and SOS1 families represent examples of two different paths in the molecular evolution of stress tolerance.

# RESULTS AND DISCUSSION

## Phylogeny of the NHX Family Exhibits Multiple Independent Duplication Events

We selected 121 genes in 33 taxa that appear, or are described, to belong to the NHX family. The distribution of *NHX* genes among the various species is shown in Table 1. Among these genes, 108 are found in 28 angiosperm species, 2 in 1 gymnosperm taxa, and 9 in 2 non-seed plant taxa. We also identified 1 sequence from the alga *C. reinhardtii*, and we used 1 sequence from the budding yeast *S. cerevisiae* as an outgroup. The NHX phylogeny agrees with the classification of these proteins in two distinct evolutionary groups. We obtained two main clades, with ~99% bootstrap support, which showed evolutionary divergence of NHX protein groups according to their protein localization, as previously reported [22, 40] (Figure 1 for more detail and unrooted tree). Proteins localized in the endosomal compartment of the Golgi and trans-Golgi network grouped in clade 1 (closer to the yeast NHX1), while proteins localized in the tonoplast grouped in clade 2.

**Table 1**: Species and number of NHX and SOS1 sequences used in this study

| | Species | Number of putative NHX sequences | Number of putative SOS1 sequences |
|---|---|---|---|
| **Yeast** | *Sacharomyces cerevisiae* | 1 | - |
| **Alga** | *Chlamydomonas reinhardtii* | 1 | - |
| **Moss** | *Physcomitrella patens* | 7 | 2 |
| **Spikemoss** | *Selaginella moellendorffii* | 2 | 2 |
| **Gymnosperm** | *Picea sitchensis* | 2 | - |
| **Angiosperms Monocots** | *Brachypodium distachyon* | 5 | 2 |
| | *Oryza sativa* | 5 | 1 |
| | *Setaria italica* | 5 | 1 |
| | *Zea mays* | 5 | 1 |
| | *Sorghum bicolor* | 4 | 1 |
| **Angiosperms Dicots** | *Aquilegia coerulea* | 3 | 2 |
| | *Mesembryanthemum crystallinum* | 3 | 1 |
| | *Atriplex gmelinii* | 1 | - |
| | *Atriplex dimorphostegia* | 1 | - |
| | *Suaeda salsa* | 1 | - |
| | *Salicornia europaea* | 1 | - |
| | *Chenopodium glaucum* | 1 | - |
| | *Kalidium foliatum* | 1 | - |
| | *Mimulus guttatus* | 6 | 1 |
| | *Vitis vinifera* | 5 | 1 |
| | *Eucalyptus grandis* | 4 | 2 |
| | *Citrus clementina* | 5 | 1 |
| | *Citrus sinensis* | 3 | 1 |
| | *Carica papaya* | 4 | 1 |
| | *Thellungiella halophila* | 1 | 2 |
| | *Arabidopsis thaliana* | 6 | 2 |
| | *Prunus persica* | 6 | 1 |
| | *Cucumis sativus* | 3 | 1 |
| | *Glycine max* | 8 | 1 |
| | *Medicago truncatula* | 3 | - |
| | *Populus trichocarpa* | 6 | 2 |
| | *Ricinus communis* | 5 | 1 |
| | *Manihot esculenta* | 7 | 2 |
| **TOTAL** | **33** | **121** | **32** |

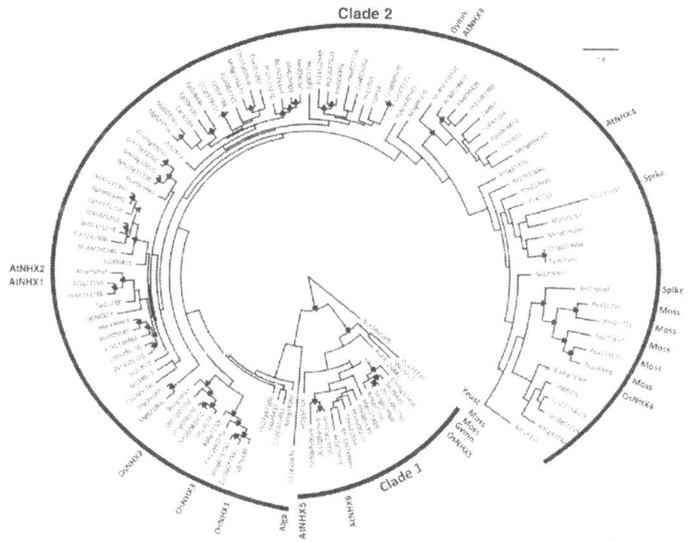


**Figure 1**: Maximum likelihood phylogeny of the NHX family. Nodes marked with black filled triangles represent nodes with bootstrap support>75%. The root was placed using the yeast NHX protein as outgroup. The two main clades and Arabidopsis, rice, gymnosperm (Gymn.), spikemoss (Spike.), moss, alga, and yeast sequences are highlighted.

We observe that multiple independent duplication events have occurred throughout the evolutionary history of the NHX family. Based on the reconciled phylogeny (Figure 2), we estimate 27 independent gene duplication and 40 gene loss events during the diversification of this gene family. Gene loss events, however, are not displayed on the phylogeny since they might not represent true losses but may be due to partial gene inventories due to incomplete whole genome sequencing data (e.g., *Thellungiella halophila*) and to some genes still remaining to be identified. In this reconciled tree, all the branches with bootstrap support inferior to a set value (here 75%) are rearranged to achieve the most parsimonious duplication and loss history for the gene family. Nevertheless our results, together with the fact that non-plant species also have multiple *NHX*genes [9, 39], suggest that multiple copies of NHX proteins were already present in a common ancestor of modern land plants (even though AtNHX1-4 and their orthologs are plant specific) [39].

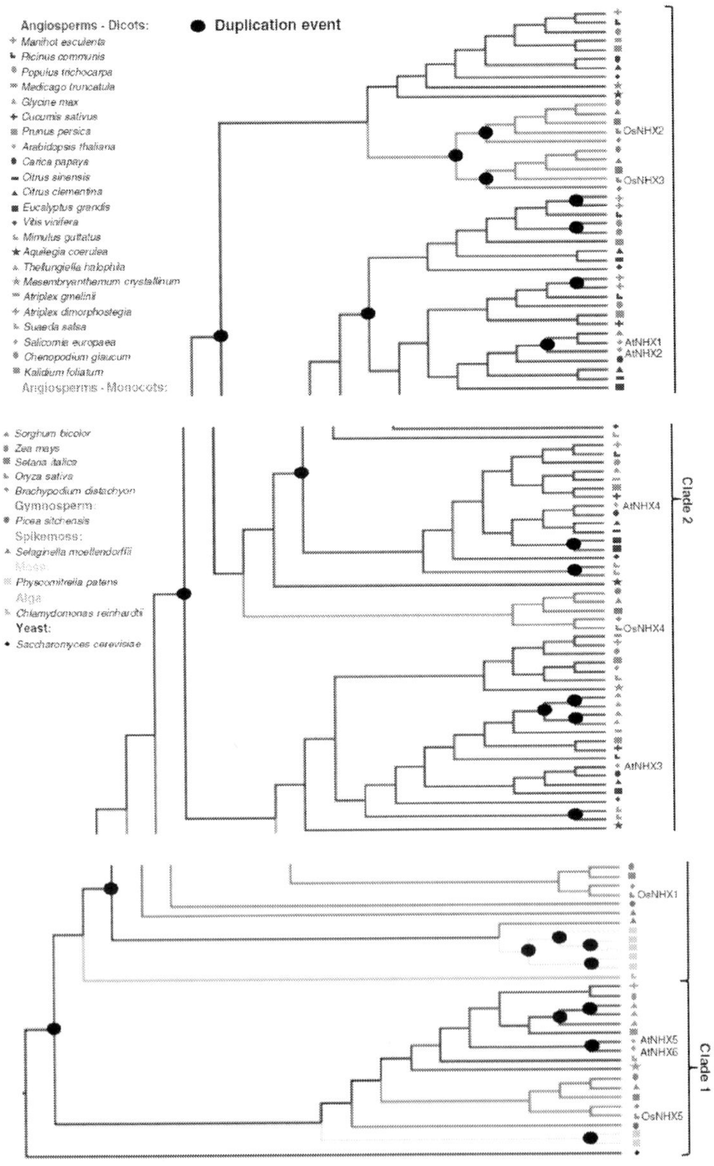


**Figure 2**: Phylogeny of the NHX family that indicates multiple independent duplication events throughout its evolutionary history. The reconciled NHX tree obtained using NOTUNG v2.6, has a duplication/loss score of 60.5 and shows 27 independent gene duplication and 40 gene loss (not shown) events.

Interestingly, orthologs of Arabidopsis NHX1 and 2 might be specific to (at least) seed plants. NHX1 and 2 proteins consistently grouped separately from proteins from basal plant lineages such as spikemoss (*Selaginella moellendorffii*), moss (*Physcomitrella patens*), and the green photosynthetic alga (*Chlamydomonas reinhardtii*) (Figure 1). Due to the limited number of gymnosperm sequences (only two NHX proteins from *Picea sitchensis*) we could not assess divergence from the more closely related species to the angiosperms. Sequences from more gymnosperm species are available as EST sequences in PlantGDB database, which coupled with greater availability of genomics resource data from other non-flowering plant taxa species, could potentially be used in the future to examine more closely the precise origins of specific gene family clades. However, our results suggest that NHX proteins grouping with AtNHX1 and 2 might be more recent and specific to seed plants. Nevertheless, bootstrap supports obtained for these branches were not very high (Figure 1) and more sequences will be necessary to be analyzed to confirm this.

## Purifying Selection on NHX Proteins

We examined the sequences of the plant *NHX* gene family to determine whether there was significant heterogeneity in selective pressure among amino acid sites. Selective pressure at the protein level can be measured as dN/dS, where dN and dS are the number of non or synonymous substitutions per non or synonymous site, respectively [41]. If non-synonymous mutations are favored by positive selection, non-synonymous mutations are fixed at a faster rate and $dN/dS > 1$ [41]. In the case of purifying selection $dN/dS < 1$. However, if selection has no effect on fitness due to neutral evolution, both mutations are fixed at the same rate and $dN/dS = 1$ [41].

For *NHX* coding region sequences from both clade 1 and 2, the average dN/dS was estimated at ~0.08 and no sites were predicted to be under positive selection. Additionally, no difference in dN/dS between *NHX* genes was identified. We believe that these results might reflect the key function of NHX proteins, which besides salt tolerance [10, 11] include normal cell functions in vesicular trafficking and protein targeting [8, 9]. Additionally, NHX proteins are predicted to have ~12 transmembrane domains [42] and are probably constrained at the structural level. Our results are consistent with the purifying selection expected for these proteins [43] and agree with results recently obtained by Hudson *et al.* [44].

## Phylogeny of the SOS1 Family Exhibits Few and Recent Duplication Events

Unlike the NHX family, plant species appear to have lower numbers of *SOS1* genes in their genomes. We have selected 32 genes in 22 taxa that appear, or are described, to belong to the SOS1 family. The distribution of *SOS1* genes among the various species is shown in Table 1. Among these genes, 22 are found in 16 angiosperm species, and 4 in 2 non-seed plant taxa. In this case we used 2 sequences from the moss *P. patens* as outgroup sequences.

We found that eight plant species, including *A. thaliana*, seem to have two SOS1-like genes (Table 2). At the time of submission of this manuscript, however, a new BLASTp search using Phytozome v8.0 (http://www.phytozome.net/, verified in June 2012) suggested that more species might possess more than one Putative *SOS1* protein. The existence of two *SOS1* genes has been described previously for *Arabidopsis thaliana*[16] and two other plant species not included in this study, namely Neptune grass (*Cymodocea nodosa*) [45] and quinoa (*Chenopodium quinoa*) [46]. Nevertheless, to confirm that these seven species have two SOS1-like proteins like in Arabidopsis, it is necessary to perform further functional studies.

As in Arabidopsis, the species with two SOS1-like proteins usually displayed differences in protein length (Table 2). However, two classes of length differences were identified - one in which both duplicates had similar lengths ($\Delta<110$ amino acid residues), and another in which the difference in duplicates length was substantially larger ($\Delta>400$ amino acid residues). Nevertheless, as in Arabidopsis, the differences in length were mainly due to shortening in the C-terminus instead of deletions within the protein primary sequence. This is important to notice, because it was shown by Quintero *et al.* [47] that it is within the C-terminus of SOS1A that lays the activation site and the auto-inhibitory domain of the protein. Additionally, SOS1-like proteins from *A. coerulea*, *E. grandis*, and *M. esculenta* have deletions in the N-terminus of the protein when compared to the Arabidopsis SOS1 proteins. This suggests that SOS1-like proteins from these three species might have a smaller transmembrane region, but further biochemical analyses need to be performed to confirm this.

The SOS1 phylogeny (Figure 3) exhibits three distinct clades: (1) one that contains the moss and spikemoss proteins (~94% bootstrap support), (2) another that has monocot proteins (~91% bootstrap support), and (3) finally, one clade with dicots SOS1 sequences. The SOS1 protein phylogeny is almost identical to the accepted species phylogeny of the study species, although there appears to be some long-branch attraction that alters the position within the dicot clade of one *Eucalyptus grandis* and one *A. thaliana* protein.

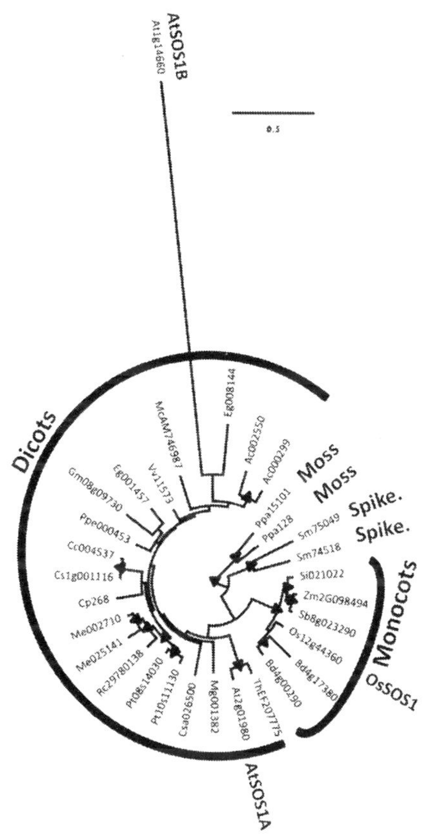


**Figure 3**: Maximum likelihood phylogeny of the SOS1 family. Nodes marked with black filled triangles represent nodes with bootstrap support>75%. The root was placed using the Physcomitrella moss SOS1 as outgroup. The monocots and dicots clades, as well as the Arabidopsis, rice, spikemoss (Spike.), and moss sequences are highlighted.

**Table 2**: Plant species predicted to have two SOS1-like genes

| **Species** | **Putative SOS1 – longer gene** | | **Putative SOS1 – shorter gene** | | **Δ Length (AA)** |
|---|---|---|---|---|---|
| | **Sequence code** | **Length (AA)** | **Sequence code** | **Length (AA)** | |
| *Physcomitrella patens* | Pp1s12_8V6.1 | 1153 | Pp1s15_101V6.1 | 1060 | 93 |
| *Selaginella moellendorffii* | 74518 | 1024 | 75049 | 940 | 84 |
| *Brachypodium distachyon* | Bradi4g00290.1 | 1138 | Bradi4g17380.1 | 719 | 419 |
| *Aquilegia coerulea* | Aca_000299m | 1150 | Aca_002550m | 588 | 562 |
| *Eucalyptus grandis* | Egrandis_v1_0.001457m | 1001 | Egrandis_v1_0.008144m | 547 | 454 |
| *Arabidopsis thaliana* | At2g01980 | 1146 | At1g14660 | 733 | 413 |
| *Populus trichocarpa* | POPTR_0010s11130.1 | 1147 | POPTR_0008s14030.1 | 1145 | 2 |
| *Manihot esculenta* | cassava4.1_025141m | 810 | cassava4.1_002710m | 703 | 107 |

Δ represents the difference in protein length.

Unlike the NHX family, the SOS1 family has undergone fewer gene duplication events. Reconciliation of the SOS1 tree with the species tree (Figure 3) estimates that 8 independent gene duplication events and one loss have occurred within the land plants (Figure 4). Once again, gene losses may simply represent unidentified genes; in fact Oh et al. [2] recently reported that *Thellungiella parvula* has three copies of the *A. thaliana* SOS1B and that these result from recent tandem duplications.

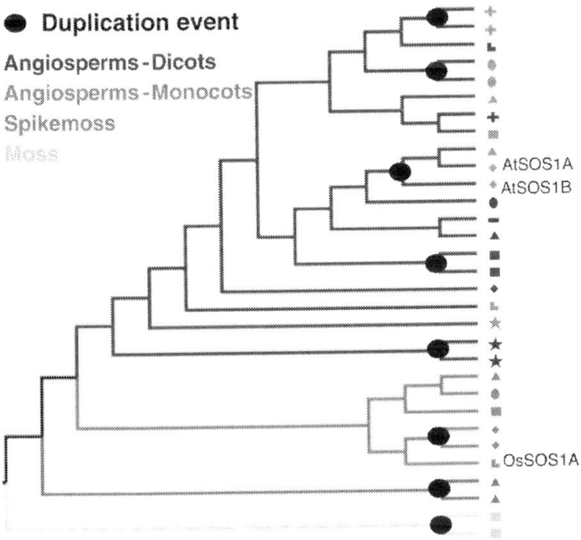


**Figure 4**: Phylogeny of the SOS1 genes indicating few recent duplication events. The reconciled SOS1 tree obtained using NOTUNG v2.6, has a duplication/loss score of 13.0 and shows 8 independent gene duplication and 1 gene loss (not shown) events. All duplication events appear towards the terminal nodes of the tree indicating that these events occurred recently.

Interestingly, gene duplication events in the SOS1 family appear toward the terminal nodes of the reconciled tree, indicating that these events were recent in evolutionary time. The analyses of some species that have a second copy of SOS1 further support this observation. For example, *Populus trichocarpa* is thought to have undergone a recent whole genome duplication event [48] and monocots such as *Brachypodium distachyon*are thought to be more prone to tandem duplications which suggests that any gene duplication in one monocot species is probably recent and not shared with the common ancestor of this group [49].

## Purifying Selection on SOS1 Proteins with few Amino Acids Under Positive Selection

Like the NHX proteins, the SOS1 proteins are generally subject to purifying selection, but our analysis indicates that some amino acid sites appear to be under positive selection. Among the land plants the SOS1 proteins had an average dN/dS of ~0.16. Using alignments that largely represented orthologous sequences, the glutamine at position 958 (Q958) in *A. thaliana* is evolving neutrally (dN/dS ~ 1.00) and two amino acid residues are under positive selection (dN/dS ~ 1.23) - a valine at position 366 (V366) and a serine at position 738 (S738) (Figure 5). Other residues (L843, C902, and P915 in *A. thaliana*) were identified in our analysis as strong candidate residues to be under neutral or positive selection, but these were not as well supported as the other three sites. Nevertheless, these other residues might be indeed under neutral or positive selection since the tests designed to detect positive selection are very stringent.

In general, structurally constrained sites should be under stronger purifying selection, while unstructured sites have higher levels of amino acid replacements [43]. According to different prediction tools available at ExPASy (http://expasy.org/tools/), however, the positively selected residue V366 is in a predicted transmembrane helix. Amino acid site Q958 is predicted to be in a transition zone between coiled and beta-sheet regions, and residue S738 is predicted to be in a random coil.

Two of these residues, V366 and S738, are in the cytoplasmic domain of AtSOS1A (Figure 5) but do not fall in the auto-inhibitory domain (amino acids 1013–1047) or near sites known to be relevant for protein activation by phosphorylation (S1138 and S1136) [47]. Additionally, mutations of these three residues under neutral or positive selection are not predicted to result in changes that affect secondary or tertiary structure, or create new sites for putative posttranslational modifications (data not shown). Nevertheless, residue S738 is in close proximity to residues S742 and V743, which were shown to affect the activity of the *A. thaliana* protein when mutated [47]. Like S742 and V743, S738 is not well conserved between species. Mutations within the transmembrane pore were also shown to affect AtSOS1A activity [47]. Thus, it may be that the roles of residues S738 and V366 in the activity of the SOS1A protein are worth study.

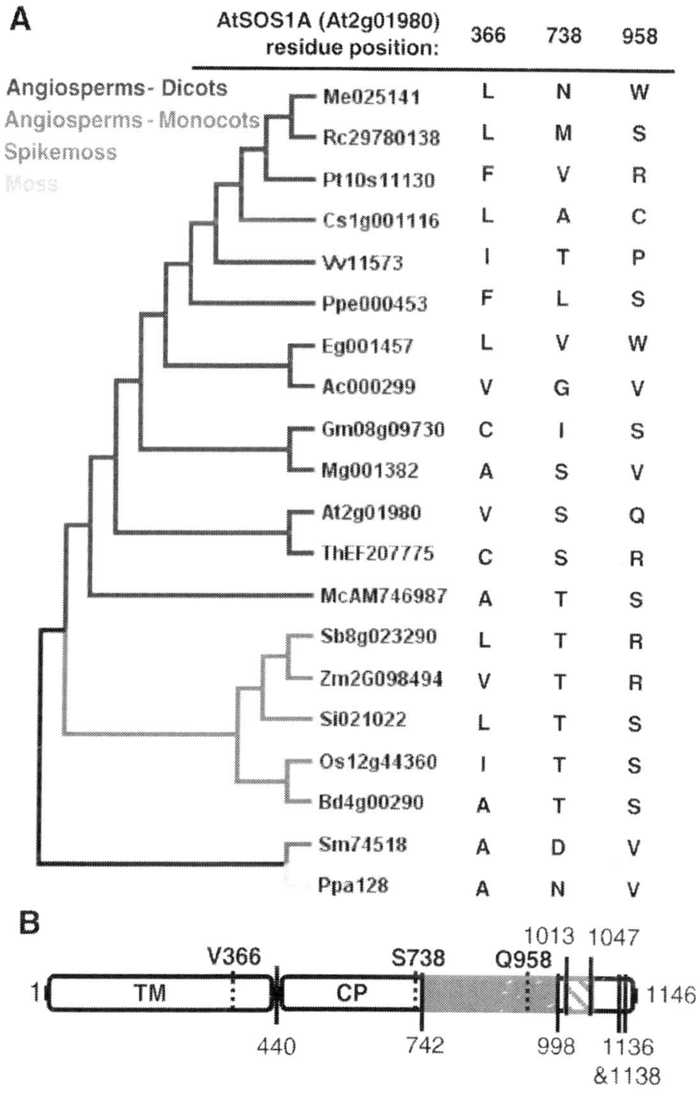


**Figure 5**: Amino acid sites under neutral or positive selection in SOS1A. SOS1 proteins are generally under purifying selection but appear to have some residues under neutral or positive selection. However, these residues do not fall in the auto-inhibitory domain or in sites important for the activation of the protein. **A** – Amino acids under neutral or positive selection for each SOS1A ortholog, which are shown in the NJ tree on the left. **B** – *A. thaliana* SOS1A protein diagram based on Figure 2A and information obtained by others [47]. Amino acid residues predicted to be under neutral or positive selection are indicated by the broken lines. The grey area (residues 742–998) corresponds to the

target-region of the auto-inhibitory domain, which is highlighted in dashed grey (residues 1013–1047). Amino acid residues 1136 and 1138 correspond to sites of recognition and phosphorylation, respectively, by the SOS2-SOS3 complex. TM: Trans-membrane region (residues 1–440). CP: Cytoplasmic region (residues 441–1146).

## Relaxation of Selection on SOS1 Gene Duplicates

The SOS1 family apparently exhibits a relaxation of purifying selective pressure upon duplication, as previously reported in other gene families [50] and, specifically for ion transporters [44]. This relaxation of selection, however, seems to be present in only one duplicated gene copy, and only in species that have two SOS1-like proteins differing in length. Although the free-ratio model [51] is not generally considered to be a good method to predict dN/dS because it has too many free parameters, the comparison between values of dN/dS is suggestive. Species which have two Putative *SOS1*-like proteins similar in length ($\Delta < 110$ amino acid residues) showed a difference in dN/dS values less than 0.1, while species that retain two putative *SOS1*-like proteins with very different length ($\Delta > 400$ amino acid residues) displayed a difference in dN/dS values always greater than 0.2. Furthermore, in the cases where the two putative *SOS1*-like proteins substantially differed in length, the shortest appeared to be evolving faster than the longer protein from the same species. We also confirmed that these results were not affected by a saturation of nucleotide substitutions on the shorter sequences, except possibly for At*SOS1B*.

More data is needed to verify if this trend is statistically significant. Nevertheless, we can speculate that the possible relation between the differences in protein length and dN/dS might be due to the relative recent occurrence of the duplications in *SOS1*, which could indicate that not enough time has elapsed for differentiation between duplicated genes to have occurred (e.g., woody plants, such as Populus, have a slower evolution rate than Arabidopsis) [48]. Another hypothesis to explain the different dN/dS values observed between *SOS1* proteins that differ in length is that the duplication of some *SOS1* genes did not result in functionally equivalent gene copies, and hence functional divergence between copies is immediately observed.

# CONCLUSIONS

## Contrasting Evolutionary Histories between the *NHX* and *SOS1* Gene Families

*NHX* and *SOS1* plant gene families exhibit markedly different evolutionary histories - while the *NHX* family expanded and developed functionally specialized members throughout the history of the land plants (Figures 1 and 2), the *SOS1* family remained a low copy gene family (Figures 3 and 4). It is clear that after gene duplication events, the NHX family members have undergone a series of protein subcellular relocalization and spatial subfunctionalization events. This can be observed by phylogenetically mapping functional and expression information collected from studies on Arabidopsis proteins (Figure 6) [6, 7, 17, 18, 19, 20, 21, 22, 23]. Rice (*Oryza sativa*) proteins also exhibit functional differentiation, although they are not always functionally equivalent to their Arabidopsis homologues. For example, both *AtNHX5* and *OsNHX5*, which are both part of clade 1 (Figure 1), have higher affinity to $K^+$ compared to $Na^+$[6, 40], although AtNHX5 appears to be unresponsive to ABA [22] while OsNHX5 responds to this hormone [40].

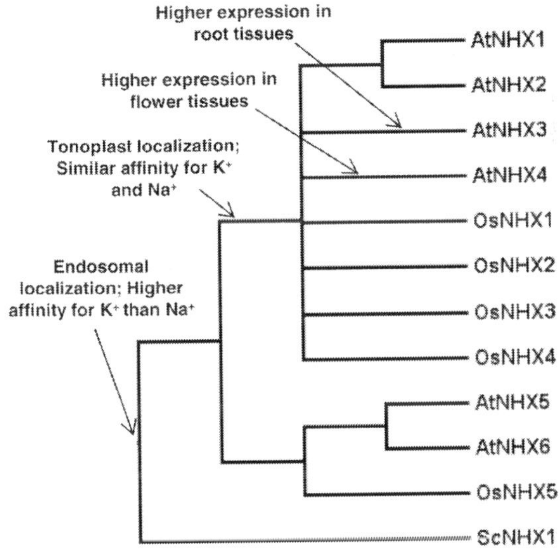


**Figure 6**: Protein subcellular relocalization and spatial subfunctionalization events occurred during the evolution of the NHX family. NHX family has

undergone a series of protein subcellular relocalization and spatial subfunctionalization events. These events are clearly observed by phylogenetically mapping functional and expression information collected from studies on Arabidopsis proteins [6, 7, 17, 18, 19, 20, 21, 22, 23]. Terminal nodes correspond to the following sequences: AtNHX1 - At5g27150; AtNHX2 - At3g05030; AtNHX3 - At3g06370; AtNHX4 - At5g55470; AtNHX5 - At1g54370; AtNHX6 - At1g79610; OsNHX1 - Os07g47100; OsNHX2 - Os05g05590; OsNHX3 - Os11g42790; OsNHX4 - Os06g21360; OsNHX5 - Os09g11450; ScNHX1 - YDR456W.

In contrast, the *SOS1* genes are found only as single- or low-copy genes in most plant species, suggesting either low duplication rates or a higher rate of gene deletion after gene duplication events for these loci. Although it is unclear why *SOS1* is constrained to be a single- or low-copy gene family, the few cases where a second SOS1-like gene has remained in the genome, it appears to have undergone neofunctionalization. For example, Arabidopsis *SOS1*B is only able to transport $Li^+$[16], and the *C. nodosa* SOS1B is described as having a different function from CnSOS1A [45]. Moreover, we found a difference in dN/dS between proteins that are diverging in protein length, thus supporting the observed neofunctionalization or possible loss of function of a second copy of *SOS1*. This difference in dN/dS seems to imply that one of the *SOS1* gene copies suffers, more frequently, a non-synonymous mutation resulting in probable loss of function, and more rarely in neofunctionalization.

In our work, we do not detect a relaxation of selective constraint after duplication in the *NHX* gene family, in agreement with a previous study [44]. Furthermore, the difference in average dN/dS between NHX (~0.08) and SOS1 proteins (~0.16) supports previous results [50, 52] indicating that in eukaryotes, old duplicated genes evolve slower than singletons, despite an initial relaxation of constraint right after duplication. Our hypothesis is that while NHX proteins can be duplicated and subfunctionalized, they must retain their basic function and thus are under stronger purifying selection. In contrast, SOS1 proteins cannot be as easily subfunctionalized and any duplicate copy seems to be preferentially lost and thus exhibiting more relaxed selective pressures.

Additionally, it may be that the evolutionary constraint on SOS1 proteins arises from fewer possibilities for subfunctionalization, in contrast to cytoplasmatic membrane proteins such as the NHXs. In a simplistic view, this might occur because intracellular membrane proteins can have multiple locations within a cell without detriment to the function it performs, while a plasma membrane protein that is not localized to the plasma membrane would be unable to perform its desired function. Moreover, after subcellular relocalization within the cell, it may be easier to vary transporter affinity towards different ions, and intracellular membrane proteins can then be further subfunctionalized according to their specific location within the cell. This is observed for the NHX family in which AtNHX1-4 have equal affinities towards $Na^+$ and $K^+$, while AtNHX5 and 6 have a higher affinity towards $K^+$ (Figure 6).

## Different means of Achieving Stress Adaptation through *NHX* and *SOS1* Duplications

NHX and SOS1 plant protein families have two contrasting evolutionary histories that seem to be related to their protein function and location within the cell. It is especially interesting to assess how the evolution of these two gene families, while different, has resulted in plant adaptation to stress conditions. Our hypothesis is that the first NHX protein was localized in the endosomal compartment of the Golgi and trans-Golgi membranes, corroborating a previous suggestion by Chanroj et al. [39]. The *NHX* duplications, however, allowed the appearance of other NHX proteins that were targeted to the tonoplast. Consequently, an organism with an improved ability to tolerate salt stress arose, since it was able to better retain $K^+$ after stress induction [36, 37].

On the other hand, the duplication followed by neofunctionalization of SOS1 has generated an increased capacity to tolerate soils with high $Li^+$ content in the case of Arabidopsis and related species. This is especially relevant for the Arabidopsis wild relative, Thellungiella spp., that lives in habitats with naturally high $Li^+$concentrations [2]; indeed, *Thellungiella parvula* appears to have three copies of *SOS1B*, probably due to environmental adaptation [2]. Altogether, our results represent two examples of different molecular evolutionary trajectories in land plant genomes that result in organismal stress adaptation through gene duplication.

# METHODS

## NHX and SOS1 Protein Sequences

We used rice protein sequences OsNHX1 (Os07g47100) and OsSOS1A (Os12g44360), in Node Consensus BLASTp of Phytozome v7.0 [53] to obtain the homologous sequences from several plant species (from mosses to eudicots). Sequences belonging to families with an e-value lower than $10^{-100}$ were used in the analysis. Independent BLASTp searches in Phytozome v7.0 and in NCBI (http://www.ncbi.nlm.nih.gov/) allowed retrieval of homologous sequences of OsNHX1 from the algae *Chlamydomonas reinhardtii* and the gymnosperm *Picea sitchensis*. In these particular cases, sequences with an e-value lower than $10^{-10}$ were used. Additionally, we also retrieved from NCBI sequences identified as being NHX- or SOS1-like from the yeast *Saccharomyces cerevisiae* (ScNHX1 – YDR456W) and from salt tolerant plant species.

In order to improve the quality of the alignments obtained, sequences shorter than 500 or longer than 940 or 1200 amino acids from the NHX and SOS1 families, respectively, were removed from the study (38 out of 158 NHX like sequences, and 6 out of 38 SOS1 like sequences, were excluded). This exclusion was performed because the alignments these sequences produced had a large number of discontinuous aligned stretches and we had low confidence on the homology assignments from these alignments.

## Phylogenetic Analysis

The phylogeny of NHX proteins was obtained from 121 sequences in 33 different species (Table 1), with the NHX1 protein from the yeast *Saccharomyces cerevisiae* as the outgroup sequence. The SOS1 phylogeny was estimated from 32 sequences in 23 plant species (Table 1), with the sequences from the moss *P. patens* used as outgroups. These outgroups were chosen because the *S. cerevisiae* sequence (NHA1) most similar to the plant SOS1 genes proved difficult to align with the rest of the plant genes.

Alignments of protein sequences were obtained using MUSCLE [54] and the alignments were refined using Gblocks (default settings) [55, 56] in order to minimize the number of positions with missing information (gaps). For each cleaned or non-cleaned alignment the best-fit amino acid substitution model was selected using ProtTest from MEGA5 [57]. Since the Jones-Taylor-Thornton model [58] together with a discrete approximation of the gamma distribution (JTT+Γ) was always within the top five best-fit models, we selected this model to obtain the phylogenies using the neighbor-joining (NJ) method [59]. We ran the NJ analysis with 1000 bootstrap replicates in MEGA5. Majority-rule consensus trees of 100 bootstrap replicates were also obtained from maximum likelihood (ML) phylogenies obtained using the JTT model plus a gamma rate distribution approximated using 5 categories. For this we used the SEQBOOT, PROML and CONSENSE programs from PHYLIP 3.69 (http://evolution.genetics.washington.edu/phylip.html). Branches with a good support (>75% bootstrap) were generally the same in the consensus trees obtained using either NJ (data not shown) or ML, although generally ML resulted in higher bootstrap support values.

## Estimation of Duplication History

Using the information on species relationships available at the Angiosperm Phylogeny Site (http://www.mobot.org/mobot/research/apweb/) we constructed a cladogram representing the species tree. Estimation of the gene duplication and loss history of both protein families was performed, using NOTUNG v2.6 [60, 61], through reconciliation of the species tree with the NHX and SOS1 ML gene trees (Figures 1 and 2). In order to obtain the most parsimonious estimation of duplication and loss events, the reconciliation of the species and gene trees was followed by rooting, and rearranging branches with less than 75% bootstrap support.

## Determination of Selective Pressure on Amino Acid Sites

NHX protein sequences were separated into two groups according to the Arabidopsis proteins with which they grouped in the estimated phylogeny. Sequences grouping with Arabidopsis NHX1-4 formed group I, and sequences grouping with Arabidopsis NHX5 and 6 formed group II. Each group was used to obtain a new alignment and a new NJ tree as previously explained. Additionally, for SOS1 sequences three other alignments were obtained in which only orthologs were included or sequences deleted at the N- or C-terminus were excluded. All these alignments were cleaned and analyzed in order to maximize the alignment positions tested. Afterwards, in order to obtain codon-cleaned alignments, we mapped the coding sequences (CDS) of *NHX* and *SOS1* genes to protein alignments.

The codon-cleaned alignments and the NJ phylogenetic trees (both rooted and unrooted), were further used in Codeml from the PAML package [62] to determine heterogeneous selective pressure on amino acid sites. Codon substitution models [41] M0, M1a, M2a, M3, M7, and M8 were applied to the alignments. Each model builds on the preceding one by adding new dN/dS classes. M0 assumes that all sites in an alignment are under purifying selection (dN/dS < 1), M1a allows for some sites to be under neutral selection (dN/dS = 1), and M2a allows for some sites to be under positive selection (dN/dS > 1). M3, M7, and M8 have increased classes of sites that are allowed to be under different selection pressures, being M8 the most complex model with 13 classes of sites. All models were tested at least twice to check for convergence problems and all alignments analyzed had an average synonymous substitution (dS) rate less than 14, which is near the lower limit of the dS range estimated to be disruptive for this type of analyses [63, 64]. Afterwards, Likelihood Ratio Tests (LRT) [65] and Akaike Information Criteria (AIC) [66] methods were used to determine the model that best described the data. Neutral or positive selected sites were accepted when model M8 was the best-fit model to the alignment and when sites came up both in model M2 and M8, with a probability ≥ 95% in model M8.

## Comparison of Non-Synonymous/Synonymous Rate Ratio between Clades

Branch-site model [67] from Codeml (PAML) was used to compare the non-synonymous/synonymous substitution rate ratio (dN/dS) between clades or sequences. Bonferroni correction [68] was used to examine significance under these multiple tests. For the SOS1 alignment, the free-ratio model [51] was also used to estimate an independent dN/dS per branch. In both cases, models were tested at least twice to check for convergence problems. Convergence might be an issue, especially for parameter rich models, when there is lack of information in the data, normally resulting from highly similar or divergent sequences (see PAML manual).

## ACKNOWLEDGEMENTS

We are grateful to Ulises Rosas (Purugganan Lab, NYU) for critical reading of the manuscript. We also thank funding by Fundação para a Ciência e a Tecnologia (FCT) – Portugal through National Funds (# PEst-OE/EQB/LA0004/2011), and in part by a grant from the NSF Plant Genome Research Program to MDP. ISP has a PhD fellowship grant (SFRH/BD/68835/2010) from FCT – Portugal. SN and IAA have postdoctoral fellowship grants (SFRH/BPD/34593/2007 and SFRH/BPD/78314/2011, respectively) also from FCT – Portugal.

## REFERENCES

1. Dassanayake M, Oh D-h, Hong H, Bohnert HJ, Cheeseman JM: Transcription strength and halophytic lifestyle. Trends Plant Sci. 2011, 16 (1): 1-3. 10.1016/j.tplants.2010.10.006.
2. Oh D-H, Dassanayake M, Bohnert HJ, Cheeseman JM: Life at the extreme: lessons from the genome. Genome Biol. 2012, 13: 241. 10.1186/gb4003.
3. Flowers TJ, Galal HK, Bromham L: Evolution of halophytes: multiple origins of salt tolerance in land plants. Funct Plant Biol. 2010, 37: 604-612. 10.1071/FP09269.
4. Zhu J-K: Plant salt tolerance. Trends Plant Sci. 2001, 6 (2): 66-71. 10.1016/S1360-1385(00)01838-0.
5. Apse MP, Blumwald E: Na+ transport in plants. FEBS Lett. 2007, 581 (12): 2247-2254. 10.1016/j.febslet.2007.04.014.
6. Venema K, Belver A, Marín-Manzano MC, Rodríguez-Rosales MP, Donaire JP: A Novel Intracellular K+/H+ Antiporter Related to Na+/H+ Antiporters Is Important for K+ Ion Homeostasis in Plants. J Biol Chem. 2003, 278 (25): 22453-22459. 10.1074/jbc.M210794200.
7. Apse MP, Sottosanto JB, Blumwald E: Vacuolar cation/H+ exchange, ion homeostasis, and leaf development are altered in a T-DNA insertional mutant of AtNHX1, the Arabidopsis vacuolar Na+/H+ antiporter. Plant J. 2003, 36: 229-239. 10.1046/j.1365-313X.2003.01871.x.

8. Bowers K, Levi BP, Patel FI, Stevens TH: The Sodium/Proton Exchanger Nhx1p Is Required for Endosomal Protein Trafficking in the Yeast Saccharomyces cerevisiae. Mol Biol Cell. 2000, 11: 4277-4294. 10.1091/mbc.11.12.4277.
9. Brett CL, Tukaye DN, Mukherjee S, Rao R: The yeast endosomal Na+(K+)/H+ exchanger Nhx1 regulates cellular pH to control vesicle trafficking. Mol Biol Cell. 2005, 16 (3): 1396-1405. 10.1091/mbc.E04-11-0999.
10. Apse MP: Salt Tolerance Conferred by Overexpression of a Vacuolar Na+/H+ Antiport in Arabidopsis. Science. 1999, 285 (5431): 1256-1258. 10.1126/science.285.5431.1256.
11. Hernandez A, Jiang X, Cubero B, Nieto PM, Bressan RA, Hasegawa PM, Pardo JM: Mutants of the Arabidopsis thaliana cation/H+ antiporter AtNHX1 conferring increased salt tolerance in yeast: the endosome/prevacuolar compartment is a target for salt toxicity. J Biol Chem. 2009, 284 (21): 14276-14285. 10.1074/jbc.M806203200.
12. Li T, Zhang Y, Liu H, Wu Y, Li W, Zhang H: Stable expression of Arabidopsis vacuolar Na+/H+ antiporter gene AtNHX1, and salt tolerance in transgenic soybean for over six generations. Chin Sci Bull. 2010, 55 (12): 1127-1134. 10.1007/s11434-010-0092-8.
13. Cao D, Hou W, Liu W, Yao W, Wu C, Liu X, Han T: Overexpression of TaNHX2 enhances salt tolerance of 'composite' and whole transgenic soybean plants. Plant Cell Tissue Organ Cult. 2011, 107 (3): 541-552. 10.1007/s11240-011-0005-9.
14. Brett CL, Donowitz M, Rao R: Evolutionary origins of eukaryotic sodium/proton exchangers. Am J Physiol Cell Physiol. 2005, 288: C223-C239. 10.1152/ajpcell.00360.2004.
15. Shi H, Ishitani M, Kim C, Zhu JK: The Arabidopsis thaliana salt tolerance gene SOS1 encodes a putative Na+/H+ antiporter. PNAS. 2000, 97 (12): 6896-6901. 10.1073/pnas.120170197.
16. An R, Chen QJ, Chai MF, Lu PL, Su Z, Qin ZX, Chen J, Wang XC: AtNHX8, a member of the monovalent cation: proton antiporter-1 family in Arabidopsis thaliana, encodes a putative Li/H antiporter. Plant J. 2007, 49 (4): 718-728. 10.1111/j.1365-313X.2006.02990.x.
17. Pardo JM, Cubero B, Leidi EO, Quintero FJ: Alkali cation

exchangers: roles in cellular homeostasis and stress tolerance. J Exp Bot. 2006, 57 (5): 1181-1199. 10.1093/jxb/erj114.

18. Bassil E, Ohto MA, Esumi T, Tajima H, Zhu Z, Cagnac O, Belmonte M, Peleg Z, Yamaguchi T, Blumwald E: The Arabidopsis intracellular Na+/H+ antiporters NHX5 and NHX6 are endosome associated and necessary for plant growth and development. Plant Cell. 2011, 23 (1): 224-239. 10.1105/tpc.110.079426.
19. Bassil E, Tajima H, Liang Y-C, Ohto M-a, Ushijima K, Nakano R, Esumi T, Coku A, Belmonte M, Blumwald E: The Arabidopsis Na+/H+ Antiporters NHX1 and NHX2 Control Vacuolar pH and K+ Homeostasis to Regulate Growth, Flower Development, and Reproduction. Plant Cell. 2011, 23 (9): 3482-3497. 10.1105/tpc.111.089581.
20. Venema K, Quintero FJ, Pardo JM, Donaire JP: The arabidopsis Na+/H+ exchanger AtNHX1 catalyzes low affinity Na+ and K+ transport in reconstituted liposomes. J Biol Chem. 2002, 277 (4): 2413-2418. 10.1074/jbc.M105043200.
21. Aharon GS, Apse MP, Duan S, Hua X, Blumwald E: Characterization of a family of vacuolar Na+/H+ antiporters in Arabidopsis thaliana. Plant Soil. 2003, 00: 245-256.
22. Yokoi S, Quintero FJ, Cubero B, Ruiz MT, Bressan RA, Hasegawa PM, Pardo JM: Differential expression and function of Arabidopsis thaliana NHX Na+/H+ antiporters in the salt stress response. Plant J. 2002, 30 (5): 529-539. 10.1046/j.1365-313X.2002.01309.x.
23. Li HT, Liu H, Gao XS, Zhang H: Knock-out of Arabidopsis AtNHX4 gene enhances tolerance to salt stress. Biochem Biophys Res Commun. 2009, 382 (3): 637-641. 10.1016/j.bbrc.2009.03.091.
24. Fukuda A, Nakamura A, Tagiri A, Tanaka H, Miyao A, Hirochika H, Tanaka Y: Function, Intracellular Localization and the Importance in Salt Tolerance of a Vacuolar Na+/H+ Antiporter from Rice. Plant Cell Physiol. 2004, 45 (2): 146-159. 10.1093/pcp/pch014.
25. Fukuda A, Nakamura A, Tanaka Y: Molecular cloning and expression of the Na+/H+ exchanger gene in Oryza sativa. Biochimica et Biophysica Acta - Gene Structure and Expression. 1999, 1446 (1–2): 149-155.
26. Li M, Lin X, Li H, Pan X, Wu G: Overexpression of AtNHX5

improves tolerance to both salt and water stress in rice (Oryza sativa L.). Plant Cell Tissue Organ Cult. 2011, 107 (2): 283-293. 10.1007/s11240-011-9979-6.

27. Xue ZY, Zhi DY, Xue GP, Zhang H, Zhao YX, Xia GM: Enhanced salt tolerance of transgenic wheat (Tritivum aestivum L.) expressing a vacuolar Na+/H+ antiporter gene with improved grain yields in saline soils in the field and a reduced level of leaf Na+. Plant Sci. 2004, 167 (4): 849-859. 10.1016/j.plantsci.2004.05.034.
28. Mahajan S, Pandey GK, Tuteja N: Calcium- and salt-stress signaling in plants: shedding light on SOS pathway. Arch Biochem Biophys. 2008, 471 (2): 146-158. 10.1016/j.abb.2008.01.010.
29. Qiu QS, Guo Y, Dietrich MA, Schumaker KS, Zhu JK: Regulation of SOS1, a plasma membrane Na+/H+ exchanger in Arabidopsis thaliana, by SOS2 and SOS3. PNAS. 2002, 99 (12): 8436-8441. 10.1073/pnas.122224699.
30. Quintero FJ, Ohta M, Shi H, Zhu JK, Pardo JM: Reconstitution in yeast of the Arabidopsis SOS signaling pathway for Na+ homeostasis. PNAS. 2002, 99 (13): 9061-9066. 10.1073/pnas.132092099.
31. Maser P, Thomine S, Schroeder JI, Ward JM, Hirschi K, Sze H, Talke IN, Amtmann A, Maathuis FJM, Sanders D, et al: Phylogenetic relationships within cation transporter families of Arabidopsis. Plant Physiol. 2001, 126: 1646-1667. 10.1104/pp.126.4.1646.
32. Guo KM, Babourina O, Rengel Z: Na(+)/H(+) antiporter activity of the SOS1 gene: lifetime imaging analysis and electrophysiological studies on Arabidopsis seedlings. Physiol Plant. 2009, 137 (2): 155-165. 10.1111/j.1399-3054.2009.01274.x.
33. Qiu QS, Guo Y, Quintero FJ, Pardo JM, Schumaker KS, Zhu JK: Regulation of vacuolar Na+/H+ exchange in Arabidopsis thaliana by the salt-overly-sensitive (SOS) pathway. J Biol Chem. 2004, 279 (1): 207-215.
34. Gaxiola RA, Rao R, Sherman A, Grisafi P, Alper SL, Fink GR: The Arabidopsis thaliana proton transporters, AtNhx1 and Avp1, can function in cation detoxification in yeast. Proc Natl Acad Sci USA. 1999, 96 (4): 1480-1485. 10.1073/pnas.96.4.1480.
35. Leidi EO, Barragan V, Rubio L, El-Hamdaoui A, Ruiz MT, Cubero B, Fernandez JA, Bressan RA, Hasegawa PM, Quintero FJ, et al:

The AtNHX1 exchanger mediates potassium compartmentation in vacuoles of transgenic tomato. Plant J. 2010, 61 (3): 495-506. 10.1111/j.1365-313X.2009.04073.x.

36. Barragan V, Leidi EO, Andres Z, Rubio L, De Luca A, Fernandez JA, Cubero B, Pardo JM: Ion exchangers NHX1 and NHX2 mediate active potassium uptake into vacuoles to regulate cell turgor and stomatal function in Arabidopsis. Plant Cell. 2012, 24 (3): 1127-1142. 10.1105/tpc.111.095273.
37. Jiang X, Leidi EO, Pardo JM: How do vacuolar NHX exchangers function in plant salt tolerance?. Plant Signal Behav. 2010, 5 (7): 792-795. 10.4161/psb.5.7.11767.
38. Hastings PJ, Lupski JR, Rosenberg SM, Ira G: Mechanisms of change in gene copy number. Nat Rev Genet. 2009, 10 (8): 551-564. 10.1038/nrg2593.
39. Chanroj S, Wang G, Venema K, Zhang MW, Delwiche CF, Sze H: Conserved and diversified gene families of monovalent cation/H+ antiporters from algae to flowering plants. Frontiers in Plant Science. 2012, 3: 1-18.
40. Fukuda A, Nakamura A, Hara N, Toki S, Tanaka Y: Molecular and functional analyses of rice NHX-type Na+/H+ antiporter genes. Planta. 2011, 233 (1): 175-188. 10.1007/s00425-010-1289-4.
41. Yang Z, Nielsen R, Goldman N, Pedersen AMK: Codon-substitution models for heterogeneous selection pressure at amino acid sites. Genetics. 2000, 155: 431-449.
42. Google Scholar
43. Sato Y, Sakaguchi M: Topogenic properties of transmembrane segments of Arabidopsis thaliana NHX1 reveal a common topology model of the Na+/H+ exchanger family. J Biochem (Tokyo). 2005, 138 (4): 425-431. 10.1093/jb/mvi132.
44. Brown CJ, Takayama S, Campen AM, Vise P, Marshall TW, Oldfield CJ, Williams CJ, Dunker AK: Evolutionary rate heterogeneity in proteins with long disordered regions. J Mol Evol. 2002, 55: 104-110. 10.1007/s00239-001-2309-6.
45. Hudson CM, Puckett EE, Bekaert M, Pires JC, Conant GC: Selection for higher gene copy number after different types of plant gene duplications. Genome Biol Evol. 2011, 3: 1369-1380. 10.1093/gbe/

evr115.

46. Garciadeblas B, Haro R, Benito B: Cloning of two SOS1 transporters from the seagrass Cymodocea nodosa. SOS1 transporters from Cymodocea and Arabidopsis mediate potassium uptake in bacteria. Plant Mol Biol. 2007, 63 (4): 479-490. 10.1007/s11103-006-9102-2.
47. Maughan PJ, Turner TB, Coleman CE, Elzinga DB, Jellen EN, Morales JA, Udall JA, Fairbanks DJ, Bonifacio A: Characterization of Salt Overly Sensitive 1 (SOS1) gene homoeologs in quinoa (Chenopodium quinoa Willd.). Genome. 2009, 52 (7): 647-657. 10.1139/G09-041.
48. Quintero FJ, Martinez-Atienza J, Villalta I, Jiang X, Kim WY, Ali Z, Fujii H, Mendoza I, Yun DJ, Zhu JK, et al: Activation of the plasma membrane Na/H antiporter Salt-Overly-Sensitive 1 (SOS1) by phosphorylation of an auto-inhibitory C-terminal domain. PNAS. 2011, 108 (6): 2611-2616. 10.1073/pnas.1018921108.
49. Smith SA, Donoghue MJ: Rates of molecular evolution are linked to life history in flowering plants. Science. 2008, 322 (5898): 86-89. 10.1126/science.1163197.
50. Duarte JM, Wall PK, Edger PP, Landherr LL, Ma H, Pires JC, Leebens-Mack J, dePamphilis CW: Identification of shared single copy nuclear genes in Arabidopsis, Populus, Vitis and Oryza and their phylogenetic utility across various taxonomic levels. BMC Evol Biol. 2010, 10: 61. 10.1186/1471-2148-10-61.
51. Jordan IK, Wolf YI, Koonin EV: Duplicated genes evolve slower than singletons despite the initial rate increase. BMC Evol Biol. 2004, 4: 22. 10.1186/1471-2148-4-22.
52. Geng S, Zhao Y, Tang L, Zhang R, Sun M, Guo H, Kong X, Li A, Mao L: Molecular evolution of two duplicated CDPK genes CPK7 and CPK12 in grass species: a case study in wheat (Triticum aestivum L.). Gene. 2011, 475 (2): 94-103. 10.1016/j.gene.2010.12.015.
53. Davis JC, Petrov DA: Preferential duplication of conserved proteins in eukaryotic genomes. PLoS Biol. 2004, 2 (3): 0318-0326.
54. Goodstein DM, Shu S, Howson R, Neupane R, Hayes RD, Fazo J, Mitros T, Dirks W, Hellsten U, Putnam N, et al: Phytozome: a comparative platform for green plant genomics. Nucleic Acids Res.

2012, 40 (Database issue): D1178-1186.

55. Edgar RC: MUSCLE: multiple sequence alignment with high accuracy and high throughput. Nucleic Acids Res. 2004, 32 (5): 1792-1797. 10.1093/nar/gkh340.

56. Castresana J: Selection of conserved blocks from multiple alignments for their use in phylogenetic analysis. Mol Biol Evol. 2000, 17: 540-552. 10.1093/oxfordjournals.molbev.a026334.

57. Talavera G, Castresana J: Improvement of phylogenies after removing divergent and ambiguously aligned blocks from protein sequence alignments. Syst Biol. 2007, 56 (4): 564-577. 10.1080/10635150701472164.

58. Tamura K, Peterson D, Peterson N, Stecher G, Nei M, Kumar S: MEGA5: molecular evolutionary genetics analysis using maximum likelihood, evolutionary distance, and maximum parsimony methods. Mol Biol Evol. 2011, 28 (10): 2731-2739. 10.1093/molbev/msr121.

59. Jones DT, Taylor WR, Thornton JM: The rapid generation of mutation data matrices from protein sequences. CABIOS. 1992, 8 (3): 275-282.

60. Saitou N, Nei M: The neighbor-joining method: a new method for reconstructing phylogenetic trees. Mol Biol Evol. 1987, 4 (4): 406-425.

61. Durand D, Halldórsson BV, Vernot B: A hybrid micro–macroevolutionary approach to gene tree reconstruction. J Comput Biol. 2006, 13 (2): 320-335. 10.1089/cmb.2006.13.320.

62. Vernot B, Stolzer M, Goldman A, Durand D: Reconciliation with non-binary species trees. J Comput Biol. 2008, 15 (8): 981-1006. 10.1089/cmb.2008.0092.

63. Yang Z: PAML: a program package for phylogenetic analysis by maximum likelihood. CABIOS. 1997, 13 (5): 555-556.

64. Yang Z: On the best evolutionary rate for phylogenetic analysis. Syst Biol. 1998, 47 (1): 125-133. 10.1080/106351598261067.

65. Anisimova M, Bielawski JP, Yang Z: Accuracy and power of the likelihood ratio test in detecting adaptive molecular evolution. Mol Biol Evol. 2001, 18 (8): 1585-1592. 10.1093/oxfordjournals.molbev.a003945.

66. Yang Z: Likelihood ratio tests for detecting positive selection and application to primate lysozyme evolution. Mol Biol Evol. 1998, 15 (5): 568-573. 10.1093/oxfordjournals.molbev.a025957.

67. Le SQ, Gascuel O: An improved general amino acid replacement matrix. Mol Biol Evol. 2008, 25 (7): 1307-1320. 10.1093/molbev/msn067.

68. Zhang J, Nielsen R, Yang Z: Evaluation of an improved branch-site likelihood method for detecting positive selection at the molecular level. Mol Biol Evol. 2005, 22: 2472-2479. 10.1093/molbev/msi237.

69. Anisimova M, Yang Z: Multiple hypothesis testing to detect lineages under positive selection that affects only a few sites. Mol Biol Evol. 2007, 24 (5): 1219-1228. 10.1093/molbev/msm042.

CHAPTER

# 13

# VACUOLAR $Na^+/H^+$ NHX-TYPE ANTIPORTERS ARE REQUIRED FOR CELLULAR $K^+$ HOMEOSTASIS, MICROTUBULE ORGANIZATION AND DIRECTIONAL ROOT GROWTH

Tyler McCubbin, Elias Bassil, Shiqi Zhang and Eduardo Blumwald

Department of Plant Sciences, University of California, Davis, CA 95616, USA

## ABSTRACT

$Na^+/H^+$ antiporters (NHXs) are integral membrane transporters that catalyze the electroneutral exchange of $K^+$ or $Na^+$ for $H^+$ and are implicated in cell expansion, development, pH and ion homeostasis and salt tolerance. *Arabidopsis* contains four vacuolar NHX isoforms (NHX1–NHX4), but only the functional roles for NHX1 and NHX2 have been assessed thus far. Colocalization studies indicated that NHX3 and NHX4 colocalize to the tonoplast. To investigate the role of all vacuolar NHX isoforms, a quadruple knockout *nhx1nhx2nhx3nhx4*, lacking all vacuolar NHXs, was generated. Seedlings of *nhx1nhx2nhx3nhx4* displayed significantly reduced growth, with markedly shorter hypocotyls. Under high $K^+$, but not $Na^+$, pronounced root skewing occurred in *nhx1nhx2nhx3nhx4*, suggesting that the organization of the cytoskeleton might be perturbed. Whole mount immunolabeling of cortical microtubules indicated that high $K^+$ caused significant microtubule reorganization in *nhx1nhx2nhx3nhx4* root cells of the elongation zone. Using microtubule stabilizing (Taxol) and destabilizing (propyzamide) drugs, we found that the effect of $K^+$ on *nhx1nhx2nhx3nhx4* root growth was antagonistic to that of Taxol, whereas elevated $K^+$exacerbated the endogenous effect of propyzamide on root skewing. Collectively, our results suggest that altered $K^+$ homeostasis leads to an increase in the dynamics of cortical microtubule reorganization in *nhx1nhx2nhx3nhx4* root epidermal cells of the elongation zone.

**Keywords:** vacuole; potassium, salt, homeostasis, pH, root, growth, microtubule; cytoskeleton, curling, skewing, NHX

## INTRODUCTION

Ion homeostasis and pH regulation are essential cellular processes necessary for plant growth, development and responses to environmental stress. $Na^+/H^+$ antiporters (NHXs) are integral membrane proteins localized to the plasma membrane and endomembrane compartments that facilitate the electroneutral exchange of $Na^+$ or $K^+$ for $H^+$, thereby maintaining both luminal pH, as well as cation homeostasis. The *Arabidopsis* genome encodes eight NHX homologs that have been grouped based on sequence similarity and function into three distinct classes: plasma membrane (NHX7/SOS1 and NHX8), endosomal/vesicular (NHX5, NHX6), as well as four that are vacuolar (NHX1, NHX2, NHX3, NHX4) [1,2]. We previously reported the vacuolar localization and function of two of the vacuolar NHX isoforms, NHX1 and NHX2 [3,4], that bear strong sequence homology to the putatively vacuolar NHX3 and NHX4. Other work has shown colocalization of AtNHX3 with vacuolar proteins δ-TIP and γ-TIP, as well as SYP22 in tobacco protoplasts, suggesting a vacuolar and PVC localization of AtNHX3 [5]. Similarly, constitutive expression of AtNHX3 in sugar beet (*Beta vulgaris* L.) has been shown to confer increased salt tolerance, but also has led to high $K^+$ accumulation in roots when grown under high salt [6]. In addition, Liu *et al.* found that *Arabidopsis nhx3* null mutants were hypersensitive to $K^+$ deficiency, further implicating NHX3 in $K^+$ homeostasis [5]. AtNHX4 localization was suggested to be similar to NHX1 at the tonoplast [7], but additional experiments are required to demonstrate localization convincingly. Given that little is known about the roles of NHX3 and NHX4 in vacuolar ion and pH homeostasis, further investigations are warranted.

Potassium transport into the vacuole has been shown to be a key regulator of cell expansion [8]. Both NHX1 and NHX2 are necessary for the accumulation of vacuolar $K^+$ and vacuolar alkalization, because the double knockout *nhx1nhx2* accumulated 70% less vacuolar $K^+$ and had significantly more acidic vacuoles than comparable wild-type (WT) plants [3]. *nhx1nhx2* plants also exhibited reduced overall growth, had smaller cells in all tissues examined, as well as aberrant flower development, such as incomplete filament elongation and anther dehiscence, which resulted in significantly reduced seed set. Others studies also showed that *nhx1nh2* double knockouts have high cytosolic $K^+$ [9], altered vacuole morphology, compromised turgor, severely impaired stomatal conductance and less daytime transpiration due to the lack of $K^+$ accumulation in guard cell vacuoles [10]. One intriguing phenotype of *nhx1nhx2* was their high sensitivity to additional $K^+$ in the growth media, with a severe reduction in overall growth [3]. This phenotype was not evident when *nhx1nhx2* plants were grown in the presence of equimolar concentrations of $Na^+$ ions. The presence of high $K^+$ also led to profound curling of seedling roots, suggesting that altered $K^+$ homeostasis, due to the lack of NHX1 and NHX2 antiporters, altered directional root growth [3].

Root growth patterning is controlled by microtubule and cytoskeleton reorganization [11]. Cellular $Na^+$ homeostasis can affect cytoskeleton-mediated root growth, and the effects of salinity on directional root growth have already been reported [12,13]. Little is known about the role(s) of cellular $K^+$homeostasis in microtubule organization. Here, we show that vacuolar NHX1 to NHX4 reside at the tonoplast. Using the quadruple knockout *nhx1nhx2nhx3nhx4*, lacking all four vacuolar NHXs, we show that vacuolar NHXs impart essential functions in $K^+$ homeostasis that affect directional root growth and cytoskeletal dynamics. The quadruple knockout exhibits more pronounced phenotypes compared to the *nhx1nhx2* double knockout, including increased sensitivity to $K^+$, suggesting that NHX3 and NHX4 have additional roles in cell expansion and $K^+$ homeostasis.

## RESULTS AND DISCUSSION

We sought to assess the contribution of all vacuolar NHXs to overall plant growth and development by generating a quadruple knockout lacking all vacuolar NHX antiporters, NHX1, NHX2, NHX3 and NHX4. Using available T-DNA insertion lines (Figure A1), we generated the quadruple knockout *nhx1nhx2nhx3nhx4* through successive crosses of single, double and triple knockouts, as described in the Methods. The level of expression of all four vacuolar NHX transcripts in the *nhx1nhx2nhx3nhx4*knockout indicated that no full-length transcripts exist (Figure A1).

### NHX3 and NHX4 are Colocalized at the Tonoplast

To determine the subcellular localization of NHX3 and NHX4, we generated constitutively expressed translational fusion proteins of both NHX3 and NHX4 with CFP and RFP, respectively. Transient co-expression of NHX3-CFP and NHX4-RFP in *Arabidopsis* cotyledons displayed overlapping expression at the periphery of the cell, as well as at transvacuolar strands (Figure 1A,C). Because NHX1 and NHX2 were shown previously to localize at the tonoplast [3], we transiently expressed NHX4-CFP in plants stably expressing NHX2-YFP and observed an overlapping expression pattern between NHX2 and NHX4 that was characteristic of a vacuolar signal, including trans-vacuolar strands and vacuolar bulbs (arrowheads in Figure 1D–F). We then expressed NHX3-RFP and NHX4-CFP in aquaporin γ-TIP [14] and the R-SNARE VAMP711 [15] expressing lines, which were previously shown to colocalize with NHX1 [3]. In cotyledon epidermal cells, an overlapping expression between NHX3-RFP and γ-TIP-GFP, including in vacuolar bulbs that are characteristic of γ-TIP, was seen (Figure 1G–I). Overlapping expression was also seen with NHX4-CFP and VAMP711-RFP (Figure 1J–L). Taken together, these results indicated that both NHX3 and NHX4 localize to the tonoplast, as suggested by other heterologous expression studies [5,7] and sequence similarity to NHX1 and NHX2 [1].

## Growth Phenotypes of nhx1nhx2nhx3nhx4 and Responses to $Na^+$, $K^+$

We used the etiolation response of dark-grown seedlings to examine the contribution of *NHX1*, *NHX2*, *NHX3* and *NHX4* to cell expansion and seedling growth. Previously, we reported that etiolated *nhx1nhx2* double knockout seedlings had profound hypocotyl and root elongation phenotypes [3]. Etiolated *nhx1nhx2nhx3nhx4* seedlings displayed significantly reduced growth of both roots and hypocotyls (Figure 2). In control media, hypocotyl lengths of *nhx1nhx2nhx3nhx4* were less than 50% the length of WT after 11 days of etiolated growth (Figure 2M). The hypocotyl elongation response of *nhx1nhx2nhx3nhx4* was highly responsive to both added $Na^+$ and $K^+$. For example, hypocotyls of *nhx1nhx2nhx3nhx4* grown in 30 mM NaCl were 31% longer than those of seedlings grown on control media (1 mM NaCl, 1 mM KCl), but hypocotyls of *nhx1nhx2nhx3nhx4* seedlings grown in 30 mM KCl were 49% shorter (Figure 2D,H,K,L) than those grown in control media. By contrast, little change in hypocotyl length was observed in the WT seedlings grown on either 30 mM NaCl or KCl (Figure 2B,F,J,L). Root growth was also markedly different between *nhx1nhx2nhx3nhx4* and WT vertically-grown seedlings, but only in response to either $K^+$ or $Na^+$. Similar to the $K^+$ and $Na^+$-dependent growth response of hypocotyls, *nhx1nhx2nhx3nhx4* roots were also significantly shorter (26%) under $K^+$ and longer (13%) under $Na^+$ as compared to *nhx1nhx2nhx3nhx4* grown in control media. In addition, the *nhx1nhx2nhx3nhx4* seedlings began to display prominent left-handed skewing (as viewed through the agar, following the convention assigned by Rutherford and Masson [16]) after four days of growth on media supplemented with 30 mM $K^+$ (Figure A2). Seedlings grown for another seven days on the same media displayed complete curling of roots (Figure 2H). By contrast, WT root growth remained unchanged under 30 mM $K^+$ even after 11 days.

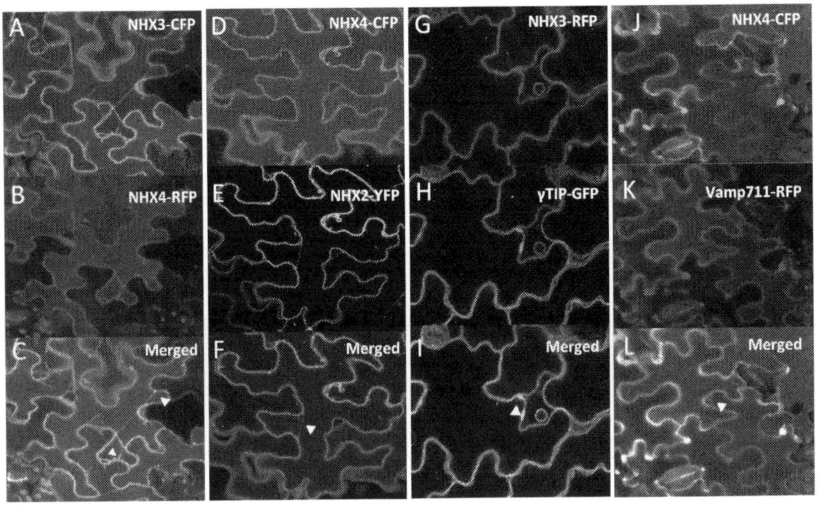


**Figure 1:** Intracellular localization of *NHX3* and *NHX4*. Transient co-expression, in *Arabidopsis* cotyledon epidermal cells, of *35S::NHX3-CFP* (A) with *35S::NHX4-RFP* (B) colocalized as shown in the merge fluorescent panel (C). Transient expression of *35S::NHX4-CFP* (D) in stable *35S::NHX2-YFP* background (E) also colocalized as shown in the merged image (F). Transient expression of *35S::NHX3-RFP* (G) in stable expressing *γTIP::γTIP-GFP* (H) colocalized as shown in the merged panel (I). Transient expression of *35S::NHX4-CFP* (J) in the stable expressing *Ub10::VAMP711-RFP*(K) colocalized as shown in the merged image (L). Arrowheads in (C,F,I,L) point to tonoplast invaginations and transvacuolar cytoplasmic strands.

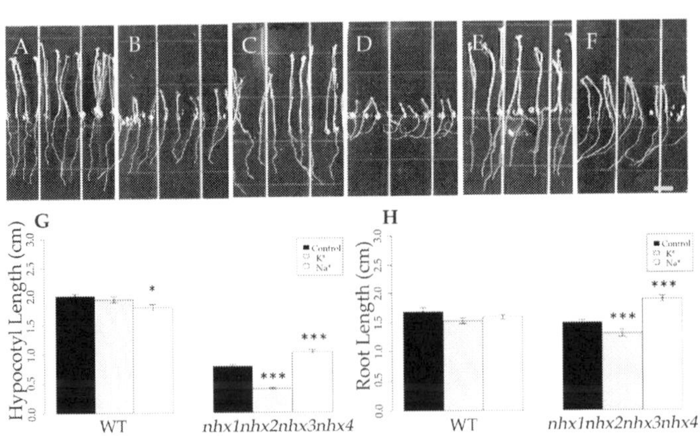


**Figure 2:** Etiolation response of the knockout *nhx1nhx2nhx3nhx4*. Etiolated wild-type (A,C,E) and *nhx1nhx2nhx3nhx4* (B,D,F) seedlings grown for 11 days

in the dark on either 30 mM $K^+$ (C,D) or 30 mM $Na^+$ media (E,F). No significant difference in hypocotyl or root growth was noted in the wild-type; however, in *nhx1nhx2nhx3nhx4*, root and hypocotyl growth was significantly inhibited in seedlings grown in 30 mM $K^+$ and increased in those grown in 30 mM $Na^+$, as compared to control media. Quantification of hypocotyl (G) and root (H) growth normalized relative to growth on control media (1 mM $Na^+$ and 1 mM $K^+$). *nhx1nhx2nhx3nhx4*displayed left-handed root skewing when grown on media containing 30 mM $K^+$ (D), but not 1 mM $K^+$ (B). Values represent the mean ± SE ($n = 29$). * $p \leq 0.05$, *** $p \leq 0.001$; *p*-values reflect significant differences between control and either $K^+$ or $Na^+$ treatments within each genotype. The yellow bar is 0.5 cm.

Previously, we reported that etiolated *nhx1nhx2* plants also displayed reduced hypocotyl growth, that was 70% the length of WT hypocotyls on media containing 1 mM $K^+$ and 50% of WT on 30 mM $K^+$ [3], while in etiolated *nhx1nhx2nhx3nhx4* seedlings, the average hypocotyl length was 40% and 20% of WT on 1 and 30 mM $K^+$, respectively. A comparison of root growth responses between *nhx1nhx2nhx3nhx4*and *nhx1nhx2* to high $K^+$ [3] indicates that root curling is also more pronounced and root growth more sensitive to high $K^+$, similar to that observed in hypocotyls. The dramatic differences in root and hypocotyl elongation between *nhx1nhx2* and *nhx1nhx2nhx3nhx4* under the higher $K^+$ suggest that NHX3 and NHX4 have additional roles in $K^+$ homeostasis and cell expansion.

To determine if the root curling observed in *nhx1nhx2nhx3nhx4* roots was specific to $K^+$, we compared the root growth of *nhx1nhx2nhx3nhx4* on media supplemented with KCl, $KNO_3$, NaCl or mannitol. Mannitol was included in order to determine whether the root curling response of *nhx1nhx2nhx3nhx4*under high $K^+$ was due to an osmotic effect; thus, 60 mM mannitol (60 mOsm, equivalent to the osmolarity of the added salt) was used. We found that *nhx1nhx2nhx3nhx4* root growth on mannitol was not different from the control (Figure A3J,B), and therefore not likely due to an osmotic effect. As shown in Figure A3, both 30 mM KCl and 30 mM $KNO_3$ supplemented media resulted in similar left-handed root skewing and curling of the root tip that was not seen when *nhx1nhx2nhx3nhx4* was grown on mannitol, suggesting that the response of *nhx1nhx2nhx3nhx4* was specific to $K^+$. Collectively, experiments comparing *nhx1nhx2nhx3nhx4* seedling growth (*i.e.*, Figure 2 and Figure A3) indicate that the knockouts' root curling response was caused by aberrant $K^+$ homeostasis.

Previous work has provided a strong link between ion stress and directional root growth. For example, *sos1(nhx7)* null mutants displayed a loss of normal gravitropic root growth when grown vertically on media supplemented with high (50–150 mM) $Na^+$, with young roots exhibiting skewing and eventual upward directed growth [17]. Moreover, roots of both WT and *sos1(nhx7)* seedlings grown on high $Na^+$ showed delayed gravitropic responses, thereby suggesting that aberrant intracellular ion homeostasis can specifically affect directional root growth. The *sos1* skewing phenotype is remarkably similar to *nhx1nhx2nhx3nhx4* grown on $K^+$ supplemented media, and although effects of high cytosolic $Na^+$ have been suggested, the effects of $K^+$ on root skewing remain unclear. Furthermore, *sos1(nhx7)* mutants were more sensitive to microtubule reorganization, with whole mount immunostaining experiments in elongating root epidermal cells demonstrating that *sos1 (nhx7)* in the presence of 50 mM $Na^+$ had helically-orientated microtubules, while WT plants maintained the normal transverse array [12]. Although microtubule remodeling has been shown to be induced by high external $Na^+$, reports of $K^+$-induced microtubule reorganization are sparse [12,13,18]. In KCl (350 mM)-treated *Zea mays*, a reorganization of the cortical microtubule array from the transverse to oblique orientation in root cells of the maturation zone occurred within 3 min after application. By contrast, treatment with sorbitol did not result in the reorganization of microtubules [19]. These results suggested that cortical microtubule organization in roots was affected by high concentrations of $Na^+$ and $K^+$. Given the phenotypes of *nhx1nhx2nhx3nhx4*, the knockout represents a useful tool to study the role of $K^+$ in cytoskeletal organization. The positive effect of $Na^+$ and the negative effect on $K^+$ on *nhx1nhx2nhx3nhx4* growth is well supported by similar results obtained in *nhx1nhx2* in which $Na^+$ was proposed to substitute for the lack of $K^+$ accumulation in vacuoles, providing the osmoticum needed to generate the turgor for cell expansion [3,10]. The sensitivity of both *nhx1nhx2* and *nhx1nhx2nhx3nhx4* to added $K^+$, as well as the aberrantly high cytosolic $K^+$ concentrations of *nhx1nhx2*, together with previously reported effects of $Na^+$ on cytoskeleton disassembly and reorganization, suggests that aberrant $K^+$ homeostasis in the *nhx1nhx2nhx3nhx4* might affect the organization of the cytoskeleton.

## nhx1nhx2nhx3nhx4 Lacks Accumulation of $K^+$ *in the Vacuole*

Previously, we quantified the concentration of $K^+$ in vacuoles of the knockout *nhx1nhx2* using the $K^+$binding dye, PBFI [3]. Using the same approach, we quantified the vacuolar [$K^+$] of *nhx1nhx2nhx3nhx4* in root tip cells of the elongation zone, where $K^+$-induced root curling was observed (Figure 2D). In *nhx1nhx2nhx3nhx4*, vacuolar $K^+$ was ~19 mM (±7 mM) and significantly lower than WT, which was 77 mM (±13 mM). These values are similar to those obtained in *nhx1nhx2*, suggesting that the lack of *NHX3*and *NHX4* in addition to *NHX1* and *NHX2* did not significantly alter root tip cell vacuolar $K^+$. In the same *nhx1nhx2* knockout, the cytosolic $K^+$ was also reported to be aberrantly high compared to WT [9]. Therefore, given that the electrochemical potential of $K^+$ favors uptake into the cell [1,3], it is possible that in *nhx1nhx2nhx3nhx4*, the lack of $K^+$ accumulation into the vacuole might lead to an aberrantly high $K^+$ accumulation in the cytosol.

## Cells of the Root Transition Zone Lack Expansion in nhx1nhx2nhx3nhx4

To investigate the significance of NHX1 to NHX4-mediated $Na^+$ and $K^+$ transport on cell expansion and root skewing, we stained roots of *nhx1nhx2nhx3nhx4* seedlings with propidium iodide and visualized cell size. Compared with WT, cells in the *nhx1nhx2nhx3nhx4* root transition zone appeared to be much shorter when grown on 30 mM KCl-supplemented media (Figure 3E,F). Cell size was similar between WT and *nhx1nhx2nhx3nhx4* in the apical meristem, but epidermal cells in the transition zone were markedly smaller in *nhx1nhx2nhx3nhx4*. Whereas in the WT, a gradual increase in cell size from the meristem to the elongation zone (Figure 3E, arrows) was apparent, the length of epidermal cells in the transition zone of *nhx1nhx2nhx3nhx4* was aberrantly small, and as a consequence, the transition zone appeared "compressed" (Figure 3F, white arrows). This phenotype was not evident in *nhx1nhx2nhx3nhx4* grown on 30 mM NaCl-supplemented media, where no difference in cell size or shape was observed between WT and *nhx1nhx2nhx3nhx4* in any root zones (Figure 3C,D).

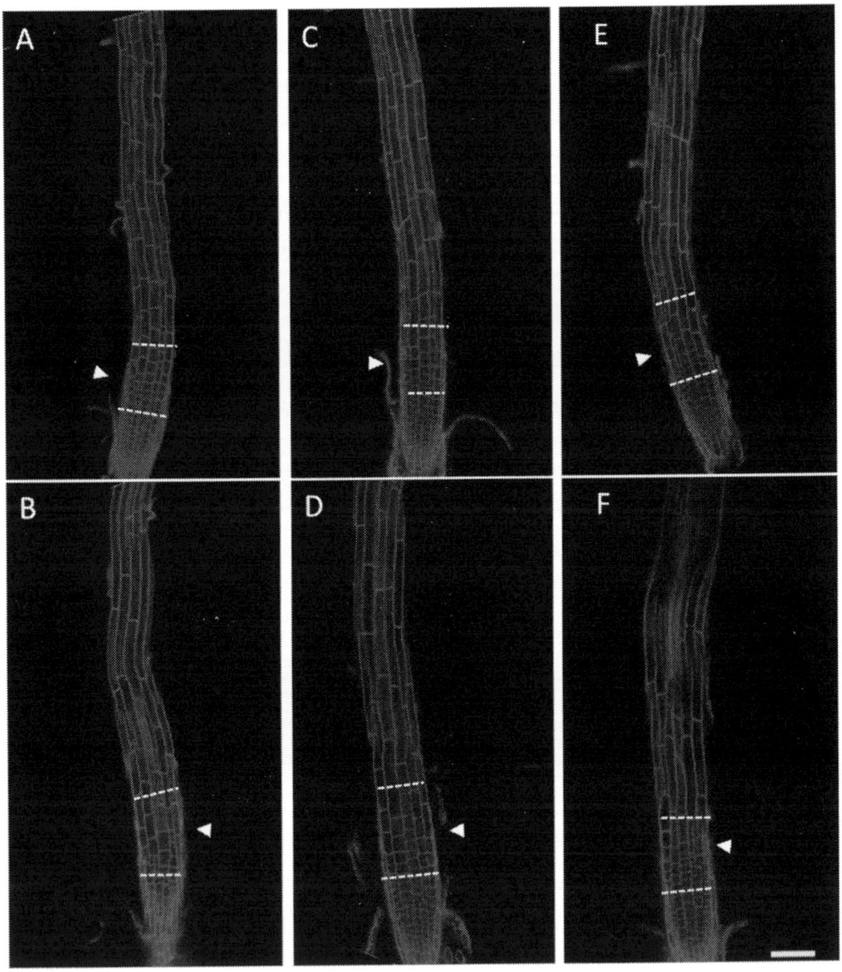


**Figure 3:** Root cell expansion in the *nhx1nhx2nhx3nhx4* root transition zone is reduced. WT (A,C,E) or *nhx1nhx2nhx3nhx4* (B,D,F) six-day-old dark-grown seedlings were stained with 10 μg·mL$^{-1}$propidium iodide. Seedlings were grown on control media (1 mM $K^+$, 1 mM $Na^+$) (A,B) or on media supplemented with 30 mM NaCl (C,D) or 30 mM KCl (E,F). The transition zone is bounded by the dashed lines. Arrowheads point to cells that show differences in cell size and are discussed in the Results. Bar is 50 μm.

### $K^+$ *Dependent Root Skewing Depends on the Dynamic Organization of the Cytoskeleton*

In order to investigate whether the root growth response of *nhx1nhx2nhx3nhx4* to elevated $K^+$concentrations was associated with dynamic changes in the organization of the cytoskeleton, we tested the effect of microtubule perturbing drugs on root morphology and directional root growth in etiolated *nhx1nhx2nhx3nhx4* seedlings as above. Propyzamide is a microtubule destabilizing drug that is known to induce right-handed root skewing (when viewed through the agar) without inhibiting overall growth and root elongation [12,20,21].

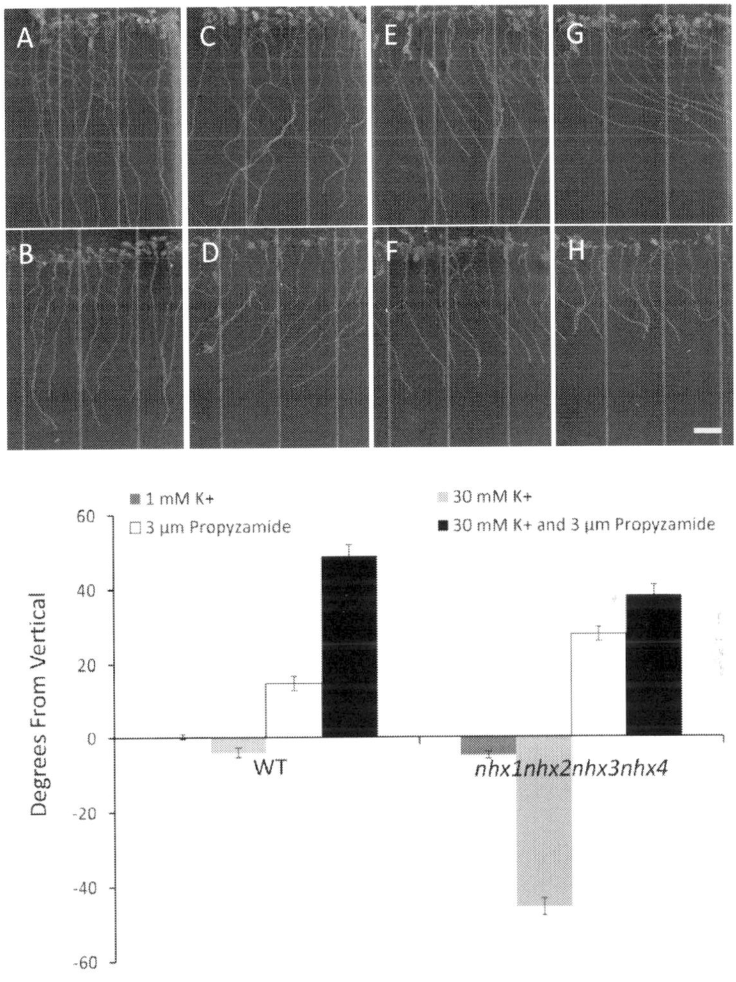


**Figure 4:** Propyzamide alters potassium-induced root skewing in the *nhx1nhx2nhx3nhx4* knockout. Wild-type (A,C,E,G) and *nhx1nhx2nhx3nhx4* (B,D,F,H) seedlings were germinated on minimal media and transferred to plates with either 30 mM $K^+$ (C,D), 3 µM propyzamide (E,F) or both 30 mM $K^+$ and propyzamide (G,H) when

four day-old and grown for an additional seven days. Root skewing is quantified as the angle of root growth in degrees from vertical (I). Values represent the mean ± SE ($n = 28$). The bar is 0.5 cm.

When applied at 3 μM, anisotropic root growth of wild-type *Arabidopsis*is moderately impaired without causing swelling of epidermal roots cells [22]. We measured root growth angles for both WT and *nhx1nhx2nhx3nhx4* vertically-grown seedlings on minimal media (containing 1 mM KCl) and an identical media supplemented with an additional 30 mM KCl. As shown in Figure 4, both the WT and *nhx1nhx2nhx3nhx4* displayed the characteristic right-handed skewing (Figure 4E,F) when grown in the presence of 3 μM propyzamide, but only *nhx1nhx2nhx3nhx4* displayed left-handed skewing in response to 30 mM $K^+$ (Figure 4D). When both elevated $K^+$ and 3 μM propyzamide were included in the media, roots from WT seedlings continued to skew towards the right, unlike the roots from *nhx1nhx2nhx3nhx4*, which displayed less pronounced skewing (Figure 4G,H).

Wang and colleagues demonstrated that WT and *sos1*(*nhx7*) seedlings exhibited left-handed root skewing when grown vertically on NaCl-enriched MS media, similar to the root skewing of *nhx1nhx2nhx3nhx4* in the presence of high KCl. Their results indicated that 50 mM $Na^+$ was sufficient to initiate root skewing that eventually resulting in strong root curling and the formation of root "loops". Interestingly, $Na^+$-dependent root skewing was exacerbated in *sos1*(*nhx7*) mutants, which displayed more prominent root skewing than the WT at lower $Na^+$ treatments, suggesting that cytosolic ion accumulation played an important role in directional root growth. NaCl-induced salt stress resulted in seedling death, which occurred more frequently in *sos1*(*nhx7*) than WT seedlings, but was partially prevented by propyzamide treatment, suggesting that microtubule reorganization was required for salt tolerance [18].

We also tested the effect of Taxol, a microtubule stabilizing drug, on $K^+$-induced root skewing in *nhx1nhx2nhx3nhx4* plants. A characteristic effect of Taxol on roots is a right-handed skewing response that is dose dependent [23]. We observed that 0.5 μM Taxol was sufficient to induce right-handed root skewing in the WT (Figure 5E), but the same concentration did not significantly change the root growth direction of *nhx1nhx2nhx3nhx4* plants (Figure 5F). It is unlikely that the lack of *nhx1nhx2nhx3nhx4* root skewing observed in the 0.5 μM Taxol treatment was caused by a lack of overall root growth, because *nhx1nhx2nhx3nhx4* roots did not vary significantly in length between 0.5 μM or 1.0 μM Taxol treatments. Indeed, the addition of 30 mM KCl was sufficient to induce left-handed skewing in *nhx1nhx2nhx3nhx4*, despite the presence of 0.5 μM Taxol (Figure 5J). In contrast, the added KCl had no effect on WT root growth, with seedlings continuing to skew to the right (Figure 5I). We observed that a threshold of 1.0 μM Taxol was sufficient to overcome the $K^+$-induced left-handed skewing in *nhx1nhx2nhx3nhx4* plants and to cause right-handed skewing, as shown (Figure 5L). Previously, it was shown that 0.5 μM Taxol counters the effect of $Na^+$-induced root skewing, which is dose dependent [18]; a result that is strikingly similar to what we observed in Taxol- and $K^+$-treated *nhx1nhx2nhx3nhx4* (Figure 5).

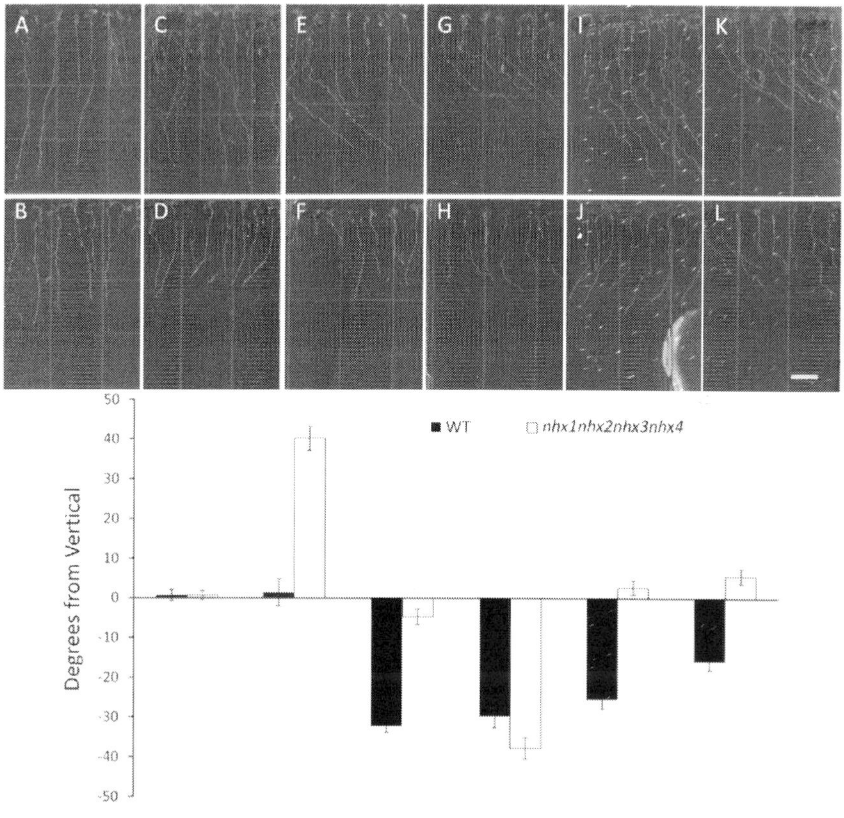


**Figure 5:** Taxol reverses directional root growth in *nhx1nhx2nhx3nhx4* grown on potassium-supplemented media. Wild-type (top) and *nhx1nhx2nhx3nhx4* (bottom) seedlings were germinated on minimal media containing 1 mM $K^+$

for three days and transferred to plates containing two concentrations of Taxol and $K^+$ (indicated below) and grown for an additional four days. (A,B) 1 mM $K^+$; (C,D) 30 mM $K^+$; (E,F) 0.5 μM Taxol and 1 mM $K^+$; (G,H) 1 μM Taxol and 1 mM $K^+$; (I,J) 0.5 μM Taxol and 30 mM $K^+$; (K,L) 1 μM Taxol and 30 mM $K^+$. (M) The angle of root growth was quantified as the degrees from vertical; values represent the mean ± SE ($n$ = 24). The bar is 0.5 cm.

To better understand the possible effect of high $K^+$ on cytoskeletal organization, we used a whole mount immunostaining approach to visualize the organization of cortical microtubules in cells of the root elongation zone, where pronounced skewing and curling occurred.

As depicted in Figure 6, no differences were observed between the microtubule organization of roots from *nhx1nhx2nhx3nhx4* and the WT when grown on media containing 1 mM $K^+$ (Figure 6A,B). By contrast, *nhx1nhx2nhx3nhx4* seedlings grown on media containing 30 mM $K^+$ displayed a helical orientation of the cortical microtubule array in elongating *nhx1nhx2nhx3nhx4* root cells (Figure 6D) that was observed only when roots exhibited left-handed skewing. Both WT and *nhx1nhx2nhx3nhx4*displayed the characteristic Taxol-induced "bundling" of microtubules [24,25] with a highly parallel microtubule organization (Figure 6E,F). However, when seedlings were grown on media with both 0.5 μM Taxol and 30 mM KCl, the *nhx1nhx2nhx3nhx4* microtubules displayed a similar spiraling pattern that was seen when grown on 30 mM KCl plates (Figure 6H,D). Higher $K^+$ had no observable effect on microtubule organization in the WT, even in the presence of Taxol (Figure 6C,G).

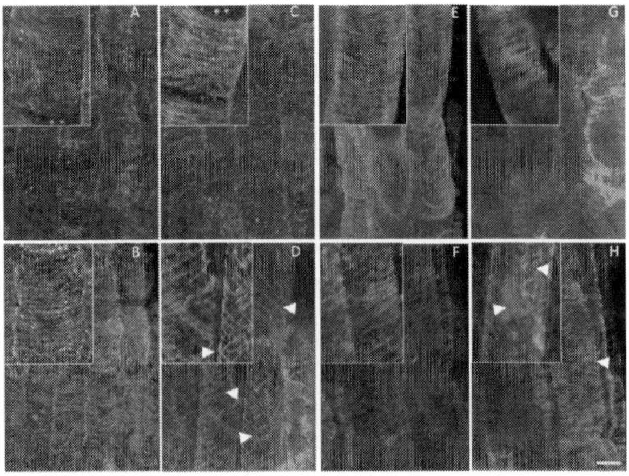

**Figure 6:** Potassium induces helical reorganization of cortical microtubules of *nhx1nhx2nhx3nhx4* epidermal cells in the root elongation zone. Immunolabeled microtubules in WT (top) and *nhx1nhx2nhx3nhx4* (bottom). Seedlings were germinated on minimal media containing 1 mM $K^+$ for three days, then transferred to plates with two concentrations of Taxol and grown for an additional five days before fixation. Normal microtubule orientation is observed in WT (A) and *nhx1nhx2nhx3nhx4* (B) when grown on media containing 1 mM $K^+$. WT seedlings grown on 30 mM $K^+$ displayed normally-orientated microtubules (C), whereas 30 mM $K^+$ induced a helical orientation of the cortical microtubules in *nhx1nhx2nhx3nhx4*(D). In the presence of 1 mM $K^+$ and 0.5 μM Taxol, both WT (E) and *nhx1nhx2nhx3nhx4* (F) display characteristic bundling of cortical microtubules. Whereas 30 mM $K^+$ had no effect on microtubule orientation in WT grown with 0.5 μM Taxol (G), *nhx1nhx2nhx3nhx4*seedlings grown with 30 mM $K^+$ and 0.5 μM Taxol (H) exhibited a helical microtubule organization similar to that seen when grown with just 30 mM $K^+$. Arrows point to helically-oriented microtubules. The bar is 10 μM. The upper left insets are enlarged 50% to highlight microtubule orientation.

In particular, cortical microtubules depolymerize and reorganize in a $Ca^{2+}$-dependent manner under salt stress; preventing this response with microtubule stabilizing drugs, such as Taxol, results in seedling death [12,13], suggesting that cytoskeletal reorganization of root cells is important during salt stress. In addition, mutations of the $Na^+/H^+$ antiporter SOS1 and associated SOS2 kinase were shown to suppress the cortical microtubule disruptions and subsequent helical growth associated with the *Arabidopsis spiral1* mutants. This study also demonstrated that even in the absence of salt stress, loss of function SOS mutants resulted in altered cortical microtubule arrays when seedlings were treated with the same dose of propyzamide as used in the present study [12].

# EXPERIMENTAL SECTION

## Plant Materials and Growth Conditions

T-DNA insertions lines for *NHX1* (At5g27150), *NHX2* (At3g05030), *NHX3* (At5g55470) and *NHX4*(At3g06370) were obtained and were *nhx1-1* [SALK_034001], *nhx2-1* [SALK_036114], *nhx3-1* [WISC_345-348F19] and *nhx4-1* [SALK_112901] [26]. All single knockouts were backcrossed twice prior to crosses for generating higher order knockouts. Subsequent crosses generated double knockouts and two triple knockouts, which were then crossed to generate a tetra *nhx* knockout. RT-PCR confirmed no DNA amplification with allele-specific primers from leaf-derived *nhx1nhx2nhx3nhx4* cDNA. Positions of T-DNA insertion sites and RT-PCR results are shown in Figure A1.

*Arabidopsis thaliana* (ecotype Columbia [Col-0]) and *nhx1nhx2nhx3nhx4* seedlings were grown at 22 °C under diurnal light conditions (8 h light and 16 h dark), unless otherwise specified. Seeds were germinated on vertically-positioned plates containing a modified Murashige and Skoog media [27] (0.5% sucrose, 1 mM $Na^+$, 1 mM $K^+$, 0.8% phytagel, pH 5.7) and supplemented as specified for each experiment. For experiments testing stress responses, this media was considered as a control and was modified by the addition of 30 mM NaCl or KCl or 60 mM mannitol. For drug treatment assays, media was supplemented with 3 μM propyzamide (Sigma cat.#45645) or either 0.5 μM or 1 μM Taxol (Sigma cat.#T7191).

## Cloning and Expression of Translational Fusion Proteins

*NHX3* and *NHX4* translational fusions were constructed using the Gateway cloning system (Invitrogen, Carlsbad, CA, USA). cDNAs without stop codons were cloned into pDONR207 to generate entry vectors (pDONR207-NHX2, pDONR207-NHX3, pDONR207-NHX4) and recombined into pEarley-Gate101 for YFP fusion and pEarleyGate102 for CFP fusion [28]. For the 35S-NHX4-RFP construct, pDONR207-NHX4 was recombined with pH7RWG2 [29]. All constructs were introduced to *A. tumefasciens* strain GV3101 for both stable and transient transformation of Col-0, VAMP711-RFP [30] and γ-TIP-GFP.

## Transient Transformation of Arabidopsis Seedlings

Transient transformation of *Arabidopsis* was adopted from Li, *et al.* [31] using 4–5-day-old seedlings grown in a sterile 24-well plate. Seedlings were co-cultivated with *Agrobacterium tumefasciens* (strain GV3101) for 18–24 h before observation. Co-transformation using two constructs (Figure 1I,L) was accomplished by co-cultivating seedlings with two bacterial cultures, each diluted to one half of the final optical density used when transforming a single construct (OD 0.5).

## Hypocotyl and Root Length Measurements

Seedlings grown on agar plates were scanned on a flatbed scanner as color TIFF files with 300-dpi resolution. Root and hypocotyl lengths were quantified using the ImageJ [32] plugin SmartRoot [33]. Thc anglcs of root growth were measured with the same software suite using gridlines etched on the rear of the petri dishes as horizontal and vertical reference points (see Figure A4).

## Quantification of Vacuolar $K^+$

Vacuolar $K^+$ was quantified using the $K^+$-specific dye PBFI-AM, as described previously [3].

## Propidium Iodide Staining

Dark-grown, six-day-old WT and *nhx1nhx2nhx3nhx4* seedlings were grown as described above. Seedlings were stained with a 10-μg·mL$^{-1}$ solution of propidium iodide (Sigma cat.#81845).

## Whole Mount Immunostaining of Arabidopsis Seedlings

Whole mount immunostaining of root cortical microtubules was modified from the protocol of Hauser *et al.* [34]. Dark-grown seedlings grown for five days after germination were fixed on agar plates in a solution of 4% p-formaldehyde, 5% DMSO, 0.025% glutaraldehyde in 1× microtubule stabilizing buffer (MTSB; 50 mM PIPES, 5 mM EGTA, 5 mM $MGSO_4$, pH 7) vacuum infiltrated at room temperature for 1 h. Seedlings were then washed three times in wash buffer (MTSB with 5% DMSO) three times for ten minutes each, and the solution was exchanged for a cell wall digestion buffer (MTSB with 1% cellulase, 0.1% pectolyase) and incubated at room temperature for 30 min. Following digestion, three washes were performed with MTSB for five minutes each, and the seedlings were incubated for ten minutes in a solution of MTSB with 1 mg/mL $NaBH_4$ for ten minutes. After three more washes with MTSB for five minutes, each seedling was incubated for ten minutes in a solution of MTSB containing 10 mM glycine and then washed once more with MTSB three times for a total of 30 min. Severed roots were transferred to microscope slides (Thermo-Scientific Superfrost plus, cat.#4951PLUS-001) and incubated in a membrane permeabilization solution (MTSB with 1% Triton X-100) at room temperature for 15 min and then in methanol at −20 °C for 20 min.

Immunostaining was carried out by an incubation with DM1A primary antibody (Sigma cat.#T9026) diluted 1:400 in a blocking solution (PBS containing 3% BSA, 0.05% Tween-20, and 0.1% thimerosal) for one hour at 37 °C. After washing, roots were incubated with FITC goat anti-mouse (Sigma cat. #F027) for one hour at 37 °C.

### Confocal Microscopy

Fluorescence microscopy was performed using a confocal laser scanning microscope (Zeiss LSM 710) equipped with a 40× water immersion objective. The excitation wavelength was 488 nm, with emission wavelengths of 500–535 nm for GFP or FITC, 525–615 nm for YFP/Venus, 450–500 nm for CFP and 565–605 nm for RFP. For imaging of GFP/RFP and CFP/YFP double reporter lines, sequential scanning on multiple channels was used to avoid crosstalk between fluorescence emission spectra.

## CONCLUSIONS

We previously reported that NHX1 and NHX2 colocalize to the vacuole and are essential for vacuolar $K^+$ accumulation and the regulation of pH [3]. Here, we demonstrate that both NHX3 and NHX4 are also localized to the tonoplast and that vacuolar cation/$H^+$ exchange by all four vacuolar NHXs are essential for normal growth and development. Compared to both WT and the *nhx1nhx2* double knockout, the quadruple knockout *nhx1nhx2nhx3nhx4* has significantly reduced vacuolar $K^+$ and growth, as well as high sensitivity to added $K^+$. *nhx1nhx2nhx3nhx4* also exhibited pronounced root skewing and curling responses not observed under equimolar $Na^+$ that were correlated with an increase in cytoskeletal dynamics and a reduction in the size of the root cells of the transition zone.

## ACKNOWLEDGMENTS

We thank Bo Liu and Yuh-Ru Julie Lee (Department of Plant Biology, University of California, Davis) and Aviah Zilberstein, (Department of Molecular Biology and Ecology of plant, Tel Aviv University) for technical assistance and helpful discussions. This work was supported in part by grants from the National Science Foundation (MCB-0343279; IOS-0820112) and the Will W. Lester Endowment, University of California, to Eduardo Blumwald.

## REFERENCES


1. Bassil, E.; Coku, A.; Blumwald, E. Cellular ion homeostasis: Emerging roles of intracellular nhx $Na^+/H^+$ antiporters in plant growth and development. *J. Exp. Bot.* 2012, *63*, 5727–5740.
2. Reguera, M.; Bassil, E.; Blumwald, E. Intracellular NHX-type cation/$H^+$ antiporters in plants. *Mol. Plant* 2013. [Google Scholar] [CrossRef]
3. Bassil, E.; Tajima, H.; Liang, Y.-C.; Ohto, M.-A.; Ushijima, K.; Nakano, R.; Esumi, T.; Coku, A.; Belmonte, M.; Blumwald, E. The arabidopsis $Na^+/H^+$ antiporters NHX1 and NHX2 control vacuolar pH and $K^+$ homeostasis to regulate growth, flower development, and reproduction. *Plant Cell* 2011, *23*, 3482–3497.
4. Apse, M.P.; Sottosanto, J.B.; Blumwald, E. Vacuolar cation/$H^+$ exchange, ion homeostasis, and leaf development are altered in a T-DNA insertional mutant of atnhx1, the *Arabidopsis* vacuolar $Na^+/H^+$ antiporter. *Plant J.* 2003, *36*, 229–239.
5. Liu, H.; Tang, R.; Zhang, Y.; Wang, C.; Lv, Q.; Gao, X.; Li, W.; Zhang, H. AtNHX3 is a vacuolar $K^+/H^+$ antiporter required for low-potassium tolerance in *Arabidopsis thaliana. Plant Cell Environ.*2010, *33*, 1989–1999.
6. Liu, H.; Wang, Q.; Yu, M.; Zhang, Y.; Wu, Y.; Zhang, H. Transgenic salt-tolerant sugar beet (*Beta vulgaris* L.) constitutively expressing an arabidopsis thaliana vacuolar $Na^+/H^+$ antiporter gene, *AtNHX3*, accumulates more soluble sugar but less salt in storage roots. *Plant Cell Environ.* 2008, *31*, 1325–1334. [Google Scholar]
7. Li, H.T.; Liu, H.; Gao, X.S.; Zhang, H.X. Knock-out of arabidopsis *AtNHX4* gene enhances tolerance to salt stress. *Biochem. Biophys. Res. Commun.* 2009, *382*, 637–641.
8. Maathuis, F.J.M.; Podar, D. Uptake, distribution, and physiological functions of potassium, calcium, and magnesium. In *The Molecular*

*and Physiological Basis of Nutrient Use Efficiency in Crops*; Wiley-Blackwell: Hoboken, NJ, USA, 2011; pp. 265–293. [Google Scholar]

9. Barragán, V.; Leidi, E.O.; Andrés, Z.; Rubio, L.; de Luca, A.; Fernández, J.A.; Cubero, B.; Pardo, J.M. Ion exchangers NHX1and NHX2 mediate active potassium uptake into vacuoles to regulate cell turgor and stomatal function in arabidopsis. *Plant Cell* 2012, *24*, 1127–1142. [Google Scholar]
10. Andrés, Z.; Pérez-Hormaeche, J.; Leidi, E.O.; Schlücking, K.; Steinhorst, L.; McLachlan, D.H.; Schumacher, K.; Hetherington, A.M.; Kudla, J.; Cubero, B.; *et al.* Control of vacuolar dynamics and regulation of stomatal aperture by tonoplast potassium uptake. *Proc. Natl. Acad. Sci. USA*2014. [Google Scholar] [CrossRef]
11. Bibikova, T.N.; Blancaflor, E.B.; Gilroy, S. Microtubules regulate tip growth and orientation in root hairs of *Arabidopsis thaliana. Plant J.* 1999, *17*, 657–665.
12. Shoji, T.; Suzuki, K.; Abe, T.; Kaneko, Y.; Shi, H.; Zhu, J.K.; Rus, A.; Hasegawa, P.M.; Hashimoto, T. Salt stress affects cortical microtubule organization and helical growth in arabidopsis. *Plant Cell Physiol.* 2006, *47*, 1158–1168.
13. Wang, S.; Kurepa, J.; Hashimoto, T.; Smalle, J.A. Salt stress-induced disassembly of *Arabidopsis*cortical microtubule arrays involves 26S proteasome-dependent degradation of spiral1. *Plant Cell*2011, *23*, 3412–3427.
14. Saito, C.; Ueda, T.; Abe, H.; Wada, Y.; Kuroiwa, T.; Hisada, A.; Furuya, M.; Nakano, A. A complex and mobile structure forms a distinct subregion within the continuous vacuolar membrane in young cotyledons of arabidopsis. *Plant J.* 2002, *29*, 245–255.
15. Uemura, T.; Ueda, T.; Ohniwa, R.L.; Nakano, A.; Takeyasu, K.; Sato, M.H. Systematic analysis of snare molecules in arabidopsis: Dissection of the post-golgi network in plant cells. *Cell Struct. Funct.* 2004, *29*, 49–65.
16. Rutherford, R.; Masson, P.H. *Arabidopsis thaliana* sku mutant seedlings show exaggerated surface-dependent alteration in root growth vector. *Plant Physiol.* 1996, *111*, 987–998.
17. Sun, F.; Zhang, W.; Hu, H.; Li, B.; Wang, Y.; Zhao, Y.; Li, K.; Liu, M.;

Li, X. Salt modulates gravity signaling pathway to regulate growth direction of primary roots in arabidopsis. *Plant Physiol.*2008, *146*, 178–188.

18. Wang, C.; Li, J.; Yuan, M. Salt tolerance requires cortical microtubule reorganization in *Arabidopsis*. *Plant Cell Physiol.* 2007, *48*, 1534–1547.
19. Blancaflor, E.B.; Hasenstein, K.H. Growth and microtubule orientation of *Zea mays* roots subjected to osmotic stress. *Int. J. Plant Sci.* 1995, *156*, 774–783.
20. Yuen, C.Y.; Sedbrook, J.C.; Perrin, R.M.; Carroll, K.L.; Masson, P.H. Loss-of-function mutations of root hair defective3 suppress root waving, skewing, and epidermal cell file rotation in arabidopsis. *Plant Physiol.* 2005, *138*, 701–714.
21. Nakamura, M.; Naoi, K.; Shoji, T.; Hashimoto, T. Low concentrations of propyzamide and oryzalin alter microtubule dynamics in *Arabidopsis* epidermal cells. *Plant Cell Physiol.* 2004, *45*, 1330–1334.
22. Naoi, K.; Hashimoto, T. A semidominant mutation in an arabidopsis mitogen-activated protein kinase phosphatase-like gene compromises cortical microtubule organization. *Plant Cell* 2004, *16*, 1841–1853.
23. Sedbrook, J.C.; Ehrhardt, D.W.; Fisher, S.E.; Scheible, W.R.; Somerville, C.R. The *Arabidopsis*SKU6/SPIRAL1 gene encodes a plus end-localized microtubule-interacting protein involved in directional cell expansion. *Plant Cell* 2004, *16*, 1506–1520.
24. Takahashi, H.; Hirota, K.; Kawahara, A.; Hayakawa, E.; Inoue, Y. Randomization of cortical microtubules in root epidermal cells induces root hair initiation in lettuce (*Lactuca sativa* L.) seedlings. *Plant Cell Physiol.* 2003, *44*, 350–359. [Google Scholar]
25. Baskin, T.I.; Wilson, J.E.; Cork, A.; Williamson, R.E. Morphology and microtubule organization in arabidopsis roots exposed to oryzalin or Taxol. *Plant Cell Physiol.* 1994, *35*, 935–942.
26. Alonso, J.M.; Stepanova, A.N.; Leisse, T.J.; Kim, C.J.; Chen, H.; Shinn, P.; Stevenson, D.K.; Zimmerman, J.; Barajas, P.; Cheuk, R.; *et al.* Genome-wide insertional mutagenesis of *Arabidopsis thaliana*. *Science* 2003, *301*, 653–657.

27. Spalding, E.P.; Hirsch, R.E.; Lewis, D.R.; Qi, Z.; Sussman, M.R.; Lewis, B.D. Potassium uptake supporting plant growth in the absence of akt1 channel activity: Inhibition by ammonium and stimulation by sodium. *J. Gen. Physiol.* 1999, *113*, 909–918.
28. Earley, K.W.; Haag, J.R.; Pontes, O.; Opper, K.; Juehne, T.; Song, K.; Pikaard, C.S. Gateway-compatible vectors for plant functional genomics and proteomics. *Plant J.* 2006, *45*, 616–629.
29. Karimi, M.; Bleys, A.; Vanderhaeghen, R.; Hilson, P. Building blocks for plant gene assembly. *Plant Physiol.* 2007, *145*, 1183–1191.
30. Geldner, N.; Devervaud-Tendon, V.; Hyman, D.L.; Mayer, U.; Stierhof, Y-D.; Chory, J. Rapid, combinatorial analysis of membrane compartments in intact plants with a multicolor marker set. *Plant J.* 2009, *59*, 169–178.
31. Li, J.F.; Park, E.; von Arnim, A.G.; Nebenfuhr, A. The fast technique: A simplified agrobacterium-based transformation method for transient gene expression analysis in seedlings of arabidopsis and other plant species. *Plant Methods* 2009, *5*, 6.
32. Schneider, C.A.; Rasband, W.S.; Eliceiri, K.W. Nih image to imagej: 25 years of image analysis. *Nat. Methods* 2012, *9*, 671–675.
33. Lobet, G.; Pagès, L.; Draye, X. A novel image-analysis toolbox enabling quantitative analysis of root system architecture. *Plant Physiol.* 2011, *157*, 29–39.
34. Hauser, M.T.; Morikami, A.; Benfey, P.N. Conditional root expansion mutants of Arabidopsis. *Development* 1995, *121*, 1237–1252.

CHAPTER

# 14

# IDENTIFICATION AND CHARACTERIZATION OF ORTHOLOGS OF ATNHX5 AND ATNHX6 IN BRASSICA NAPUS

Brett A. Ford, Joanne R. Ernest and Anthony R. Gendall*

Department of Botany, La Trobe University, Melbourne, VIC, Australia

Improving crop species by breeding for salt tolerance or introducing salt tolerant traits is one method of increasing crop yields in saline affected areas. Extensive studies of the model plant species *Arabidopsis thaliana* has led to the availability of substantial information regarding the function and importance of many genes involved in salt tolerance. However, the identification and characterization of *A. thaliana* orthologs in species such as *Brassica napus* (oilseed rape) can prove difficult due to the significant genomic changes that have occurred since their divergence approximately 20 million years ago (MYA). The recently released *Brassica rapa*genome provides an excellent resource for comparative studies of *A. thaliana* and the cultivated *Brassica* species, and facilitates the identification of *Brassica* species orthologs which may be of agronomic importance. Sodium hydrogen antiporter (NHX) proteins transport a sodium or potassium ion in exchange for a hydrogen ion in the other direction across a membrane. In *A. thaliana* there are eight members of the NHX family, designated *AtNHX1-8*, that can be sub-divided into three clades, based on their subcellular localization: plasma membrane (PM), intracellular class I (IC-I) and intracellular class II (IC-II). In plants, many NHX proteins are primary determinants of salt tolerance and act by transporting $Na^+$ out of the cytosol where it would otherwise accumulate to toxic levels. Significant work has been done to determine the role of both PM and IC-I clade members in salt tolerance in a variety of plant species, but relatively little analysis has been described for the IC-II clade. Here we describe the identification of *B. napus* orthologs of *AtNHX5* and *AtNHX6*, using the *B. rapa* genome sequence, macro- and micro-synteny analysis, comparative expression and promoter motif analysis, and highlight the value of these multiple approaches for identifying true orthologs in closely related species with multiple paralogs.

## INTRODUCTION

It is estimated that more than 800 million hectares of land worldwide and approximately 20% of irrigated farmland are negatively impacted by salinity (FAO, 2009). While there are many different salts that contribute to the salinization of a landscape, by far the most abundant and damaging is NaCl (Tester and Davenport, 2003). NaCl in the soil inhibits plant growth by causing an initial osmotic stress, followed by the accumulation of $Na^+$ ions in the plant to toxic levels (Munns and Tester, 2008). Improved crop yield under saline conditions can be achieved by identifying genes which confer salt tolerance, and introducing them to crop species by traditional breeding or transgenesis.

In plants, members of the monovalent cation/proton antiporter (CPA1) gene family exchange a sodium, potassium or lithium ion for a hydrogen ion across a cellular membrane. They can be classified into two distinct sub families, the plasma membrane (PM) localized NHAP family and the intracellular localized (IC) NHX family (Brett et al., 2005a; Chanroj et al., 2012). These proteins are important in maintaining pH and ion homeostasis, and are conserved across all phyla and kingdoms (Brett et al., 2005a). In *Arabidopsis thaliana*, there are two members of the NHAP family (SOS1/AtNHX7 and AtNHX8) and six members of the IC-NHX family, designated AtNHX1-6 (Brett et al., 2005a; Chanroj et al., 2012). The IC-NHX family can be further sub-divided into two clades based on their cellular localization: the IC-I clade (*AtNHX1-4*), localized to the tonoplast; or the IC-II clade (*AtNHX5-6*), localized to endosomal compartments (Brett et al., 2005a; Bassil et al., 2011). In plants, NHX proteins are important determinants of $Na^+$ tolerance, detoxifying cytosolic $Na^+$ by transportation out of the cell to the apoplastic space, or by sequestration into subcellular compartments (Apse et al., 1999; Shi et al., 2003; Rodriguez-Rosales et al., 2008). Members of the PM (*AtNHX7/SOS1*) and IC-I (*AtNHX1*) clades have been shown to be upregulated in response to NaCl, and to confer salt tolerance upon over-expression in *A. thaliana* (Apse et al., 1999; Shi et al., 2002, 2003; Shi and Zhu, 2002). Of the IC-II clade, *AtNHX5* and the rice ortholog *OsNHX5* are both up regulated in response to NaCl (Yokoi et al., 2002; Fukuda et al., 2011) and over expressing *AtNHX5* in a variety of plant species including rice (Li et al., 2011) and tomato (Rodriguez-Rosales et al., 2008) has been shown to increase their tolerance to NaCl. Recent work has shown that *AtNHX5* and *AtNHX6* are functionally redundant, are important for normal plant growth and development, are endosome associated and have an important role in protein trafficking to the vacuole (Bassil et al., 2011). A similar role in protein trafficking has been demonstrated for the yeast ortholog ScNHX1 (Bowers et al., 2000; Brett et al., 2005b). Interestingly, comparatively little work has been undertaken investigating the role of the Class II NHX proteins from important crop species in NaCl tolerance.

*Brassica napus* is a member of the Brassicaceae family and is an important oil crop species. *B. napus*cultivation has increased significantly over recent years and it is now the second largest crop for the production of oil seeds and oil meals, and the third largest crop used in the production of vegetable oil (USDA, 2009). As both *A. thaliana* and *B. napus* are members of the family Brassicaceae and therefore relatively closely related, a comparative study could be undertaken to try to identify the homologs of *AtNHX5* and *AtNHX6* to assess their importance in salt tolerance. However, there have been extensive large- and small-scale genome changes in the *A. thaliana* and *B. napus* genomes since their divergence approximately 13–17 million years ago (MYA) (Figure 1). *B. napus* is an allopolyploid species (AACC, 2*n* = 38) arising from the hybridization of *B. rapa* (AA, 2*n* = 20) and *Brassica oleracea* (CC, 2*n* = 18) (Figure 1). Additionally, all *Brassica* species appear to have undergone a whole genome triplication (WGT, Br—α) event since they diverged from *A. thaliana* (Figure 1). The *A. thaliana* genome has undergone extensive duplications, deletions, re-arrangements and a reduction in chromosome number even since its divergence from its close relative *Arabidopsis lyrata* only 5 MYA (Figure 1) (Hu et al., 2011). These confounding factors make direct comparative studies and homolog identification in *B. napus* based solely on *A. thaliana* genomic information problematic.

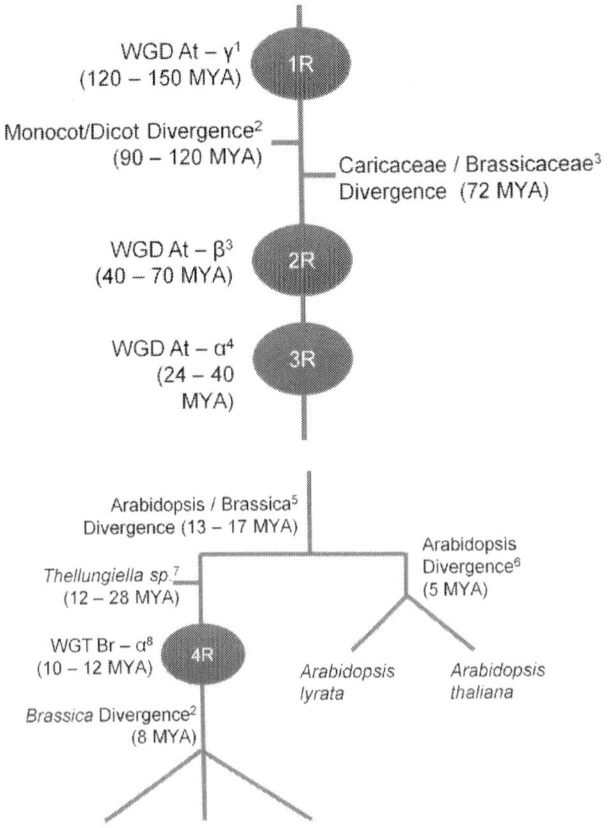

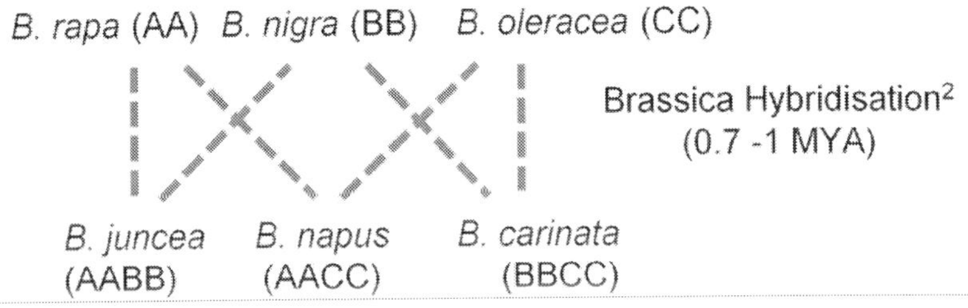


**Figure 1**: Genome evolution of *B. napus* **and** *A. thaliana.* Prior to their divergence approximately 13–17 MYA *B. napus* and *A. thaliana* share a complex genome evolution including three (denoted 1R, 2R, and 3R) whole genome duplication (WGD) events. The last two WGD events post date the divergence of the Moncots/Dicots and the Caricaceae/Brassicaceae. Since their divergence the ancestor of extant Brassica species underwent a whole genome triplication event (WGT, denoted 4R) with the allopolyploid *B. napus* arising from the hybridization of *B. rapa* and *B. oleracea*. Superscript numbers relate to references for divergence times. For an in depth analysis of evolutionary history of Brassicaceae please refer to the excellent review by Franzke et al. (2011). [1](Soltis et al., 2009); [2](Mun et al., 2009); [3](Bell et al., 2010); [4](Henry et al., 2006); [5](Yang et al., 1999); [6](Koch et al., 2000); [7](Couvreur et al., 2010); [8](Lysak et al., 2005).

To overcome this problem, Schranz et al. (2006), building on the work of Parkin et al. (2005), have proposed that the comparative studies of *A. thaliana* and the *Brassica* species should be based upon an ancient karyotype of $n = 8$, that is observed in at least 37% of the Brassicaceae species. The genome of this ancient karyotype can be further segmented into 24 conserved syntenic blocks (designated A–X) that can be identified in both the *A. thaliana* genome and the *B. rapa* (A genome) component of *B. napus* (Parkin et al., 2005; Schranz et al., 2006).

While this approach greatly aided comparative studies between *A. thaliana* and *B. napus*, the lack of high quality *Brassica* sequence still inhibited homolog identification. The recent release of the fully sequenced *B. rapa* genome (Wang et al., 2011) has enabled in-depth comparative studies between *A. thaliana* and *B. napus*.

Analysis of the *B. rapa* genome reveals that there has been significant gene loss since the most recent WGT (Wang et al., 2011). The triplicated *B. rapa* genome can be divided into three sub genomes based on their level of gene loss relative to *A. thaliana*, which have been designated the least fractionated genome (LF) with 30% gene loss, the medium fractionated genome (MF1) with 54% gene loss and the most fractionated genome (MF2) with 64% gene loss (Wang et al., 2011). However, analysis of the *B. rapa* genome shows that genes associated with response to environmental factors such as salt were over-retained (Wang et al., 2011) and as a result it is possible that there are three homologs of *AtNHX5* and three homologs of *AtNHX6* in *B. rapa*. As *B. napus* has both a *B. rapa* triplicated genome, and a triplicated *B. oleracea* C genome component there may be up to six *AtNHX5* and six *AtNHX6* homologs. Here we describe the use of the *B. rapa* genomic sequence to aid in the clear identification of the *B. napus*homologs of *AtNHX5* and *AtNHX6* as a first step in evaluating their potential to improve salt tolerance.

## MATERIALS AND METHODS

### Identification and Reannotation of *B. rapa NHX* Genes

The initial identification of the complete set of *NHX* genes in *B. rapa* used representative *A. thaliana* NHX protein sequences (See Table 1 for accession numbers) as query sequences in TBLASTN searches of the *B. rapa* genomic database in Phytozome version 8.0 (http://www.phytozome.org; Goodstein et al., 2012). Initial searches were performed using AtNHX1, AtNHX6, and AtSOS1/AtNHX7 amino acid sequences. Hits with an *E*-value of less than $10^{-45}$ were classified as putative NHXs and examined further. Preliminary alignments of *A. thaliana* NHX proteins to the annotated *B. rapa* proteins revealed some minor sequence dissimilarities in annotations that often corresponded to the lack of one or more exons in the *B. rapa* gene models, or from fusion of two adjacent genes into a single mis-annotated *BrNHX* gene. To refine these predictions, two approaches were used. Expressed sequence tag (EST) or cDNA sequences were identified by using the NHX amino acid sequences as the query for a TBLASTN search of all EST databases restricting the search to the taxid *Brassica*. The EST or cDNA sequences were then assigned to the relevant *NHX* gene based on the regions of maximum match and sequence alignment to the corresponding *B. rapa* genome. These sequences were aligned with the corresponding genomic region using Spidey (http://www.ncbi.nlm.nih.

gov/IEB/Research/Ostell/Spidey/index.html), using the "Plant" genomic sequence option and other default options. The identified exons were then incorporated into a final re-annotated open reading frame model. If no EST or cDNA support existed, amino acid sequences from the *A. thaliana* or other *B. rapa* ortholog corresponding to the mis-annotated regions were used as queries in a low stringency TBLASTN approach (expected *E*-value of 1.0), which allowed the correction of the original annotations. Gene models for *BrNHX6.1* and *BrNHX6.2* were generated using the FancyGene 1.4 gene model drawing program (http://www.bio.ieo.eu/fancygene Rambaldi and Ciccarelli, 2009). *B. rapa NHX* gene locations were determine by using the Brassica Database (http://brassicadb.org/brad).

**Table 1**: Chromosomal locations of *B. rapa NHX* **genes**.

| A. thaliana gene name | AGI | A. thaliana genomic segmenta | Corresponding B. rapa segment locationsb | Br gene name | B. rapa (scaffold, fractionated genomec) | Location (ATG-STOP) | Brassica sp. ESTs |
|---|---|---|---|---|---|---|---|
| NHX1 | AT5G27150 | Q | 2, 6, 9 | BrNHX1.1 | Bra020599 (A02, MF2) | 24295626–24298768 | B. juncea HQ848296, HQ848294, HQ848295 |
| – | – | – | – | BrNHX1.2 | Bra036110 (A09, MF1) | 2662958–2665986 | B. rapa EX098258.1; B. napus: GU192449.1 |
| – | – | – | – | BrNHX1.3 | Bra009975 (A06, LF) | ~18416170–~18415490 | None (and likely pseudogene) |
| NHX2 | AT3G05030 | F | 1, 3, 5 | BrNHX2 | Bra039469 (A05, LF) | 23480049–23483092 | B. napus AY189676.1 |
| NHX3 | AT5G55470 | C | 5, 6, 8 | BrNHX3 | Bra002905 (A10, LF) | 6853381–6850500 | None |
| NHX4 | AT3G06370 | F | 1, 3, 5 | BrNHX4 | Bra020755 (A02, LF) | 23272879–23269821 | None |
| NHX5 | AT1G54370 | C | 5, 6, 8 | – | Not detected | – | None |
| NHX6 | AT1G79610 | E | 2, 7, 7 | BrNHX6.1 | Bra035130 (A07, LF) | 22302738–22298998 | B. oleracea DK554651.1; DK541204.1; DK548975.1; DK489391.1 |
| – | – | – | – | BrNHX6.2 | Bra003601 (A07, MF2) | 14001071–14007892 | B. napus FG553917.1; ES983056.1: CD827425.1; B. oleraceaEE531575.1; EE530829.1; DK495073.1; DK481606.1 |
| NHX7/ SOS1 | AT2G01980 | K | 2, 6, 9 | BrSOS1 | Bra017430 (A09, MF2) | 15259302–15253531 | B. rapa HQ848290, B. napus EU487184.1; B. juncea HQ848289,HQ949287, HQ848288, EF206779 |
| NHX8 | AT1G14660 | A | 6, 8, 9, 10 | BrNHX8 | Bra026197 (A06, LF) | 5641695–5637360 | None |

## Identification of *Thellungiella parvula NHX* Genes

The identification of the intra-cellular *NHX* genes in *T. parvula* used representative Arabidopsis NHX protein sequences (See Table 1 for accession numbers) as query sequences in BLASTP searches of the *T. parvula* predicted gene models—protein version 2.0 database in the *Thellungiella* website [http://www.phytozome.org; (Dassanayake et al., 2011; Goodstein et al., 2012)]. Initial searches were performed using AtNHX1 and AtNHX6 amino acid sequences. Hits with an E value of less than $10^{-45}$ were classified as putative NHXs and examined further by phylogentic analysis.

## Phylogenetic Analysis

The predicted amino acid sequences of all predicted *B. rapa, T. parvula* and *A. thaliana* NHXs were imported into MEGA5 (Tamura et al., 2011), and aligned using ClustalW with default parameters, and the corresponding amino acid alignments used for subsequent analyses. A neighbor-joining analysis was performed, using the Poisson model of amino acid substitution, with uniform rate among sites and pairwise deletion, and the phylogeny tested with 1000 bootstrap replications.

NHX6 nucleotide sequences from *B. rapa*, *B. oleracea* and *B. napus* and *Brassica* NHX6 ESTs from Genbank were imported into MEGA5 (Tamura et al., 2011) and aligned using Clustal W with default parameters. A neighbor-joining analysis was performed using the maximum composite likelihood method with uniform rate among sites and complete deletion, and the phylogeny tested with 1000 bootstrap replications.

## Microsynteny of *AtNHX5* and *AtNHX6* and their co-linearity in *B. rapa*

*AtNHX5* and *AtNHX6* sequences were used as the query to search for homeologous regions in the three sub genomes in *B. rapa* using the syntenic paralogs search tool in the Brassica Database (http://brassicadb.org/brad). The region of *A. thaliana* chromosome 1 containing *AtNHX6* and eight flanking genes upstream and downstream was then aligned to the three co-linear genomic segments from *B. rapa*. The region of *A. thaliana* chromosome 1 surrounding *AtNHX5* was aligned in a similar manner, however due to the more complex nature of the co-linearity between this region and the corresponding *B. rapa* genomic segments the region of analysis was expanded to include 14 genes downstream and 12 genes upstream of *AtNHX5*.

## *BrNHX6* Promoter Analysis

Nucleotide sequences of *AtNHX6*, *BrNHX6.1*, and *BrNHX6.2*, including 1119 bp upstream of the predicted ATG start codon, and 500 bp downstream of the predicted stop codon were obtained from Phytozome. Sequences were aligned and subjected to sliding window analysis of homology using VISualization Tool for Alignments (VISTA) web-based software (Frazer et al., 2004). Regions identified as highly conserved were analysed for conserved *cis*-regulatory elements (CRE) motifs using the Plant cis-acting regulatory DNA elements (PLACE) database (Higo et al., 1999).

### *Plant Growth Conditions*

Seeds of *B. rapa* (cv Chinese cabbage), *B. oleracea* (cv White cabbage) and *B. napus* (cv Westar) were sown direct to soil in seed raising mix and vermiculite (3:1) and watered twice a week. A general purpose fertilizer (Osmocote) was applied in liquid form every two weeks and plants were grown in a 16 h light: 8 h dark photoperiod at 22°C.

For material used for qRT-PCR and isolation of cDNA, *B. napus* (cv Westar) seeds were surface sterilized by soaking in 70% ethanol for 5 min and 1% commercial bleach for 10 min. Seeds were washed three times in sterile double distilled water and plated onto half strength MS plates (Murashige and Skoog, 1962) with 0 mM or 200 mM NaCl. Three replicate plates were made for each treatment. Seeds were stratified for 72 h and grown in a 16 h light: 8 h dark photoperiod at 22°C.

## Cloning of Brassica NHX6 ORF Sequences

Genomic DNA was extracted from *B. rapa* (cv Chinese cabbage) as described (Herrmann and Frischauf, 1987). RNA was extracted from eight week old leaf material of *B. napus* (cv Westar) and *B. oleracea* (cv white cabbage) and from two week old whole seedlings *B. napus* (cv Westar) using the Qiagen RNeasy Kit (Qiagen #74104) as per manufacturer›s instructions. Approximately 500 ng of RNA was used to synthesize cDNA using Superscript III (Life Technologies #18080) following the manufacturer›s recommendations, and the cDNA was diluted to a final volume of 100 μL with sterile double distilled water.

Full length *Brassica* NHX6 ORF sequences were amplified by PCR from cDNA derived from eight week old leaf material of *B. oleracea* (Forward 5′-CACCATGTCGGAGATTTCGCCG-3′ and Reverse 5′-TAACCGGGGGCTAAATTTCTGA-3′) and *B. napus* (Foward 5′-CACCATGTCGGAGATTTCGCCG-3′ and Reverse 5′-TAACC-GGGGGCTAAATTTCTGA-3′). The PCR product was then cloned into the pENTR D/TOPO vector (Life Technologies #K2435-20) as per manufacturer›s instructions. Approximately 10 independent clones from each ligation were sequenced with one unique coding sequence being identified for both *B. napus* (Genbank JX082291) and *B. oleracea* (Genbank JX082292).

## Identification and Cloning of Partial *Brassica* NHX6 Sequences

### *Primer design*

The *B. rapa NHX6.1* and *NHX6.2* genes, the full length *B. napus* (JX082291) and *B. oleracea* (JX082292) ORF sequences as well as other *B. napus*. *B. rapa* and *B. oleracea* EST sequences that were available were aligned in Vector NTI v11.5 (Invitrogen). From this alignment regions of 100% nucleotide conservation between all divergent *Brassica* species sequences were identified and primers (Forward 5′-GCTTGAAGCCCTAGAGGTTGT-3′ and Reverse 5′-CGT TATTACTTGTGAAGAACGTGTT-3′) designed within these regions to amplify all potential A and C genome *B. napus* NHX6 homologs.

Using these primers, partial *Brassica* NHX6 sequences were amplified by PCR from cDNA of 2-week-old whole seedlings in *B. napus* and from genomic DNA in *B. rapa*. DNA from the PCR was purified using the Promega Wizard PCR clean up kit (Promega #A9282) and then cloned into pGEM T-easy vector (Promega #A1360) as per manufacturer's instructions. Approximately 10 independent clones from each ligation were sequenced using the M13_F sequencing primer.

## Gene Expression Analysis

Total RNA was extracted from whole seedlings of *B. napus* (cv Westar) two weeks after germination using the Qiagen RNeasy Kit (Qiagen

#74,104) as per manufacturer›s instructions. Total RNA (2 µg) was DNase treated with the Promega RQ1 DNase kit (Promega #M6101) as per manufacturer›s instructions to remove any genomic DNA contamination.

Primers were designed for the *BnNHX6.1* gene (Forward 5′-CATCCTTTTCTCATTCTGTTCATCG-3′ and Reverse 5′-TCGAAGTCCACTGTACCAAAG-3′) and for the *BnNHX6.2* gene (Forward 5′-GATAGCCGTGATACATCCCTTG-3′ and Reverse 5′-AGTTCTGAAAATGACTTTGCGC-3′) based on the corresponding BrNHX6.1 or BrNHX6.2 open reading frame sequences identified from the *B. rapa*genome.

Quantitative real-time PCR was performed using the Bio-Rad iCycler and the iScript One Step RT-PCR kit with SYBR green (Bio-Rad #170-8892) as per manufacturer's instructions. PCR conditions consisted of a reverse transcription step at 50°C for 10 min, a reverse transcription inactivation step at 95°C for 5 min and 40 cycles of 95°C for 10 s followed by either 61°C (*BnNHX6.1*) or 60°C (*BnNHX6.2*) for 30 s. Results were visualized using BioRad iQ5 Optical System Software. A dissociation analysis was performed and the PCR products were sequenced to confirm PCR specificity to the *BnNHX6* transcript. Expression of the *BnNHX6* genes was normalized relative to the expression of BnUBC_21 (Forward 5′-CCTCTGCAGCCTCCTCAAGT-3′ and Reverse 5′-CATATCTCCCCTGTCTTGAAATGC-3′) previously validated as a reference gene for qRT-PCR in *B. napus* (Chen et al., 2010). Transcript abundance was calculated using the Pfaffl model for relative quantification with efficiency correction (Pfaffl, 2004). All experiments were completed in triplicate on three independent biological replicates.

# RESULTS

## Identification and Phylogenetic Analysis of *B. rapa* *NHX* Genes

As a first step to identify the *B. napus AtNHX5* and *AtNHX6* homologs, we searched the now fully sequenced *B. rapa* genome. As *AtNHX5* and *AtNHX6* are very closely related to the other *A. thaliana*NHX members *AtNHX1-4* and *AtNHX7-8*, unmabiguous identification of the corresponding *B. rapa*homologs may be difficult.

To clearly identify and unambiguously assign the *AtNHX5* and *AtNHX6* *B. rapa* homologs, we identified all the *NHX* genes in the genome of *B. rapa*. To determine the complete complement of *NHX* genes, we searched the *B. rapa* genome sequence and identified ten *NHX* genes (hereafter called *BrNHX;* see Table 1 for complete details). A phylogenetic analysis of the corresponding predicted amino acid sequences of the encoded proteins revealed that *B. rapa* is likely to express five intracellular class one (IC-I) genes, two intracellular class two (IC-II) genes and two PM clade members (Figure 2A). Additional evidence of one pseudogene (*BrNHX1.3*) could also be detected. In some cases, multiple *B. rapa* genes could be detected that corresponded to a single *NHX* gene in *A. thaliana*, but surprisingly, in only one case could three *B. rapa* orthologs corresponding to a *AtNHX* gene be detected (*AtNHX1* and *Br.NHX1.1, Br NHX1.2* and *BrNHX1.3*; Table 1) suggesting substantial gene loss of *NHX*genes in *B. rapa* following the most recent WGT. To further investigate this gene loss, we examined the chromosomal location of the *B. rapa NHX* genes, and compared these to the described fractionated genomes (Wang et al., 2011). Six of the 10 *B. rapa* NHXs were retained on the LF sub-genome, one on the MF1 and three on the MF2 (Table 1). The proteins encoded by the two class II-IC *NHX* genes identified in *B. rapa* are clearly differentiated from other members of the *A. thaliana* NHX family (Figure 2A) and are likely to be the true orthologs of *AtNHX5* and *AtNHX6*. Interestingly, both of the identified class II-IC *NHX* genes are more closely related to *AtNHX6* than to *AtNHX5* (Figure 2A) indicating that there may not be a *B. rapa* ortholog of *AtNHX5*.

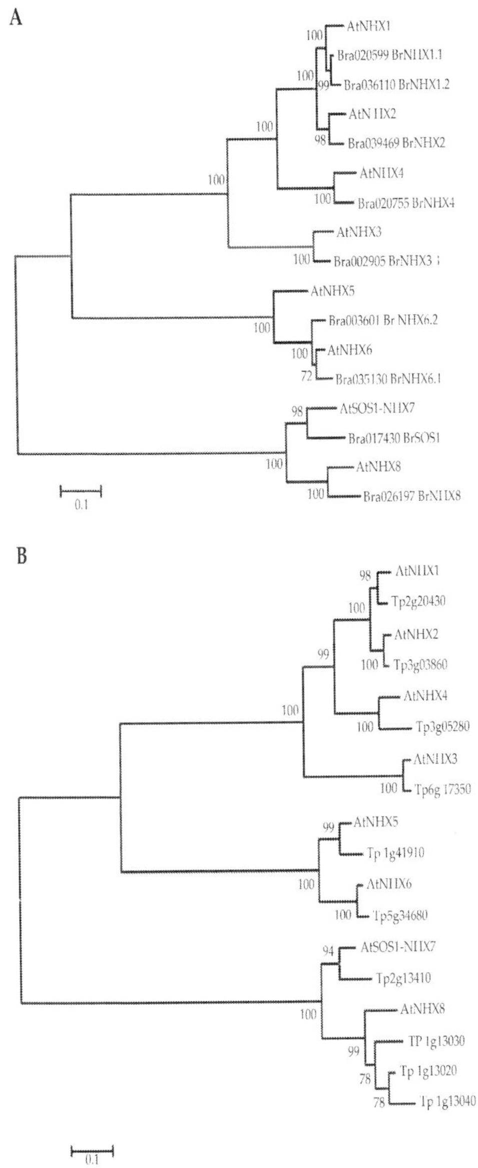


**Figure 2: Phylogenetic analysis of *B. rapa* and *T. parvula* NHX proteins.** Amino acid sequences of NHX proteins from *B. rapa* **(A)**, *T. parvula* **(B)** and *A. thaliana* were aligned using Clustal W, and a Neighbor-joining tree generated. Numbers at nodes correspond to bootstrap support percentages following 1000 replications, and the scale is in amino acid changes per site.

To further investigate the apparent absence of a *AtNHX5* ortholog in *B. rapa*, we decided to try to identify the intra-cellular *NHX* genes in the closely related species *T. parvula. T. parvula* diverged from the *Brassica* species after the divergence from *A. thaliana* but before the Br—α WGT event (Figure 1). Six intra-cellular *NHX* genes were identified and the phylogentic analysis showed that there was one unique ortholog of each of the *A. thaliana NHX1-6* genes (Figure 2B), indicating that there was an *AtNHX5*ortholog present in the ancient *Brassica* ancestor species before the Br—α WGT.

## Macro- and Micro-Synteny of *AtNHX5 and AtNHX6*

To be confident that the *BrNHX6.1* and *BrNHX6.2* genes identified are both orthologs of *AtNHX6* and not *AtNHX5*, we analysed the macrosynentic relationship between these genes based on the 24 conserved syntenic blocks between *A. thaliana* and *B. rapa* (Schranz et al., 2006). Analysis of the conserved co-linearity in *A. thaliana* and *B. rapa* shows that *BrNHX6.1*, *BrNHX6.2*, and *AtNHX6* are all located in the retained conserved genomic block E and are likely to be true orthologs, whereas *AtNHX5* is retained within the conserved genomic Block C on the *A. thaliana* genome (Table 1), and appears to have no clear ortholog in *B. rapa*. We also analysed the gene models for *AtNHX5, AtNHX6, BrNHX6.1*, and *BrNHX6.2*(Figure 3). This showed that *AtNHX6, BrNHX6.1*, and *BrNHX6.2* all have a common 22 exon/21 intron gene structure, whereas *AtNHX5* has only 21 exons (Figure 3) supporting the notion that the *BrNHX6.1*and *BrNHX6.2* genes are both orthologous to *AtNHX6* and not *AtNHX5*.

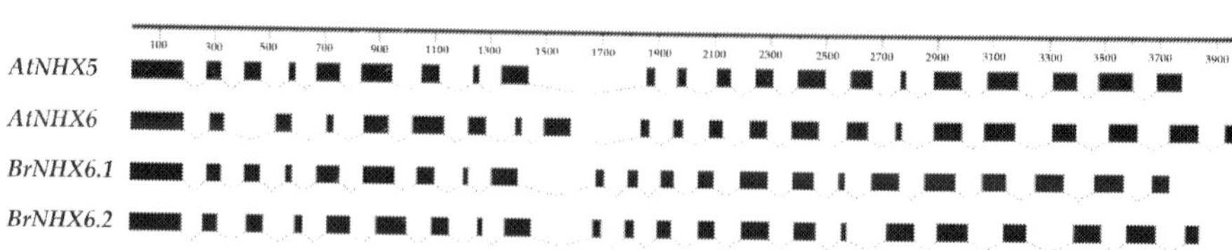


**Figure 3**: Predicted gene models for AtNHX5, AtNHX6, BrNHX6.1 and BrNHX6.2.Predicted gene models for AtNHX5, AtNHX6, BrNHX6.1, and BrNHX6.2. Exons are shown as black boxes and introns are shown as dotted lines. Gene models were generated using FancyGene 1.4 (http://www.bio.ieo.eu/fancygene).

Although it is clear that there was an AtNHX5 ortholog present prior to the Br—α WGT event, no *AtNHX5*homolog was identified in the *B. rapa* genome. To determine the cause of this discrepancy we decided to carefully investigate the microsynteny around *AtNHX5* (Figure 4A). The microsynteny in the *A. thaliana*genome, including fourteen genes downstream and twelve genes upstream of *AtNHX5*, was compared to the corresponding co-linear sub genome regions in the *B. rapa* genome (Figure 4A). Surprisingly, the corresponding co-linear regions of the three *B. rapa* sub genomes contain only five (LF), three (MF1) and nine (MF2) genes respectively, and only four (LF), three (MF1) and five (MF2) of these genes were homologous to *A. thaliana* genes (Figure 4A). The colinearity between *A. thaliana* and the *B. rapa* sub genomes is relatively well conserved from the At1G54220 gene to At1G54290 (Figure 4A). However, the next 14 genes upstream of At1G54290 in the *A. thaliana* genome including *AtNHX5* are completely absent from all three co-linear *B. rapa* sub genomes (Figure 4A). Co-linear genes are again found between the *A. thaliana* genome and the *B. rapa* LF sub genome from At1G54490 and the MF2 sub genomes from At1G5440, however, no homologous genes can be detected in the MF1 sub genome (Figure 4A). Interestingly, upstream of At1G54440 the co-linear region on the LF sub genome is found on chromosome BrA09, whereas the co-linear region of the LF sub genome downstream of At1G54290 was found on chromosome BrA06 (Figure 4A). Similarly, the region of co-linearity upstream of At1G54440 on the MF2 sub genome is found on chromosome BrA01 and the co-linear region of the MF2 sub genome downstream of At1G54290 was found on chromosome BrA08 (Figure 4A). This indicates that the syntenic regions around *AtNHX5* in the LF and MF2 sub genomes have probably undergone an inter-chromosomal translocation event. It also appears that there has been a large fragment deletion event in the MF1 sub genome in the syntenic region around *AtNHX5*. These chromosomal re-arrangments and deletions would account for the absence of any *AtNHX5* orthologs in *B. rapa* and may be the result of the genome reduction processes post the Br—α WGT event. This evidence, combined with the inability to identify any *AtNHX5* orthologs when searching the *B. rapa* genome, strongly suggest that there are no *AtNHX5*orthologs present in *B. rapa*.

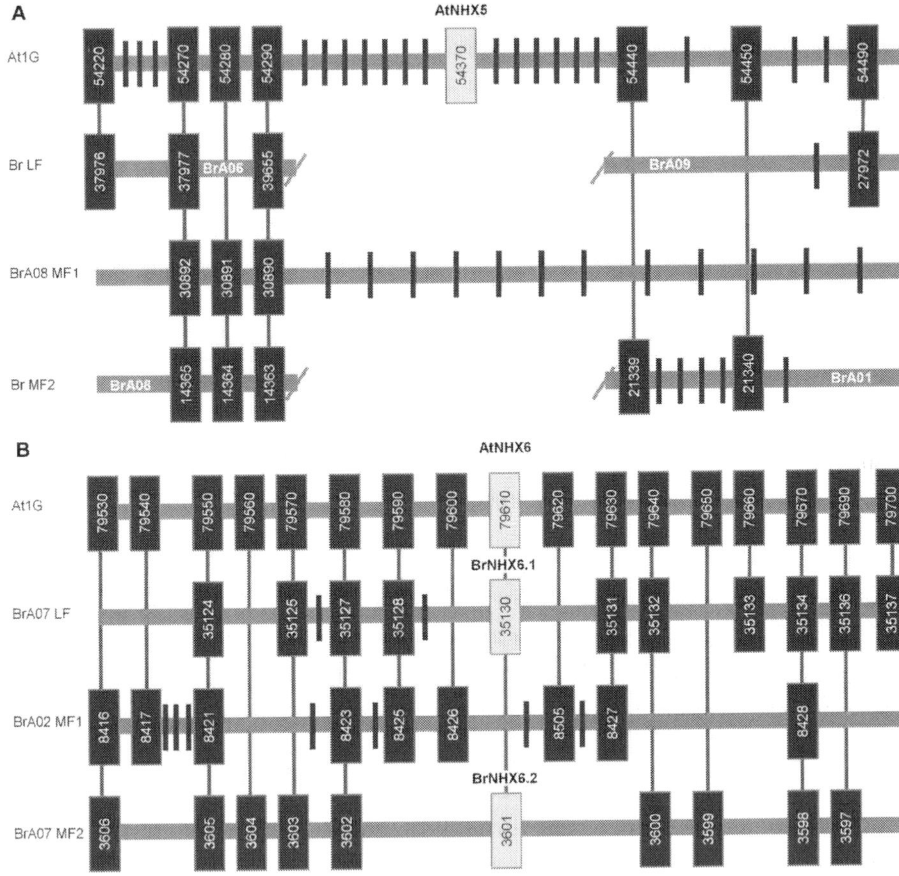


**Figure 4**: Microsynteny of AtNHX5 and AtNHX6 and conserved co-linearity in *B. rapa*.Comparison of co-linearity between AtNHX5 **(A)** and AtNHX6 **(B)** (highlighted in yellow) in the Arabidopsis genome and the least fractionated (LF), medium fractionated (MF1) and most fractionated (MF2) sub-genomes in *B. rapa*. The boxes and bars denote annotated genes (boxes contain gene accession number) and the red lines connect syntenic homologous genes (Note corresponding genomic regions are not drawn to scale).

The microsynteny in the *A. thaliana* genome, including eight genes downstream and eight genes upstream of *AtNHX6*, was compared to the corresponding co-linear sub genome regions in the *B. rapa* genome (Figure 4B). The corresponding three co-linear *B. rapa* sub genomes contain 13 (LF), 16 (MF1), and 10 (MF2) genes, respectively, with 11 (LF), nine (MF1), and 10 (MF2) homologous genes to the co-linear region in the *A. thaliana* genome (Figure 4B). The *BrNHX6.1* gene is located on the LF sub genome and the *BrNHX6.2* gene is located of the

MF2 sub genome, but there is no *AtNHX6* ortholog present on the MF1 sub genome (Figure 4B). Importantly, all of the 17 *A. thaliana* genes in the region surrounding *AtNHX6* have at least one homolog in one of the corresponding *B. rapa* sub genomes and the order of the genes in *A. thaliana* is maintained within *B. rapa* (Figure 4B). The high degree of co-linearity between the region in *A. thaliana* genome containing *AtNHX6* and the corresponding regions in the *B. rapa* genome strongly support the notion that there are only two *AtNHX6* orthologs in *B. rapa* with the third ortholog likely to have been lost from the MF1 sub genome since the last whole genome triplication event.

## PROMOTER ANALYSIS OF BRNHX6.1, BRNHX6.2 AND ATNHX6

Although the *B. rapa* genome does not appear to contain a homolog to *AtNHX5*, there is a high level of genetic conservation between the *A. thaliana* and *B. rapa NHX6.1* and *NHX6.2* orthologs. To explore the potential for differential functionality between these two genes, we investigated the level of sequence similarity between their respective promoter regions and the conservation of specific *cis*-acting regulatory motifs.

To identify conserved regions within the promoters of *B. rapa NHX6* orthologs, we compared nucleotide sequences of *AtNHX6*, *BrNHX6.1*, and *BrNHX6.2* (Figure 5A). The gene sequences, including 1119 bp upstream and 500 bp downstream, were subjected to sliding-widow analysis of homology. It has previously been shown that requiring 70% identity within a 25-bp window returns the greatest number of regulatory elements with acceptable specificity (Guo and Moose, 2003). Using these parameters, sliding window analysis revealed a highly conserved region of approximately 300 bp immediately upstream of the start codon, including the 5′UTR (Figure 5B).

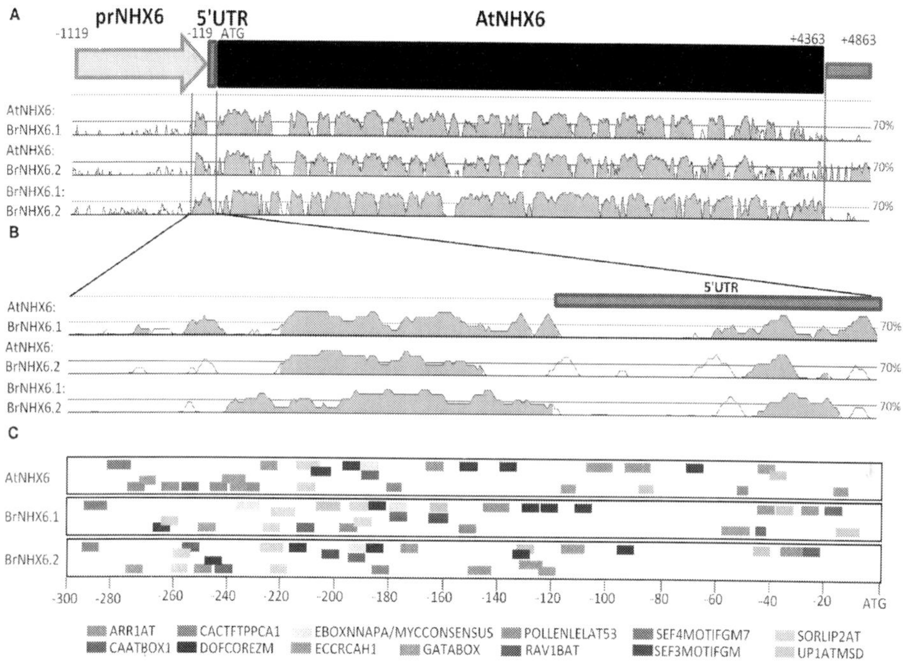


**Figure 5**: Conservation of promoter regions of NHX6 homologues in *B. rapa***.** Graphic output for comparisons of *Arabidopsis* and *B. rapa* IC II NHX genome region **(A)** and putative promoter region **(B)** using VISTA, with parameters of at least 70% identity within a window of 25 b. Regions with >70% identity shared by the three orthologs are shaded in pink. **(C)** Positions of known plant *cis*-acting regulatory elements which are common to all three genes are shown as colored bars.

Evaluation of this highly conserved promoter region (−300 bp) using the PLACE database SIGNALSCAN tool identified 10 *cis*-acting regulatory elements (CREs) which were common to all three genes: ARR1AT (5′-NGATT-3′), CAATBOX1 (5′-CAAT-3′), CACTFTPPCA1 (5′-YACT-3′), DOFCOREZM (5′-AAAG-3′), EBOXNNAPA (5′-CANNAG-3′), EECCRCAH1 (5′-GANTTNC-3′), GATABOX (5′-GATA-3′), MYCCONSENSUS (5′-CANNAG-3′), POLLENLELAT53 (5′-AGAAA-3′), RAV1BAT (5′-CACCTG-3′), SORLIP2AT (5′-GCCAC-3′), and UP1ATMSD (5′-GGCCCAWWW-3′). Additio-nally, the functionally similar elements SEF4MOTIFGM7S (5′-RTTTTTR-3′) and SEF3MOTIFGM (5′-AACCCA-3′) were conserved between *AtNHX6*:*BrNHX6.1* and *BrNHX6.1*:*BrNHX6.2*, respectively. The position of these motifs in the promoter sequence relative to the start codon is depicted in Figure 5C. Interestingly, the position of the DOFCOREZM (5′-AAAG-3′) and MYCCONSENSUS/EBOXNNAPA (5′-CANNAG-3′)/RAV1BAT (5′-CACCAG-3′) elements are conserved across the three promoter regions.

## Identification *B. napus* NHX6 Homologs

*B. napus* is an agriculturally important member of the Brassicaceae family. Due to the significance of salinity in *B. napus* cultivation, we sought to identify the complement of expressed *B. napus NHX6*homologs and assess their role in salt tolerance. To identify *NHX6* homologs in *B. napus* without the advantage of a complete *B. napus* genome sequence, we designed primers in absolutely conserved regions present in all available nucleotide sequences from *B. napus*, *B. rapa*, and *B. oleracea*. These primers were first tested on *B. rapa* genomic DNA to ensure they could identify both *BrNHX6.1* and *BrNHX6.2*. Using these "universal" *Brassica* NHX6 primers, four unique sequences were identified in *B. rapa.* A phylogenetic analysis of these sequences revealed two distinct groups including either the *BrNHX6.1* or the *BrNHX6.2* sequences from the *B. rapa* genome (Figure 6A). The presence of two unique sequences grouping with each *BrNHX6* sequence is most likely the result of heterozygous *BrNHX6* alleles present in the *B. rapa* genomic DNA tested. Importantly, however, the primers used were able to amplify both the *BrNHX6* homologs.

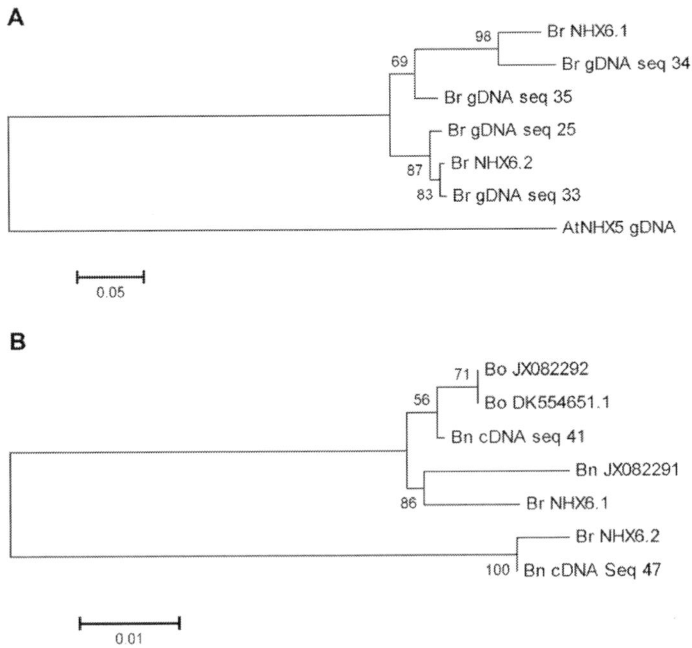


**Figure 6**: Phylogenetic analysis of *B. rapa*, *B. oleracea*, and *B. napus* **NHX6 partial sequences.** *B. rapa* gDNA **(A)** and *B. rapa*, *B. oleracea* and *B. napus* mRNA **(B)** sequences were aligned using Clustal W, and a Neighbor-joining tree generated. Numbers at nodes correspond to bootstrap support percentages following 1000 replications.

The complement of expressed *AtNHX6* homologs in *B. napus* was then investigated using the same approach, but using mRNA isolated from *B. napus* seedlings. Sequencing the cloned amplicons from *B. napus* cDNA identified only two unique sequences. A phylogenetic analysis was performed including *BrNHX6.1* and *BrNHX6.2* sequences from the *B. rapa* genome, the *B. napus* (JX082291) and *B. oleracea*(JX082292) ORF sequences and a *B. oleracea* (DK554651) EST sequence. The phylogenetic analysis showed two major groupings, each containing either the *BrNHX6.1* or the *BrNHX6.2* sequence (Figure 6B). Interestingly, only one sequence grouped with *BrNHX6.2* while four unique sequences grouped with *BrNHX6.1* (Figure 6B). The sequences that grouped with *BrNHX6.1* were divided into two sub-groups—one group has a *B. napus* sequence more closely related to the *BrNHX6.1* sequence, while the other group includes only *B. oleracea* and *B. napus* sequences, possibly indicating both an A and C genome version of *BrNHX6.1* (Figure 6B). From this analysis, the *B. napus* sequences that are clearly closely related to *BrNHX6.1* and *BrNHX6.2* will be hereafter designated as *BnNHX6.1* and *BnNHX6.2*, respectively.

## Expression Analysis of *BnNHX6.1* and *BnNHX6.2*

Having identified two *NHX6* genes in *B. napus*, we examined the relative expression of *BnNHX6.1* and *BnNHX6.2* in the presence and absence of salt stress. Firstly, to optimise the concentration of NaCl required to induce salt stress, we compared the germination percentage and fresh weight of *B. napus* cv. Westar seedlings grown in the presence and the absence of NaCl. The NaCl treatment had a marked effect on the germination and growth of *B. napus* seedlings (Figure 7A). Germination in the presence of NaCl was severely inhibited with only 41% of seeds germinating compared with 92% of seeds sown on plates lacking additional NaCl (Figure 7B). The mean fresh weight per seedling grown in the presence of NaCl was also greatly reduced compared to those grown in the absence of NaCl (Figure 7C). These results demonstrate that the NaCl treatment was sufficient to induce a severe stress impacting on the germination and growth of *B. napus* seedlings. We then examined the differential expression of both the *BnNHX6.1*and the *BnNHX6.2* genes in the presence and absence of the NaCl stress. The *BnNHX6.2* gene showed no significant difference in relative transcript abundance in response to the NaCl stress, whereas there was a significant increase in relative transcript abundance of the *BnNHX6.1* gene in response to the NaCl stress (Figure 7D). Interestingly, the relative expression of *BnNHX6.1* was extremely low compared to *BnNHX6.2* both in the presence and absence of NaCl (Figure 7D) suggesting that there may be some differential regulation of these two genes.

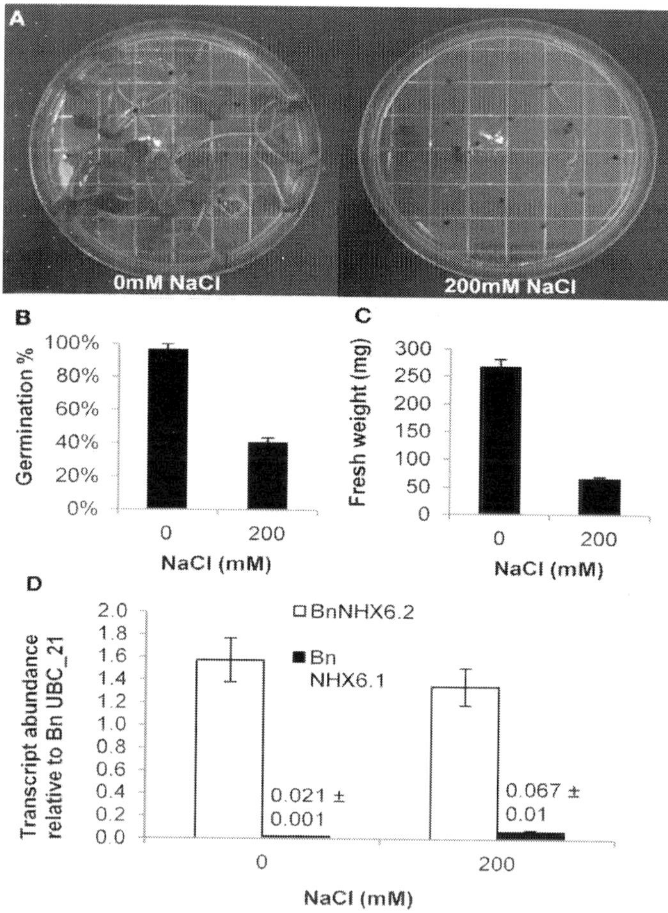


**Figure 7**: Differential expression of BnNHX6.1 and BnNHX6.2 in *B. napus* **seedlings in response to** NaCl. *B. napus* (cv Westar) seeds were sown on ½ strength MS media with or without 200 mM NaCl **(A)** and the germination percentage **(B)** and fresh weight per seedling **(C)** were compared between the two treatments. RNA was extracted from seedlings and the transcript abundance of BnNHX6.1 and BnNHX6.2 was assayed relative to the BnUBC_21 reference gene **(D)**. Data are the means of three biological replicates (±s.e.).

# DISCUSSION

## Ortholog Identification in *B. napus*

In this paper we have attempted to identify the *B. napus* orthologs of two *A.*

*thaliana* proteins, AtNHX5 and AtNHX6. Although *A. thaliana* and *B. napus* are relatively closely related, precise ortholog identification is complicated by the recent WGT event in the *Brassica* species and the amphidiploid nature of the *B. napus* (AACC 2*n* = 38) genome. Consequently, there are potentially six *B. napus* homologs of *AtNHX5* and *AtNHX6* comprising three triplicated paralogs from both the A and C parental genomes.

Identification of *A. thaliana* homologs in the *Brassica* species has traditionally relied upon comparison to incomplete genomic sequences, generally in the form of bacterial artificial chromosome (BAC) clones. For example identification of *FLOWERING LOCUS C* (*FLC*) homologs in *B. rapa* entailed probing *B. rapa*BAC libraries to identify BAC clones containing co-linear regions (Yang et al., 1999). While successful in this case, this methodology was reliant on having all the appropriate co-linear region of the *B. rapa*genome present in the BAC library. The recent release of the *B. rapa* genome provides a complete *Brassica*genomic resource allowing *A. thaliana* gene orthologs to be identified with a high level of confidence.

Our search of the *B. rapa* genome identified only two *AtNHX6* orthologs. Analysis of the *B. rapa* genome has shown that there has been significant loss of gene paralogs in the three sub genomes since the WGT event (Wang et al., 2011), which is likely to be the reason that there is no *AtNHX6* ortholog on the MF1 sub genome. Interestingly, we could not find any evidence of an *AtNHX5* homolog in the *B. rapa* genome. While this came as a surprise, recent evidence shows that *AtNHX5* and *AtNHX6* act redundantly in *A. thaliana* (Bassil et al., 2011), and therefore the the *Brassica* species may not require a functional *AtNHX5*ortholog. Our analysis of the microsynteny around *AtNHX5* and its co-linearity with the three corresponding sub genomes in *B. rapa* indicated that a large block of *A. thaliana* genes including *AtNHX5*were missing from all three co-linear *B. rapa* sub genomes as a result of inter-chromosomal translocations in the LF and MF2 sub genomes and a fragment deletion in the MF1 sub-genome. The identification of an *AtNHX5* ortholog in the closely related *T. parvula* species indicates that there was an *AtNHX5* ortholog present in the ancient *Brassica* ancestral species prior to the Br—α WGT event. The loss of *AtNHX5* in the *Brassica* species could only have occurred either due to a localized deletion event occurring in the ancestral *Brassica* genome after the *T. parvula-Brassica* divergence, but before the *Brassica* WGT event or due to the loss of all three *AtNHX5* genes following the WGT event. It seems more likely that the *AtNHX5* orthologs were lost after the Br—α WGT event due to the differential disruption in the syntentic region around *AtNHX5* in the three *B. rapa* sub-genomes. If *AtNHX5* had been lost prior to the Br—α WGT event it would be reasonable to expect that the syntenic region in all three *B. rapa* sub-genomes would display a common chromosomal disruption that was then triplicated as part of the Br—α WGT event.

Sequencing of orthologous BAC clones of the *B. rapa* A genome, the *B. oleracea* C genome and the A and C genome components of *B. napus* have shown essentially complete co-linearity between the protein coding genes in the A and C genomes and sequence identity of 94–97% (Cheung et al., 2009). It could, therefore, be anticipated that the *B. oleracea* C genome is also likely to only have two *AtNHX6* orthologs. Our sequence analysis in *B. oleracea* and *B. napus* could only identify two distinct *AtNHX6* homologs that were also homologous to the *BrNHX6.1* and *BrNHX6.2* genes. It is important to note however, that all of the *B. oleracea* and *B. napus* sequences analysed in this phylogeny were derived from mRNA; it is therefore possible that additional *B. oleracea NHX6* homologs are present in the genome but have not been detected in the limited mRNA sequencing described here. The complete complement of *B. oleracea NHX6* homologs could only be resolved with a completed *B. oleracea* genome. The phylogenetic analysis of homologous *Brassica NHX6.1* sequences shows two distinct groupings, one including *B. rapa NHX6.1*and a *B. napus* sequence, and the other containing only *B. oleracea* and *B. napus* sequences. This provides evidence of possible A and C genome versions of *Brassica NHX6.1*. Unfortunately, we do not have a corresponding *B. oleracea NHX6.2* sequence available, making it difficult to clearly identify the A and C genome versions of *NHX6.2*. It is therefore most likely that there are two *AtNHX6* homologs in both *B. rapa* and *B. oleracea* and, as a consequence, four *AtNHX6* homologs in *B. napus*.

## Expression Patterns of *BnNHX6.1* and *BnNHX6.2*

The analysis of the *B. rapa* genome showed that of the eight Arabidopsis *NHX* genes only *AtNHX1* and *AtNHX6* have retained multiple paralogs since the most recent triplication event. Gene duplications arising from polyploidy events are thought to be a major source of novel variation and gene function (Moore and Purugganan, 2003) and duplicated genes may be lost, develop new functions or share the functions of the original gene through differential expression or regulation (Blanc and Wolfe, 2004). It may be the case that the two orthologs of *AtNHX6* may have different functions or expression patterns. As the coding regions of all the identified *Brassica* NHX6 genes are highly homologous with *AtNHX6* and each other, it is likely that their functions ($Na^+/K^+–H^+$ antiporting) remain identical. The analysis of the *BrNHX6.1* and *BrNHX6.2* promoter regions highlighted a significant region of similarity 300 bp upstream of the ATG. The conserved MYCCONSENSUS element is often associated with genes responsive to abiotic stress, such as the stress hormone abscisic acid (ABA), dehydration-responsive RD22, and cold responsive genes CBF, DREB1, and ICE1 (Chinnusamy et al., 2004). This is highly interesting given the potential role for *NHX* proteins in improving salt tolerance. Additionally, the equivalent EBOXNNAPA motif was found to be an essential promoter element for high expression of the napA storage protein in *B. napus* seeds (Stalberg et al., 1996). The conservation of these regulatory elements suggests conservation of functionality between AtNHX6 and its *B. rapa* homologs.

The expression of the *B. napus NHX6.2* gene is significantly higher than the *B. napus NHX6.1* gene in two week old seedlings, indicating that in this tissue type at least there does appear to be some differential expression. While there is no significant difference in expression of *BnNHX6.2* in response to NaCl, the expression of *BnNHX6.1* increased approximately three-fold in response to NaCl again highlighting differences in expression. It should be noted that the relative expression of *BnNHX6.1* was 20 times lower than the expression of *BnNHX6.2* in the NaCl treatment, and is suggestive of a more prominent role for *BnNHX6.2*. This is in contrast to the situation in *A. thaliana*, where the largest difference in expression observed is a 3.5 increase of AtNHX5 transcript in leaf tissue (Bassil et al., 2011). The two Brassica NHX6 isoforms do not appear to have different functions, but the expression of BnNHX6.1 is significantly lower than that of *BnNHX6.2*. This may indicate that the *BnNHX6.1* gene is simply sitting in the genome waiting to be lost, like most of the other triplicated paralogs of the *BrNHX* gene family. It would be of great interest to examine the expression patterns of *BnNHX6.1* and *BnNHX6.2* in a variety of tissues types in the presence and absence of NaCl to better elucidate differential expression patterns.

In this study we have demonstrated the use of the *B. rapa* genome in the successful identification of two *B. rapa* NHX6 orthologs, and used this information to identify two unique *B. napus* NHX6 orthologs. Our analysis also strongly suggests that there are four *B. napus* orthologs, comprising an A and C genome version of *BrNHX6.1* and *BrNHX6.2*. This study also indicates that there may be some differential expression of the *BnNHX6.1* and *BnNHX6.2*.

## REFERENCES


1. Apse, M. P., Aharon, G. S., Snedden, W. A., and Blumwald, E. (1999). Salt tolerance conferred by overexpression of a vacuolar Na+/H+ antiport in Arabidopsis. *Science* 285, 1256–1258.
2. Bassil, E., Ohto, M. A., Esumi, T., Tajima, H., Zhu, Z., Cagnac, O., Belmonte, M., Peleg, Z., Yamaguchi, T., and Blumwald, E. (2011). The *Arabidopsis* intracellular Na+/H+ antiporters *NHX5* and *NHX6* are endosome associated and necessary for plant growth and development. *Plant Cell* 23, 224–239.
3. Bell, C. D., Soltis, D. E., and Soltis, P. S. (2010). The age and diversification of the angiosperms re-revisited. *Am. J. Bot.* 97, 1296–1303.
4. Blanc,G.,andWolfe,K.H.(2004).Functionaldivergenceofduplicated genes formed by polyploidy during *Arabidopsis* evolution. *Plant Cell* 16, 1679–1691.
5. Bowers, K., Levi, B. P., Patel, F. I., and Stevens, T. H. (2000). The sodium/proton exchanger Nhx1p is required for endosomal protein trafficking in the yeast *Saccharomyces cerevisiae*. *Mol. Biol. Cell* 11, 4277–4294.
6. Brett, C. L., Donowitz, M., and Rao, R. (2005a). Evolutionary origins of eukaryotic sodium/proton exchangers. *Am. J. Physiol. Cell Physiol* 288, C223–C239.
7. Brett, C. L., Tukaye, D. N., Mukherjee, S., and Rao, R. J. (2005b). The yeast endosomal Na+(K+)/H+ exchanger Nhx1 regulates cellular pH to control vesicle trafficking. *Mol. Biol. Cell* 16, 1396–

1405.

8. Chanroj, S., Wang, G., Venema, K., Zhang, M. W., Delwiche, C. F., and Sze, H. (2012). Conserved and diversified gene families of monovalent cation/$H^{(+)}$ antiporters from algae to flowering plants. *Front. Plant Science* 3:25. doi: 10.3389/fpls.2012.00025
9. Chen, X., Truksa, M., Shah, S., and Weselake, R. J. (2010). A survey of quantitative real-time polymerase chain reaction internal reference genes for expression studies in *Brassica napus*. *Anal. Biochem*. 405, 138–140.
10. Cheung, F., Trick, M., Drou, N., Lim, Y. P., Park, J. Y., Kwon, S. J., Kim, J. A., Scott, R., Pires, J. C., Paterson, A. H., Town, C., and Bancroft, I. (2009). Comparative analysis between homoeologous genome segments of *Brassica napus* and its progenitor species reveals extensive sequence-level divergence. *Plant Cell* 21, 1912–1928.
11. Chinnusamy, V., Schumaker, K., and Zhu, J. K. (2004). Molecular genetic perspectives on cross-talk and specificity in abiotic stress signalling in plants. *J. Exp. Bot*. 55, 225–236.
12. Couvreur, T. L., Franzke, A., Al-Shehbaz, I. A., Bakker, F. T., Koch, M. A., and Mummenhoff, K. (2010). Molecular phylogenetics, temporal diversification, and principles of evolution in the mustard family (Brassicaceae). *Mol. Biol. Evol*. 27, 55–71.
13. Dassanayake, M., Oh, D. H., Haas, J. S., Hernandez, A., Hong, H., Ali, S., Yun, D. J., Bressan, R. A., Zhu, J. K., Bohnert, H. J., and Cheeseman, J. M. (2011). The genome of the extremophile crucifer *Thellungiella parvula*. *Nat. Genet*. 43, 913–918.
14. Food, and Agriculture Organization of the United Nations. (2009). "Increasing crop production sustainably. The perspective biological processes," ed *Department of Economic and Social Development* (Rome: FAO).
15. Franzke, A., Lysak, M. A., Al-Shehbaz, I. A., Koch, M. A., and Mummenhoff, K. (2011). Cabbage family affairs: the evolutionary history of Brassicaceae. *Trends Plant Sci*. 16, 108–116.
16. Frazer, K. A., Pachter, L., Poliakov, A., Rubin, E. M., and Dubchak, I. (2004). VISTA: computational tools for comparative genomics. *Nucleic Acids Res*. 32, W273–W279.

17. Fukuda, A., Nakamura, A., Hara, N., Toki, S., and Tanaka, Y. (2011). Molecular and functional analyses of rice NHX-type $Na^+$/$H^+$antiporter genes. *Planta* 233, 175–188.

18. Goodstein, D. M., Shu, S., Howson, R., Neupane, R., Hayes, R. D., Fazo, J., Mitros, T., Dirks, W., Hellsten, U., Putnam, N., and Rokhsar, D. S. (2012). Phytozome: a comparative platform for green plant genomics. *Nucleic Acids Res.* 40, D1178–D1186.

19. Guo, H., and Moose, S. P. (2003). Conserved noncoding sequences among cultivated cereal genomes identify candidate regulatory sequence elements and patterns of promoter evolution. *Plant Cell* 15, 1143–1158.

20. Henry, Y., Bedhomme, M., and Blanc, G. (2006). History, protohistory and prehistory of the *Arabidopsis thaliana* chromosome complement. *Trends Plant Sci.* 11, 267–273.

21. Herrmann, B. G., and Frischauf, A. M. (1987). Isolation of genomic DNA. *Methods Enzymol.* 152, 180–183.

22. Higo, K., Ugawa, Y., Iwamoto, M., and Korenaga, T. (1999). Plant cis-acting regulatory DNA elements (PLACE) database: 1999. *Nucleic Acids Res.* 27, 297–300.

23. Hu, T. T., Pattyn, P., Bakker, E. G., Cao, J., Cheng, J. F., Clark, R. M., Fahlgren, N., Fawcett, J. A., Grimwood, J., Gundlach, H., Haberer, G., Hollister, J. D., Ossowski, S., Ottilar, R. P., Salamov, A. A., Schneeberger, K., Spannagl, M., Wang, X., Yang, L., Nasrallah, M. E., Bergelson, J., Carrington, J. C., Gaut, B. S., Schmutz, J., Mayer, K. F., Van De Peer, Y., Grigoriev, I. V., Nordborg, M., Weigel, D., and Guo, Y. L. (2011). The *Arabidopsis lyrata* genome sequence and the basis of rapid genome size change. *Nat. Genet.* 43, 476–481.

24. Koch, M. A., Haubold, B., and Mitchell-Olds, T. (2000). Comparative evolutionary analysis of chalcone synthase and alcohol dehydrogenase loci in *Arabidopsis*, *Arabis*, and related genera (Brassicaceae). *Mol. Biol. Evol.* 17, 1483–1498.

25. Li, M., Li, Y., Li, H., and Wu, G. (2011). Overexpression of *AtNHX5* improves tolerance to both salt and drought stress in *Broussonetia papyrifera* (L.) Vent. *Tree Physiol.* 31, 349–357.

26. Lysak, M. A., Koch, M. A., Pecinka, A., and Schubert, I. (2005). Chromosome triplication found across the tribe Brassiceae. *Genome*

*Res*. 15, 516–525.

27. Moore, R. C., and Purugganan, M. D. (2003). The early stages of duplicate gene evolution. *Proc. Natl. Acad. Sci. U.S.A*. 100, 15682–15687.
28. Mun, J. H., Kwon, S. J., Yang, T. J., Seol, Y. J., Jin, M., Kim, J. A., Lim, M. H., Kim, J. S., Baek, S., Choi, B. S., Yu, H. J., Kim, D. S., Kim, N., Lim, K. B., Lee, S. I., Hahn, J. H., Lim, Y. P., Bancroft, I., and Park, B. S. (2009). Genome-wide comparative analysis of the *Brassica rapa* gene space reveals genome shrinkage and differential loss of duplicated genes after whole genome triplication. *Genome Biol*. 10, R111.
29. Munns, R., and Tester, M. (2008). Mechanisms of salinity tolerance. *Annu. Rev. Plant Biol*. 59, 651–681.
30. Murashige, T., and Skoog, F. (1962). A revised medium for rapid growth and bio-assays with tobacco tissue cultures. *Physiol. Plant* 15, 473–497.
31. Parkin, I. A., Gulden, S. M., Sharpe, A. G., Lukens, L., Trick, M., Osborn, T. C., and Lydiate, D. J. (2005). Segmental structure of the *Brassica napus* genome based on comparative analysis with *Arabidopsis thaliana*. *Genetics* 171, 765–781.
32. Pfaffl, M. (2004). "Quantification strategies in real-time PCR," in *A-Z of quantitative PCR*, ed S. Bustin (La Jolla, CA: International University Line), 87–112.
33. Rambaldi, D., and Ciccarelli, F. D. (2009). FancyGene: dynamic visualization of gene structures and protein domain architectures on genomic loci. *Bioinformatics* 25, 2281–2282.
34. Rodriguez-Rosales, M. P., Jiang, X., Galvez, F. J., Aranda, M. N., Cubero, B., and Venema, K. (2008). Overexpression of the tomato $K^+/H^+$ antiporter LeNHX2 confers salt tolerance by improving potassium compartmentalization. *New Phytol*. 179, 366–377.
35. Schranz, M. E., Lysak, M. A., and Mitchell-Olds, T. (2006). The ABC's of comparative genomics in the Brassicaceae: building blocks of crucifer genomes. *Trends Plant Sci*. 11, 535–542.
36. Shi, H., Lee, B. H., Wu, S. J., and Zhu, J. K. (2003). Overexpression of a plasma membrane $Na^+/H^+$ antiporter gene improves salt tolerance in *Arabidopsis thaliana*. *Nat. Biotechnol*. 21, 81–85.

37. Shi, H., Quintero, F. J., Pardo, J. M., and Zhu, J. K. (2002). The putative plasma membrane Na(+)/H(+) antiporter SOS1 controls long-distance Na(+) transport in plants. *Plant Cell* 14, 465–477.

38. Shi, H., and Zhu, J. K. (2002). *SOS4*, a pyridoxal kinase gene, is required for root hair development in Arabidopsis. *Plant Physiol.* 129, 585–593.

39. Soltis, D. E., Albert, V. A., Leebens-Mack, J., Bell, C. D., Paterson, A. H., Zheng, C., Sankoff, D., Depamphilis, C. W., Wall, P. K., and Soltis, P. S. (2009). Polyploidy and angiosperm diversification. *Am. J. Bot.* 96, 336–348.

40. Stalberg, K., Ellerstom, M., Ezcurra, I., Ablov, S., and Rask, L. (1996). Disruption of an overlapping E-box/ABRE motif abolished high transcription of the *napA* storage-protein promoter in transgenic *Brassica napus* seeds. *Planta* 199, 515–519.

41. Tamura, K., Peterson, D., Peterson, N., Stecher, G., Nei, M., and Kumar, S. (2011). MEGA5, molecular evolutionary genetics analysis using maximum likelihood, evolutionary distance, and maximum parsimony methods. *Mol. Biol. Evol.* 28, 2731–2739.

42. Tester, M., and Davenport, R. (2003). $Na^+$ tolerance and $Na^+$ transport in higher plants. *Ann. Bot. (Lond.)* 91, 503–527.

43. United States Department of Agriculture. (2009). "Oilseeds: world market and trade," ed *Foreign Agricultural Service Circular Series 11-09 - November*, (Washington, DC: USDA).Wang, X., Wang, H., Wang, J., Sun, R., Wu, J., Liu, S., Bai, Y., Mun, J. H., Bancroft, I., Cheng, F., Huang, S., Li, X., Hua, W., Freeling, M., Pires, J. C., Paterson, A. H., Chalhoub, B., Wang, B., Hayward, A., Sharpe, A. G., Park, B. S., Weisshaar, B., Liu, B., Li, B., Tong, C., Song, C., Duran, C., Peng, C., Geng, C., Koh, C., Lin, C., Edwards, D., Mu, D., Shen, D., Soumpourou, E., Li, F., Fraser, F., Conant, G., Lassalle, G., King, G. J., Bonnema, G., Tang, H., Belcram, H., Zhou, H., Hirakawa, H., Abe, H., Guo, H., Jin, H., Parkin, I. A., Batley, J., Kim, J. S., Just, J., Li, J., Xu, J., Deng, J., Kim, J. A., Yu, J., Meng, J., Min, J., Poulain, J., Hatakeyama, K., Wu, K., Wang, L., Fang, L., Trick, M., Links, M. G., Zhao, M., Jin, M., Ramchiary, N., Drou, N., Berkman, P. J., Cai, Q., Huang, Q., Li, R., Tabata, S., Cheng, S., Zhang, S., Sato, S., Sun, S., Kwon, S. J., Choi, S. R., Lee, T. H.,

Fan, W., Zhao, X., Tan, X., Xu, X., Wang, Y., Qiu, Y., Yin, Y., Li, Y., Du, Y., Liao, Y., Lim, Y., Narusaka, Y., Wang, Z., Li, Z., Xiong, Z., and Zhang, Z. (2011). The genome of the mesopolyploid crop species *Brassica rapa*. *Nat. Genet.* 43, 1035–1039.

44. Yang, Y. W., Lai, K. N., Tai, P. Y., and Li, W. H. (1999). Rates of nucleotide substitution in angiosperm mitochondrial DNA sequences and dates of divergence between *Brassica* and other angiosperm lineages. *J. Mol. Evol.* 48, 597–604.
45. Yokoi, S., Quintero, F. J., Cubero, B., Ruiz, M. T., Bressan, R. A., Hasegawa, P. M., and Pardo, J. M. (2002). Differential expression and function of *Arabidopsis thaliana* NHX $Na^+/H^+$ antiporters in the salt stress response. *Plant J.* 30, 529–539.

CHAPTER

15

# SPAHA1 AND SPSOS1 COORDINATE IN TRANSGENIC YEAST TO IMPROVE SALT TOLERANCE

Yang Zhou[1,2,3], Xiaochang Yin[3], Ruijun Duan[1], Gangping Hao[4], Jianchun Guo[1,2], Xingyu Jiang[3]

[1]Institute of Tropical Bioscience and Biotechnology, Chinese Academy of Tropical Agricultural Sciences, Haikou, China,

[2]College of Plant Science & Technology, Huazhong Agricultural University, Wuhan, China,

[3]Hainan Key Laboratory for Sustainable Utilization of Tropical Bioresources/College of Agriculture, Hainan University, Haikou, China,

[4]Department of Biochemistry, Taishan Medical University, Tai'an, China

## ABSTRACT

In plant cells, the plasma membrane Na+ /H+ antiporter SOS1 (salt overly sensitive 1) mediates Na+ extrusion using the proton gradient generated by plasma membrane H+ - ATPases, and these two proteins are key plant halotolerance factors. In the present study, two genes from Sesuvium portulacastrum, encoding plasma membrane Na+ /H+ antiporter (SpSOS1) and H+ -ATPase (SpAHA1), were cloned. Localization of each protein was studied in tobacco cells, and their functions were analyzed in yeast cells. Both SpSOS1 and SpAHA1 are plasma membrane-bound proteins. Real-time polymerase chain reaction (PCR) analyses showed that SpSOS1 and SpAHA1 were induced by salinity, and their expression patterns in roots under salinity were similar. Compared with untransformed yeast cells, SpSOS1 increased the salt tolerance of transgenic yeast by decreasing the Na+ content. The Na+ /H+ exchange activity at plasma membrane vesicles was higher in SpSOS1-transgenic yeast than in the untransformed strain. No change was observed in the salt tolerance of yeast cells expressing SpAHA1 alone; however, in yeast transformed with both SpSOS1 and SpAHA1, SpAHA1 generated an increased proton gradient that stimulated the Na+ /H+ exchange activity of SpSOS1. In this scenario, more Na+ ions were transported out of cells, and the yeast cells co-expressing SpSOS1 and SpAHA1 grew better than the cells transformed with only SpSOS1 or SpAHA1. These findings demonstrate that the plasma membrane Na+ /H+ antiporter SpSOS1 and H+ -ATPase SpAHA1 can function in coordination. These results provide a reference for developing more salt-tolerant crops via co-transformation with the plasma membrane Na+ /H+ antiporter and H+ -ATPase.

## INTRODUCTION

Sodium is a trace element in plants. However, excess sodium in the cytoplasm is toxic to some important physiological and biochemical processes, and efflux of $Na^+$ out of the cytoplasm is an important mechanism of salt tolerance. Excess $Na^+$ in the cytoplasm can be transported into the external medium across the plasma membrane or into the vacuole across the tonoplast, such that the $Na^+$ in the cytoplasm is maintained at a non-toxic concentration. The tonoplast $Na^+/H^+$ antiporter NHX compartmentalizes $Na^+$ into vacuoles using the proton gradient generated by vacuolar $H^+$-ATPase and $H^+$-pyrophosphatase ($H^+$-PPase), whereas the plasma membrane $Na^+/H^+$ antiporter SOS1, which is driven by the proton gradient generated by plasma membrane $H^+$-ATPases, transports $Na^+$ out of cells [1].

NHX proteins from more than 60 plant species have been reported to date [1]. *Arabidopsis*AtNHX1 was the first NHX protein to be described, and its overexpression can significantly improve the salt tolerance of transgenic yeast or plants [2, 3]. Many lines of evidence have indicated that tonoplast-bound NHXs function as $Na^+/H^+$ antiporters and mediate plant salt tolerance by sequestering $Na^+$ into vacuoles [4]. As $Na^+$ influx into vacuoles via NHXs is energized by the $H^+$ gradient created by vacuolar $H^+$-ATPase and $H^+$-PPase, these two $H^+$-pumps should play important roles in plant salt tolerance. Indeed, the salt tolerance of transgenic *Arabidopsis* harboring *AVP1* (vacuolar $H^+$-PPase 1) was stronger than that of wild type [5]. Moreover, the co-expression of NHX and AVP conferred stronger salt tolerance to transgenic plants than NHX or AVP alone, a finding that further demonstrates the relationship between NHX and the $H^+$-pump, with plant NHX proteins functioning as $Na^+/H^+$ antiporters using the force from the proton gradient produced by the vacuolar $H^+$-PPase [6–8]. SOS1 is the only $Na^+$ efflux protein at the plant plasma membrane reported to date and functions as a $Na^+/H^+$ antiport system that is driven by the plasma membrane $H^+$-pump $H^+$-ATPase [9–11], and the salt tolerance of *Arabidopsis* transformed with the *AtSOS1* gene is significantly higher than that of wild type [12]. In addition, plasma membrane $H^+$-ATPase activity is involved in plant adaptation to salinity [13]. These findings suggest that the salt tolerance efficiency of the simultaneous expression of SOS1 and plasma membrane $H^+$-ATPase genes should be further enhanced, just as the co-expression of the vacuolar $H^+$-PPase and $Na^+/H^+$ antiporter results in even higher salt tolerance. However, this hypothesis has yet to be tested. Therefore, the aim of the present investigation was to determine whether the combination of the plasma membrane $Na^+/H^+$ antiporter with $H^+$-ATPase effectively increases salt tolerance in transgenic organisms. Herein, a plasma membrane $Na^+/H^+$ antiporter gene, *SpSOS1*, and a plasma membrane $H^+$-ATPase gene, *SpAHA1*, were cloned from *Sesuvium portulacastrum*, a halophyte with optimal growth at 200–300 mM NaCl [14, 15]. Functional analyses indicated that SpSOS1 could transport $Na^+$ out of cells using the force of the proton gradient generated by SpAHA1. In particular, the co-expression of *SpSOS1* and *SpAHA1* conferred stronger salt tolerance to transgenic yeast cells than the expression of either gene alone. These results may be helpful in developing more salt-tolerant crops through the co-expression of SOS1 and $H^+$-ATPase.

# MATERIALS AND METHODS

## Plant Material and Cultivation

*Sesuvium portulacastrum* (L.) L. plants, dicotyledonous halophytes belonging to the Aizoaceae family, were collected from the seashore of Haikou, China, and propagated through cuttings (S1 Text). The collection site in the seashore is a wasteland, so no one is responsible for this field. Furthermore, *Sesuvium portulacastrum* grows worldwide and is not specifically exploited or an endangered or protected species. Therefore, no specific permission was required for the collection of the plants. Three stem internodes were cut from the mother plants, incubated in Hoagland solution and allowed to grow at room temperature. After 3 weeks, uniform seedlings were treated with 0, 400, 600 or 800 mM NaCl; three replications for each treatment were performed. The roots and leaves were harvested at various times (0, 3, 6, 9, 12, 24 and 48 h) after NaCl treatment and immediately stored in liquid nitrogen.

## Cloning of *SpSOS1* and *SpAHA1*

Total RNA isolation and cDNA synthesis were performed using a reverse transcription polymerase chain reaction (RT-PCR) kit (Invitrogen, Carlsbad, CA, USA) according to the manufacturer's protocol. Degenerate primers were designed for *SpSOS1* according to the conserved sequences of plant *SOS1* genes: 5'-ATGAATGATGGGAC(G/T)GC(A/T)AT-3', 5'-TATAG(A/T/C/G)CCCAA(A/T)GT(A/G)CTTCCATG-3'. To clone the *SpSOS1* gene, the cDNA was used as the template in PCR with the degenerate primers. The PCR product, 1,850 bp in length, is homologous with other plant *SOS1* genes. To obtain the full-length *SpSOS1*sequence, 3'-RACE (rapid amplification of cDNA ends) and 5'-RACE cDNA syntheses were performed with kits according to the manufacturer's instructions (Invitrogen, Carlsbad, CA, USA). The clones obtained were sequenced, and the overlapping region with the first clone was confirmed. After deducing the open reading frame, gene primers were designed according to the open reading frame (ORF) and including E*co*R I and S*al* I restriction sites (underlined): Forward: 5'-gGAATTCATGGCGGCGTT GACTGAT-3'; Reverse: 5'-ataGTCGACTCATGGTGCAGGGCGG-3'. A fragment containing the ORF was synthesized via PCR from cDNA and sequenced for further confirmation. The *SpSOS1* ORF was cloned into the yeast expression vector p416 GPD to produce the plasmid p416-SpSOS1. Regarding the cloning of *SpAHA1*, an *SpAHA1* EST fragment (GenBank Accession No. DQ925817) was first re-obtained via PCR from cDNA and sequenced for further confirmation. Elongation of the 5'- and 3'- ends of *SpAHA1* was performed using 5'-RACE and 3'-RACE techniques with two pairs of nested primers and cDNA as the template. One pair of primers deduced from the polylinker sequence was from the RACE kit (Invitrogen, Carlsbad, CA, USA), and gene-specific primers were designed according to EST sequences. The 3'-RACE nested PCR primers were 5'-ATTGTCGGTATGACTGGAGATGGTGTG-3' and 5'-CGTCTCTATTACCATTCGTA TTGTGCTC-3', and the 5'-RACE nested PCR primers were 5'-TTGGAAAATAGCCCTACTGGTAAGCAC-3' and 5'-TGAGCATAAATCCGAGCACAATAC-3'. The fragments were cloned into pMD-19T and sequenced. Full-length *SpAHA1* was deduced by overlap of the above two fragments and was obtained via PCR with cDNA as the template and two gene-specific primers (5'-CGGGATCCATGGCCAAGGCCATCAATC-3'; 5'-CGGAATTCCCTTAGACGGTGTAGTTCTGTTGG-3') and cloned into the yeast expression vector p414 to produce the plasmid p414-SpAHA1. The complete cDNA sequences of *SpSOS1*(accession No. JX674067) and *SpAHA1* (accession No. JX628604) were deposited in GenBank.

## Quantitative real-time PCR

Real-time PCR was performed using a Fast Real-Time PCR System (ABI 7900HT, TaKaRa, Kyoto, Japan). The primers 5'-TTGGCATCGTTGAGGGTCT-3' and 5'-CAGTGGGAACACGGAAAGC-3' were used to amplify the housekeeping gene *GAPDH*. Gene-specific primers were designed for a 116-bp fragment of *SpSOS1*; the primers were 5'-GAGCATCTTTGGCAGCACG-3' in the sense direction and 5'-CTGAGTTTCGGGGATTTTAGG-3' in the anti-sense direction. A forward primer (5'-GATGGGCTGGTGTCGTTTG-3') and reverse primer (5'-CCGTAATCCTTCTTGGTGGTG-3') were designed for a 148-bp fragment of *SpAHA1*. The amplification conditions were as follows: 95°C for 30 s, followed by 40 cycles at 95°C for 5 s and

60°C for 30 s. A dissociation curve from 60°C to 95°C was generated to verify primer specificity, and data analysis was performed using the SDS plate utility software, version 2.4. The mRNA abundances of the *SpSOS1* and *SpAHA1* genes are expressed based on the values measured at 0 h of salt treatment.

## Subcellular Localization of SpSOS1 and SpAHA1 Proteins

The full-length genes *SpSOS1*, *SpAHA1* or *Arabidopsis SOS1* (*AtSOS1*) without the stop codon at the 3' end were obtained via PCR and fused in-frame to GFP in the modified binary vector pCAMBIA1300. Primers were designed for *SpSOS1*, *SpAHA1* and *AtSOS1*, as follows: SpSOS1–1300F: 5'-ataGTCGACATGGCGGCGTTGACTGATT-3', SpSOS1–1300R: 5'-cggGGTACCTGGTGCAGGGCGGGGGAAAG-3'; SpAHA1–1300F: 5'-gACTAGTATGGCCAAGGCCATCAATC-3', SpAHA1–1300R: 5'-cgGGATCCGACGGTGTAGTTCTGTTGG-3'; AtSOS1–1300F: 5'- gcTCTAGAATGACGACTGTAATCGACGCG-3', AtSOS1–1300R: 5'- cggGGTACCTAGATCGTTCCTGAAAACG-3'. The resulting constructs were transferred into *Agrobacterium tumefaciens* strain LBA4404. For transient expression in plants, different combinations of strains containing these constructs were co-infiltrated into *Nicotiana tabacum*leaves [16]. After 5 days, protoplasts were isolated from the infiltrated leaves with a buffer containing 0.4 M mannitol, 20 mM KCl, 20 mM MES (pH 5.7), 10 mM $CaCl_2$, 3.5 mM β-mercaptoethanol, 0.1% bovine serum albumin, 1.5% cellulase R10, and 0.4% macerozyme R10 [17]. GFP fluorescence in the leaf epidermal cells and mesophyll protoplasts was detected using a confocal laser scanning microscope (FV3000; Olympus Corporation, Tokyo, Japan).

## Yeast Strains and Incubation

*Saccharomyces cerevisiae* strain AXT3K (Δena1::HIS3::Δena4, Δnha1::LEU2, and Δnhx1::KanMX4), which is deficient in the main endogenous $Na^+$ transporters, is a derivative of W303 (MAT ura3–1 leu2–3,112 his3–11,15 trp1–1 ade2–1 can1–100) [10]. To assess the functionality of the SpSOS1 and SpAHA1 proteins, p416-SpSOS1, p414-SpAHA1 and p416-SpSOS1 and p414-SpAHA1 were transformed into

AXT3K. The wild-type yeast strain W303 was used as a positive control [18]. W303, AXT3K, and AXT3K transformed with p416-SpSOS1, AXT3K transformed with p414-SpAHA1, and AXT3K co-transformed with p416-SpSOS1 and p414-SpAHA1 were grown in AP medium (10 mM arginine, 8 mM phosphoric acid, 2% glucose, 2 mM $MgSO_4$, 1 mM KCl, 0.2 mM $CaCl_2$, and trace minerals and vitamins) [19]. When these precultures grew to saturation, they were diluted 10-fold, 100-fold, 1,000-fold, and 10,000-fold, and then 10 μl of each serial dilution was spotted onto AP plates with the indicated concentration of NaCl. After incubation for 3–5 days at 30°C, growth was imaged and analyzed.

## Measurement of $Na^+$ Content in Yeast Cells

Yeast cells were grown in AP medium in the presence or absence of 30 mM NaCl at 30°C. The $Na^+$ content in the cells was measured using an atomic absorption spectrophotometer (AA-670, Shimadzu Corporation, Kyoto, Japan), as previously described [19].

## Plasma Membrane Isolation and Measurements of $Na^+$/$H^+$ Exchange and $H^+$-ATPase Activity

Yeast cells were first cultivated to an OD660 of 1.5 in AP medium. A total of 50 mM NaCl (final concentration) was added to further ensure the activation of SpAHA1 and SpSOS1. At one hour after NaCl treatment, the yeast cells were harvested via centrifugation, and the plasma membrane was isolated using an aqueous two-phase system [20, 21]. The plasma membrane protein content was measured using the Bradford method with BSA as a standard [22]. The activities of $H^+$-ATPase and $Na^+$/$H^+$ exchange at the plasma membrane were determined as described previously using a fluorescence spectrophotometer (Hitachi F-2500, Hitachi, Ltd., Tokyo, Japan) [23]. The reaction mixture consisted of 5 μM quinacrine, 50 mM BTP-HCl (pH 7.5), 25 mM BTP-HEPES (pH 7.5), 50 mM $KNO_3$, 250 mM mannitol, 4 mM $MgSO_4$ and 50 μg of plasma membrane protein. The assay was initiated by the addition of 3 mM ATP (pH 7.5); when the ΔpH reached a steady state, 50 mM $Na_2SO_4$ was added to the reaction mixture. Finally, the reaction was stopped with 25 mM $(NH_4)_2SO_4$. $H^+$-ATPase activity was assayed by measuring the ΔpH across the membrane, which is expressed as the change in fluorescence quenching during the first 30 s after the addition of ATP (ΔF/min/mg protein). $Na^+$/$H^+$ exchange activity was calculated as the change in the initial rate of fluorescence recovery during the first 30 s after the addition of $Na_2SO_4$, and this value is expressed as a percentage of the maximal fluorescence (ΔF (%)/min/mg protein). The reaction was controlled at excitation and emission wavelengths of 430 nm and 500 nm, respectively.

### Statistical Analysis

A two-tailed Student's t-test was used to analyze the data, which are presented as the means ± SE.

## RESULTS

### Cloning and Molecular Description of *SpSOS1* and *SpAHA1*

To clone the genes encoding the plasma membrane $Na^+/H^+$ antiporter and $H^+$-ATPase from *S. portulacastrum*, we utilized homologs of other plant plasma membrane SOS1s or $H^+$-ATPases and designed degenerate primers. We obtained a fragment of cDNA encoding a putative homolog of the $Na^+/H^+$ antiporter *SOS1* gene, and the complete cDNA was then obtained using the rapid amplification of cDNA end (RACE) method and sequenced (GenBank accession number: JX674067). The homology-based computer search identified a 396-bp EST in *S. portulacastrum* (GenBank accession number: DQ925817) encoding a putative plant plasma membrane $H^+$-ATPase. The EST was isolated and confirmed via sequencing, and the cDNA was cloned using RACE. The *SpSOS1* gene encodes a protein of 1,155 amino acids. The SpSOS1 protein shares 61.53%, 60.58%, 67.67%, 58.94% and 60.39% amino acid identity with similar proteins from *Arabidopsis thaliana* (AtSOS1), rice (OsSOS1), poplar (PtSOS1), wheat (TaSOS1) and *Thellungiella halophila* (ThSOS1), respectively [11, 19, 23–25] (Fig 1A). TMHMM software analysis revealed that SpSOS1 contains 12 putative transmembrane domains at the N-terminus and a long hydrophilic C-terminus (http://www.cbs.dtu.dk/services/TMHMM/), and a conserved phosphorylation site (1143DSPS1146) is also present at the C-terminus. These findings suggest that SpSOS1 may function as a plasma membrane $Na^+/H^+$ antiporter that is involved in salt resistance in *S. portulacastrum*. The 3,394-bp *SpAHA1* cDNA fragment for a putative $H^+$-ATPase from *S. portulacastrum* was cloned using the RACE method and then sequenced (GenBank accession number: JX628604). This fragment contains a 2,862-bp ORF encoding a protein of 953 amino acids. An alignment of the amino acid sequences of *Arabidopsis* $H^+$-ATPases showed that SpAHA1 shares significant similarity with *Arabidopsis*AtAHAs and is closest to AtAHA1 (85.83%) (Fig 1B), indicating that SpAHA1 may function as an $H^+$-ATPase at the plasma membrane and assist in energy generation in *S. portulacastrum*by catalyzing ATP.

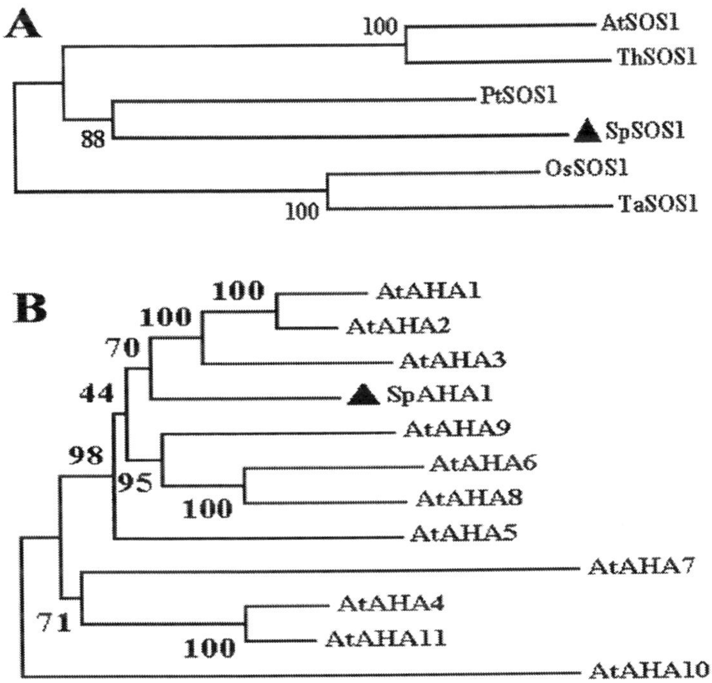


**Figure 1:** Structural analysis of SpSOS1 and SpAHA1 proteins.

(A) Phylogenetic tree of plasma membrane $Na^+/H^+$ antiporters from plants. SpSOS1 (JX674067), AtSOS1 (NM_126259), OsSOS1 (AY785147), PtSOS1 (XM_002315801), TaSOS1 (FN356232) and ThSOS1 (EF207775) indicate $Na^+/H^+$ antiporters from *Sesuvium portulacastrum*, *Arabidopsis thaliana*, *Oryza sativa*, *Populus trichocarpa*, *Triticum aestivum* and *Thellungiella halophila*, respectively. (B) Phylogenetic tree of plasma membrane $H^+$-ATPases from plants. SpAHA1 (JX628604) indicates *Sesuvium portulacastrum* $H^+$-ATPase.

AtAHA1 (NM_127453), AtAHA2 (NM_001203941), AtAHA3 (NM_001203630), AtAHA4 (NM_114664), AtAHA5 (NM_128013), AtAHA6 (NM_126721), AtAHA7 (NM_115897), AtAHA8 (NM_114131), AtAHA9 (NM_001198521), AtAHA10 (NM_101587), and AtAHA11 (NM_125662) indicate *Arabidopsis thaliana* $H^+$-ATPases. The phylogenetic trees were constructed using the MEGA5.0 software based on multiple alignments of full-length protein sequences. Evolutionary distances were computed using the neighbor joining method. Bootstrap values are indicated at the nodes of the tree and are expressed as percentages.

## Subcellular Localization of SpSOS1 and SpAHA1 Proteins

To analyze the localizations of SpSOS1 and SpAHA1 in plant cells, their ORFs fused with the *GFP* gene were each cloned into the plasmid pCAMBIA1300. AtSOS1, a plasma membrane-localized $Na^+$/$H^+$ antiporter [9], was used as a control. GFP-tagged SpSOS1, SpAHA1, AtSOS1 or GFP alone was transiently expressed in tobacco leaves. Confocal imaging showed that when GFP was expressed alone, a fluorescent signal was clearly observed in the cytosol and nucleus (Fig 2), which agrees with the previous observation indicating that GFP localizes to both the cytosol and nucleus [26, 27]. The GFP-fused AtSOS1 protein localized to the cell membrane (Fig 2). The fluorescence images of the SpSOS1-GFP or SpAHA1-GFP fusion protein precisely coincided with the pattern observed for the plasma membrane-localized protein AtSOS1 in tobacco leaf epidermal cells and mesophyll protoplasts, in contrast to the typical cytosolic and nuclear fluorescence distribution of GFP (Fig 2). These results indicate that both SpSOS1 and SpAHA1 are plasma membrane-bound proteins.

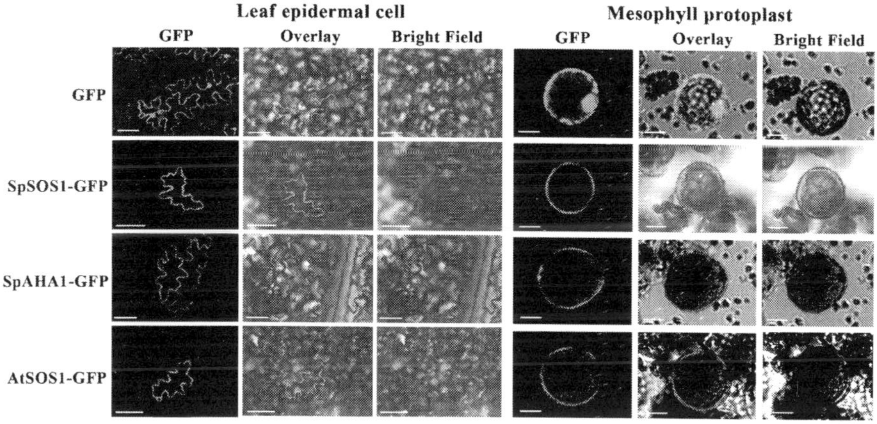


**Figure 2:** SpSOS1 and SpAHA1 proteins are localized at the plasma membrane.

SpSOS1-GFP, SpAHA1-GFP and AtSOS1-GFP fusion proteins were produced as described in the Methods and Materials section. Tobacco leaves were transformed with the vector pCAMBIA1300 containing GFP alone (GFP), pCAMBIA1300-SpSOS1-GFP (SpSOS1-GFP), pCAMBIA1300-SpAHA1-GFP (SpAHA1-GFP) or pCAMBIA1300-AtSOS1-GFP (AtSOS1-GFP). GFP signals from epidermal cells (left panel; scale bars = 50 μm) and mesophyll protoplasts (right panel; scale bars = 10 μm) of tobacco leaves transiently expressing either GFP alone, SpSOS1-GFP, SpAHA1-GFP or AtSOS1-GFP were recorded using confocal microscopy.

## Expression of SpSOS1 and SpAHA1 in *S. Portulacastrum*

*Sesuvium portulacastrum* is a halophyte with optimal growth at 200–300 mM NaCl [14]; thus, 400 mM, 600 mM and 800 mM NaCl were chosen for treatments to assess the expression of *SpSOS1* and *SpAHA1* in *S. portulacastrum* seedlings under salt stress. *SpSOS1* expression in roots exhibited a consistent pattern under salt stress: all of the treatment groups exhibited a transient increase in *SpSOS1* expression in the roots that peaked at approximately 3–6 hours, after which *SpSOS1* expression gradually decreased to basal levels at 48 h (Fig 3A). In contrast, the change in *SpSOS1* expression in leaves was not significant under salinity conditions (Fig 3B). The expression pattern of *SpAHA1* in roots exposed to salt stress was similar to that of *SpSOS1* (Fig 3C). However, compared to *SpSOS1*, *SpAHA1* had a more complex leaf expression pattern under salt stress, whereby expression was initially up-regulated and then decreased (Fig 3D).

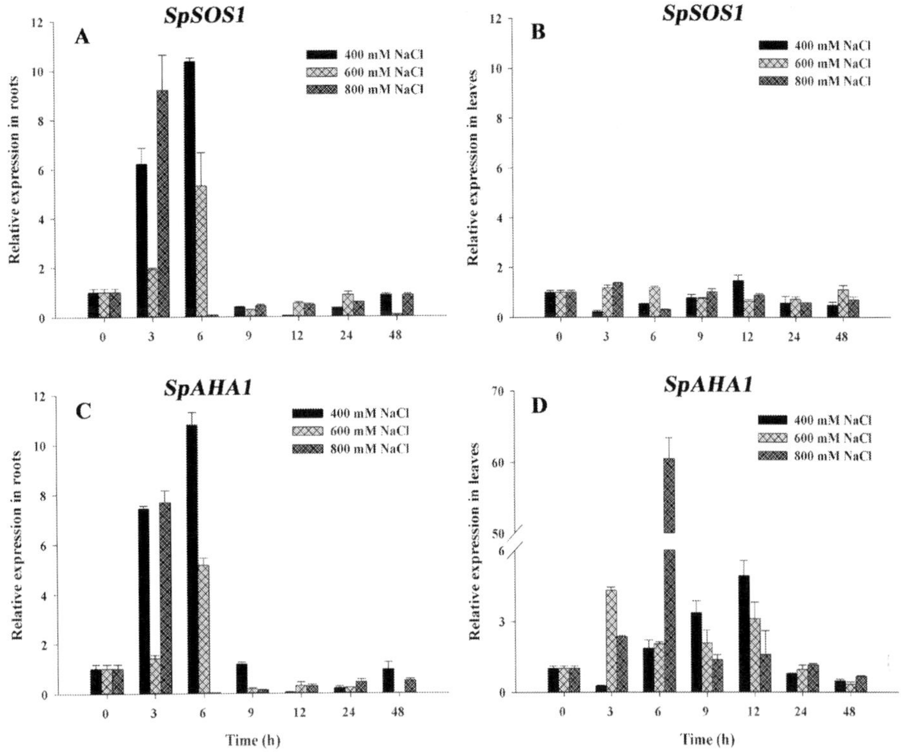


**Figure 3**: Changes in *SpSOS1* **and** *SpAHA1* **mRNAs in** *S. portulacastrum* **exposed to salinity.**

Seedlings were treated with 400, 600 or 800 mM NaCl. The expression of *SpSOS1* (A, B) and *SpAHA1* (C, D) was analyzed in roots (A, C) and leaves (B, D) at different time intervals (0, 3, 6, 9, 12, 24 and 48 h) under salt treatment using real-time PCR. The expression value at the initial time (0 h) was set to 1, and the expression levels at the other time points under salinity treatment were calculated as the fold of the value at 0 h. Values are expressed as the means±SE (n = 3).

## Effect of Salt stress on the Growth of Yeast Cells

To investigate the roles of SpSOS1 and SpAHA1 in halotolerance, the two genes were inserted into the yeast expression vectors p416 GPD and p414 and then transformed individually or co-transformed into AXT3K. The growth of the untransformed and transgenic strains was similar under normal conditions. Untransformed yeast cells grew very poorly in the presence of NaCl, and these cells could not grow on AP medium containing 75 mM NaCl. The growth of *SpAHA1*-transgenic AXT3K was similar to that of the control strain under salt stress, whereas *SpSOS1*expression increased the salt tolerance of AXT3K. Interestingly, although the expression of *SpAHA1* alone did not influence the response of transgenic yeast cells to salt stress, AXT3K cells co-transformed with *SpSOS1* and *SpAHA1* grew better than with *SpSOS1* alone under salinity conditions. Indeed, the growth of yeast cells expressing *SpSOS1* and *SpAHA1* together was the best among the control and transgenic yeast cells and nearly reached the salt tolerance of the wild-type yeast strain W303 (Fig 4).

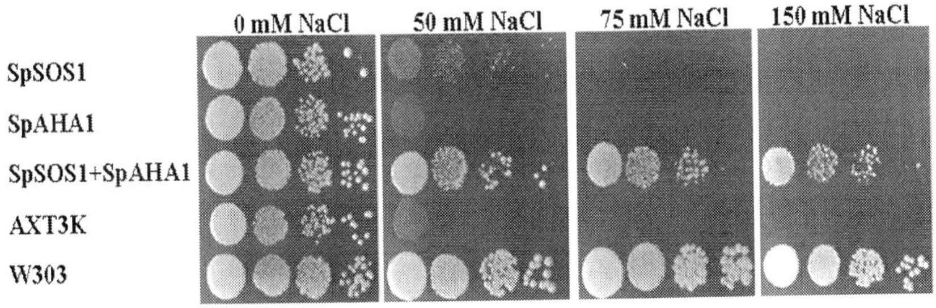


**Figure 4**: Effect of salt stress on the growth of yeast cells.

The wild-type strain (W303), mutant cells (AXT3K) and recombinant strains (SpSOS1, SpAHA1 and SpSOS1+SpAHA1) were grown to saturation, and then 10-μL serial decimal dilutions were spotted onto AP plates supplemented with 0, 50, 75, or 150 mM NaCl. SpSOS1: the AXT3K strain transformed with plasmid p416-SpSOS1; SpAHA1: the AXT3K strain transformed with plasmid p414-SpAHA1; SpSOS1+SpAHA1: mutant strain AXT3K co-transformed with plasmids p416-SpSOS1 and p414-AHA1; AXT3K: untransformed mutant strain AXT3K; W303: wild-type yeast strain W303.

## Change in $Na^+$ content

The $Na^+$ content in untransformed and transgenic cells was analyzed using atomic absorption spectrophotometry (Fig 5). Although the $Na^+$ content did not differ between the untransformed strain and various transgenic yeast cells grown under normal conditions, NaCl treatment significantly increased the $Na^+$ content in all of the yeast strains. A comparative analysis showed that the $Na^+$ content increased to a higher level in the untransformed cells than in the *SpSOS1*-transgenic cells, whereas *SpAHA1* expression did not induce a significant change in the $Na^+$ concentration in transgenic cells compared with untransformed cells. Interestingly, the $Na^+$ content in yeast cells co-expressing *SpSOS1* and *SpAHA*1 was the lowest among all yeast strains.

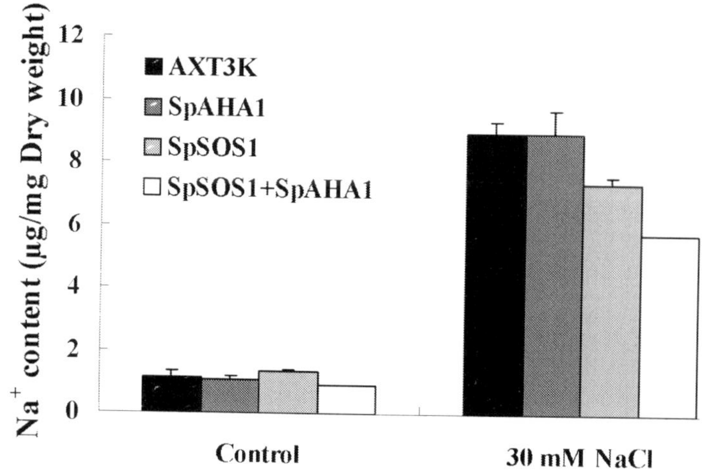


**Figure 5**: $Na^+$ **content in yeast cells.**

The $Na^+$ content in unstressed yeast cells and yeast cells treated with 30 mM NaCl was measured. Values are expressed as the means±SE (n = 3). AXT3K: the non-transformed yeast strain; SpAHA1: the AXT3K strain transformed with plasmid p414-SpAHA1; SpSOS1: the AXT3K strain transformed with plasmid p416-SpSOS1; SpSOS1+SpAHA1: the mutant strain AXT3K co-transformed with plasmids p416-SpSOS1 and p414-AHA1.

## $Na^+/H^+$ Exchange and $H^+$-ATPase Activity

The yeast strain AXT3K, which is nearly devoid of background plasma membrane $Na^+/H^+$exchange activity, has been used to investigate the $Na^+/H^+$ exchange of some plant SOS1 proteins [19, 23, 25]. SpSOS1 suppressed the salt sensitivity of AXT3K (Fig 4). Subcellular localization showed that SpSOS1 is a plasma membrane-bound protein (Fig 2); thus,

AXT3K should be a good tool for studying the $Na^+/H^+$ exchange activity of SpSOS1 at the plasma membrane. To determine whether SpSOS1 functions as a $Na^+/H^+$ exchanger in a manner similar to its counterparts AtSOS1, TaSOS1 and ThSOS1 in transgenic yeast cells [11, 19, 25], we isolated plasma membrane vesicles from AXT3K expressing the *SpSOS1* gene and from untransformed yeast cells. Little background $Na^+/H^+$ exchange activity was detected for the plasma membrane vesicles from AXT3K, which contains a mutation in the plasma membrane $Na^+/H^+$ antiporter (Fig 6), whereas the plasma membrane $Na^+/H^+$ exchange activity of yeast expressing *SpSOS1* was higher than that in untransformed yeast cells, indicating that SpSOS1 functions as a $Na^+/H^+$ antiporter. As shown in Fig 6, the generation of a transmembrane pH gradient is a prerequisite for the $Na^+$-dependent movement of $H^+$ out of vesicles, suggesting that if the transmembrane proton gradient increases, the $Na^+$ transport capacity of SOS1 should be stronger. To test this hypothesis, the $H^+$-ATPase gene *SpAHA1* was transformed into yeast cells expressing *SpSOS1*, and comparative studies were performed for $Na^+/H^+$ exchange and $H^+$-ATPase activities at the plasma membrane. $H^+$-pump activity was the highest for plasma membrane vesicles from the yeast cells co-expressing *SpSOS1* and *SpAHA1*, followed by that of the yeast cells individually transgenic for *SpAHA1* or *SpSOS1*; the untransformed strain displayed the lowest activity (Fig 6A and 6B). The co-transformed strain showed the highest $Na^+/H^+$ exchange activity among these tested yeast strains, and the strain transgenic for *SpSOS1* had a higher $Na^+/H^+$ exchange activity than the untransformed strain. Although plasma membrane $H^+$-ATPase activity in the *SpAHA1*-transformed strain was higher than that in the untransformed cells, inconclusive results were obtained for differences in the $Na^+/H^+$exchange activity between the *SpAHA1*-transformed strain and untransformed cells (Fig 6A and 6C).

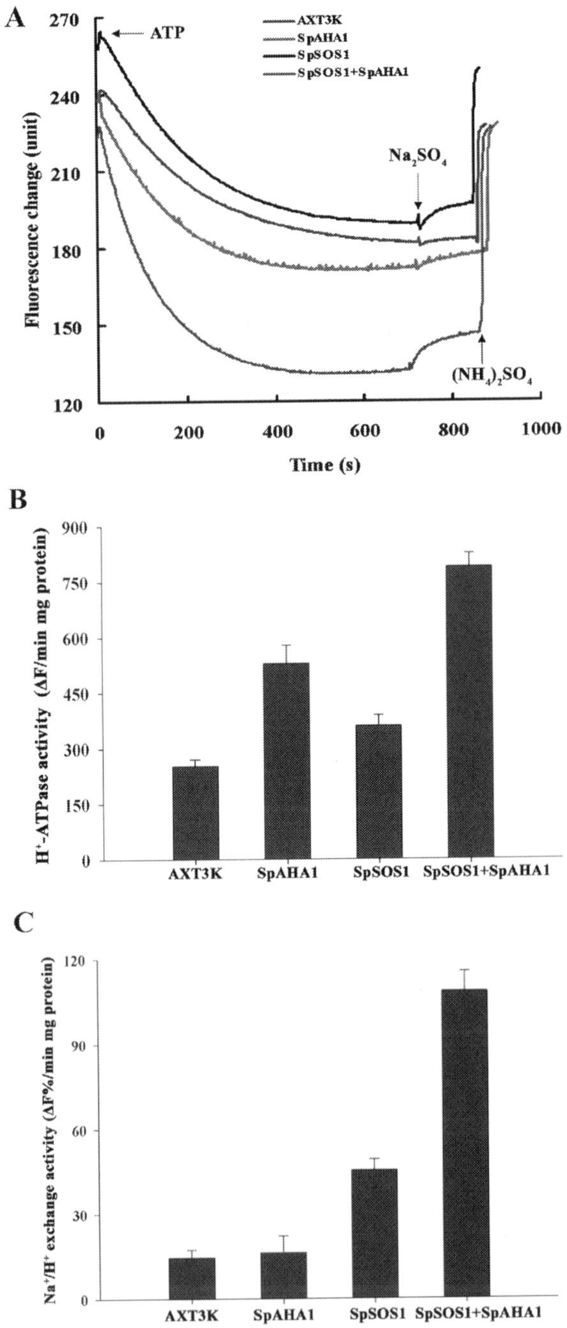


**Figure 6**: $H^+$-ATPase and $Na^+/H^+$ antiport activities at plasma membrane vesicles.

Plasma membranes were isolated from mutant cells (AXT3K) and related transgenic strains (SpSOS1, SpAHA1 and SpSOS1+SpAHA1) using the aqueous two-phase system. Fluorescent quenching of quinacrine was used to monitor the acidification of plasma membrane vesicles from these yeast cells. Formation of a pH gradient, indicating that $H^+$ was pumped into the vesicles, was initiated with ATP. Once fluorescence was stabilized, 50 mM $Na_2SO_4$ was added to the cuvette, and fluorescence recovery, indicating $Na^+/H^+$ exchange, was monitored. The reaction was terminated by adding 25 mM $(NH_4)_2SO_4$, which dissipated the pH gradient. The change in fluorescence is expressed as arbitrary units (A). $H^+$-ATPase activity was analyzed by measuring ΔpH across the membrane; it is expressed as the change in fluorescence quenching per minute per milligrams of membrane protein (ΔF/min/mg protein) (B). $Na^+/H^+$ exchange activity is given as the proportion of dissipation of the preformed pH gradient per minute per milligram of membrane protein (ΔF%/ min/mg protein) (C). Values are expressed as the means±SE of three replicates.

## DISCUSSION

One key mechanism of salinity tolerance is the ability to remove $Na^+$ ions from the cytosol via plasma membrane $Na^+$ transporters. SOS1 is the only plant plasma membrane $Na^+$ efflux protein reported to date; thus, the protein is likely a key halotolerance factor in plants. SpSOS1 shows similarities with certain reported plasma membrane $Na^+/H^+$ antiporters, and subcellular localization analysis showed that SpSOS1 is distributed at the plasma membrane in plant cells, suggesting that SpSOS1 can regulate $Na^+$ efflux and is involved in salt resistance in *S. portulacastrum*. SpAHA1 is a plasma membrane-bound protein and has high similarity to the *Arabidopsis* AHA family, indicating that SpAHA1 functions as a $H^+$-ATPase at the plasma membrane and provides energy for metabolic processes in plants.

Real-time PCR indicated that the *SpSOS1* expression in roots increased sharply within the first 3–6 h and then decreased to basal levels under salt stress. In leaves, no obvious change in *SOS1* transcriptional abundance was found during NaCl treatment (up to 48 h). The salt-induced expression pattern of *SpSOS1* is similar to that of rice *SOS1*, which shows mRNA accumulation in roots but not in shoots under salinity [23]. In wheat, the *SOS1* expression level in leaves was not up-regulated under salinity treatment, but mRNA abundance in leaves was also substantially lower than that in roots [19]. These findings suggest that more active SOS1s can transport excess $Na^+$ out of plants through the roots; correspondingly, $Na^+$ transport into shoots from roots was decreased, supporting an important salt tolerance mechanism (i.e., retarding $Na^+$ transport from roots to shoots). Salt stress induced the up-regulation of *SpAHA1*in both roots and leaves, indicating that the $H^+$-ATPase can provide more energy for the plant to resist salinity. Interestingly, the expression patterns of *SpAHA1* and *SpSOS1* in roots are similar, suggesting that they can work in coordination. Additionally, SpSOS1 functions as a $Na^+/H^+$ antiporter using the proton gradient produced by SpAHA1.

The expression of *SpSOS1* increased the salt tolerance of transgenic AXT3K, a strain that is very sensitive to salt treatment because $Na^+$ efflux transporter deficiency at the plasma membrane. Cells expressing *SpSOS1* showed higher $Na^+$ efflux activity than the untransformed yeast strain, which is congruent with the lower intracellular sodium levels in the NaCl-treated yeast strain expressing *SpSOS1* compared to untransformed cells. Thus, the association between $Na^+$ transport activity, intracellular $Na^+$ concentration and halotolerance suggests that SpSOS1 may function as a sodium extrusion system. The plasma membrane $H^+$-ATPase is very important for energizing sodium efflux, and the plasma membrane $H^+$-ATPase is involved in salt tolerance in tomato [13]. In the present study, yeast cells expressing *SpAHA1* exhibited significantly increased $H^+$-pump activity compared with non-transformed cells, but there were no obvious differences in either salt tolerance or $Na^+$ content between *SpAHA1*-transgenic cells and non-transformed cells under salt stress. As a hypothesis, although enhanced plasma membrane $H^+$ pumping in transgenic yeast cells provides extra energy to support $Na^+$ export mediated by the $Na^+/H^+$ antiporter at the plasma membrane, excess $Na^+$ in the cytosol cannot be transported out of cells because the $Na^+$ transporter utilizing the driving force is deficient at the plasma membrane of AXT3K. This hypothesis is supported by the measurement of the $Na^+/H^+$ exchange activity of plasma membrane vesicles, showing that $Na^+/H^+$ antiport activity did not markedly increase at the plasma membrane of *SpAHA1*-transgenic cells compared with the non-transformed strain. Although excess salt is toxic to physiological and biochemical processes in plants, some species can grow in environments containing high levels of salts because they have evolved the unique ability to resist salinity [28]. One key mechanism in salinity tolerance is the ability to remove excessive $Na^+$ from the cytosol via tonoplast $Na^+/H^+$ antiporters (NHXs) and the plasma membrane $Na^+/H^+$ antiporter (SOS1). The proton motive force utilized by vacuolar $Na^+/H^+$ antiporters is generated by $H^+$-translocating enzymes in the tonoplast, such as V-ATPase and $H^+$-inorganic PPase. Recent studies have demonstrated that the vacuolar $Na^+/H^+$antiporter can utilize the electrochemical $H^+$ gradient generated by tonoplast $H^+$-pumps to sequester $Na^+$ into vacuoles via the simultaneous overexpression of the two tonoplast proteins in plants. The salt tolerance of rice co-expressing *SsNHX1* and *AtAVP1* was higher than that achieved by single-gene transformation [6], and rice plants engineered to overexpress native *NHX* and $H^+$-PPase genes showed higher salt tolerance than plants expressing *OsNHX1* or *OsVP1* alone [7]. In addition, Gouiaa et al. reported that tobacco plants transformed with both *TNHXS1* and *TVP1* grew better than the single gene-transformed lines under salinity [8]. The plasma membrane $Na^+/H^+$ antiporter can also transport excess $Na^+$ in the cytoplasm out of cells using the driving force of the $H^+$ gradient produced by plasma membrane $H^+$-pumps. *Arabidopsis* plants transformed with the *AtSOS1* gene showed higher salt tolerance than wild-type plants [12]. Although plasma membrane $H^+$-ATPase activity correlates with the adaptation of plants to salt stress [13], the overexpression of plasma membrane $H^+$-ATPase has not been reported to confer strong salt tolerance to transgenic plants. It is difficult to engineer organisms with vacuolar $H^+$-ATPase because it comprises multiple protein subunits located in the tonoplast, and there is no experimental evidence for enhanced salt tolerance in transgenic organisms with a vacuolar $H^+$-ATPase. In the present investigation, SpAHA1 produced an increased electrochemical gradient of protons to promote the $Na^+/H^+$ exchange activity of SpSOS1, with more $Na^+$ ion being transported out of cells. Correspondingly, yeast cells co-expressing SpSOS1 and SpAHA1 grew better than cells transformed with SpSOS1 or SpAHA1 alone, demonstrating that the plasma membrane $Na^+/H^+$ antiporter SpSOS1 and plasma membrane $H^+$-ATPase SpAHA1 could work in coordination in transgenic yeast cells. These results provide a reference for the development of more salt-tolerant crops via co-transformation of the plasma membrane $Na^+$/$H^+$ antiporter and $H^+$-ATPase.

## ACKNOWLEDGMENTS

We thank American Journal Experts (AJE) for English language editing.

## REFERENCES

1. Pardo JM, Cubero B, Leidi EO, Quintero FJ. Alkali cation exchangers: roles in cellular homeostasis and stress tolerance. J Exp Bot. 2006; 57(5): 1181–1199. pmid:16513813
2. Apse MP, Aharon GS, Snedden WA, Blumwald E. Salt tolerance conferred by overexpression of a vacuolar Na+/H+ antiport in Arabidopsis. Science. 1999; 285(5431):1256–1258. pmid:10455050
3. Gaxiola RA, Rao R, Sherman A, Grisafi P, Alper SL, Fink GR. The Arabidopsis thaliana proton transporters, AtNhx1 and Avp1, can function in cation detoxification in yeast. Proc Natl Acad Sci USA. 1999; 96(4): 1480–1485. pmid:9990049
4. Rodriguez-Rosales MP, Galvez FJ, Huertas R, Aranda MN, Baghour M, Cagnac O, et al. Plant NHX cation/proton antiporters. Plant Signal Behav. 2009; 4(4): 265–276. pmid:19794841
5. Gaxiola RA, Li J, Undurraga S, Dang LM, Allen GJ, Alper SL, et al. Drought- and salt-tolerant plants result from overexpression of the AVP1 H+-pump. Proc Natl Acad Sci USA. 2001; 98: 11444–11449. pmid:11572991
6. Zhao FY, Zhang XJ, Li PH, Zhao YX, Zhang H. Co-expression of the Suaeda salsa SsNHX1 and Arabidopsis AVP1 confer greater salt tolerance to transgenic rice than the single SsNHX1. Mol Breeding. 2006; 17(4): 341–353.
7. Liu SP, Zheng LQ, Xue YH, Zhang Q, Wang L, Shou HX. Overexpression of OsVP1 and OsNHX1 increases tolerance to drought and salinity in rice. J Plant Biol. 2010; 53: 444–452.
8. Gouiaa S, Khoudi H, Leidi EO, Pardo JM, Masmoudi K. Expression of wheat Na+/H+ antiporter TNHXS1 and H+-pyrophosphatase TVP1 genes in tobacco from a bicistronic transcriptional unit improves salt tolerance. Plant Mol Biol. 2012; 79: 137–155. pmid:22415161

9. Qiu QS, Guo Y, Dietrich MA, Schumaker KS, Zhu JK. Regulation of SOS1, a plasma membrane Na+/H+ exchanger in Arabidopsis thaliana, by SOS2 and SOS3. Proc Nat Acad Sci USA. 2002; 99: 8436–8441. pmid:12034882
10. Quintero FJ, Ohta M, Shi H, Zhu JK, Pardo JM. Reconstitution in yeast of the Arabidopsis SOS signalling pathway for Na+ homeostasis. Proc Nat Acad Sci USA. 2002; 99: 9061–9066. pmid:12070350
11. Quintero FJ, Martı´nez-Atienza J, Caros I, Jiang XY, Kim WY, Ali Z, et al. Activation of the plasma membrane Na/H antiporter SOS1 by phosphorylation of an auto-inhibitory C-terminal domain. Proc Nat Acad Sci USA. 2011; 108: 2611–2616. pmid:21262798
12. Yang Q, Chen ZZ, Zhou XF, Yin HB, Li X, Xin XF, et al. Overexpression of SOS (Salt Overly Sensitive) genes increases salt tolerance in transgenic Arabidopsis. Mol Plant. 2009; 2: 22–31. pmid:19529826
13. Kerkeb L, Donaire JP, Venema K, Rodriguez-Rosales MP. Tolerance to NaCl induces changes in plasma membrane lipid composition, fluidity and H+-ATPase activity of tomato calli. Physiol Plant. 2001; 113: 217–224. pmid:12060299
14. Yi XP, Sun Y, Yang Q, Guo AP, Chang LL, Wang D, et al. Quantitative proteomics of Sesuvium portulacastrum leaves revealed that ion transportation by V-ATPase and sugar accumulation in chloroplast played crucial roles in halophyte salt tolerance. J Proteomics. 2014; 99: 84–100. pmid:24487036
15. Yang CL, Duan RJ, Li RM, Hu XW, Fu SP, Guo JC. The physiological characteristics of salt-tolerance in Sesuvium portulacastrum L. Acta Ecologica Sinica. 2010; 30(17): 4617–4627.
16. Waadt R, Schmidt LK, Lohse M, Hashimoto K, Bock R, Kudla J. Multicolor bimolecular fluorescence complementation reveals simultaneous formation of alternative CBL/CIPK complexes in planta. Plant J. 2008; 56: 505–516. pmid:18643980
17. Yoo SD, Cho YH, Sheen J. Arabidopsis mesophyll protoplasts: a versatile cell system for transient gene expression analysis. Nat Protocols. 2007; 2(7): 1565–1572. pmid:17585298
18. Xu YY, Zhou Y, Hong S, Xia ZH, Cui DQ, Guo JC, et al. Functional

characterization of a wheat NHX antiporter gene TaNHX2 that encodes a K+/H+ exchanger. Plos One. 2013; 8(11): e78098. pmid:24223765

19. Xu HX, Jiang XY, Zhan KH, Cheng XY, Chen XJ, Pardo JM, et al. Functional characterization of a wheat plasma membrane Na+/H+ antiporter in yeast. Arch Biochem Biophys. 2008; 473: 8–15. pmid:18316035
20. Menendez A, Larsson C, Ugalde U. Purification of functionally sealed cytoplasmic side-out plasma membrane vesicles from Saccharomyces cerevisias. Anal Biochem. 1995; 230(2): 308–314. pmid:7503423
21. Hong S, Cong XL, Jing HY, Xia ZH, Huang X, Hu XW, et al. Characterization of Ca2+/H+ exchange in the plasma membrane of Saccharomyces cerevisiae. Arch Biochem Biophys. 2013; 537: 125–132. pmid:23871844
22. Brandford MM. A rapid and sensitive method for the quantitation of microgram quantities of protein utilizing the principle of protein-dye binding. Anal Biochem. 1976; 72: 248–254. pmid:942051
23. Martinez-Atienza J, Jiang XY, Garciadeblas B, Mendoza I, Zhu JK, Pardo JM, et al. Conservation of the salt overly sensitive pathway in rice. Plant Physiol. 2007; 143: 1001–1012. pmid:17142477
24. Tang RJ, Liu H, Bao Y, Lv QD, Yang L, Zhang HX. The woody plant poplar has a functionally conserved salt overly sensitive pathway in response to salinity stress. Plant Mol boil. 2010; 74: 367–380.
25. Oh DH, Leidi E, Zhang Q, Hwang SM, Li YZ, Quintero FJ, et al. Loss of halophytism by interference with SOS1 expression. Plant Physiol. 2009; 151: 210–222. pmid:19571313
26. Kim BG, Waadt R, Cheong YH, Pandey GK, Dominguez-Solis JR, Schultke S, et al. The calcium sensor CBL10 mediates salt tolerance by regulating ion homeostasis in Arabidopsis. Plant J. 2007; 52: 473–484. pmid:17825054
27. D'Angelo C, Weinl S, Batistic O, Pandey GK, Cheong YH, Schultke S, et al. Alternative complex formation of the Ca-regulated protein kinase CIPK1 controls abscisic acid-dependent and independent stress responses in Arabidopsis. Plant J. 2006; 48: 857–872. pmid:17092313

28. Flowers TJ, Colmer TD. Salinity tolerance in halophytes. New Phytol. 2008; 179: 945–963. pmid:18565144

CHAPTER

# 16

# MN-EUVERING MANGANESE: THE ROLE OF TRANSPORTER GENE FAMILY MEMBERS IN MANGANESE UPTAKE AND MOBILIZATION IN PLANTS

Amanda L. Socha and Mary Lou Guerinot

Department of Biological Sciences, Dartmouth College, Hanover, NH, USA

**Keywords**: manganese, metal transport, Arabidopsis, rice, synchrotron x-ray fluorescence

Manganese (Mn), an essential trace element, is important for plant health. In plants, Mn serves as a cofactor in essential processes such as photosynthesis, lipid biosynthesis and oxidative stress. Mn deficient plants exhibit decreased growth and yield and are more susceptible to pathogens and damage at freezing temperatures. Mn deficiency is most prominent on alkaline soils with approximately one third of the world's soils being too alkaline for optimal crop production. Despite the importance of Mn in plant development, relatively little is known about how it traffics between plant tissues and into and out of organelles. Several gene transporter families have been implicated in Mn transport in plants. These transporter families include NRAMP (natural resistance associated macrophage protein), YSL (yellow stripe-like), ZIP (zinc regulated transporter/iron-regulated transporter [ZRT/IRT1]-related protein), CAX (cation exchanger), CCX (calcium cation exchangers), CDF/MTP (cation diffusion facilitator/metal tolerance protein), P-type ATPases and VIT (vacuolar iron transporter). A combination of techniques including mutant analysis and Synchrotron X-ray Fluorescence Spectroscopy can assist in identifying essential transporters of Mn. Such knowledge would vastly improve our understanding of plant Mn homeostasis.

## INTRODUCTION

With an increasing human population, there is a growing demand for improvements in crop production. Mineral nutrients such as iron (Fe), copper (Cu), and zinc (Zn) are essential for both plants and animals, but maintaining optimal micronutrient levels presents challenges in both the plant and animal kingdoms. To combat human nutrient deficiency, efforts are being made to increase the bioavailability of nutrients in the edible portions of plants (Schroeder et al., 2013). Another concern is contamination of the soil with non-essential metal(oid)s such as arsenic (As), lead (Pb), cadmium (Cd), and mercury (Hg) (Salt et al., 1998). These elements are detrimental to both plants (through direct uptake from the soil) and humans (through consumption of contaminated plant products). One strategy for remediating soil contaminated with metals is using hyperaccumulator plants, which can tolerate and store remarkably high levels of metals in the aerial portion of the plant (Kramer, 2010). Mn is an example of an element that is both required in humans but can potentially be toxic. In excess, Mn can induce neurological symptoms that resemble Parkinson's disease (Martinez-Finley et al., 2013). Therefore, reducing human exposure to high levels of Mn is a major worldwide health concern. By better understanding the molecular mechanisms, specifically which transport proteins allow plants to take up and store metals, steps can be taken toward improved crop growth and the engineering of biofortified plants for improved human nutrition.

This review will focus on the micronutrient Mn. Despite its importance in cellular processes, such as photosynthesis and protecting cells against reactive oxygen species (ROS), little is known about transporters essential for Mn uptake and storage in the cell. We describe common and emerging techniques for assaying Mn localization and accumulation within the plant. In addition, we detail the protein families implicated in Mn transport. Particular emphasis will be placed on the model plant species *Arabidopsis thaliana* and *Oryza sativa* (rice), on which the majority of characterization work has been carried out. The transporter families will be organized based on their putative function of transporting Mn into or out of the cytoplasm. Table 1 summarizes the transporters discussed in this review. The subcellular localization (in *A. thaliana* and rice) and tissue localization (in *A. thaliana*) of the transporters is shown in Figures 1 and 2, respectively.

**Table 1**: Putative Mn transporters referenced in the text.

| Transporter | Tissue expression | Subcellular localization | Transcript response to Mn deficiency | Transcript response to Mn toxicity | Other proposed substrates | References |
|---|---|---|---|---|---|---|
| NRAMP FAMILY | | | | | | |
| AtNRAMP1 | Root (all tissues) > shoot | PM | Up in root | - - | Fe, Cd | Curie et al., 2000; Thomine et al., 2000; Cailliatte et al.,2010 |
| AtNRAMP3 | Shoot and root vasculature, developing seed | VM | None | - - | Fe, Cd | Thomine et al., 2000 |
| AtNRAMP4 | Shoot vasculature > root vasculature, developing seed | VM | None | - - | Fe, Zn, Cd | Thomine et al., 2000 |
| OsNRAMP3 | Nodes > root and leaf vasculature, panicle, husk, flower | PM | None | None (PTM) | | Yamaji et al., 2013 |
| OsNRAMP5 | Root exodermis and endodermis | PM | None | – | Cd | Ishimaru et al , 2012; Sasaki et al., 2012 |
| DMTI | Root, leaf, stems | PM/peri-bacteroid membrane | - - | - - | Fe, Zn, Cu | Kaiser et al., 2003 |
| LeNRAMP1 | Root | VM | - - | - - | | Bereczky et al., 2003 |
| LeNRAMP3 | Root > shoot | VM | - - | - - | | Bereczky et al., 2003 |
| YSL FAMILY | | | | | | |
| AtYSL4 | Shoot, silique, root, flower, developing seed | VM/EM/CM | None | - - | Ni, Fe** | Conte et al., 2013; Divol et al., 2013 |
| AtYSL6 | Shoot, flower, silique, developing seed | VM/EM/CM | None | - - | Ni, Fe** | Conte et al., 2013; Divol et al., 2013 |
| OsYSL2 | Leaf and leaf sheath phloem, root phloem, developing seed | PM | None | - - | Fe | Koike et al., 2004 |
| OsYSL6 | Root, shoot | Un-determined | None | - - | - - | Sasaki et al., 2011 |
| ZmYS1 | Root epidermis, leaf mesophyll | PM | - - | - - | Fe, Zn, Cu, Ni, Cd | Roberts et al., 2004; Schaaf et al., 2004; Ueno et al., 2009 |
| ZIP FAMILY | | | | | | |
| AtIRT1 | Root epidermis, flower | PM | None | - - | Fe, Zn, Cu, Cd, Co | Eide et al., 1996; Korshunova et al., 1999; Vert et al., 2002 |
| HvIRT1 | Root | PM, ER | None | - - | Fe, Zn | Pedas et al., 2008 |
| PsIRT1 | Root | PM | - - | - - | Fe, Zn, Cd | Cohen et al., 2004 |

| LeIRT1 | Root, flowers | - - | - - | - - | Fe, Zn, Cd | Eckhardt et al., 2001 |
|---|---|---|---|---|---|---|
| LeIRT2 | Root | - - | - - | - - | Fe, Zn, Cd | Eckhardt et al., 2001 |
| AtZIP1 | Root vasculature > shoot vasculature | VM | Up in shoot | - - | Zn | Milner et al., 2013 |
| AtZIP2 | Root vasculature > shoot vasculature | PM | Down in shoot | - - | Zn | Milner et al., 2013 |
| AtZIP5 | Root > shoot | - - | - - | - - | | Milner et al., 2013 |
| AtZIP6 | Root > shoot | - - | - - | - - | | Milner et al., 2013 |
| AtZIP7 | Shoot > root | - - | - - | - - | Zn, Fe | Milner et al., 2013 |
| AtZIP9 | Root, shoot | - - | - - | - - | | Milner et al., 2013 |
| MtZIP4 | Leaf, root* | MM (predicted) | Down in leaf | - - | | Lopez-Millan et al., 2004 |
| MtZIP7 | Leaf | PM (predicted) | None | - - | | Lopez-Millan et al., 2004 |
| CAX FAMILY | | | | | | |
| AtCAX2 | Root, shoot and flower vasculature, fruit, stem | VM | - - | None | Ca, Cd, Zn | Hirschi et al., 2000; Schaaf et al., 2002; Shigaki et al.,2003; Pittman et al., 2004; Edmond et al., 2009 |
| AtCAX4 | Root, leaf, stem, flower, silique | VM | - - | Up | Cd | Cheng et al., 2002 |
| AtCAX5 | Root, stem, fruit, flower, leaf | VM | - - | Up | | Edmond et al., 2009 |
| OsCAX1a | Root, shoot, flower | - - | - - | - - | Ca | Kamiya and Maeshima, 2004; Kamiya et al., 2005 |
| OsCAX3 | Root, shoot, flower | - - | - - | - - | Ca | Kamiya and Maeshima, 2004; Kamiya et al., 2005 |
| LeCAX2 | Leaf, fruit | - - | - - | - - | Ca | Edmond et al., 2009 |
| HvCAX2 | Root, shoot, seed | - - | - - | None | Ca | Edmond et al., 2009 |
| CCX FAMILY | | | | | | |
| AtCCX3 | Flowers, stem, leaf, root | VM, EM | - - | Up in root and flowers | Na, K | Morris et al., 2008 |
| CDF/MTP FAMILY | | | | | | |
| AtMTP11 | Leaf hydathodes, guard cells, root tip | Golgi/PVC | - - | None | | Delhaize et al., 2007; Peiter et al., 2007 |
| OsMTP8.1 | Shoot | VM | None | Up in shoot | | Chen et al., 2013 |
| ShMTP8 | - - | Internal organelle | - - | - - | - - | Delhaize et al., 2003 |
| PtMTP11.1 | - - | GLC | - - | - - | | Peiter et al., 2007 |
| PtMTP11.2 | - - | GLC | - - | - - | | Peiter et al., 2007 |
| BmMTP10 | Root, shoot | Golgi | - - | Up in root and shoot | | Erbasol et al., 2013 |
| BmMTP11 | Root, shoot | Golgi | - - | None | | Erbasol et al., 2013 |
| P-type ATPase FAMILY | | | | | | |
| AtECA1 | Root vasculature, flower, leaf vasculature, stem, silique | ER | - - | - - | Ca, Zn, Ni | Wu et al., 2002 |
| AtECA3 | Root vasculature and tip, Leaf vasculature, hydathodes, guard cells, flower, stem, silique | Golgi/EM | None | - - | Ca | Mills et al., 2008 |
| LeLCA1 | - - | - - | - - | - - | Ca | Johnson et al., 2009 |
| VIT/CCC1-LIKE FAMILY | | | | | | |
| AtVIT1 | Developing seed, vasculature | VM | - - | - - | Fe | Kim et al., 2006 |
| OsVIT1 | Leaf > root, stem, panicle, embryo | VM | - - | - - | Fe, Zn | Zhang et al., 2012 |
| OsVIT2 | Leaf > root, stem, panicle, embryo | VM | - - | - - | Fe, Zn | Zhang et al., 2012 |

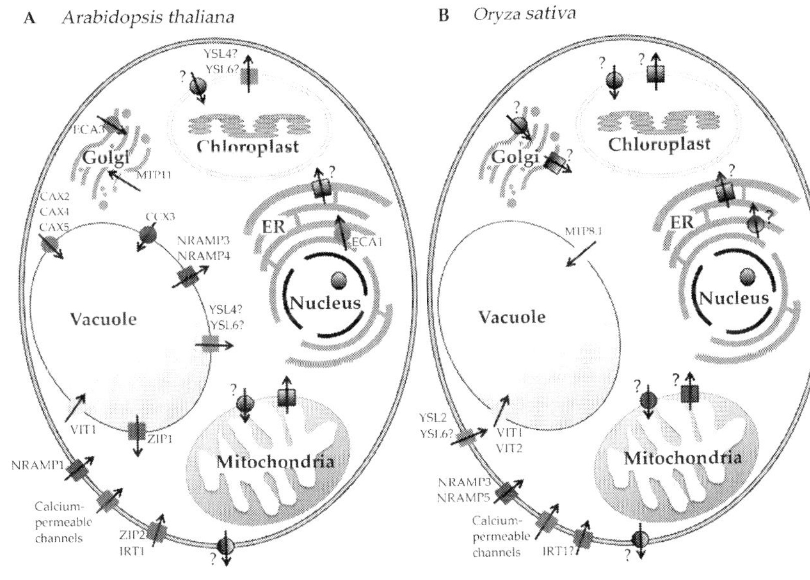


**Figure 1:** Subcellular localization of putative Mn transporters. A diagram of a plant cell showing the Mn transport pathways in **(A)** *A. thaliana* and **(B)** *O. Sativa*. Squares, import into the cytosol; Circles, export out of the cytosol; Gray, unknown; Red, ZIP family; Magenta, Calcium-permeable channels; Orange, YSL family; Light blue, CAX family; Dark Blue, CCX family; Yellow, VIT family; Purple, NRAMP family; Brown, P2A-Type ATPase family; Green, CDF/MTP family.

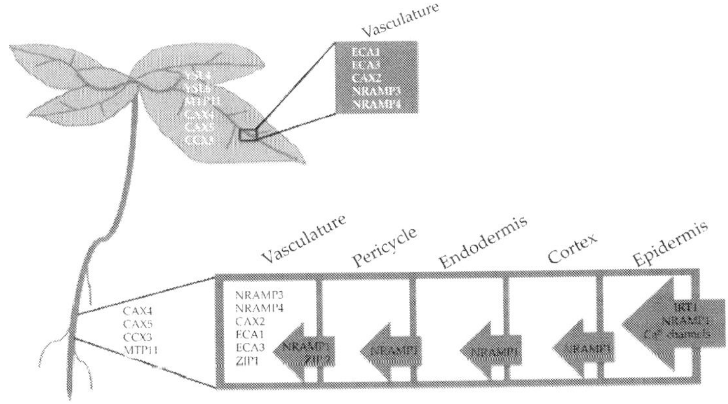


**Figure 2:** Tissue localization of Mn transporters. The probable role of transporter family members in translocating Mn from the soil into the aerial portion of the

plant in *A. thaliana*. The transporters listed to the left of the cells are not yet localized to a specific tissue.

## SYMPTOMS OF MN DEFICIENCY AND MN TOXICITY

Mn is a micronutrient element required for normal plant growth and development. It is essential for most photosynthetic organisms as a component of the oxygen-evolving complex in photosystem II (PSII), which catalyzes the water-splitting reaction to produce oxygen and provides electrons for the photosynthetic electron transport chain (Goussias et al., 2002; Nickelsen and Rengstl, 2013). Mn is also required for multiple steps in carbohydrate, lipid and lignin biosynthesis in plants (Marschner, 2012). Mn acts as a direct cofactor of a variety of enzymes (~35 in plants), which include decarboxylases of the TCA cycle, RNA polymerases and numerous glycosyl transferases. In some enzymes, other metals can replace Mn as an enzymatic cofactor (Hebbern et al., 2009). Typically magnesium (Mg) replaces Mn because it is 50–100 times more abundant in the cell. Conversely, when Mn is in excess, it can replace Mg, which can have detrimental effects on the cellular processes in which Mg is involved. There are some enzymes that specifically require Mn such as those involved in cellular redox reactions. Mn is an indispensable component of Mn superoxide dismutase (MnSOD), a principal antioxidant enzyme of the mitochondria. A recent study analyzing the effect of Mn deficiency on Chlamydomonas showed that MnSOD activity decreases before that of PSII (Allen et al., 2007). This finding suggests intracellular regulation of Mn to support PSII function in the chloroplast in preference to MnSOD function in the mitochondria.

Plant uptake of Mn is a function of the Mn oxidation state in the soil. While Mn can exist in a range of oxidation states ($Mn^{1+}$, $Mn^{2+}$, $Mn^{3+}$, $Mn^{4+}$, $Mn^{6+}$, and $Mn^{7+}$), the most commonly found forms in biological systems are $Mn^{2+}$, $Mn^{3+}$ and $Mn^{4+}$with $Mn^{4+}$ being the least stable (Marschner, 2012). The most soluble species in soil is the divalent cation, $Mn^{2+}$, which is also the form of Mn that is most efficiently accumulated in plants (Marschner, 2012). Soil pH is a major determinant of Mn oxidation state in soil. At neutral or higher pH, $Mn^{3+}$ and $Mn^{4+}$ predominate and insoluble Mn oxides will form (Rengel, 2000; Marschner, 2012). Mn solubility is also influenced by microorganisms, which can either reduce or oxidize Mn, thereby affecting its availability to the plant (Lovley et al., 2011; Geszvain et al., 2012).

Despite its necessity, Mn is required in relatively small amounts (~20–40 milligrams per kilogram of dry weight in most crop species) (He et al., 2005; Jiang, 2006; Marschner, 2012). However, it is one of the most prevalent trace element deficiencies seen in cereals including wheat and barley (Jiang, 2006). Fertilizers containing Mn sulphate can be added to the soil, however this is costly and the added Mn can be oxidized making it unavailable for plant acquisition. A more cost-effective and eco-friendly alternative is to preferentially breed plants that are able to thrive in low Mn soil. There has been a significant effort to identify crop species that are considered Mn efficient and can tolerate growth on low Mn, either by storing more Mn or by having an increased ability to absorb Mn from the soil (Jiang, 2006; Pedas et al., 2008). The genetic basis of Mn efficiency is not well understood.

Mn-deficient plants exhibit inhibited growth and decreased biomass (Marschner, 2012). Interveinal chlorosis due to decreases in net photosynthesis and chlorophyll content are common. This is most likely because the Mn-complex is needed to stabilize the PSII reaction center protein, D1 (Krieger et al., 1998; Allen et al., 2007; Yanykin et al., 2010). Mn-deficient plants are also characterized by tissue necrosis due to a decrease in MnSOD levels and an increase in oxygen free radicals (Allen et al., 2007; Marschner, 2012). Because Mn is a cofactor in the biosynthesis of cinnamic acid and its polymerization into lignin, decreased lignin concentration is especially prominent in the roots of Mn deficient plants (Marschner, 2012; Salvador et al., 2013). This is believed to be a contributing factor to an increased susceptibility to damage by freezing temperatures and root-infecting pathogens (Marschner, 2012). Mn deficiency can result from multiple factors including high concentrations of other minerals in the soil [i.e., Fe, Mg, calcium (Ca), phosphorus (P)] that can interfere with Mn absorption as well as soil alkalinity ($pH > 7.5$) (Lynch and St. Clair, 2004; Marschner, 2012). Highly calcareous soils and soils found in arid and semi-arid regions are described as too alkaline for vegetative growth (Lynch and St. Clair, 2004).

Mn toxicity can also be detrimental to plants. Toxicity can occur in poorly drained acidic soils ($pH < 5.5$), where there is a predominance of

$Mn^{2+}$ (Lynch and St. Clair, 2004). Excess Mn can also prevent the uptake and translocation of other essential elements such as Ca, Mg, Fe, and P, presumably due to the similarity in ionic radius or binding strength for ligands (Marschner, 2012; Millaleo et al., 2013). In addition, toxicity can inhibit PSII and increase the accumulation of oxidized Mn and oxidized phenolic compounds in the leaf apoplast (Fecht-Christoffers et al., 2003; Marschner, 2012). As a result, common symptoms of Mn toxicity include interveinal chlorosis and tissue necrosis, which manifests as brown spots on mature leaves of plants, ultimately resulting in reduced plant biomass (Marschner, 2012). There are multiple mechanisms proposed to combat the deleterious effects of excess Mn. For example, plants can sequester Mn in the apoplast or vacuole (Horst and Maier, 1999; Hirschi et al., 2000; Schaaf et al., 2002). In fact, ectopic expression of vacuolar Mn transporters can increase the plant's tolerance to excess Mn (Delhaize et al., 2003). Free Mn ions can be chelated in metabolically inactive $Mn^{2+}$-organic acid complexes (Horst and Maier, 1999; Pittman, 2005; Fernando et al., 2010). In addition, $Mn^{2+}$ can be mobilized into the endoplasmic reticulum (ER), thereby reducing cytoplasmic Mn (Wu et al., 2002).

## TECHNIQUES FOR STUDYING METAL TRANSPORT AND ACCUMULATION IN PLANTS

The range of metals transported by a particular transport protein can be determined by expressing the gene in a simpler model, usually yeast. Some of the yeast strains commonly used to test metal transport capabilities are *smf1*, *fet3fet4*, *ctr1*, and *zrt1zrt2*, which are unable to transport Mn, Fe, Cu, and Zn, respectively, across the plasma membrane (Dancis et al., 1994; Dix et al., 1994; Supek et al., 1996; Zhao and Eide, 1996a,b). Specific metal transport by a protein of interest is indicated by rescued growth of these yeast strains in a low Mn, Fe, Cu or Zn media or in the presence of a divalent metal chelator. For example, the *smf1* mutant strain is unable to grow on media containing the divalent cation chelator EGTA. Another commonly used yeast strain to assay metal transport is *pmr1*. The $P_2$-type Ca-ATPase, PMR1 (plasma membrane ATPase related 1) pumps both $Ca^{2+}$ and $Mn^{2+}$ in the Golgi for detoxification purposes or for use as a cofactor for Golgi-localized proteins (Rudolph et al., 1989; Durr et al., 1998). When PMR1 is defective, yeast are more sensitive to high concentrations of $Mn^{2+}$ (Durr et al., 1998). Therefore, complementation of *pmr1* yeast with a Mn efflux transporter should restore growth when Mn in the media is high. Indirect studies using yeast, such as competition assays, are sometimes used to determine if a transporter has broad specificity (examples can be found in Grotz et al., 1998; Kaiser et al., 2003). However, further studies are necessary to confirm biological function.

Xenopus oocytes, immature eggs of an aquatic frog, are used to study the physiological function of a transporter. Electrophysiological measurements can be recorded in this system as well as uptake of radioactively labeled metals like $^{54}Mn$. This system also allows for the addition of any potential metal chelators necessary for Mn translocation across the membrane. While it is not clear whether Mn-specific metallochaperones exist in plants, Mn can complex with nicotianamine (NA), phytosiderophores (PS), phytate and organic acids (Koike et al., 2004; Haydon and Cobbett, 2007; Fernando et al., 2010).

Mn content in plant tissues or in yeast expressing a plant transporter is measured to demonstrate difference in Mn transport efficiency. To date, the most accurate and sensitive method to measure metal content in a sample is Inductively Coupled Plasma Mass Spectrometry (ICP-MS) (Baxter et al., 2008; Donner et al., 2012). However, because ICP-MS requires the total digestion of a sample, it does not collect spatial information about an element *in vivo*. For some metals, ion specific fluorophores exist, which, in conjunction with confocal microscopy, can image the subcellular localization of a metal (Burdette et al., 2001; Miller et al., 2006, 2008; Yoon et al., 2007), but there is currently no Mn-specific fluorescent probe. Therefore, other techniques are needed to spatially resolve Mn localization in a cell.

Quantitative *in vivo* cryo-scanning electron microscopy (SEM)/energy dispersive X-ray analysis (EDAX) has been used to provide detailed electron micrographs of tissue from hyperaccumulator plants along with energy dispersive X-ray spectra from regions of interest (Fernando et al., 2006b).The samples are prepared by rapidly freezing them in liquid nitrogen, which preserves the metal location during processing and microbeam exposure (Fernando et al., 2013). Further analysis using Particle-Induced X-ray Emission induced by a focused ion beam (μ PIXE) was used to confirm the cryo-SEM/EDAX results (Fernando et al., 2006a). Synchrotron X-Ray Fluorescence (SXRF) technology is a method used to localize metals *in vivo* at resolutions down to 250 nm. For a review on how SXRF can be used to study gene function see Punshon et al. (2013). It is important to recognize that no one method stands alone in determining the role of a protein in Mn translocation and multiple methods must be used in parallel.

## MANGANESE LOCALIZATION *IN PLANTA*

Intracellular Mn is found in multiple locations in the cell including the chloroplast, cell wall, mitochondria and Golgi apparatus (Pittman, 2005). Mn is also located in the vacuole, an organelle that is critical for cellular metal homeostasis, where it serves as an intracellular sink when metals are in excess and as a source when metals are limited (Pittman, 2005; Fernando et al., 2006a; Lanquar et al., 2010). Much of the Mn imaging in plant tissues has been performed on Mn hyperaccumulators, due to their unique metabolism and ease of imaging Mn-enriched tissues (Leitenmaier and Kupper, 2013). To date, there are around 22 Mn hyperaccumulators identified around the world (Fernando et al., 2013). The majority of hyperaccumulator plants' primary sequestration sites are in non-photosynthetic tissues such as trichomes (leaf hairs) and epidermal tissue. However, Mn hyperaccumulators can sequester excess Mn in photosynthetic tissues as well as non-photosynthetic tissues. In previous studies of five different Mn hyperaccumulators, excess foliar Mn was concentrated in the double layer of chloroplast containing palisade mesophyll cells in the leaf (Fernando et al., 2006a,b). In contrast, the Mn hyperaccumulator, *Maytenus founieri*, exhibited increased Mn accumulation in the non-photosynthetic leaf epidermal tissues, which is reminiscent of other metal hyperaccumulators such as the Zn/Cd hyperaccumulator *A. halleri* (Fernando et al., 2008).

SXRF can be used to spatially localize and quantify Mn in *A. thaliana* seeds at a resolution high enough to resolve metals at a subcellular level (Donner et al., 2012). Recent studies using μPIXE confirmed SXRF data that Mn is localized to spongy mesophyll on the abaxial side of the embryonic leaf (Kim et al., 2006; Schnell Ramos et al., 2013) (Figure 3). Donner et al. (2012) pointed out that, due to its requirement for photosynthesis, Mn needs to relocate to the palisade mesophyll cells during germination.

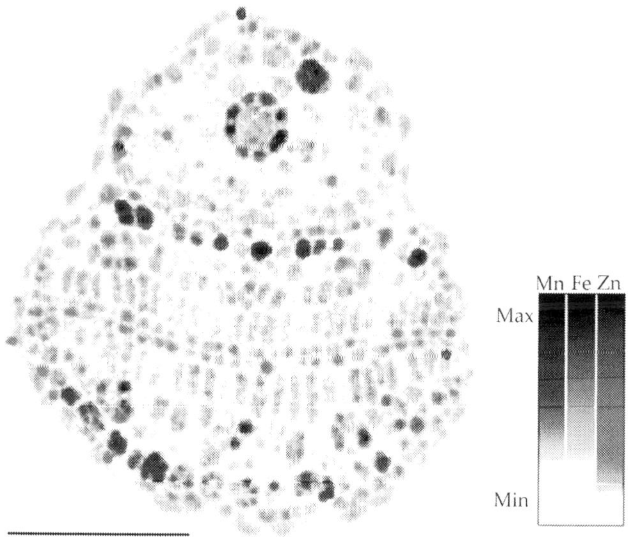


**Figure 3**: Elemental distribution of Mn, Zn and Fe in *A. thaliana* **seeds collected using Synchrotron X-Ray Florescence.** Two dimensional map of a cross-section of an *A. thaliana* seed collected at 1 micron resolution at the Advanced Photon Source. Mn (blue), Zn (magenta), and Fe (green) are shown.

## MANGANESE TRANSPORTERS

Studying the metal transport mechanisms of other species can greatly assist our understanding of how plants take up and distribute Mn. While Mn transport is relatively understudied in plants, it is well understood in bacterial and yeast systems (Jkubovics and Jenkinson, 2001; Culotta et al., 2005; Papp-Wallace and Maguire, 2006; Reddi et al., 2009). In the yeast, *Saccharaomyces cerevisiae*, Mn is transported into the cell through the high affinity $Mn^{2+}$ transporter SMF1 of the NRAMP family (Supek et al., 1996) and the high affinity phosphate transporter, PHO84 (Jensen et al., 2003). PHO84-type phosphate transporters prefer neutral metal-phosphate complexes; therefore the substrate is presumed to be $MnHPO_4$ (Jensen et al., 2003). The photosynthetic alga *Chlamydomonas* also encodes an NRAMP protein that is predicted to be involved in high affinity $Mn^{2+}$ influx (Allen et al., 2007). Photosynthetic autotrophs such as cyanobacteria require Mn for PSII assembly and a gene product with sequence similarity to an ABC (ATP binding cassette)-type transporter is implicated in Mn uptake in this organism (Bartsevich and Pakrasi, 1995).

There are few Mn-only transporters identified in plants. One possible explanation is that Mn shares many of the same transporters as other divalent cations such as Fe. One study found that it might not be possible to uncouple Mn and Fe transport in one high affinity Fe transporter after systematically introducing point mutations into the metal binding domain of IRT1 (Rogers et al., 2000). In fact, the majority of transporters implicated in Mn translocation have broad specificity for several divalent cations including Fe, Zn, Cu, Cd, Ca, Co (cobalt), Ni (nickel). The transporter families discussed below are organized based on transport of Mn into or out of the cytosol.

## TRANSPORT INTO THE CYTOSOL

### *NRAMP Family*

NRAMP1 was first characterized in mice as a gene involved in resistance to intracellular pathogens (Nevo and Nelson, 2006). Like many of the transporter families implicated in Mn transport, the NRAMPs transport a broad range of metals and can transport $Mn^{2+}$ as well as $Fe^{2+}$, $Zn^{2+}$, $Cd^{2+}$, $Cu^{2+}$, $Ni^{2+}$, $Co^{2+}$, and $Al^{3+}$ (Nevo and Nelson, 2006; Xia et al., 2010). The plant NRAMP proteins are predicted to be ion transporters based on sequence and predicted structural homology to the NRAMPs characterized in organisms such as humans and yeast. NRAMPs have between 10 and 12 transmembrane domains and a consensus transport sequence between transmembrane domains 8 and 9 that is shared with other ion channels and transporters (Cellier et al., 1995).

NRAMP proteins have been characterized in a number of plant species including *Solanum lycopersicum*, *Glycine max*, *Malas baccata*, *Thlaspi japonicum* and hyperaccumulators *A. halleri* and *Thlaspi caerulescens* (Bereczky et al., 2003; Kaiser et al., 2003; Mizuno et al., 2005; Xiao et al., 2008; Oomen et al., 2009). There are 6 and 7 NRAMP transporter proteins in *A. thaliana* and *Oryza sativa*, respectively. The first NRAMP genes cloned from plants were from rice in 1997 (Belouchi et al., 1997). However, not all are functionally characterized (Bennetzen, 2002; Nevo and Nelson, 2006).

Functional characterization of AtNRAMP1, AtNRAMP3, and AtNRAMP4 was first reported in 2000 (Thomine et al., 2000).

AtNRAMP1 is believed to be a high affinity Mn transporter in the root due to the transcriptional upregulation of *AtNRAMP1* in the root under Mn deficiency and its PM localization (Cailliatte et al., 2010). In addition, *AtNRAMP1* expression increases in Fe deficiency and the protein complements yeast defective in Fe or Mn transport (Curie et al., 2000). Studies show that Arabidopsis *Atnramp1* knockout lines are more susceptible to Mn deficiency, accumulate less Mn in the shoot when grown without Mn and have decreased Mn in the root when grown under Mn replete conditions (Cailliatte et al., 2010). Plants overexpressing *AtNRAMP1* are more resistant to Mn deficiency. Elemental images of Mn in *Atnramp1–1* mutant seeds shows wild type distribution of Mn, therefore it is unlikely that these genes are necessary for Mn loading into the embryonic cotyledons (Donner et al., 2012).

AtNRAMP3 and AtNRAMP4 are also able to complement yeast deficient in either Mn or Fe uptake. *Atnramp3–1* mutants show increased Mn in the roots of plants grown in Fe limited conditions while Mn accumulation decreased when *AtNRAMP3* is expressed in the cell under control of the strong 35S promoter (Thomine et al., 2003). This phenotype is possibly due to the vacuolar localization of this transporter. Expression data and transcriptional fusions to the promoters of *AtNRAMP3* and *AtNRAMP4*show that the transporters are expressed in the root stele and the vasculature of leaves and cotyledons and are highly upregulated during Fe deficiency (Thomine et al., 2000, 2003; Lanquar et al., 2005). AtNRAMP4 was identified as part of the vacuolar proteome in *A. thaliana* mesophyll cells (Carter et al., 2004). Translational fusions of both AtNRAMP3 and AtNRAMP4 proteins show that they are localized to the vacuolar membrane; therefore they are thought to be important for metal remobilization from the vacuole, *in planta* (Thomine et al., 2003).

AtNRAMP3 and AtNRAMP4 transporters are functionally redundant; only *Atnramp3nramp4* double mutants exhibit a strong phenotype in response to metal deficiency. Both AtNRAMP3 and AtNRAMP4 are considered necessary for Fe remobilization during early germination due to their high expression levels at this growth stage (Lanquar et al., 2005). Lanquar et al. (2010) demonstrated that AtNRAMP3 and AtNRAMP4 are necessary for Mn export from the vacuole in the mesophyll cells of adult plant leaves. Double mutants have increased Mn accumulation in vacuoles. Under Mn deficiency, double mutants exhibit stunted growth and chlorosis when compared to wild type. Expression of either AtNRAMP3 or AtNRAMP4 in the double mutant can restore growth in Mn deficiency. Interestingly, it has been shown in other photosynthetic organisms that MnSOD activity decreases before there is a loss of function PSII in response to Mn deficiency (Allen et al., 2007). However, in Arabidopsis *Atnramp3nramp4* double mutants, the level of active PSII decreases while there is no loss of MnSOD activity (Lanquar et al., 2010). These results suggest intra-organellar crosstalk between the chloroplast and vacuole that mediate proper Mn distribution in the cell.

To date, it is not known whether AtNRAMP2, AtNRAMP5 or AtNRAMP6 are important for Mn homeostasis.

Of the 7 NRAMPs in rice, only 4 have been functionally characterized. *O. sativa* plants utilize OsNRAMP3 to respond to environmental changes in Mn availability. *OsNRAMP3* is constitutively expressed in the node of rice, which connects the vasculature of the root to aerial tissues of the plant (including the leaves, panicles and stems) (Yamaji et al., 2013). Under Mn deficiency, the PM localized OsNRAMP3 transports Mn from the transpiration stream in the xylem of enlarged vascular bundles to the younger tissues and panicles to meet minimal growth requirements. In contrast, when Mn is in excess, OsNRAMP3 is internalized in vesicles and rapidly degraded. Then, Mn is preferentially loaded into the older leaves, which are directly connected to the xylem-enlarged vascular bundles, thereby protecting developing tissue from Mn toxicity. This study demonstrates the importance of post-translational regulation in response to environmental nutrient availability.

*OsNRAMP5*, on the other hand, does not respond to varying levels of environmental Mn, although gene expression increases slightly in the roots when plants are under Fe or Zn deficiency (Sasaki et al., 2012). Expression is seen in the mature zones of the root, more specifically at the PM of the exodermal and endodermal layers. Knockdown and RNAi lines accumulate less Mn in the roots and shoots and growth is stunted in response to Mn deficiency (Ishimaru et al., 2012; Sasaki et al., 2012). This phenotype is specific to Mn and not to other metal deficiencies tested, such as Fe-deficiency (Sasaki et al., 2012). Supplementing the plant with excess Mn does not rescue the growth defect. Therefore, it is likely that OsNRAMP5 is essential for Mn uptake from the soil in rice.

## *YSL Family*

The YSL transporters belong to a poorly characterized family of oligopeptide transporters (OPTs), which can transport amino acid-containing compounds and their derivatives (Yen et al., 2001). They are not found in all kingdoms, rather they are only found in plants, bacteria, fungi and archaea (Yen et al., 2001). The family is named after maize YS1 (ZmYS1), which accumulates in the root PM when Fe is scarce (Von Wiren et al., 1994; Roberts et al., 2004; Schaaf et al., 2004; Ueno et al., 2009). The characteristic interveinal chlorosis in the leaf of *ys1* knockdown mutants prompted the name "yellow stripe." Transport studies in both yeast and oocytes suggest that it can transport metals such as Mn, Zn, Cu, Ni, Cd as well as Fe (Schaaf et al., 2004). Therefore, it is reasonable to hypothesize that YSLs can transport Mn in other plant species.

YSL proteins mediate cellular uptake of metals complexed to non-proteinogenic amino acids: PS or its biosynthetic precursor NA (Yen et al., 2001). There is evidence that PS and NA are ligands for Mn (Schaaf et al., 2004; Haydon and Cobbett, 2007). Graminaceous plants (grasses) that include many crop species, use PS and NA for metal uptake and translocation, respectively (Palmer and Guerinot, 2009). PS is not produced by non-graminaceous plants (such as *A. thaliana*). Therefore, YSLs in non-grasses most likely use NA for intercellular and intracellular metal transport (Bashir et al., 2011).

There are eight predicted *YSL* genes in *A. thaliana* based on sequence similarity to YS1 (Curie et al., 2001). Knocking out a single *AtYSL* gene does not produce a strong visible phenotype (Didonato et al., 2004; Waters et al., 2006; Curie et al., 2009; Conte et al., 2013; Divol et al., 2013). However, it was discovered that if crosses are made between *Atysl* knockouts whose genes are in the same AtYSL subclass, a visible phenotype is observed. Examples include *Atysl1ysl3, Atysl4ysl6*, and *Atysl5ysl7* (Curie et al., 2009; Conte et al., 2013; Divol et al., 2013). Proteomics studies of the Arabidopsis vacuole predicted AtYSL4 and AtYSL6 to be vacuolar-localized (Jaquinod et al., 2007). Evidence shows that they are either located at the vacuole/endocompartments (shown via translational fusions) or the chloroplast (shown via immunofluorescence) (Conte et al., 2013; Divol et al., 2013). Data suggests that these proteins may be effluxers of heavy metals from an internal cellular compartment. Conte et al. (2013) observed that in comparison to wild type and single mutants, the double mutant is more resistant to prolonged growth on excess Mn. This phenotype was not reported by Divol et al. (2013). Thus far, the physiological role of AtYSL4 and AtYSL6 in Mn homeostasis remains to be clarified.

In rice, there are 18 putative *YSL* genes (Koike et al., 2004; Conte et al., 2013). OsYSL2 and OsYSL6 are implicated in Mn homeostasis. Electrophysiological and localization studies suggest the OsYSL2 is involved in lateral movement of $Mn^{2+}$-NA complexes via the phloem and loading into developing seeds (Koike et al., 2004). OsYSL6 is required for Mn detoxification when soil Mn is in excess (Sasaki et al., 2011). OsYSL6 is expressed constitutively in all cells of the roots and the shoots, particularly in senescing leaves. GFP-fusions of the protein disrupt localization, but due to successful complementation of the *smf1*yeast mutant, it is hypothesized to localize to the PM. When expressed in yeast, OsYSL6 translocates $Mn^{2+}$-NA. In *Osysl6* knockouts, plants exhibit symptoms of toxicity and excess Mn accumulates in the apoplast of the roots and shoots. When $Mn^{2+}$ accumulates in the apoplast, it is oxidized into $Mn^{3+}$, which in turn will oxidize proteins and lipids (Fecht-Christoffers et al., 2003). Therefore, it is probable that when plants are exposed to high levels of Mn, OsYSL6 translocates Mn from the apoplast to the symplast where it is properly sequestered.

## ZIP Family

The *ZIP* gene family members are known to transport a broad range of metals including $Fe^{2+}$, $Zn^{2+}$, $Cd^{2+}$, $Co^{2+}$, and $Mn^{2+}$ in both eukaryotes and prokaryotes (Korshunova et al., 1999; Guerinot, 2000). Structurally, they are composed of eight transmembrane domains with the N- and C- termini facing outside the PM. Their length and amino acid composition varies within the cytoplasmic variable region between transmembrane domains 3 and 4. The variable region contains a conserved histidine motif that is believed to be important for metal binding and point mutations of inter-membrane domains can alter metal selectivity (Guerinot, 2000; Rogers et al., 2000).

The ZIP family of metal transporters is named for its founding members in both *S. cerevisae* and *A. thaliana*, respectively (Grotz et al., 1998). Of the 16 ZIP transporters in Arabidopsis, few have been functionally characterized (Maser et al., 2001; Rampey et al., 2013).

IRT1 (Iron Regulated Transporter 1) is a high affinity Fe transporter that also has low affinity for other metals including Mn (Eide et al., 1996; Korshunova et al., 1999; Vert et al., 2002; Yang et al., 2008). It is upregulated in Fe deficient roots. In addition, exogenous Mn cannot rescue the chlorotic phenotype of *irt1* mutants, which only grow in the presence of excess Fe (Vert et al., 2002). However, *irt1* mutant plants accumulate less Mn in their tissues during Fe deficiency, suggesting that IRT1 is a major pathway for $Mn^{2+}$ uptake when Fe is scarce (Connolly et al., 2002). Because Mn uptake is still required for the plant during Fe replete conditions, there is likely another PM membrane localized transporter on the root epidermis that can transport Mn. Interestingly, under $Mn^{2+}$ deficiency, IRT1 transcript levels decrease and Fe transport increases as evidenced by high Fe concentrations in the plant (Yang et al., 2008).

A more recent study found that in addition to AtIRT1, six *A. thaliana* ZIPs could restore growth to the Mn uptake defective *smf1* mutant: AtZIP1, AtZIP2, AtZIP5, AtZIP6, AtZIP7, and AtZIP9 (Milner et al., 2013). AtZIP3, AtZIP4, AtZIP10, AtZIP11, AtZIP12 did not. *AtZIP8* is a pseudogene in Arabidopsis and this did not complement any of the yeast strains used in the study. Under normal conditions, AtZIP1, AtZIP2, and AtZIP6 are more highly expressed in the root while AtZIP7 transcript is more highly concentrated in the shoot. *AtZIP5* and *AtZIP9* transcripts are not very abundant and can be found in the root and shoot. The physiological roles of AtZIP6, AtZIP7, and AtZIP9 in Mn homeostasis *in planta* remain to be explored.

*AtZIP1* is most highly expressed in the root stele and is localized to the vacuole (Milner et al., 2013). AtZIP1 is believed to remobilize Mn from vacuoles into the cytoplasm to allow for their translocation to the shoot. This function is supported by the phenotype of *Atzip1* T-DNA insertion lines, which are more sensitive to low Mn and accumulate more Mn in the root. AtZIP2 is proposed to supply Mn to the aerial tissues in the plant via the root vasculature and not through direct uptake from the soil based on it localization to the PM in the root stele. In addition, mutant plants lacking functional AtZIP2 are more tolerant to Mn toxicity and less tolerant to Mn-deficiency, suggesting their importance in Mn translocation to the shoot for sequestration of excess Mn or normal plant health. However, it is unlikely that AtZIP2 is the primary transporter of

Mn in the root because gene expression decreases during Mn deficiency.

To date, the literature does not detail any OsZIPs that are able to transport Mn. However, *OsZIP5*transcript was slightly induced in the root upon Mn deficiency (Lee et al., 2010). Yet, it was not able to rescue *smf1* mutants and is therefore unlikely to be a direct transporter of Mn. ZIPs are described in other plant species, aside from Arabidopsis, that transport Mn. These species include *Solanum lycopersicum*, *Pisum sativa*, *Hordeum vulgare*, and *Medicago truncatula* (Eckhardt et al., 2001; Bereczky et al., 2003; Cohen et al., 2004; Lopez-Millan et al., 2004; Pedas et al., 2008). Tomato LeIRT1 and LeIRT2 along with PsIRT1, HvIRT1, MtZIP4, and MtZIP7 can restore growth to *smf1* mutants in Mn limited media. In barley, Mn efficiency correlates with increased expression of HvIRT1, an OsIRT1 ortholog, in roots (Pedas et al., 2008).

## Calcium-Permeable Channels

Calcium-permeable channels located in the PM of the root cells are a $Ca^{2+}$ influx pathway that may be permeable to $Mn^{2+}$ (White et al., 2002). Initial studies of these channels used biochemical and electrophysiological experiments to characterize their role in catalyzing cation influx because the genes encoding these channels had not been identified. A subset of cation channels that may be permeable to $Mn^{2+}$ (as well as a broad range of other cations) are found in the apical PM of Arabidopsis root hairs (Very and Davies, 2000), protoplasts from the endodermis, cortex and root elongation zone and the epidermis of a developing root tip (Kiegle et al., 2000). They are also identified in the stele of maize roots (White et al., 2002). In addition, *LCT1*, a gene in wheat roots that is potentially a $Ca^{2+}$-permeable channel, was cloned and heterologously expressed in a yeast mutant where the transport of $Ca^{2+}$ is blocked by the addition of $Mn^{2+}$ to the media (Clemens et al., 1998). There is further evidence based on competition assays that $Ca^{2+}$-permeable channels transport $Mn^{2+}$ in *A. thaliana* root hair tips (Wymer et al., 1997) and maize roots (Marshall et al., 1994).

## Export From the Cytosol

### *CAX Family*

The CAX proteins are one of five transporter families that constitute the $Ca^{2+}$/cation antiporters (CaCA) superfamily (Shigaki and Hirschi, 2006; Emery et al., 2012). Although the CAX family members were originally identified as $Ca^{2+}$ transporters, further study revealed their ability to transport a wide array of ions, hence their name modification from "calcium exchanger" to "cation exchanger." Typically, the CAXs contain 11 transmembrane domains and are found in plants, fungi and bacteria and in lower vertebrates. They facilitate the redistribution of cations across a membrane using electrochemical energy generated by a proton pump in order to maintain optimal ionic concentrations in the cell.

There are six CAX proteins in *A. thaliana*, which are split into two distinct phylogenetic groups: AtCAX1, AtCAX3, and AtCAX4 are part of the Type IA subgroup and AtCAX2, AtCAX5, and AtCAX6 are part of the Type 1B subgroup (Shigaki et al., 2006). The biological significance of the subgroups is unclear. In plants, the CAXs mediate efflux of ions into the vacuole. AtCAX2 contains a three-amino acid $Mn^{2+}$ binding domain (Cys-Ala-Phe) between transmembrane domain 4 and transmembrane domain 5 (Shigaki et al., 2003). Expression of AtCAX2 in yeast can confer tolerance to $Mn^{2+}$ toxicity (Schaaf et al., 2002; Shigaki et al., 2003). Furthermore, ectopic expression of AtCAX2 in tobacco (*Nicotiana tabacum*) can mediate vacuolar sequestration of Mn and confer resistance to high Mn stress (Hirschi et al., 2000), indicating a Mn transport function. In *A. thaliana*, *AtCAX2* is expressed at low levels in all tissues and does not respond to changes in Mn availability (Hirschi et al., 2000). However, Arabidopsis *Atcax2* mutants accumulate significantly less Mn in the vacuole compared to wild type (Pittman et al., 2004). The *Atcax2*knockout mutants display no obvious phenotype, even under Mn stress, and some Mn does continue to accumulate in the vacuole, suggesting that there are other vacuolar transporters that can compensate for the absence of AtCAX2 (Pittman et al., 2004). Candidates include AtCAX4 and AtCAX5 along with the AtCCX3 and AtVIT1 transporters, which will be discussed below.

AtCAX4 and AtCAX5 are likely involved in $Mn^{2+}/H^{+}$ antiport activity. Like AtCAX2, AtCAX4, and AtCAX5 are located at the vacuolar membrane and are constitutively expressed at low levels in all tissues (Cheng et al., 2002; Edmond et al., 2009). AtCAX5 is most highly expressed in the stem and root and AtCAX4 is more highly expressed in the roots (determined via qRT-PCR) and plays a role in root growth under ion stress, such as $Mn^{2+}$ toxicity (Edmond et al., 2009; Mei et al., 2009). Interestingly, AtCAX4 and AtCAX5 RNA levels increase when plants are exposed to high $Mn^{2+}$(Cheng et al., 2002; Edmond et al., 2009; Mei et al., 2009). AtCAX4 increases Mn stress tolerance when expressed in tobacco and AtCAX5 can rescue Mn-sensitive yeast, indicating their ability to transport Mn (Korenkov et al., 2007; Edmond et al., 2009).

Thus far, single *Atcax* mutants do not exhibit strong phenotypes, even when exposed to Mn stress. This is possibly due to functional redundancy of the AtCAX proteins. A subset of double mutants in Arabidopsis are reported in the literature: *Atcax1cax3*, *Atcax2cax3*, and *Atcax1cax2* (Connorton et al., 2012; Punshon et al., 2012). AtCAX1 appears to be important for determining the phenotype of a double mutant plant. A study by Connorton et al. (2012), found that *Atcax2* and *Atcax2cax3* mutants are both more sensitive to high Mn than wild type. However, the *Atcax1cax2* double mutant does not display any signs of Mn toxicity, suggesting that by deleting *Atcax1* from the *Atcax2* mutants, Mn is once again able to accumulate in the vacuole. The phenotype of the *Atcax1cax2* mutant is consistent with the phenotype of the plants lacking a functional *AtCAX1* gene, which are also more tolerant to Mn toxicity. Therefore, it will be important to understand the apparent cross-talk that occurs between the AtCAX proteins, which confers their Mn transport capabilities.

There are five CAX proteins in the rice genome, all of which have been cloned and their transport specificities assessed in yeast (Kamiya et al., 2005). OsCAX1a and OsCAX3 confer Mn tolerance in yeast, therefore they are potentially $Mn^{2+}/H^{+}$ exchangers *in planta* (Kamiya and Maeshima, 2004; Kamiya et al., 2005). It is important to note that CAX proteins can also transport Mn in other plant species. Examples include HvCAX2 in barley and LeCAX2 in tomato (Edmond et al., 2009). Future work is needed to confirm that they are also localized to the vacuole *in planta*, where they mediate Mn tolerance.

## CCX Family

CCXs are one of five families of transporters, (along with the CAXs) which make up the CaCA superfamily (Shigaki et al., 2006; Emery et al., 2012). There are five CCX proteins (CCX1-5), which were previously identified as CAX7-11. They were reclassified due to their higher sequence homology to the mammalian PM $K^+$-dependent $Na^+$/$Ca^{2+}$ exchangers (NCXs). Expression of AtCCX3 in yeast rescues mutants defective in either PM or vacuolar import of $Mn^{2+}$ (Morris et al., 2008). However, GFP and HA tagged versions of the protein support localization to intracellular compartments. Wild type yeast expressing AtCCX3 had nearly double the Mn concentration when compared to yeast expressing the control vector. Also, when ectopically expressed in tobacco, plant $Mn^{2+}$ concentration significantly increased as leaves matured, which led to tissue necrosis. This phenotype is likely due to the defects in ion homeostasis in the plant resulting in increased ROS as evidenced by higher oxidation of proteins compared to controls. In plants, treatment with Mn induces *AtCCX3* expression in roots and flowers (Morris et al., 2008). Like *Atcax*single mutants, there are no apparent growth defects in Arabidopsis *Atccx3* mutant plants, which could be due to functional redundancy. Interestingly, AtCCX3 can transport both monovalent and divalent cations. However, the only divalent cation that can be transported by AtCCX3 is $Mn^{2+}$. Whether other AtCCXs can translocate $Mn^{2+}$ remains to be studied.

## CDF/MTP Family

The CDF family, also known as MTP, is ubiquitous among all kingdoms of life (Maser et al., 2001; Montanini et al., 2007). The CDFs act as proton antiporters, which efflux metals such as $Zn^{2+}$, $Fe^{2+}$, $Co^{2+}$, $Ni^{2+}$, $Cd^{2+}$, and $Mn^{2+}$ out of the cytoplasm or into subcellular compartments (Gustin et al., 2011). However, one CDF protein is reported to be responsible for Zn uptake into the cytoplasm (Cragg et al., 2002). The plant CDFs are clustered into three functional groups based on phylogenetic analysis: Zn-CDFs, Fe/Zn-CDFs, and Mn-CDFs (Montanini et al., 2007; Gustin et al., 2011). Metal selectivity of the proteins is inferred from metal transport activity (either confirmed or hypothesized) of the respective members of the three subgroups. Mn-CDFs also contain amino acid residues that may predict metal specificity. The CDFs typically have six transmembrane domains and Mn-CDFs contain the highly conserved consensus sequence DXXXD (where $X$ = any amino acid) in transmembrane domains 2 and 4, which is not found in the Zn- or Fe/Zn-CDFs (Montanini et al., 2007). Studies show that even single point mutations within key structural sites can alter metal ion specificity of the MTPs. For example, single point mutations in AtMTP1 (Podar et al., 2012) and OsMTP1 (Menguer et al., 2013) allow Mn uptake, which is not observed with the wild type protein.

In plants, the CDFs are identified as MTPs due to their role in detoxification of heavy metals (Ricachenevsky et al., 2013). The first characterized Mn-CDF transporter was identified in the Mn hyperaccumulating tropical legume, *Stylosanthes hamata* (Delhaize et al., 2003). ShMTP8 (previously annotated ShMTP1) can confer $Mn^{2+}$ tolerance when ectopically expressed in Arabidopsis or yeast by sequestering excess metal in the vacuole. Of the twelve identified CDFs in *A. thaliana*, there are four in the Mn-CDF subgroup, which are also highly similar to ShMTP8, suggesting similar function. These proteins are AtMTP8, AtMTP9, AtMTP10, AtMTP11. The Mn-CDFs are further divided into subgroup 8 (including AtMTP8 and ShMTP8) and subgroup 9 (including AtMTP9, AtMTP10, and AtMTP11). To date, AtMTP11 is the only functionally characterized Arabidopsis Mn-CDF. Yeast transformed with AtMTP11are more tolerant to $Mn^{2+}$ (Delhaize et al., 2007; Peiter et al., 2007). In addition, increased $Mn^{2+}$ dependent proton-transport activity was recorded in yeast microsomal vesicles prepared from yeast expressing AtMTP11 in comparison to vesicles prepared from the control yeast strain (Delhaize et al., 2007). These data further support the role of AtMTP11 as an $Mn^{2+}/H^{+}$ antiporter. Two independent studies localized AtMTP11 to the pre-vacuolar and Golgi-like compartments. Surprisingly, *AtMTP11* is most highly expressed in the leaf hydathodes and the root tip rather than in tissues that typically accumulate excess Mn such as the trichomes (Peiter et al., 2007). Hydathodes are involved in secretion of water containing salts and metals from the leaf. Therefore, Peiter et al. (2007) hypothesized that AtMTP11 is involved in vesicular trafficking and exocytosis of excess Mn at secretory tissues where Mn will be excreted rather than stored. This hypothesis is supported by the phenotype of both knockout mutants and plants overexpressing AtMTP11. Arabidopsis *Atmtp11* mutants are hypersensitive to high Mn while overexpression lines are hypertolerant to high Mn (Delhaize et al.,

2007; Peiter et al., 2007). Also, the *Atmtp11* mutants accumulate more Mn in the shoot and root, presumably due to their inability to secrete Mn from tissues (Peiter et al., 2007).

In rice there are five Mn-CDFs: OsMTP8 and OsMTP8.1 from Group 8 and OsMTP9, OsMTP11, and OsMTP11.1 from group 9. Recently, OsMTP8.1 was identified in a screen of rice shoot cDNAs that conferenced Mn tolerance in yeast (Chen et al., 2013). *pmr1* yeast mutants expressing OsMTP8.1 also accumulated more Mn. In rice, OsMTP8.1 is mainly expressed in shoots under all conditions tested, and expression increases in response to excess Mn. *Osmtp8.1* knockdown and knockout mutants exhibited toxicity symptoms in the presence of elevated Mn and Mn accumulation in the roots and shoot compared to wild type was reduced. The protein is localized to the vacuole, resulting in the hypothesis that OsMTP8.1 is important for Mn detoxification by sequestering Mn into the vacuole in rice plants.

Two MTP11 orthologs (PtMTP11.1 and PtMTP11.2) in poplar are believed to have a similar function as AtMTP11 due to their localization to Golgi-like compartments and ability to complement Arabidopsis *Atmtp11* plants (Peiter et al., 2007). Another study in beets (*Beta vulgaris*) identified two genes, *BmMTP10* and *BmMTP11*, which were named after their orthologs in *A. thaliana*. Much like AtMTP11, these proteins are associated with the Golgi and are hypothesized to efflux excess Mn via the secretory pathway. The conservation of Mn-CDF function among multiple plant species supports the importance of these transporters in Mn detoxification and cellular homeostasis.

## P-type ATPase Family

The endomembrane system is essential for coordinating ion homeostasis in the cell. Several transporters that localize to the vacuolar membrane are described to facilitate metal uptake and release. However, there are also transporters in other endomembrane compartments that are equally important. Currently, there are two characterized proteins in Arabidopsis, AtECA1 (ER-type calcium ATPases) and AtECA3, which are localized to the ER and Golgi Apparatus, respectively (Liang et al., 1997; Wu et al., 2002; Li et al., 2008; Mills et al., 2008). AtECA1 and AtECA3 function as $Mn^{2+}$ pumps that remove $Mn^{2+}$ from the cytosol and deliver it into their respective endomembrane compartment.

The ECAs belong to the $Ca^{2+}$-ATPase subfamily within the P-type ATPase superfamily of transporters, which use energy from ATP hydrolysis to catalyze the translocation of cations across membranes (Baxter et al., 2003; Huda et al., 2013). Plant $Ca^{2+}$-ATPases are categorized into $P_{2A}$ and $P_{2B}$-types, both of which are generally described as $Ca^{2+}$ pumps (Evans and Williams, 1998; Mills et al., 2008). There are four predicted $P_{2A}$-type ECA proteins in *A. thaliana* (AtECA1-4) and three in rice (OsECA1-3) (Baxter et al., 2003). $P_{2A}$-type ATPases show sequence homology to the sarcoplasmic/ER $Ca^{2+}$-ATPases (SERCA type) found in mammals. SERCAs, PMCAs (plasma membrane $Ca^{2+}$-ATPases) and SPCAs (secretory pathway $Ca^{2+}$-ATPases) comprise three distinct subfamilies of $P_2$-type Ca-ATPases in animal cells (Pittman et al., 1999; Mills et al., 2008). PMCAs are homologous to $P_{2B}$-type PMCAs, while there are no known SPCAs in plants.

Both AtECA1 and AtECA3 are able to restore the growth of *pmr1* yeast when Mn is high (Wu et al., 2002; Li et al., 2008; Mills et al., 2008). Proteomics analysis of Arabidopsis organelles, translational fusions and co-localization with marker proteins supports the ER localization of AtECA1 as well as the Golgi localization of AtECA3 (Wu et al., 2002; Dunkley et al., 2006; Mills et al., 2008; Nikolovski et al., 2012). AtECA1 is expressed in all major organs (especially in the root and flower) where it is believed to play a major role in managing Mn toxicity in the cell (Wu et al., 2002). *Ateca1* mutants appear wild type when grown under standard conditions, yet are smaller and chlorotic in the presence of high Mn. Two different *Ateca3* mutants exhibit opposite phenotypes to Mn stress. *Ateca3-2* mutants are more susceptible to Mn deficiency (Mills et al., 2008). When grown without Mn, they display stunted growth and leaf chlorosis, which can be rescued by the addition of even trace amounts of Mn. The *Ateca3-4* mutant allele is more sensitive to Mn toxicity (Li et al., 2008). The phenotypes of plants carrying these alleles suggest that AtECA3 is important for pumping Mn into the Golgi for proper plant nutrition as well as detoxification via the Golgi compartments. Further investigation is needed to understand the variation between the alleles that uncouples the proposed dual function of the AtECA3 transporter.

There is no evidence that the Arabidopsis AtECA2 or AtECA4 transporters play a role in Mn homeostasis. In fact, *Ateca2* mutants appear similar to wild type when grown under high or low Mn (Mills et al., 2008). A tomato ECA, LCA1 (Lycopersicon esculentum $Ca^{2+}$-ATPases), shows high amino acid sequence similarity to ECA2 (77%) in comparison to ECA1 and ECA3 (Pittman et al., 1999). LCA1 can also complement the *pmr1* yeast mutant, suggesting that it is also an endocompartment-localized $Mn^{2+}$ pump (Johnson et al., 2009). Additional study of ECAs *in A. thaliana*, rice and other crop species is required to clarify their role in Mn homeostasis *in planta*.

### *VIT/CCC1-Like Family*

CCC1 ($Ca^{2+}$-sensitive cross complementer 1), which transports Fe and Mn into the vacuole in yeast (Li et al., 2001), has six orthologs in Arabidopsis referred to as CCC1-like (Rampey et al., 2006; Gollhofer et al., 2011). To date, AtVIT1 is the best characterized CCC1-like transporter and is likely involved in Mn uptake into the vacuole. As hypothesized, *ccc1* mutant yeast, when expressing AtVIT1, could accumulate Mn in the vacuoles (Kim et al., 2006). High-resolution SXRF images of *Atvit1* mutant seeds display highly localized Mn in the Arabidopsis embryo in a pattern identical to that seen in wild type. This is not wholly surprising considering the multitude of transporters that are implicated in Mn transport in the absence of AtVIT1. Therefore, although AtVIT1 is a potential route for Mn transport in planta, it does not appear to be essential for localization of Mn in the seed. The functional orthologs of AtVIT1 in rice (OsVIT1 and OsVIT2) are localized to the vacuole. They transport Mn, Fe and Zn in yeast, however physiological studies of mutants suggest that they are only Fe and Zn transporters *in planta* (Zhang et al., 2012).

## CONCLUSIONS AND FUTURE PERSPECTIVES

Mn, although essential for plant survival, can be toxic to plants. Over the past few decades, several transporter families have been identified that play a role in Mn homeostasis. Interestingly, many of the transporters that translocate Mn have broad specificity for other ions, particularly divalent cations. For example, transporters such as IRT1, originally described as an Fe transporter in the root, can also transport Mn into the plant. In addition, CAX family members along with ECA proteins were originally thought to transport Ca; it was only subsequent research that discovered their roles in Mn transport as well. Despite the advancements in identifying Mn transporters, there is still a lot unknown regarding the molecular mechanisms controlling Mn homeostasis in plants. Many Mn transporters are not differentially regulated at the transcriptional level in response to Mn stress. Further research can explore the possibility that many of the transporters are under post-transcriptional or post-translational control. The degradation of OsNRAMP3 in response to Mn toxicity in rice may be a common mechanism used by plants to tolerate stress conditions (Yamaji et al., 2013).

Further characterization of orthologs of Mn transporters found in *A. thaliana* in crop species is a necessary step toward agronomic advancement. Research involving identification of genes involved in Mn homeostasis can be exploited to generate plants that are able to thrive in suboptimal soil conditions which would increase crop production and guarantee food security. For example, if we breed plants with high Mn uptake efficiency in the root, we can grow plants in highly alkaline soils where plants usually succumb to Mn deficiency. In conjunction, developing plants with increased Mn storage capability can assist in either soil Mn detoxification or improved growth of plants in acidic soils.

## ACKNOWLEDGMENTS

We thank members of the Guerinot laboratory for helpful discussions; we also thank the many laboratories that have contributed to this field of investigation. In addition we would like to thank Tracy Punshon (Dartmouth College) for critical reading of the manuscript. Work in our laboratory is supported by grants from the National Science Foundation (IOS-091994; DBI 0701119), the National Institutes of Health (RO1 GM 078536), the Department of Energy (DE-FG-2-06ER15809) and the National Institute of Environmental Health Sciences (5 P42 ES007373) to Mary Lou Guerinot and a Fellowship from Sigma Delta Epsilon-Graduate Women in Science to Amanda L. Socha. Amanda L. Socha has also been partially supported by a training grant from the Department of Education's Graduate Assistance in Areas of National Need (GAANN) program. The SXRF image was obtained at The Advanced Photon Source. The Advanced Photon Source, an Office of Science User Facility operated for the U.S. Department of Energy (DOE) Office of Science by Argonne National Laboratory, was supported by the U.S. DOE under Contract No. DE-AC02-06CH11357.

## REFERENCES

1. Allen, M. D., Kropat, J., Tottey, S., Del Campo, J. A., and Merchant, S. S. (2007). Manganese deficiency in Chlamydomonas results in loss of photosystem II and MnSOD function, sensitivity to peroxides, and secondary phosphorus and iron deficiency. *Plant Physiol.* 143, 263–277. doi: 10.1104/pp.106.088609
2. Bartsevich, V. V., and Pakrasi, H. B. (1995). Molecular identification of an ABC transporter complex for manganese: analysis of a cyanobacterial mutant strain impaired in the photosynthetic oxygen evolution process. *EMBO J.* 14, 1845–1853.
3. Bashir, K., Ishimaru, Y., Shimo, H., Nagasaka, S., Fujimoto, M., Takanashi, H., et al. (2011). The rice mitochondrial iron transporter is essential for plant growth. *Nat. Commun.* 2, 322. doi: 10.1038/ncomms1326
4. Baxter, I., Muthukumar, B., Park, H. C., Buchner, P., Lahner, B., Danku, J., et al. (2008). Variation in molybdenum content across broadly distributed populations of *Arabidopsis thaliana* is controlled by a mitochondrial molybdenum transporter (MOT1). *PLoS Genet.* 4:e1000004. doi: 10.1371/journal.pgen.1000004
5. Baxter, I., Tchieu, J., Sussman, M. R., Boutry, M., Palmgren, M. G., Gribskov, M., et al. (2003). Genomic comparison of P-type ATPase ion pumps in Arabidopsis and rice. *Plant Physiol.* 132, 618–628. doi: 10.1104/pp.103.021923
6. Belouchi, A., Kwan, T., and Gros, P. (1997). Cloning and characterization of the OsNramp family from *Oryza sativa*, a new family of membrane proteins possibly implicated in the transport of metal ions. *Plant Mol. Biol.* 33, 1085–1092. doi: 10.1023/A:1005723304911
7. Bennetzen, J. (2002). The rice genome. Opening the door to comparative plant biology. *Science* 296, 60–63. doi: 10.1126/science.1071402
8. Bereczky, Z., Wang, H. Y., Schubert, V., Ganal, M., and Bauer, P. (2003). Differential regulation of *NRAMP* and *IRT* metal transporter genes in wild type and iron uptake mutants of tomato. *J. Biol. Chem.* 278, 24697–24704. doi: 10.1074/jbc.M301365200
9. Burdette, S. C., Walkup, G. K., Spingler, B., Tsien, R. Y., and Lippard,

S. J. (2001). Fluorescent sensors for $Zn^{(2+)}$ based on a fluorescein platform: synthesis, properties and intracellular distribution. *J. Am. Chem. Soc.* 123, 7831–7841. doi: 10.1021/ja010059l

10. Cailliatte, R., Schikora, A., Briat, J. F., Mari, S., and Curie, C. (2010). High-affinity manganese uptake by the metal transporter NRAMP1 is essential for Arabidopsis growth in low manganese conditions. *Plant Cell* 22, 904–917. doi: 10.1105/tpc.109.073023
11. Carter, C., Pan, S., Zouhar, J., Avila, E. L., Girke, T., and Raikhel, N. V. (2004). The vegetative vacuole proteome of *Arabidopsis thaliana*reveals predicted and unexpected proteins. *Plant Cell* 16, 3285–3303. doi: 10.1105/tpc.104.027078
12. Cellier, M., Prive, G., Belouchi, A., Kwan, T., Rodrigues, V., Chia, W., et al. (1995). Nramp defines a family of membrane proteins. *Proc. Natl. Acad. Sci. U.S.A.* 92, 10089–10093. doi: 10.1073/pnas.92.22.10089
13. Chen, Z., Fujii, Y., Yamaji, N., Masuda, S., Takemoto, Y., Kamiya, T., et al. (2013). Mn tolerance in rice is mediated by MTP8.1, a member of the cation diffusion facilitator family. *J. Exp. Bot.* 64, 4375–4387. doi: 10.1093/jxb/ert243
14. Cheng, N. H., Pittman, J. K., Shigaki, T., and Hirschi, K. D. (2002). Characterization of CAX4, an Arabidopsis $H^{(+)}$/cation antiporter. *Plant Physiol.* 128, 1245–1254. doi: 10.1104/pp.010857
15. Clemens, S., Antosiewicz, D. M., Ward, J. M., Schachtman, D. P., and Schroeder, J. I. (1998). The plant cDNA LCT1 mediates the uptake of calcium and cadmium in yeast. *Proc. Natl. Acad. Sci. U.S.A.* 95, 12043–12048. doi: 10.1073/pnas.95.20.12043
16. Cohen, C. K., Garvin, D. F., and Kochian, L. V. (2004). Kinetic properties of a micronutrient transporter from *Pisum sativum* indicate a primary function in Fe uptake from the soil. *Planta* 218, 784–792. doi: 10.1007/s00425-003-1156-7
17. Connolly, E. L., Fett, J. P., and Guerinot, M. L. (2002). Expression of the IRT1 metal transporter is controlled by metals at the levels of transcript and protein accumulation. *Plant Cell* 14, 1347–1357. doi: 10.1105/tpc.001263
18. Connorton, J. M., Webster, R. E., Cheng, N., and Pittman, J. K. (2012). Knockout of multiple Arabidopsis cation/$H^{(+)}$ exchangers

suggests isoform-specific roles in metal stress response, germination and seed mineral nutrition. *PLoS ONE* 7:e47455. doi: 10.1371/journal.pone.0047455

19. Conte, S. S., Chu, H. H., Rodriguez, D. C., Punshon, T., Vasques, K. A., Salt, D. E., et al. (2013). *Arabidopsis thaliana* Yellow Stripe1-Like4 and Yellow Stripe1-Like6 localize to internal cellular membranes and are involved in metal ion homeostasis. *Front. Plant Sci.* 4:283. doi: 10.3389/fpls.2013.00283
20. Cragg, R. A., Christie, G. R., Phillips, S. R., Russi, R. M., Kury, S., Mathers, J. C., et al. (2002). A novel zinc-regulated human zinc transporter, hZTL1, is localized to the enterocyte apical membrane. *J. Biol. Chem.* 277, 22789–22797. doi: 10.1074/jbc.M200577200
21. Culotta, V. C., Yang, M., and Hall, M. D. (2005). Manganese transport and trafficking: lessons learned from *Saccharomyces cerevisiae*. *Eukaryot. Cell* 4, 1159–1165. doi: 10.1128/EC.4.7.1159-1165.2005
22. Curie, C., Alonso, J. M., Le Jean, M., Ecker, J. R., and Briat, J. F. (2000). Involvement of NRAMP1 from *Arabidopsis thaliana* in iron transport. *Biochem. J.* 347, 749–755. doi: 10.1042/0264-6021:3470749
23. Curie, C., Cassin, G., Couch, D., Divol, F., Higuchi, K., Le Jean, M., et al. (2009). Metal movement within the plant: contribution of nicotianamine and yellow stripe 1-like transporters. *Ann. Bot.* 103, 1–11. doi: 10.1093/aob/mcn207
24. Curie, C., Panaviene, Z., Loulergue, C., Dellaporta, S. L., Briat, J. F., and Walker, E. L. (2001). Maize yellow stripe1 encodes a membrane protein directly involved in Fe(III) uptake. *Nature* 409, 346–349. doi: 10.1038/35053080
25. Dancis, A., Haile, D., Yuan, D. S., and Klausner, R. D. (1994). The *Saccharomyces cerevisiae* copper transport protein (Ctr1p). Biochemical characterization, regulation by copper, and physiologic role in copper uptake. *J. Biol. Chem.* 269, 25660–25667.
26. Delhaize, E., Gruber, B. D., Pittman, J. K., White, R. G., Leung, H., Miao, Y., et al. (2007). A role for the *AtMTP11* gene of Arabidopsis in manganese transport and tolerance. *Plant J.* 51, 198–210. doi: 10.1111/j.1365-313X.2007.03138.x

27. Delhaize, E., Kataoka, T., Hebb, D. M., White, R. G., and Ryan, P. R. (2003). Genes encoding proteins of the cation diffusion facilitator family that confer manganese tolerance. *Plant Cell* 15, 1131–1142. doi: 10.1105/tpc.009134

28. Didonato, R. J. Jr. Roberts, L. A., Sanderson, T., Eisley, R. B., and Walker, E. L. (2004). Arabidopsis *Yellow Stripe-Like2 (YSL2)*: a metal-regulated gene encoding a plasma membrane transporter of nicotianamine-metal complexes. *Plant J.* 39, 403–414. doi: 10.1111/j.1365-313X.2004.02128.x

29. Divol, F., Couch, D., Conejero, G., Roschzttardtz, H., Mari, S., and Curie, C. (2013). The Arabidopsis Yellow Stripe LIKE4 and 6 transporters control iron release from the chloroplast. *Plant Cell* 25, 1040–1055. doi: 10.1105/tpc.112.107672

30. Dix, D. R., Bridgham, J. T., Broderius, M. A., Byersdorfer, C. A., and Eide, D. J. (1994). The *FET4* gene encodes the low affinity Fe(II) transport protein of *Saccharomyces cerevisiae. J. Biol. Chem.* 269, 26092–26099.

31. Donner, E., Punshon, T., Guerinot, M. L., and Lombi, E. (2012). Functional characterisation of metal(loid) processes in planta through the integration of synchrotron techniques and plant molecular biology. *Anal. Bioanal. Chem.* 402, 3287–3298. doi: 10.1007/s00216-011-5624-9

32. Dunkley, T. P., Hester, S., Shadforth, I. P., Runions, J., Weimar, T., Hanton, S. L., et al. (2006). Mapping the Arabidopsis organelle proteome. *Proc. Natl. Acad. Sci. U.S.A.* 103, 6518–6523. doi: 10.1073/pnas.0506958103

33. Durr, G., Strayle, J., Plemper, R., Elbs, S., Klee, S. K., Catty, P., et al. (1998). The medial-Golgi ion pump Pmr1 supplies the yeast secretory pathway with $Ca^{2+}$ and $Mn^{2+}$ required for glycosylation, sorting, and endoplasmic reticulum-associated protein degradation. *Mol. Biol. Cell* 9, 1149–1162. doi: 10.1091/mbc.9.5.1149

34. Eckhardt, U., Mas Marques, A., and Buckhout, T. J. (2001). Two iron-regulated cation transporters from tomato complement metal uptake-deficient yeast mutants. *Plant Mol. Biol.* 45, 437–448. doi: 10.1023/A:1010620012803

35. Edmond, C., Shigaki, T., Ewert, S., Nelson, M. D., Connorton, J. M.,

Chalova, V., et al. (2009). Comparative analysis of CAX2-like cation transporters indicates functional and regulatory diversity. *Biochem J.* 418, 145–154. doi: 10.1042/BJ20081814

36. Eide, D., Broderius, M., Fett, J., and Guerinot, M. L. (1996). A novel iron-regulated metal transporter from plants identified by functional expression in yeast. *Proc. Natl. Acad. Sci. U.S.A.* 93, 5624–5628. doi: 10.1073/pnas.93.11.5624
37. Emery, L., Whelan, S., Hirschi, K. D., and Pittman, J. K. (2012). Protein phylogenetic analysis of $Ca^{(2+)}$/cation Antiporters and insights into their evolution in plants. *Front. Plant Sci.* 3:1. doi: 10.3389/fpls.2012.00001
38. Erbasol, I., Bozdag, G. O., Koc, A., Pedas, P., and Karakaya, H. C. (2013). Characterization of two genes encoding metal tolerance proteins from Beta vulgaris subspecies maritima that confers manganese tolerance in yeast. *Biometals* 26, 795–804. doi: 10.1007/s10534-013-9658-7
39. Evans, D. E., and Williams, L. E. (1998). P-type calcium ATPases in higher plants - biochemical, molecular and functional properties. *Biochim. Biophys. Acta* 1376, 1–25. doi: 10.1016/S0304-4157(97)00009-9
40. Fecht-Christoffers, M. M., Braun, H. P., Lemaitre-Guillier, C., Vandorsselaer, A., and Horst, W. J. (2003). Effect of manganese toxicity on the proteome of the leaf apoplast in cowpea. *Plant Physiol.* 133, 1935–1946. doi: 10.1104/pp.103.029215
41. Fernando, D. R., Bakkaus, E. J., Perrier, N., Baker, A. J., Woodrow, I. E., Batianoff, G. N., et al. (2006a). Manganese accumulation in the leaf mesophyll of four tree species: a PIXE/EDAX localization study. *New Phytol.* 171, 751–757. doi: 10.1111/j.1469-8137.2006.01783.x
42. 
43. Fernando, D. R., Batianoff, G. N., Baker, A. J., and Woodrow, I. E. (2006b). *In vivo* localization of manganese in the hyperaccumulator *Gossia bidwillii* (Benth.) N. Snow & Guymer (Myrtaceae) by cryo-SEM/EDAX. Plant *Cell Environ.* 29, 1012–1020. doi: 10.1111/j.1365-3040.2006.01498.x
44. Fernando, D. R., Marshall, A., Baker, A. J., and Mizuno,

T. (2013). Microbeam methodologies as powerful tools in manganese hyperaccumulation research: present status and future directions. *Front. Plant Sci.* 4:319. doi: 10.3389/fpls.2013.00319

45. Fernando, D. R., Mizuno, T., Woodrow, I. E., Baker, A. J., and Collins, R. N. (2010). Characterization of foliar manganese (Mn) in Mn (hyper)accumulators using X-ray absorption spectroscopy. *New Phytol.* 188, 1014–1027. doi: 10.1111/j.1469-8137.2010.03431.x
46. Fernando, D. R., Woodrow, I. E., Jaffre, T., Dumontet, V., Marshall, A. T., and Baker, A. J. (2008). Foliar manganese accumulation by *Maytenus founieri* (Celastraceae) in its native New Caledonian habitats: populational variation and localization by X-ray microanalysis. *New Phytol.* 177, 178–185. doi: 10.1111/j.1469-8137.2007.02253.x
47. Geszvain, K., Butterfield, C., Davis, R. E., Madison, A. S., Lee, S. W., Parker, D. L., et al. (2012). The molecular biogeochemistry of manganese(II) oxidation. *Biochem. Soc. Trans.* 40, 1244–1248. doi: 10.1042/BST20120229
48. Gollhofer, J., Shlawicke, C., Jungnick, N., Schmidt, W., and Buckhout, T. J. (2011). Members of a small family of nodulin-like genes are regulated under iron deficiency in roots of *Arabidopsis thaliana. Plant Physiol. Biochem.* 49, 557–564. doi: 10.1016/j.plaphy.2011.02.011
49. Goussias, C., Boussac, A., and Rutherford, A. W. (2002). Photosystem II and photosynthetic oxidation of water: an overview. *Philos. Trans. R. Soc. Lond. B Biol. Sci.* 357, 1369–1381. discussion: 1419–1320. doi: 10.1098/rstb.2002.1134
50. Grotz, N., Fox, T., Connolly, E., Park, W., Guerinot, M. L., and Eide, D. (1998). Identification of a family of zinc transporter genes from Arabidopsis that respond to zinc deficiency. *Proc. Natl. Acad. Sci. U.S.A.* 95, 7220–7224. doi: 10.1073/pnas.95.12.7220
51. Guerinot, M. L. (2000). The ZIP family of metal transporters. *Biochim. Biophys. Acta* 1465, 190–198. doi: 10.1016/S0005-2736(00)00138-3
52. Gustin, J. L., Zanis, M. J., and Salt, D. E. (2011). Structure and evolution of the plant cation diffusion facilitator family of ion transporters. *BMC Evol. Biol.* 11:76. doi: 10.1186/1471-2148-11-76

53. Haydon, M. J., and Cobbett, C. S. (2007). Transporters of ligands for essential metal ions in plants. *New Phytol.* 174, 499–506. doi: 10.1111/j.1469-8137.2007.02051.x

54. He, Z. L., Yang, X. E., and Stoffella, P. J. (2005). Trace elements in agroecosystems and impacts on the environment. *J. Trace Elem. Med. Biol.* 19, 125–140. doi: 10.1016/j.jtemb.2005.02.010

55. Hebbern, C. A., Laursen, K. H., Ladegaard, A. H., Schmidt, S. B., Pedas, P., Bruhn, D., et al. (2009). Latent manganese deficiency increases transpiration in barley (*Hordeum vulgare*). *Physiol. Plant* 135, 307–316. doi: 10.1111/j.1399-3054.2008.01188.x

56. Hirschi, K. D., Korenkov, V. D., Wilganowski, N. L., and Wagner, G. J. (2000). Expression of arabidopsis CAX2 in tobacco. Altered metal accumulation and increased manganese tolerance. *Plant Physiol.* 124, 125–133. doi: 10.1104/pp.124.1.125

57. Horst, W. J., and Maier, P. (1999). "Compartmentalization of manganese in the vacuoles and in the apoplast of leaves in relation to genotypic manganese leaf-tissue tolerance in *Vigna unguiculata* (L.) Walp," in *Plant Nutrition- Molecular Biology and Genetics*, eds G. Gissel-Nielsen and A. Jensen (Dordrecht: Kluwer Academic Publishers), 222–234.

58. Huda, K. M., Banu, M. S., Tuteja, R., and Tuteja, N. (2013). Global calcium transducer P-type $Ca^{(2)(+)}$-ATPases open new avenues for agriculture by regulating stress signalling. *J. Exp. Bot.* 64, 3099–3109. doi: 10.1093/jxb/ert182

59. Ishimaru, Y., Takahashi, R., Bashir, K., Shimo, H., Senoura, T., Sugimoto, K., et al. (2012). Characterizing the role of rice NRAMP5 in manganese, iron and cadmium transport. *Sci. Rep.* 2:286. doi: 10.1038/srep00286

60. Jkubovics, N. S., and Jenkinson, H. F. (2001). Out of the iron age: new insights into the critical role of manganese homeostasis in bacteria. *Microbiology* 147, 1709–1718.

61. Jaquinod, M., Villiers, F., Kieffer-Jaquinod, S., Hugouvieux, V., Bruley, C., Garin, J., et al. (2007). A proteomics dissection of *Arabidopsis thaliana* vacuoles isolated from cell culture. *Mol. Cell Proteomics* 6, 394–412. doi: 10.1074/mcp.M600250-MCP200

62. Jensen, L. T., Ajua-Alemanji, M., and Culotta, V. C. (2003).

The *Saccharomyces cerevisiae* high affinity phosphate transporter encoded by *PHO84* also functions in manganese homeostasis. *J. Biol. Chem.* 278, 42036–42040. doi: 10.1074/jbc.M307413200

63. Jiang, W. (2006). Mn use efficiency in different wheat cultivars. *Environ. Exp. Bot.* 57, 41–50. doi: 10.1016/j.envexpbot.2005.04.008
64. Johnson, N. A., Liu, F., Weeks, P. D., Hentzen, A. E., Kruse, H. P., Parker, J. J., et al. (2009). A tomato ER-type $Ca^{2+}$-ATPase, LCA1, has a low thapsigargin-sensitivity and can transport manganese. *Arch. Biochem. Biophys.* 481, 157–168. doi: 10.1016/j.abb.2008.11.010
65. Kaiser, B. N., Moreau, S., Castelli, J., Thomson, R., Lambert, A., Bogliolo, S., et al. (2003). The soybean NRAMP homologue, GmDMT1, is a symbiotic divalent metal transporter capable of ferrous iron transport. *Plant J.* 35, 295–304. doi: 10.1046/j.1365-313X.2003.01802.x
66. Kamiya, T., Akahori, T., and Maeshima, M. (2005). Expression profile of the genes for rice cation/$H^+$ exchanger family and functional analysis in yeast. *Plant Cell Physiol.* 46, 1735–1740. doi: 10.1093/pcp/pci173
67. Kamiya, T., and Maeshima, M. (2004). Residues in internal repeats of the rice cation/$H^+$ exchanger are involved in the transport and selection of cations. *J. Biol. Chem.* 279, 812–819. doi: 10.1074/jbc.M309726200
68. Kiegle, E., Gilliham, M., Haseloff, J., and Tester, M. (2000). Hyperpolarisation-activated calcium currents found only in cells from the elongation zone of *Arabidopsis thaliana* roots. *Plant J.* 21, 225–229. doi: 10.1046/j.1365-313x.2000.00659.x
69. Kim, S. A., Punshon, T., Lanzirotti, A., Li, L., Alonso, J. M., Ecker, J. R., et al. (2006). Localization of iron in Arabidopsis seed requires the vacuolar membrane transporter VIT1. *Science* 314, 1295–1298. doi: 10.1126/science.1132563
70. Koike, S., Inoue, H., Mizuno, D., Takahashi, M., Nakanishi, H., Mori, S., et al. (2004). OsYSL2 is a rice metal-nicotianamine transporter that is regulated by iron and expressed in the phloem. *Plant J.* 39, 415–424. doi: 10.1111/j.1365-313X.2004.02146.x
71. Korenkov, V., Hirschi, K., Crutchfield, J. D., and Wagner, G. J.

(2007). Enhancing tonoplast Cd/H antiport activity increases Cd, Zn, and Mn tolerance, and impacts root/shoot Cd partitioning in *Nicotiana tabacum* L. *Planta* 226, 1379–1387. doi: 10.1007/s00425-007-0577-0

72. Korshunova, Y. O., Eide, D., Clark, W. G., Guerinot, M. L., and Pakrasi, H. B. (1999). The IRT1 protein from *Arabidopsis thaliana* is a metal transporter with a broad substrate range. *Plant Mol. Biol.* 40, 37–44. doi: 10.1023/A:1026438615520
73. Kramer, U. (2010). Metal hyperaccumulation in plants. *Annu. Rev. Plant Biol.* 61, 517–534. doi: 10.1146/annurev-arplant-042809-112156
74. Krieger, A., Rutherford, A. W., Vass, I., and Hideg, E. (1998). Relationship between activity, D1 loss, and Mn binding in photoinhibition of photosystem II. *Biochemistry* 37, 16262–16269. doi: 10.1021/bi981243v
75. Lanquar, V., Lelievre, F., Bolte, S., Hames, C., Alcon, C., Neumann, D., et al. (2005). Mobilization of vacuolar iron by AtNRAMP3 and AtNRAMP4 is essential for seed germination on low iron. *EMBO J.* 24, 4041–4051. doi: 10.1038/sj.emboj.7600864
76. Lanquar, V., Ramos, M. S., Lelievre, F., Barbier-Brygoo, H., Krieger-Liszkay, A., Kramer, U., et al. (2010). Export of vacuolar manganese by AtNRAMP3 and AtNRAMP4 is required for optimal photosynthesis and growth under manganese deficiency. *Plant Physiol.* 152, 1986–1999. doi: 10.1104/pp.109.150946
77. Lee, S., Jeong, H. J., Kim, S. A., Lee, J., Guerinot, M. L., and An, G. (2010). OsZIP5 is a plasma membrane zinc transporter in rice. *Plant Mol. Biol.* 73, 507–517. doi: 10.1007/s11103-010-9637-0
78. Leitenmaier, B., and Kupper, H. (2013). Compartmentation and complexation of metals in hyperaccumulator plants. *Front. Plant Sci.* 4:374. doi: 10.3389/fpls.2013.00374
79. Li, L., Chen, O. S., McVey Ward, D., and Kaplan, J. (2001). CCC1 is a transporter that mediates vacuolar iron storage in yeast. *J. Biol. Chem.* 276, 29515–29519. doi: 10.1074/jbc.M103944200
80. Li, X., Chanroj, S., Wu, Z., Romanowsky, S. M., Harper, J. F., and Sze, H. (2008). A distinct endosomal $Ca^{2+}/Mn^{2+}$ pump affects root growth through the secretory process. *Plant Physiol.* 147, 1675–

1689. doi: 10.1104/pp.108.119909

81. Liang, F., Cunningham, K. W., Harper, J. F., and Sze, H. (1997). ECA1 complements yeast mutants defective in $Ca^{2+}$ pumps and encodes an endoplasmic reticulum-type $Ca^{2+}$-ATPase in Arabidopsis thaliana. *Proc. Natl. Acad. Sci. U.S.A.* 94, 8579–8584. doi: 10.1073/pnas.94.16.8579
82. Lopez-Millan, A. F., Ellis, D. R., and Grusak, M. A. (2004). Identification and characterization of several new members of the ZIP family of metal ion transporters in *Medicago truncatula. Plant Mol. Biol.* 54, 583–596. doi: 10.1023/B:PLAN.0000038271.96019.aa
83. Lovley, D. R., Ueki, T., Zhang, T., Malvankar, N. S., Shrestha, P. M., Flanagan, K. A., et al. (2011). Geobacter: the microbe electric's physiology, ecology, and practical applications. *Adv. Microb. Physiol.* 59, 1–100. doi: 10.1016/B978-0-12-387661-4.00004-5
84. Lynch, J. P., and St. Clair, S. B. (2004). Mineral stress: the missing link in understanding how global climate change will affect plants in real world soils. *Field Crop Res.* 90, 101–115. doi: 10.1016/j.fcr.2004.07.008
85. Marschner, P. (2012). *Marschner's Mineral Nutrition of Higher Plants*. Boston, MA: Academic Press.
86. Marshall, J., Corzo, A., Leigh, R. A., and Sanders, D. (1994). Membrane potential-dependent calcium transport in right-side-out plasma membrane vesicles from *Zea mays* L. roots. *Plant J.* 5, 683–694. doi: 10.1111/j.1365-313X.1994.00683.x
87. Martinez-Finley, E. J., Gavin, C. E., Aschner, M., and Gunter, T. E. (2013). Manganese neurotoxicity and the role of reactive oxygen species. *Free Radic. Biol. Med.* 62, 65–75. doi: 10.1016/j.freeradbiomed.2013.01.032
88. Maser, P., Thomine, S., Schroeder, J. I., Ward, J. M., Hirschi, K., Sze, H., et al. (2001). Phylogenetic relationships within cation transporter families of Arabidopsis. *Plant Physiol.* 126, 1646–1667. doi: 10.1104/pp.126.4.1646
89. Mei, H., Cheng, N. H., Zhao, J., Park, S., Escareno, R. A., Pittman, J. K., et al. (2009). Root development under metal stress in *Arabidopsis thaliana* requires the $H^+$/cation antiporter CAX4. *New Phytol.* 183,

95–105. doi: 10.1111/j.1469-8137.2009.02831.x

90. Menguer, P. K., Farthing, E., Peaston, K. A., Ricachenevsky, F. K., Fett, J. P., and Williams, L. E. (2013). Functional analysis of the rice vacuolar zinc transporter OsMTP1. *J. Exp. Bot.* 64, 2871–2883. doi: 10.1093/jxb/ert136
91. Millaleo, R., Reyes-Diaz, M., Alberdi, M., Ivanov, A. G., Krol, M., and Huner, N. P. (2013). Excess manganese differentially inhibits photosystem I versus II in *Arabidopsis thaliana*. *J. Exp. Bot.* 64, 343–354. doi: 10.1093/jxb/ers339
92. Miller, E. W., He, Q., and Chang, C. J. (2008). Preparation and use of Leadfluor-1, a synthetic fluorophore for live-cell lead imaging. *Nat. Protoc.* 3, 777–783. doi: 10.1038/nprot.2008.43
93. Miller, E. W., Zeng, L., Domaille, D. W., and Chang, C. J. (2006). Preparation and use of Coppersensor-1, a synthetic fluorophore for live-cell copper imaging. *Nat. Protoc.* 1, 824–827. doi: 10.1038/nprot.2006.140
94. Mills, R. F., Doherty, M. L., Lopez-Marques, R. L., Weimar, T., Dupree, P., Palmgren, M. G., et al. (2008). ECA3, a Golgi-localized P2A-type ATPase, plays a crucial role in manganese nutrition in Arabidopsis. *Plant Physiol.* 146, 116–128. doi: 10.1104/pp.107.110817
95. Milner, M. J., Seamon, J., Craft, E., and Kochian, L. V. (2013). Transport properties of members of the ZIP family in plants and their role in Zn and Mn homeostasis. *J. Exp. Bot.* 64, 369–381. doi: 10.1093/jxb/ers315
96. Mizuno, T., Usui, K., Horie, K., Nosaka, S., Mizuno, N., and Obata, H. (2005). Cloning of three ZIP/Nramp transporter genes from a Ni hyperaccumulator plant *Thlaspi japonicum* and their $Ni^{2+}$-transport abilities. *Plant Physiol. Biochem.* 43, 793–801. doi: 10.1016/j.plaphy.2005.07.006
97. Montanini, B., Blaudez, D., Jeandroz, S., Sanders, D., and Chalot, M. (2007). Phylogenetic and functional analysis of the Cation Diffusion Facilitator (CDF) family: improved signature and prediction of substrate specificity. *BMC Genomics* 8:107. doi: 10.1186/1471-2164-8-107
98. Morris, J., Tian, H., Park, S., Sreevidya, C. S., Ward, J. M., and

Hirschi, K. D. (2008). AtCCX3 is an Arabidopsis endomembrane $H^+$ -dependent $K^+$ transporter. *Plant Physiol.* 148, 1474–1486. doi: 10.1104/pp.108.118810

99. Nevo, Y., and Nelson, N. (2006). The NRAMP family of metal-ion transporters. *Biochim. Biophys. Acta* 1763, 609–620. doi: 10.1016/j.bbamcr.2006.05.007
100. Nickelsen, J., and Rengstl, B. (2013). Photosystem II assembly: from cyanobacteria to plants. *Annu. Rev. Plant Biol.* 64, 609–635. doi: 10.1146/annurev-arplant-050312-120124
101. Nikolovski, N., Rubtsov, D., Segura, M. P., Miles, G. P., Stevens, T. J., Dunkley, T. P., et al. (2012). Putative glycosyltransferases and other plant Golgi apparatus proteins are revealed by LOPIT proteomics. *Plant Physiol.* 160, 1037–1051. doi: 10.1104/pp.112.204263
102. Oomen, R. J., Wu, J., Lelievre, F., Blanchet, S., Richaud, P., Barbier-Brygoo, H., et al. (2009). Functional characterization of NRAMP3 and NRAMP4 from the metal hyperaccumulator *Thlaspi caerulescens*. *New Phytol.* 181, 637–650. doi: 10.1111/j.1469-8137.2008.02694.x
103. Palmer, C. M., and Guerinot, M. L. (2009). Facing the challenges of Cu, Fe and Zn homeostasis in plants. *Nat. Chem. Biol.* 5, 333–340. doi: 10.1038/nchembio.166
104. Papp-Wallace, K. M., and Maguire, M. E. (2006). Manganese transport and the role of manganese in virulence. *Annu. Rev. Microbiol.* 60, 187–209. doi: 10.1146/annurev.micro.60.080805.142149
105. Pedas, P., Ytting, C. K., Fuglsang, A. T., Jahn, T. P., Schjoerring, J. K., and Husted, S. (2008). Manganese efficiency in barley: identification and characterization of the metal ion transporter HvIRT1. *Plant Physiol.* 148, 455–466. doi: 10.1104/pp.108.118851
106. Peiter, E., Montanini, B., Gobert, A., Pedas, P., Husted, S., Maathuis, F. J., et al. (2007). A secretory pathway-localized cation diffusion facilitator confers plant manganese tolerance. *Proc. Natl. Acad. Sci. U.S.A.* 104, 8532–8537. doi: 10.1073/pnas.0609507104
107. Pittman, J. K. (2005). Managing the manganese: molecular mechanisms of manganese transport and homeostasis. *New Phytol.* 167, 733–742. doi: 10.1111/j.1469-8137.2005.01453.x

108. Pittman, J. K., Mills, R. F., O'connor, C. D., and Williams, L. E. (1999). Two additional type IIA $Ca^{(2+)}$-ATPases are expressed in *Arabidopsis thaliana*: evidence that type IIA sub-groups exist. *Gene* 236, 137–147. doi: 10.1016/S0378-1119(99)00242-5

109. Pittman, J. K., Shigaki, T., Marshall, J. L., Morris, J. L., Cheng, N. H., and Hirschi, K. D. (2004). Functional and regulatory analysis of the *Arabidopsis thaliana* CAX2 cation transporter. *Plant Mol. Biol.* 56, 959–971. doi: 10.1007/s11103-004-6446-3

110. Podar, D., Scherer, J., Noordally, Z., Herzyk, P., Nies, D., and Sanders, D. (2012). Metal selectivity determinants in a family of transition metal transporters. *J. Biol. Chem.* 287, 3185–3196. doi: 10.1074/jbc.M111.305649

111. Punshon, T., Hirschi, K., Yang, J., Lanzirotti, A., Lai, B., and Guerinot, M. L. (2012). The role of CAX1 and CAX3 in elemental distribution and abundance in Arabidopsis seed. *Plant Physiol.* 158, 352–362. doi: 10.1104/pp.111.184812

112. Punshon, T., Ricachenevsky, F. K., Hindt, M. N., Socha, A. L., and Zuber, H. (2013). Methodological approaches for using synchrotron X-ray fluorescence (SXRF) imaging as a tool in ionomics: examples from *Arabidopsis thaliana*. *Metallomics* 5, 1133–1145. doi: 10.1039/c3mt00120b

113. Rampey, R. A., Baldridge, M. T., Farrow, D. C., Bay, S. N., and Bartel, B. (2013). Compensatory mutations in predicted metal transporters modulate auxin conjugate responsiveness in Arabidopsis. *G3* 3, 131–141. doi: 10.1534/g3.112.004655

114. Rampey, R. A., Woodward, A. W., Hobbs, B. N., Tierney, M. P., Lahner, B., Salt, D. E., et al. (2006). An Arabidopsis basic helix-loop-helix leucine zipper protein modulates metal homeostasis and auxin conjugate responsiveness. *Genetics* 174, 1841–1857. doi: 10.1534/genetics.106.061044

115. Reddi, A. R., Jensen, L. T., and Culotta, V. C. (2009). Manganese homeostasis in *Saccharomyces cerevisiae*. *Chem. Rev.* 109, 4722–4732. doi: 10.1021/cr900031u

116. Rengel, Z. (2000). Manganese uptake and transport in plants. *Met. Ions Biol. Syst.* 37, 57–87.

117. Ricachenevsky, F. K., Menguer, P. K., Sperotto, R. A., Williams,

L. E., and Fett, J. P. (2013). Roles of plant metal tolerance proteins (MTP) in metal storage and potential use in biofortification strategies. *Front. Plant Sci.* 4:144. doi: 10.3389/fpls.2013.00144

118. Roberts, L. A., Pierson, A. J., Panaviene, Z., and Walker, E. L. (2004). Yellow stripe1. Expanded roles for the maize iron-phytosiderophore transporter. *Plant Physiol.* 135, 112–120. doi: 10.1104/pp.103.037572
119. Rogers, E. E., Eide, D. J., and Guerinot, M. L. (2000). Altered selectivity in an Arabidopsis metal transporter. *Proc. Natl. Acad. Sci. U.S.A.* 97, 12356–12360. doi: 10.1073/pnas.210214197
120. Rudolph, H. K., Antebi, A., Fink, G. R., Buckley, C. M., Dorman, T. E., Levitre, J., et al. (1989). The yeast secretory pathway is perturbed by mutations in PMR1, a member of a $Ca^{2+}$ ATPase family. *Cell* 58, 133–145. doi: 10.1016/0092-8674(89)90410-8
121. Salt, D. E., Smith, R. D., and Raskin, I. (1998). Phytoremediation. *Annu. Rev. Plant Physiol. Plant Mol. Biol.* 49, 643–668. doi: 10.1146/annurev.arplant.49.1.643
122. Salvador, V. H., Lima, R. B., Dos Santos, W. D., Soares, A. R., Bohm, P. A., Marchiosi, R., et al. (2013). Cinnamic acid increases lignin production and inhibits soybean root growth. *PLoS ONE* 8:e69105. doi: 10.1371/journal.pone.0069105
123. Sasaki, A., Yamaji, N., Xia, J., and Ma, J. F. (2011). OsYSL6 is involved in the detoxification of excess manganese in rice. *Plant Physiol.* 157, 1832–1840. doi: 10.1104/pp.111.186031
124. Sasaki, A., Yamaji, N., Yokosho, K., and Ma, J. F. (2012). Nramp5 is a major transporter responsible for manganese and cadmium uptake in rice. *Plant Cell* 24, 2155–2167. doi: 10.1105/tpc.112.096925
125. Schaaf, G., Catoni, E., Fitz, M., Schwacke, R., Von Wiren, N., and Frommer, W. B. (2002). A putative role for the vacuolar calcium/manganese proton antiporter AtCAX2 in heavy metal detoxification. *Plant Biol.* 4, 612–618. doi: 10.1055/s-2002-35432
126. Schaaf, G., Ludewig, U., Erenoglu, B. E., Mori, S., Kitahara, T., and Von Wiren, N. (2004). ZmYS1 functions as a proton-coupled symporter for phytosiderophore- and nicotianamine-chelated metals. *J. Biol. Chem.* 279, 9091–9096. doi: 10.1074/jbc.M311799200

127. Schnell Ramos, M., Khodja, H., Mary, V., and Thomine, S. (2013). Using muPIXE for quantitative mapping of metal concentration in *Arabidopsis thaliana* seeds. *Front. Plant Sci.* 4:168. doi: 10.3389/fpls.2013.00168

128. Schroeder, J. I., Delhaize, E., Frommer, W. B., Guerinot, M. L., Harrison, M. J., Herrera-Estrella, L., et al. (2013). Using membrane transporters to improve crops for sustainable food production. *Nature* 497, 60–66. doi: 10.1038/nature11909

129. Shigaki, T., and Hirschi, K. D. (2006). Diverse functions and molecular properties emerging for CAX cation/$H^+$ exchangers in plants. *Plant Biol.* 8, 419–429. doi: 10.1055/s-2006-923950

130. Shigaki, T., Pittman, J. K., and Hirschi, K. D. (2003). Manganese specificity determinants in the Arabidopsis metal/$H^+$ antiporter CAX2. *J. Biol. Chem.* 278, 6610–6617. doi: 10.1074/jbc.M209952200

131. Shigaki, T., Rees, I., Nakhleh, L., and Hirschi, K. D. (2006). Identification of three distinct phylogenetic groups of CAX cation/proton antiporters. *J. Mol. Evol.* 63, 815–825. doi: 10.1007/s00239-006-0048-4

132. Supek, F., Supekova, L., Nelson, H., and Nelson, N. (1996). A yeast manganese transporter related to the macrophage protein involved in conferring resistance to mycobacteria. *Proc. Natl. Acad. Sci. U.S.A.* 93, 5105–5110. doi: 10.1073/pnas.93.10.5105

133. Thomine, S., Lelievre, F., Debarbieux, E., Schroeder, J. I., and Barbier-Brygoo, H. (2003). AtNRAMP3, a multispecific vacuolar metal transporter involved in plant responses to iron deficiency. *Plant J.* 34, 685–695. doi: 10.1046/j.1365-313X.2003.01760.x

134. Thomine, S., Wang, R., Ward, J. M., Crawford, N. M., and Schroeder, J. I. (2000). Cadmium and iron transport by members of a plant metal transporter family in Arabidopsis with homology to *Nramp* genes. *Proc. Natl. Acad. Sci. U.S.A.* 97, 4991–4996. doi: 10.1073/pnas.97.9.4991

135. Ueno, D., Yamaji, N., and Ma, J. F. (2009). Further characterization of ferric-phytosiderophore transporters ZmYS1 and HvYS1 in maize and barley. *J. Exp. Bot.* 60, 3513–3520. doi: 10.1093/jxb/erp191

136. Vert, G., Grotz, N., Dedaldechamp, F., Gaymard, F., Guerinot, M. L., Briat, J. F., et al. (2002). IRT1, an Arabidopsis transporter essential for iron uptake from the soil and for plant growth. *Plant Cell* 14, 1223–1233. doi: 10.1105/tpc.001388

137. Very, A. A., and Davies, J. M. (2000). Hyperpolarization-activated calcium channels at the tip of Arabidopsis root hairs. *Proc. Natl. Acad. Sci. U.S.A.* 97, 9801–9806. doi: 10.1073/pnas.160250397

138. Von Wiren, N., Mori, S., Marschner, H., and Romheld, V. (1994). Iron inefficiency in maize mutant *ys1* (*Zea mays* L. cv Yellow-Stripe) is caused by a defect in uptake of iron phytosiderophores. *Plant Physiol.* 106, 71–77.

139. Waters, B. M., Chu, H. H., Didonato, R. J., Roberts, L. A., Eisley, R. B., Lahner, B., et al. (2006). Mutations in Arabidopsis yellow stripe-like1 and yellow stripe-like3 reveal their roles in metal ion homeostasis and loading of metal ions in seeds. *Plant Physiol.* 141, 1446–1458. doi: 10.1104/pp.106.082586

140. White, P. J., Bowen, H. C., Demidchik, V., Nichols, C., and Davies, J. M. (2002). Genes for calcium-permeable channels in the plasma membrane of plant root cells. *Biochim. Biophys. Acta* 1564, 299–309. doi: 10.1016/S0005-2736(02)00509-6

141. Wu, Z., Liang, F., Hong, B., Young, J. C., Sussman, M. R., Harper, J. F., et al. (2002). An endoplasmic reticulum-bound $Ca^{(2+)}/Mn^{(2+)}$ pump, ECA1, supports plant growth and confers tolerance to $Mn^{(2+)}$ stress. *Plant Physiol.* 130, 128–137. doi: 10.1104/pp.004440

142. Wymer, C. L., Bibikova, T. N., and Gilroy, S. (1997). Cytoplasmic free calcium distributions during the development of root hairs of *Arabidopsis thaliana*. *Plant J.* 12, 427–439. doi: 10.1046/j.1365-313X.1997.12020427.x

143. Xia, J., Yamaji, N., Kasai, T., and Ma, J. F. (2010). Plasma membrane-localized transporter for aluminum in rice. *Proc. Natl. Acad. Sci. U.S.A.* 107, 18381–18385. doi: 10.1073/pnas.1004949107

144. Xiao, H., Yin, L., Xu, X., Li, T., and Han, Z. (2008). The iron-regulated transporter, MbNRAMP1, isolated from *Malus baccata* is involved in Fe, Mn and Cd trafficking. *Ann. Bot.* 102, 881–889. doi: 10.1093/aob/mcn178

145. Yamaji, N., Sasaki, A., Xia, J. X., Yokosho, K., and Ma, J. F. (2013).

A node-based switch for preferential distribution of manganese in rice. *Nat. Commun.* 4:2442. doi: 10.1038/ncomms3442

146. Yang, T. J., Perry, P. J., Ciani, S., Pandian, S., and Schmidt, W. (2008). Manganese deficiency alters the patterning and development of root hairs in Arabidopsis. *J. Exp. Bot.* 59, 3453–3464. doi: 10.1093/jxb/ern195
147. Yanykin, D. V., Khorobrykh, A. A., Khorobrykh, S. A., and Klimov, V. V. (2010). Photoconsumption of molecular oxygen on both donor and acceptor sides of photosystem II in Mn-depleted subchloroplast membrane fragments. *Biochim. Biophys. Acta* 1797, 516–523. doi: 10.1016/j.bbabio.2010.01.014
148. Yen, M. R., Tseng, Y. H., and Saier, M. H. Jr. (2001). Maize Yellow Stripe1, an iron-phytosiderophore uptake transporter, is a member of the oligopeptide transporter (OPT) family. *Microbiology* 147, 2881–2883.
149. Pubmed Abstract | Pubmed Full Text
150. Yoon, S., Miller, E. W., He, Q., Do, P. H., and Chang, C. J. (2007). A bright and specific fluorescent sensor for mercury in water, cells, and tissue. *Angew. Chem. Int. Ed. Engl.* 46, 6658–6661. doi: 10.1002/anie.200701785
151. Zhang, Y., Xu, Y. H., Yi, H. Y., and Gong, J. M. (2012). Vacuolar membrane transporters OsVIT1 and OsVIT2 modulate iron translocation between flag leaves and seeds in rice. *Plant J.* 72, 400–410. doi: 10.1111/j.1365-313X.2012.05088.x
152. Zhao, H., and Eide, D. (1996a). The yeast *ZRT1* gene encodes the zinc transporter protein of a high-affinity uptake system induced by zinc limitation. *Proc. Natl. Acad. Sci. U.S.A.* 93, 2454–2458. doi: 10.1073/pnas.93.6.2454
153. Zhao, H., and Eide, D. (1996b). The *ZRT2* gene encodes the low affinity zinc transporter in *Saccharomyces cerevisiae*. *J. Biol. Chem.* 271, 23203–23210. doi: 10.1074/jbc.271.38.23203

# CITATION

## CHAPTER 2

Nieves-Cordones M, Martínez V, Benito B and Rubio F (2016) Comparison between Arabidopsis and Rice for Main Pathways of K+ and Na+ Uptake by Roots. Front. Plant Sci. 7:992. doi: 10.3389/fpls.2016.00992

## CHAPTER 3

Wang, M., Zheng, Q., Shen, Q., Guo, S. (2013) The critical role of potassium in plant stress response. Int J Mol Sci. 14, 7370-7390

## CHAPTER 4

Yamaguchi T, Hamamoto S and Uozumi N (2013) Sodium transport system in plant cells. Front. Plant Sci. 4:410. doi: 10.3389/fpls.2013.00410

## CHAPTER 5

Alemán F, Caballero F, Ródenas R, Rivero RM, Martínez V and Rubio F (2014) The F130S point mutation in the Arabidopsis high-affinity $K^+$ transporter AtHAK5 increases $K^+$ over $Na^+$ and $Cs^+$ selectivity and confers $Na^+$ and $Cs^+$ tolerance to yeast under heterologous expression. *Front. Plant Sci.* **5**:430. doi: 10.3389/fpls.2014.00430

## CHAPTER 6

Sharma T, Dreyer I and Riedelsberger J (2013) The role of $K^+$ channels in uptake and redistribution of potassium in the model plant *Arabidopsis thaliana*. *Front. Plant Sci.* **4**:224. doi: 10.3389/fpls.2013.00224

## CHAPTER 7

Eisenach C., Papanatsiou M., Hillert E.K., Blatt M.R. (2014) Clustering of the $K^+$ channel GORK of Arabidopsis parallels its gating by extracellular $K^+$. Plant J. 78, 203-214

## CHAPTER 8

Kugler, A., Köhler, B., Palme, K., Wolff, P., Dietrich, P. (2013) Salt-dependent regulation of a CNG channel subfamily in Arabidopsis. BMC Plant Biol. 9, 140

## CHAPTER 9

Ni J., Yu Z., Du G., Zhang Y., Taylor J.L., Shen C., Xu J., Liu X., Wang Y., Wu Y. (2016) Heterologous expression and functional analysis of rice GLUTAMATE RECEPTOR-LIKE family indicates its role in glutamate triggered calcium flux in rice roots. Rice. 9, 9

## CHAPTER 10

Waters, S., Gilliham, M., Hrmova, M. (2013) Plant High-Affinity Potassium (HKT) Transporters involved in salinity tolerance: structural insights to probe differences in ion selectivity. Int J Mol Sci. 14, 7660-7680

## CHAPTER 11

Pittman J.K., Hirschi K.D. (2016) Plant High-Affinity Potassium (HKT) Phylogenetic analysis and protein structure modelling identifies distinct $Ca^{2+}$/Cation antiporters and conservation of gene family structure within Arabidopsis and rice species. Rice. 9, 3

## CHAPTER 12

Pires I.S., Negrão S., Pentony M.M., Abreu I.A., Oliveira M.M., Purugganan M.D. (2013) Different evolutionary histories of two cation/proton exchanger gene families in plants. BMC Plant Biol. 13, 97

## CHAPTER 13

McCubbin T., Bassil E., Zhang S., Blumwald E. (2014) Vacuolar $Na^+$/$H^+$ NHX-type antiporters are required for cellular $K^+$ homeostasis, microtubule organization and directional root growth. Plants. 3, 409-426

## CHAPTER 14

Ford BA, Ernest JR and Gendall AR (2012) Identification and characterization of orthologs of AtNHX5 and AtNHX6 in Brassica napus. Front. Plant Sci. 3:208. doi: 10.3389/fpls.2012.00208

## CHAPTER 15

Zhou Y., Yin X., Duan R., Hao G., Guo J., Jiang X. (2015) SpAHA1 and SpSOS1 coordinate in transgenic yeast to improve salt tolerance. PLoS One. 10, e0137447

## CHAPTER 16

Socha AL and Guerinot ML (2014) Mn-euvering manganese: the role of transporter gene family members in manganese uptake and mobilization in plants. Front. Plant Sci. 5:106. doi: 10.3389/fpls.2014.00106

# INDEX

## H



## I



## K



## L



## M



## N